A Users Guide to Vacuum Technology

A Users Guide to Vacuum Technology

Fourth Edition

John F. O'Hanlon
Emeritus Professor of Electrical and Computer Engineering
University of Arizona
Tucson, Arizona, USA

Timothy A. Gessert
Gessert Consulting, LLC
Conifer, Colorado, USA

Copyright © 2024 by John Wiley & Sons, Inc. All rights reserved.

Published by John Wiley & Sons, Inc., Hoboken, New Jersey.
Published simultaneously in Canada.

No part of this publication may be reproduced, stored in a retrieval system, or transmitted in any form or by any means, electronic, mechanical, photocopying, recording, scanning, or otherwise, except as permitted under Section 107 or 108 of the 1976 United States Copyright Act, without either the prior written permission of the Publisher, or authorization through payment of the appropriate per-copy fee to the Copyright Clearance Center, Inc., 222 Rosewood Drive, Danvers, MA 01923, (978) 750-8400, fax (978) 750-4470, or on the web at www.copyright.com. Requests to the Publisher for permission should be addressed to the Permissions Department, John Wiley & Sons, Inc., 111 River Street, Hoboken, NJ 07030, (201) 748-6011, fax (201) 748-6008, or online at http://www.wiley.com/go/permission.

Trademarks: Wiley and the Wiley logo are trademarks or registered trademarks of John Wiley & Sons, Inc. and/or its affiliates in the United States and other countries and may not be used without written permission. All other trademarks are the property of their respective owners. John Wiley & Sons, Inc. is not associated with any product or vendor mentioned in this book.

Limit of Liability/Disclaimer of Warranty
While the publisher and author have used their best efforts in preparing this book, they make no representations or warranties with respect to the accuracy or completeness of the contents of this book and specifically disclaim any implied warranties of merchantability or fitness for a particular purpose. No warranty may be created or extended by sales representatives or written sales materials. The advice and strategies contained herein may not be suitable for your situation. You should consult with a professional where appropriate. Further, readers should be aware that websites listed in this work may have changed or disappeared between when this work was written and when it is read. Neither the publisher nor authors shall be liable for any loss of profit or any other commercial damages, including but not limited to special, incidental, consequential, or other damages.

For general information on our other products and services or for technical support, please contact our Customer Care Department within the United States at (800) 762-2974, outside the United States at (317) 572-3993 or fax (317) 572-4002.

Wiley also publishes its books in a variety of electronic formats. Some content that appears in print may not be available in electronic formats. For more information about Wiley products, visit our web site at www.wiley.com.

Library of Congress Cataloging-in-Publication Data
Names: O'Hanlon, John F., 1937– author. | Gessert, Timothy A., author.
Title: A users guide to vacuum technology / John F. O'Hanlon, Emeritus Professor of Electrical and Computer Engineering, University of Arizona, Tucson, Arizona, USA, Timothy A. Gessert, Gessert Consulting, LLC, Conifer, Colorado, USA.
Description: 4th edition. | Hoboken, New Jersey : John Wiley & Sons, Inc., [2024] | Includes index.
Identifiers: LCCN 2023024446 (print) | LCCN 2023024447 (ebook) | ISBN 9781394174133 (hardback) | ISBN 9781394174140 (adobe pdf) | ISBN 9781394174225 (epub)
Subjects: LCSH: Vacuum technology–Handbooks, manuals, etc.
Classification: LCC TJ940 .O37 2024 (print) | LCC TJ940 (ebook) | DDC 621.5/5–dc23/eng/20230630
LC record available at https://lccn.loc.gov/2023024446
LC ebook record available at https://lccn.loc.gov/2023024447

Cover image: Wiley
Cover design: Courtesy of NASA

Set in 9.5/12.5pt STIXTwoText by Straive, Pondicherry, India

For Jean, Carol, Paul, and Amanda
and
For Janet, Rachael, Kathryn, and Benjamin

Contents

Preface

A Users Guide to Vacuum Technology, Fourth Edition, focuses on the operation, understanding, and selection of equipment for processes used in semiconductor, optics, renewable energy, and related emerging technologies. It emphasizes subjects not adequately covered elsewhere, while avoiding in-depth treatments of topics of interest only to the vacuum system designer or vacuum historian. The discussions of gauges, pumps, and materials present a required prelude to the later discussion of fully integrated vacuum systems. System design options are grouped according to their function and include both single- and multichamber systems and how details of each design are determined by specific requirements of a production or research application.

During the twenty years since the publication of the third edition, the needs of vacuum technology users have evolved considerably. For example, in 2003, when the third edition was published, the minimum feature width for a typical semiconductor fabrication facility was on the order of 200 nm: the "200-nm node." Approximately ten years later in 2013, production at the 20-nm node was becoming available, and its related lithography tools began to require UV exposure in vacuum because any gas ambient detrimentally absorbed or scattered UV light. Presently, the 2-nm node is being tested for advanced integrated circuit processes, and their (ultra-) UV light sources require even more advanced vacuum systems, as well as related equipment with increasingly tightened specification regarding particle, film, and gas-phase contamination. Few (if any) historic vacuum textbooks include these topics to the extent required by today's technologists.

The past two decades have also featured an unprecedented increase in the use of sophisticated vacuum-based processes for mass producing consumer products, such as low-cost eyeglass reflective coatings, durable cookware coatings, secure bank notes, RFID tags, and coated plastic films.

Since the publication of the third edition, the authors of this book have collectively taught several thousands of students at academic institutions, in high-technology companies, and at professional society meetings. Through their experience, the authors have acquired an unusually diverse and unique exposure to industries involved in present vacuum technology processes, future directions, and the related problems they are facing. Much of this experience has been incorporated into this fourth edition, with the goal of assisting users with insight needed for success in both their present and future activities.

Although it is expected that academic students will continue to find this book a valuable reference in their pursuit of advanced degrees, the primary audience of this fourth edition is expected to be vacuum technologists and scientists already working in vacuum technology. However, it is expected that initiatives to expand semiconductor production and developments related to quantum computing both will require the type of advanced guidance presented in this book. Finally, vacuum users in other technology fields are also expected to find this book a valuable resource, e.g., space simulation, fusion research, renewable energy, and medical devices.

In addition to including new requirements and related equipment changes within these technology sectors, another enhancement in this fourth edition includes expanded discussions on vacuum technology Best Practices. This type of general guidance would have been acquired historically through mentoring by experienced colleagues; however, the authors have seen rapid developments in many high-technology sectors, as well as frequent career changes or added management responsibilities, have left many vacuum technologists in greater need of this type of reliable yet succinct guidance. It is hoped that this edition of *A Users Guide to Vacuum Technology* can fill some of the education gap resulting from this loss of historic "mentoring," as well as assist senior technologists in appreciating some of the more advanced vacuum concepts and descriptions.

The authors thank countless personal colleagues, students, and other researchers who over many years have provided numerous questions and practical solutions to the vacuum topics that have been included in this book. At the risk of many unintentional omissions, the authors would like to particularly thank Bruce Kendal for many discussions that continue to remain highly relevant to this book, Frank Zimone for the idea of incorporating "Best Practices," and Howard Patton for the original development of the AVS Short Course, Controlling Contamination in Vacuum Systems, on which Chapter 21 of this fourth edition is broadly based.

John F. O'Hanlon
Tucson, Arizona, USA

Timothy A. Gessert
Conifer, Colorado, USA

Symbols

Symbol	Quantity	Units
A	Area	m^2
B	Magnetic field strength	T (tesla)
C	Conductance (gas)	L/s
C′	vena contracta	
D	Diffusion constant	m^2/s
E_o	Heat transfer	$J\text{-}s^{-1}\text{-}m^{-2}$
F	Force	N (newton)
G	Electron multiplier gain	
H	Heat flow	J/s
K	Compression ratio (gas)	
K_p	Permeability constant (gas)	m^2/s
Kn	Knudsen's number	
K_R	Radiant heat conductivity	$J\text{-}s^{-1}\text{-}m^{-1}\text{-}K^{-1}$
K_T	Thermal conductivity	$J\text{-}s^{-1}\text{-}m^{-1}\text{-}K^{-1}$
M	Molecular weight	
N	Number of molecules	
N_o	Avogadro's number	$(kg\text{-}mol)^{-1}$
P	Pressure	Pa (pascal)
Q	Gas flow	$Pa\text{-}m^3/s$
R	Gas constant	$J\text{-}(kg\text{-}mol)^{-1}\text{-}s^{-1}$
R	Reynolds' number	
S	Pumping speed	L/s
S'	Gauge sensitivity	Pa^{-1}

(*Continued*)

Symbol	Quantity	Units
S_C	Critical saturation ratio	
T	Absolute temperature	K
U	Average gas stream velocity	m/s
U	Mach number	
V	Volume	m^3
V_a	Acceleration potential	V
V_b	Linear blade velocity	m/s
V_o	Normal specific volume of an ideal gas	m^3/(kg-mol)
W	Ho coefficient	
a	Transmission probability	
b	Turbopump blade chord length, or length dimension	m
c	Condensation coefficient	
c_p	Specific heat at constant pressure	J-(kg-mol)$^{-1}$-K^{-1}
c_v	Specific heat at constant volume	J-(kg-mol)$^{-1}$-K^{-1}
d	Diameter dimension	m
d_o	Molecular diameter	m
d'	Average molecular spacing	m
e	Length dimension	m
i_e	Emission current	A
i_p	Plate current	A
k	Boltzmann constant	J/K
l	Length dimension	m
m	Mass	kg
n	Gas density	m^{-3}
q	Outgassing rate	Pa-m/s
q_k	Permeation rate	Pa-m/s
r	Radius	m
s	Turbomolecular pump blade spacing	m
s_r	Turbomolecular pump blade speed ratio	
u	Local gas stream velocity	m/s
v	Average particle velocity	m/s
w	Length dimension	m
Γ	Particle flux	m^{-2}-s^{-1}

Symbol	Quantity	Units
Δ	Free molecular heat conductivity	$J\text{-}s^{-1}\text{-}m^{-2}\text{-}K^{-2}\text{-}Pa^{-1}$
α	Accommodation coefficient	
β	Molecular slip constant	
γ	Specific heat ratio c_p/c_v	
δ	Kronecker delta function	
ε	Emissivity	
η	Dynamic viscosity	Pa-s
λ	Mean free path	m
ξ	Volume to surface area ratio	
π	Pi	
ρ	Mass density	kg/m^3
τ	Vacuum system time constant	s
φ	Angle	deg
ω	Angular frequency; (heat transfer rate)	rad/s; (m/s)

Part I

Its Basis

An understanding of how vacuum components and systems function begins with an understanding of the behavior of gases at low pressures. Chapter 1 discusses the nature of vacuum technology. Chapter 2 reviews basic gas properties. Chapter 3 describes the complexities of gas flow at near-atmosphere and reduced pressures, and Chapter 4 discusses a most important topic: how gases evolve from and within material surfaces. Together, these chapters form the understanding of gauges, pumps, and systems that form the mainstay of vacuum technology as we know it today.

A Users Guide to Vacuum Technology, Fourth Edition. John F. O'Hanlon and Timothy A. Gessert.
© 2024 John Wiley & Sons, Inc. Published 2024 by John Wiley & Sons, Inc.

1

Vacuum Technology

Torricelli is credited with the conceptual understanding of the vacuum within a mercury column by the year 1643. It is written that his good friend Viviani actually performed the first experiment, perhaps as early as 1644 [1,2]. His discovery was followed in 1650 by Otto von Guericke's piston vacuum pump. Interest in vacuum remained at a low level for more than 200 years, when a period of rapid discovery began with McLeod's invention of the compression gauge. In 1905, Gaede, a prolific inventor, designed a rotary pump sealed with mercury. The thermal conductivity gauge, diffusion pump, ion gauge, and ion pump soon followed, along with processes for liquefying helium and refining organic pumping fluids. They formed the basis of a technology that has made possible everything from incandescent light bulbs to space exploration. The significant discoveries of this early period of vacuum science and technology have been summarized in a number of historical reviews [2,3,4,5,6,7].

The gaseous state can be divided into two fundamental regions. In one region, the distances between adjacent particles are exceedingly small compared to the size of the vessel in which they are contained. We call this the viscous state because gas properties are primarily determined by interactions between nearby particles. The rarefied gas state is a space in which molecules are widely spaced and rarely collide with one another. Instead, they collide with their confining walls. Figure 1.1 sketches this behavior. This is an extremely important distinction that will appear in many discussions throughout this material.

A vacuum is a space from which air or other gas has been removed. Of course, it is impossible to remove all gas from a container. The amount removed depends on the application and is done for many reasons. At atmospheric pressure, molecules constantly bombard surfaces. They can bounce from surfaces, attach themselves to surfaces, and even chemically react with surfaces. Air or other surrounding gas can quickly contaminate a clean surface. A clean surface, e.g., a freshly cleaved crystal,

A Users Guide to Vacuum Technology, Fourth Edition. John F. O'Hanlon
and Timothy A. Gessert.
© 2024 John Wiley & Sons, Inc. Published 2024 by John Wiley & Sons, Inc.

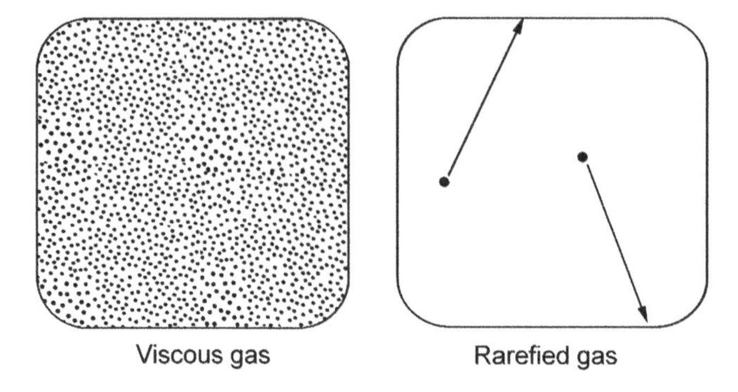

Fig. 1.1 View of a viscous gas and a rarefied gas.

will remain clean in an ultrahigh vacuum chamber for long periods of time, because the rate of molecular bombardment is low.

Molecules are crowded closely together at atmospheric pressure and travel in every direction much like people in a crowded plaza. It is impossible for molecules to travel from one wall of a chamber to another without myriad collisions with others. By reducing the pressure to a suitably low value, molecules can travel from one wall to another without collision. Many things become possible if they can travel long distances without collisions. Metals can be evaporated from pure sources without reacting in transit. Molecules or atoms can be accelerated to a high energy and sputter away or be implanted in a surface. Electrons or ions can be scattered from surfaces and be collected. The energy changes they undergo on scattering or release from a surface are used to probe or analyze surfaces and underlying layers.

For convenience the sub-atmospheric pressure scale has been divided into several ranges that are listed in Table 1.1. The ranges in this table are not so arbitrary; rather, they are a concise statement of the materials, methods, and equipment necessary to achieve the degree of vacuum needed for a given vacuum process.

The required degree of vacuum depends on the application. Reduced pressure epitaxy and laser etching of metals are two processes that are performed in the low vacuum range. Sputtering, plasma etching and deposition, low-pressure chemical vapor deposition, ion plating, and gas filling of encapsulated heat transfer modules are examples of processes performed in the medium vacuum range.

Pressures in the high vacuum range are needed for the manufacture of low- and high-tech devices such as microwave, power, cathode ray and photomultiplier tubes, light bulbs, architectural and automotive glass, decorative packaging, and processes including degassing of metals, vapor deposition, and ion implantation. A number of medium technology applications including medical, microwave susceptors, electrostatic dissipation films, and aseptic packaging use films fabricated

Table 1.1 ISO Definition of Vacuum Pressure Ranges and Descriptions

Pressure Ranges	**Definition**	**The reasoning for the definition of the ranges is as follows (typical circumstances):**
Prevailing atm. pressure (31–110 kPa) to 100 Pa (232–825 to 0.75 Torr)	Low (rough) vacuum	Pressure can be achieved by simple materials (e.g., regular steel) and positive displacement vacuum pumps; viscous flow regime for gases
<100 to 0.1 Pa ($0.75–7.5 \times 10^{-5}$ Torr)	Medium (fine) vacuum	Pressure can be achieved by elaborate materials (e.g., stainless steel) and positive displacement vacuum pumps; transitional flow regime for gases
$<0.1–1 \times 10^{-6}$ Pa ($7.5 \times 10^{-5}–7.5 \times 10^{-9}$ Torr)	High vacuum (HV)	Pressure can be achieved by elaborate materials (e.g., stainless steel), elastomer sealings, high vacuum pumps; molecular flow regime for gases
$<1 \times 10^{-6}$ Pa–1×10^{-9} Pa ($7.5 \times 10^{-9}–7.5 \times 10^{-12}$ Torr)	Ultrahigh vacuum (UHV)	Pressure can be achieved by elaborate materials (e.g., low-carbon stainless steel), metal sealings, special surface preparations and cleaning, bake-out, and high vacuum pumps; molecular flow regime for gases
$<1 \times 10^{-9}$ Pa ($<7.5 \times 10^{-12}$ Torr)	Extreme-high vacuum (EHV)	Pressure can be achieved by sophisticated materials (e.g., vacuum-fired low-carbon stainless steel, aluminum, copper–beryllium, and titanium), metal sealings, special surface preparations and cleaning, bake-out, and additional getter pumps; molecular flow regime for gases

Note 1: While there has been some variation in the selection of limits for these intervals, the above list gives typical ranges for which the limits are to be considered approximations.
Note 2: The prevailing atmospheric pressure on ground depends on weather conditions and altitude and ranges from 31 kPa (altitude of Mount Everest, weather condition: “low”) up to 110 kPa (altitude of Dead Sea, weather condition: “high”).
Source: © ISO. This material is reproduced from ISO 3529-1:2019 with permission of the American National Standards Institute (ANSI) on behalf of the International Organization for Standardization. All rights reserved.

in a vacuum environment [8]. Retail security, bank note security, and coated laser and inkjet papers are now included in this group.

The background pressure must be reduced to the very high vacuum range for electron microscopy, mass spectrometry, crystal growth, X-ray and electron beam lithography, and storage media production. For ease of reading, we call the very high vacuum region "high vacuum" and its associated pumps "high vacuum pumps."

Pressures in the ultrahigh vacuum range were formerly the domain of the surface analyst, materials researcher, or accelerator technologist. Today, critical high-volume production applications, such as semiconductor devices, thin-film media heads, and extreme UV lithography, require ultrahigh vacuum base pressures to reduce gaseous impurity contamination.

Yet another category of process takes place in the medium vacuum region using pure process gases and ultrahigh vacuum chamber starting conditions to maintain purity. Additionally, these processes must be free of particles. We call these systems ultraclean vacuum systems.

A vacuum system is a combination of pumps, valves, and pipes that creates a region of low pressure. It can be anything from a simple mechanical pump or aspirator for exhausting a vacuum storage container to a complex system such as an underground accelerator with miles of piping that must be held at an ultrahigh vacuum.

Removal of air at atmospheric pressure is usually done with a displacement pump, i.e., a pump that removes air from the chamber and expels it into the atmosphere. Rotary vane pumps are often used for this "rough pumping" purpose. Liquid nitrogen sorption pumps are used to rough pump ultraclean systems, such as those used for molecular beam epitaxy (MBE) that cannot tolerate even minute amounts of organic contamination. These pumps have finite gas sorption and require periodic regeneration.

Sorption pumps, as well as rotary vane and similar mechanical pumps, have low-pressure limits in the range 10^{-1}–10^{-3} Pa. Pumps that will function in a rarefied atmosphere are required to operate below this pressure range. The diffusion pump was the first high vacuum momentum transfer pump. Its outlet pressure is below atmosphere. The turbomolecular pump, a system of high-speed rotating turbine blades, can also pump gas at low pressures. The outlet pressures of these two pumps need to be kept in the range 0.5–50 Pa, so they must exhaust into a mechanical "backing" pump, or "fore" pump. If the diffusion or turbomolecular pump exhaust gas flow would otherwise be too great, a lobe blower would be placed between the exhaust of the diffusion or turbo pump and the inlet of the rotary pump to pump gas at an increased speed in this intermediate pressure region.

Capture pumps can effectively remove gas from a chamber at low pressure. They do so by freezing molecules on a wall (cryogenic), chemically reacting with the molecules (gettering), or accelerating the molecules to a high velocity and burying them in a metal wall (ion pumping). Capture pumps are useful, efficient, and clean high vacuum pumps.

Air is the most important gas to understand because it is in every vacuum system. It contains at least a dozen constituents, whose major components are described in Table 1.2. The differing ways in which pumps remove air, and gauges measure its pressure, can be understood in terms of the partial pressures of its components. The concentrations listed in Table 1.2 are those of dry atmospheric air at sea level whose total pressure is 101,325 Pa (760 Torr). The partial pressure of water vapor is not given in this table, because it is not constant. At 20°C a relative humidity of 50% corresponds to a partial pressure of 1165 Pa (8.75 Torr), making it the third largest component of air. The total pressure changes rapidly with altitude, as shown in Fig. 1.2, whereas its proportions change slowly but significantly. In outer space, the atmosphere is mainly H_2 with some He [10].

In the pressure region below 10 Pa, gases evolving from material surfaces contribute more flux to the total gas load than do the gases originally filling the chamber. A quality pump is not the only requirement to reach low pressures. The materials of construction, techniques for joining components, surface cleaning techniques, and operational procedures are all critically important. In the remaining chapters, the

Table 1.2 Components of Dry Atmospheric Air

	Quantity		
Constituent	**(vol%)**	**(ppm)**	**Pressure (Pa)**
N_2	78.084 ± 0.004		79,117
O_2	20.946 ± 0.002		21,223
Ar	0.934 ± 0.001		946.357
CO_2[a]		420	42.0
CH_4		2	0.203
Ne		18.18 ± 0.04	1.842
He		5.24 ± 0.004	0.51
Kr		1.14 ± 0.01	0.116
H_2		0.5	0.051
N_2O		0.5 ± 0.1	0.051
Xe		0.087 ± 0.001	0.009

[a] Carbon dioxide data from NOAA Global Monitoring Laboratory, Mauna Loa, Hawaii, August 2023. Data since 1955 are publicly available at: https://gml.noaa.gov/ccgg/trends
Source: R.C. Weast [9]/Reproduced with permission of Taylor & Francis.

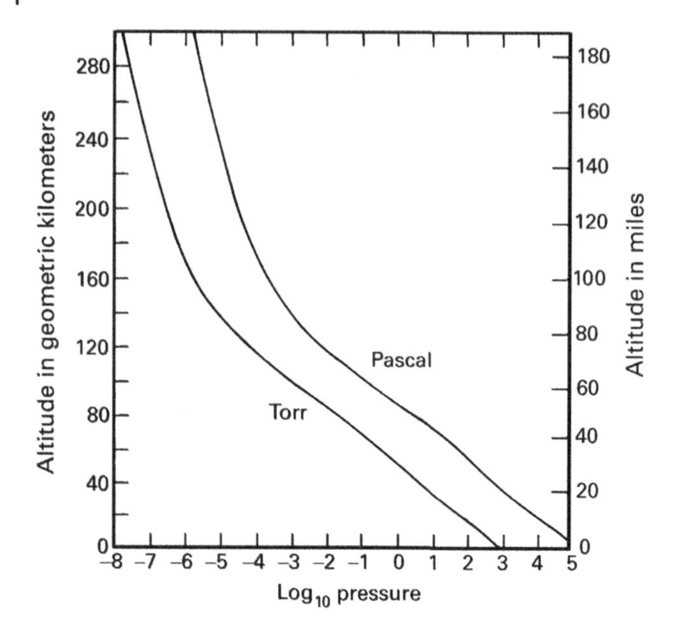

Fig. 1.2 Relation between the atmospheric pressure and the geometric altitude. R.C. Weast [9]/Reproduced with permission of Taylor & Francis.

pumps, gauges, materials of construction, and operational techniques are described in terms of fundamental gas behavior. The focus is on the understanding and operation of vacuum systems and developing best practices for many applications.

1.1 Units of Measurement

Units of measurement present problems in many disciplines, and vacuum technology is no exception. Système International, more commonly known by its abbreviation, SI, is in worldwide use. (SI base units are given in Appendix A.2.) With the exception of the United Kingdom and the United States, the use of SI vacuum units is standard. However, noncoherent units such as the mbar (United Kingdom) and Torr (United States) remain in use.

The meter-kilogram-second (MKS) system was first introduced almost 75 years ago and became commonplace only after a decade or more of classroom education by instructors committed to change. In a similar manner, those who teach vacuum technique should lead the way and promote routine use of SI units. Metrology tools are manufactured for a global economy, and their readings can be displayed easily in several formats. The ease of displaying noncoherent units does not encourage change. The advantages of using a coherent unit system are manifold. Calculations become straightforward and logical, and the chance for error is reduced. Incoherent units are cumbersome, to say the least. One example is the permeation constant, which is the volume of gas (at standard temperature and pressure) per material thickness per material area per second. Additionally, noncoherent permeation units mask their relation to solubility and diffusion. Ultimately, SI units will be routinely used, but that will take time. To assist with this change, dual labels have been added throughout the text. SI units for pressure (Pa), time (s), and length (m) will be assumed in all formulas, unless noted differently within a formula statement.

References

1 Middleton, W.E.K., *The History of the Barometer*, Johns Hopkins Press, Baltimore, 1964.
2 Redhead, P.A., *Vacuum* **53**, 137 (1999).
3 Madey, T.E., *J. Vac. Sci. Technol., A* **2**, 110 (1984).
4 Hablanian, M.H., *J. Vac. Sci. Technol., A* **2**, 118 (1984).
5 Singleton, J.H., *J. Vac. Sci. Technol., A* **2**, 126 (1984).
6 Redhead, P.A., *J. Vac. Sci. Technol., A* **2**, 132 (1984).
7 Madey, T.E. and Brown, W.C., Eds., *History of Vacuum Science and Technology*, American Institute of Physics, New York, 1984.
8 Johansen, P.R., *J. Vac. Sci. Technol., A* **8**, 2798 (1990).
9 Weast, R.C., Ed., *The Handbook of Chemistry and Physics*, 59th ed., Copyright 1978, The Chemical Rubber Publishing Co., CRC Press, Inc., West Palm Beach, FL. p. F151.
10 Santeler, D.J., Jones, D.W., Holkeboer, D.H., and Pagone, F., *Vacuum Technology and Space Simulation*, NASA SP-105, National Aeronautics and Space Administration, Washington, DC, 1966, p. 34.

2

Gas Properties

In this chapter we discuss the properties of gases at atmospheric and reduced pressures. The properties developed here are based on the kinetic picture of a gas. Kinetic theory has its limitations, but with it we are able to describe particle motion, pressure, effusion, viscosity, diffusion, thermal conductivity, and thermal transpiration of ideal gases. We will use these ideas as the starting point for understanding gas flow, surface reactions, gauges, pumps, and systems.

2.1 Kinetic Picture of a Gas

The kinetic picture of a gas is based on four assumptions: (1) The volume of gas under consideration contains a large number of molecules. For example, a cubic meter of gas at a pressure of 10^5 Pa (760 Torr) and a temperature of 22 °C contains 2.48×10^{25} molecules, whereas at a pressure of 10^{-7} Pa (10^{-5} Torr), a very high vacuum, it contains 2.5×10^{13} molecules. Indeed, any volume and pressure normally used in the laboratory will contain a large number of molecules. (2) Adjacent molecules are separated by distances that are large compared with their individual diameters. If we could stop all molecules instantaneously and place them on the coordinates of a grid, their average spacing would be about 3.4×10^{-9} m at atmospheric pressure. The diameter of most molecules is of order $2\text{–}6\times10^{-10}$ m. At this pressure, their separation distance is about 6–15 times their diameter. At very low pressure, say, 10^{-7} Pa (10^{-5} Torr), the separation distance is about 3×10^{-5} m. (3) Molecules are in a constant state of motion. All directions of motion are equally likely, and all velocities are possible, although not equally probable. (4) Molecules exert no force on one another except when they collide. If this is true, then molecules will travel in straight lines until they collide with a wall or with one another.

A Users Guide to Vacuum Technology, Fourth Edition. John F. O'Hanlon and Timothy A. Gessert.
© 2024 John Wiley & Sons, Inc. Published 2024 by John Wiley & Sons, Inc.

Using these assumptions, many interesting properties of ideal gases have been derived, some of which are reviewed here.

2.1.1 Velocity Distribution

As the individual molecules move, they collide elastically. Particles change energy after collision, but the pair conserves energy and momentum. We stated that all velocities are possible, but not with equal probability. Based on these four assumptions, Maxwell and Boltzmann independently calculated the distribution of particle velocities.

$$\frac{dn}{dv} = \frac{2N}{\pi^{1/2}}\left(\frac{m}{(2kT)}\right)^{3/2} v^2 e^{-mv^2/(2kT)} \tag{2.1}$$

The particle mass is m, T is Kelvin temperature, $T(\text{K}) = 273.16 + T(°\text{C})$, N is the total number of particles, and k is Boltzmann's constant. Figure 2.1 illustrates Eq. (2.1) for air at three temperatures. It is a plot of the relative number of molecules between velocity v and $v + dv$. We see that there are no molecules with zero or infinite velocity, and that the peak or most probable velocity v_p, is a function of the average gas temperature. The particle velocity also depends on the molecular mass. The arithmetic mean, or average velocity v_{ave} is useful when describing particle flow.

$$v_{ave} = \left(\frac{8kT}{\pi m}\right)^{1/2}; v_p = \left(\frac{2kT}{m}\right)^{1/2}; v_{rms} = \left(\frac{3kT}{m}\right)^{1/2} \tag{2.2}$$

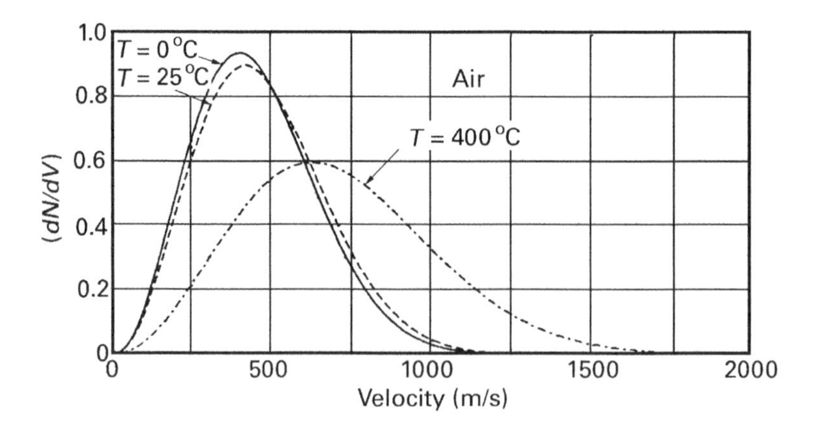

Fig. 2.1 Relative velocity distribution of air at 0°C, 25°C, and 400°C.

The average velocities of several gas and vapor molecules are given in Appendix B.2. Two other velocities, peak, and rms are displayed in Eq. (2.2). The rms velocity is the square root of the mean of each particle velocity squared times the number of particles with that velocity. For M–B statistics, the average velocity is 1.128 times the peak velocity, and the rms velocity is 1.225 times the peak. Figure 2.1 illustrates the temperature dependence of the velocity distribution for an ensemble of particles with identical mass. As the temperature increases, the peak broadens and shifts to a higher velocity. Equation 2.1 may be plotted for different gases having the same temperature. Figure 2.2 illustrates this for a collection of six gases. Figure's 2.1 and 2.2 illustrate two concepts. First, an increase in temperature or decrease in mass causes an increase in gas velocity and gas flux walls. Second, not all the particles in a distribution have the same velocity. The M–B distribution is quite broad—over 5% of the molecules travel at velocities greater than twice the average.

2.1.2 Energy Distribution

Maxwell and Boltzmann also derived an energy distribution, which is based on the same assumptions as the velocity distribution.

$$\frac{dn}{dE} = \frac{2N}{\pi^{1/2}} \frac{E^{1/2}}{(kT)^{3/2}} e^{-E/(kT)} \tag{2.3}$$

From this expression, the average energy can be calculated to be $E_{ave} = (3/2)kT$, and the most probable energy $E_p = (1/2)kT$. Notice that neither the energy distribution nor the average energy is a function of the molecular mass. Each is only a

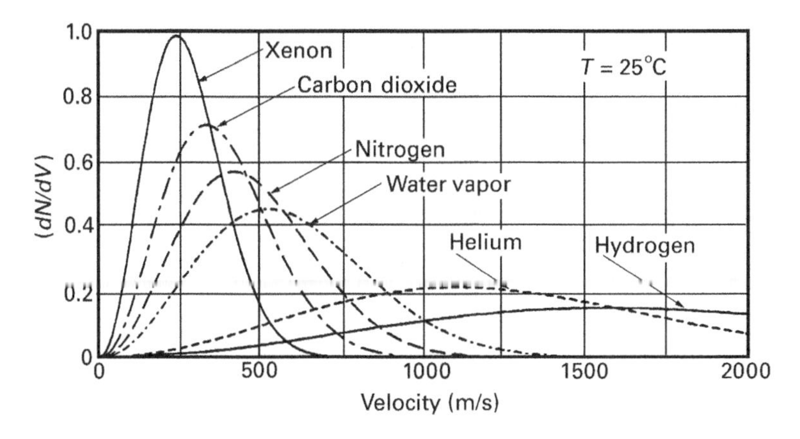

Fig. 2.2 Relative velocity distribution of several gases at 25°C.

function of temperature. The gases depicted in Fig. 2.2 have the same energy distribution, because they all have the same average temperature.

2.1.3 Mean Free Path

The fact molecules are randomly distributed and move with different velocities implies that each travels a different straight-line distance, known as a *free path* before suffering a collision. The free paths sketched in Fig. 2.3, are not all equal. The average, or mean, of the free paths, denoted by λ, is found from kinetic theory:

$$\lambda = \frac{1}{2^{1/2}\pi d_o^2 n} \tag{2.4}$$

where d_o is the molecular diameter in meters, and n is the gas density in molecules per cubic meter. The mean free path is clearly gas density-dependent. For air at room temperature the mean free path is most easily remembered by one of the following two expressions:

$$\lambda(\text{cm}) = \frac{0.67}{P(\text{Pa})} \quad \text{or} \quad \lambda(\text{cm}) = \frac{0.005}{P(\text{Torr})} \tag{2.5}$$

where λ has units of cm, and P is the pressure in Pascal or Torr, respectively. A simple mnemonic for mean free path is that $\lambda = 1$ cm at a pressure of 5 mTorr, or a pressure of 0.67 Pa. From that point, λ scales inversely with pressure, e.g., when the pressure drops by a factor of 10, λ increases by a factor of 10.

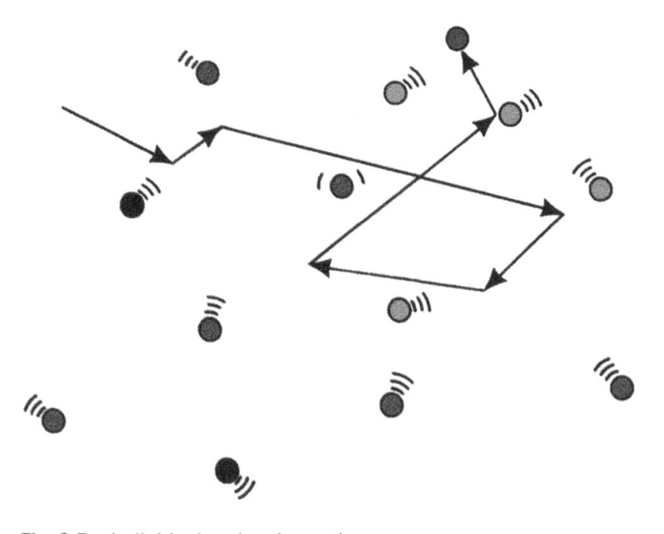

Fig. 2.3 Individual molecular paths.

Kinetic theory describes the distributions of free paths as:

$$N = N'e^{-x/\lambda} \tag{2.6}$$

N' is the number of molecules in the volume and N is the number of molecules that traverse a distance x before suffering a collision. Equation 2.6 states that 63% of the collisions occur in a distance $0 \leq x \leq \lambda$, whereas about 37% of the collisions occur in range $\lambda \leq x \leq 5\lambda$. Only about 0.6% of the particles travel distances greater than 5λ without suffering a collision.

For the case of two gases, a and b, the mean free path is:

$$\lambda_a = \frac{1}{\left[2^{1/2}\pi n_a d_a^2 + \left(1 + \frac{v_b^2}{v_a^2}\right)^{1/2} n_b \frac{\pi}{4}\left(d_a + d_b\right)^2\right]} \tag{2.7}$$

2.1.4 Particle Flux

The concept of particle flux is helpful in understanding gas flow, pumping speed, and evaporation. According to kinetic theory, the flux Γ of an ideal gas striking a unit surface or crossing an imaginary plane of unit area from one side is:

$$\Gamma\left(\text{particles-m}^2\text{-s}^{-1}\right) = nv/4 \tag{2.8}$$

where n is the particle density and v is the average velocity. Substituting the average velocity from Eq. (2.2) yields:

$$\Gamma = n\left(\frac{kT}{2\ m}\right)^{1/2} \tag{2.9}$$

The particle flux is directly proportional to particle density and square root of temperature, and inversely proportional to square root of molecular mass.

2.1.5 Monolayer Formation Time

The time to saturate a surface with one layer of molecules is a function of the molecular arrival rate Γ and molecular size. Assuming each molecule sticks and occupies surface area d_o^2, the time to form a monolayer is:

$$t_{ml} = \frac{1}{\Gamma d_o^2} = \frac{4}{nvd_o^2} \tag{2.10}$$

At ambient temperature, a monolayer of air ($d_o = 0.372$ nm, $\nu = 467$ m/s) will form in about 2.5 s at a pressure of 10^{-4} Pa. The formation time will be longer if the sticking coefficient is less than unity.

2.1.6 Pressure

The absolute pressure on a surface is defined as the rate at which momentum $m\nu$ is imparted to a unit surface. A molecule incident on a surface at an angle φ from the normal will impart a total impulse or pressure of $2m\nu\cos\varphi$. By integrating over all angles in the half-plane, we find the pressure is:

$$P = \frac{1}{3} n m v_{rms}^2 \tag{2.11}$$

However, the total energy of a molecule is proportional to its temperature:

$$E = \frac{m v_{rms}^2}{2} = \frac{3}{2} kT \tag{2.12}$$

Combining Eq's (2.11) and (2.12) yields the Ideal Gas Law.

$$P = nkT \tag{2.13}$$

In Eq. (2.13), n is given in units of m^{-3}, k in Joules/Kelvin, and T in Kelvins. The resulting pressure has units of Pascal (Pa). A Pascal is a Newton per square meter and is the basic SI pressure unit. The beauty of using SI units for calculations is the absence of annoying conversion constants required to calculate pressure in units of Torr with temperature in Celsius, etc.

When using SI, one simply divides by 133.32 to obtain units of Torr or by 100 to obtain units of millibar. For purposes of estimating, one may simply divide SI pressure by 100 to obtain an approximate value in units of Torr. Appendix A.3 gives conversion factors for myriad useful variables. Pressure-dependent values of particle density, average molecular spacing, mean free path, and surface flux are tabulated in Table 2.1 for air at room temperature. The pressure dependence of the mean free path for several other gases is depicted in Appendix B.1.

2.2 Gas Laws

The Ideal Gas Law, Eq. (2.13), summarizes all the earlier experimentally determined gas laws. However, it is useful to review these laws, because they are especially helpful to those with no experience in gas kinetics. When using kinetic theory, we need to remember that the primary assumption of a gas at rest in

Table 2.1 Low-Pressure Properties of Air

Pressure (Pa)	n (m^{-3})	d' (m)	λ (m)	Γ (m^{-2}-s^{-1})
1.01×10^{5} (760 Torr)	2.48×10^{25}	3.43×10^{-9}	6.5×10^{-8}	2.86×10^{27}
100 (0.75 Torr)	2.45×10^{22}	3.44×10^{-8}	6.6×10^{-5}	2.83×10^{24}
1 (7.5 mTorr)	2.45×10^{20}	1.6×10^{-7}	6.6×10^{-3}	2.83×10^{22}
10^{-3} (7.5×10^{-6} Torr)	2.45×10^{17}	1.6×10^{-6}	6.64	2.83×10^{19}
10^{-5} (7.5×10^{-8} Torr)	2.45×10^{15}	7.41×10^{-6}	664	2.83×10^{17}
10^{-7} (7.5×10^{-10} Torr)	2.45×10^{13}	3.44×10^{-5}	6.6×10^{4}	2.83×10^{15}

Particle density n; average molecular spacing d'; mean free path λ; and particle flux on a surface Γ, for $T=22°C$.

thermal equilibrium with its container is not always valid in some practical situations. For example, a pressure gauge close to and facing a high vacuum cryogenic pumping surface will register a lower pressure than when it is close to and facing a warm surface in the same vessel [1]. This and other nonequilibrium situations will be discussed as required. First, let us examine the individual gas laws in equilibrium.

2.2.1 Boyle's Law

In 1662, Robert Boyle demonstrated that the volume occupied by a fixed quantity of gas varied inversely with its pressure when the gas temperature remained constant.

$$P_1V_1 = P_2V_2 \quad (N,T\ \text{constant}) \tag{2.14}$$

This is easily derived from the general law by multiplying both sides of the Ideal Gas Law by the volume and recognizing that the total number is equal to the product of density and volume.

2.2.2 Amontons' Law

Guillaume Amontons discovered the pressure in a confined chamber increased as the temperature increased. Since the gas is in a confined chamber, its particle number and volume are constant. His law can be expressed as:

$$\frac{P_1}{T_1} = \frac{P_2}{T_2} \quad (N,V\ \text{constant}) \tag{2.15}$$

In 1703 he constructed an air thermometer based on this relationship. This later came to be known as the law of Gay-Lussac.

2.2.3 Charles' Law

The French chemist Jacques Charles found in 1787 that gases expanded and contracted to the same extent under the same changes of temperature provided that no change in pressure occurred. Again, by substitution in Eq. (2.13) we obtain:

$$\frac{V_1}{T_1} = \frac{V_2}{T_2} \quad (N, P \text{ constant}) \tag{2.16}$$

2.2.4 Dalton's Law

John Dalton discovered in 1801 that the total pressure of a mixture of gases was equal to the sum of the forces per unit area of each gas taken individually. By the same methods for a mixture of gases, we can develop the relation:

$$P_t = nkT = n_1kT + n_2kT + n_3kT + \cdots = P_t = P_1 + P_2 + P_3 + \cdots \tag{2.17}$$

Equation 2.17, illustrated in Fig. 2.4, is called Dalton's Law of partial pressures and is valid for pressures below atmospheric [2].

2.2.5 Avogadro's Law

In 1811, Amedeo Avogadro observed that the ratio of pressure to number of molecules was proportional for a given temperature and volume:

$$\frac{P_1}{N_1} = \frac{P_2}{N_2} \quad (T, V \text{ constant}) \tag{2.18}$$

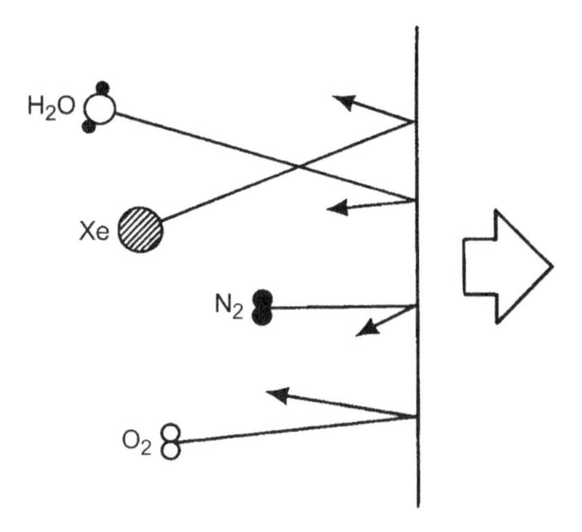

Fig. 2.4 Dalton's law: the total pressure is the sum of the partial pressures.

Two terms, "standard temperature and pressure" (STP) and "Mole" often cause confusion. STP refers to a gas with a temperature of 0°C, and pressure of 1 atmosphere, where 1 atmosphere is currently defined as 101,326 Pa. However, in Chapter 6, we will observe that many manufacturers of thermal mass flow controllers assume a temperature of 20°C when calibrating flow meters; this condition is referred to as "normal temperature and pressure" (NTP). In SI, Avogadro's number of any gas species is $N_0 = 6.02252 \times 10^{23}$ and occupies a molar volume of $V_0 = 0.0224136\,\text{m}^3$ (22.41 L). Avogadro's number of molecules is called a Mole. In SI, the unit of volume is cubic meter, and the unit of mass is kilogram. To avoid confusion, we will use kg-mol when necessary. For example, one kg-mol of oxygen contains 6.02252×10^{26} molecules and weighs 32 kg. Its density at STP is therefore $32\,\text{kg}/22.4136\,\text{m}^3$ or $1.45\,\text{kg/m}^3$.

2.2.6 Graham's Law

In the nineteenth century, Thomas Graham studied the rate of effusion of gases through very small holes in porous membranes. He observed the rate of effusion to be inversely proportional to the square root of the density of the gas provided that the pressure and temperature were held constant. Since the density of a gas is proportional to its molecular weight, Grahams' law can be stated as:

$$\frac{\text{effusion rate}_a}{\text{effusion rate}_b} = \left(\frac{\text{M}_b}{\text{M}_a}\right)^{1/2} \quad \left(P,T \;\; \text{constant}\right) \tag{2.19}$$

Grahams' law explains why a helium-filled balloon loses its gas more quickly than an air-filled balloon.

2.3 Elementary Gas Transport Phenomena

In this section, approximate views of viscosity, thermal conductivity, diffusion, and thermal transpiration are discussed. We state results from kinetic theory without derivation.

2.3.1 Viscosity

A viscous force is present in a gas when it is undergoing shear. Figure 2.5 illustrates two plane surfaces, one fixed and the other traveling in the x direction with a uniform velocity. The coefficient of absolute viscosity η is defined by:

$$\frac{F_x}{A_{xz}} = \eta \frac{du}{dy} \tag{2.20}$$

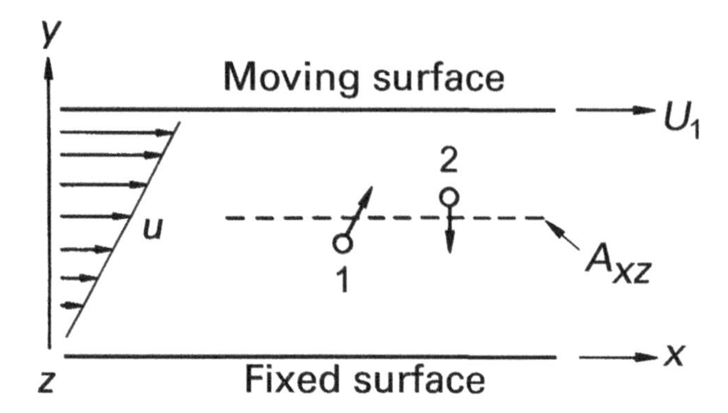

Fig. 2.5 Origin of the viscous force in a gas.

where F_x is the force in the x direction, A_{xz} is the surface area in the x–z plane, and du/dy is the rate of change of the gas velocity at this position between the two surfaces. Because the gas stream velocity increases as the moving plate is approached, those molecules crossing the plane A_{xz} from below (particle 1 in Fig. 2.5) will transport less momentum across the plane than will those crossing the same plane from above (particle 2 in Fig. 2.5). The result is that molecules crossing from below the plane will, on the average, reduce the momentum of the molecules from above the plane, in the same manner molecules crossing from above the plane will increase the momentum of those molecules below the plane. To an observer this viscous force appears to be frictional; actually, it is merely the result of momentum transfer between the separated plates by successive molecular collisions. Again, from kinetic theory, the coefficient of viscosity is:

$$\eta = \frac{1}{3} nmv\lambda \tag{2.21}$$

When gas density is measured in units of m^{-3}, molecular mass in kg, velocity in m/s, and the mean free path in m, η will have SI units of (N-s)/m^2, or Pa-s (recall 1 N/m^2 = 1 Pa). One Pa-s is equal to 10 Poise. A more rigorous treatment of viscosity yields a result with the slightly different numerical coefficient of 0.499 instead of 1/3 [3]. Using this result, we obtain:

$$\eta = \frac{0.499(4mkT)^{1/2}}{\pi^{3/2} d_o^2} \tag{2.22}$$

Kinetic theory predicts that viscosity should increase as square root of mass and temperature and decrease as the square of molecular diameter. An interesting result of this simple theory is that viscosity is independent of gas density or pressure. Observe that particle density does not appear in Eq. (2.22). This theory is valid only in a limited pressure range. If there were a perfect vacuum between the two plates, there would be no viscous force because there would be no mechanism for transferring momentum. This understanding leads to the conclusion that Eq. (2.22) is valid as long as the distance between the plates is greater than the mean free path, e.g., in a viscous gas.

In a rarefied gas, where the ratio of the mean free path to plate separation is much greater than 1, the viscous force can be expressed as:

$$\frac{F}{A_{xz}} = \left(\frac{Pmv}{4kT}\right)\frac{U_1}{C} \tag{2.23}$$

The term in parentheses is referred to as the free-molecular viscosity. The free-molecular viscosity is directly proportional to the molecular density ($n = P/kT$) that is available to transfer momentum between the plates. It is valid in the region $\lambda \gg y$. The constant C in Eq. (2.25), the Cunningham Correction Factor [4], is related to molecular slip. For $\lambda \gg y$, $C \sim 1$.

Figure 2.6 illustrates the magnitude of the viscous force generated by air at 22°C between two plates moving with a relative velocity of 100 m/s for three plate separations. Equations 2.20 and 2.22 were used to calculate the asymptotic value of the viscous drag at high pressures and Eq. (2.23) was used to calculate the free molecular limit. A more complete treatment of the intermediate or viscous slip region is given elsewhere [5]. The viscous shear force is independent of the plate spacing as long as

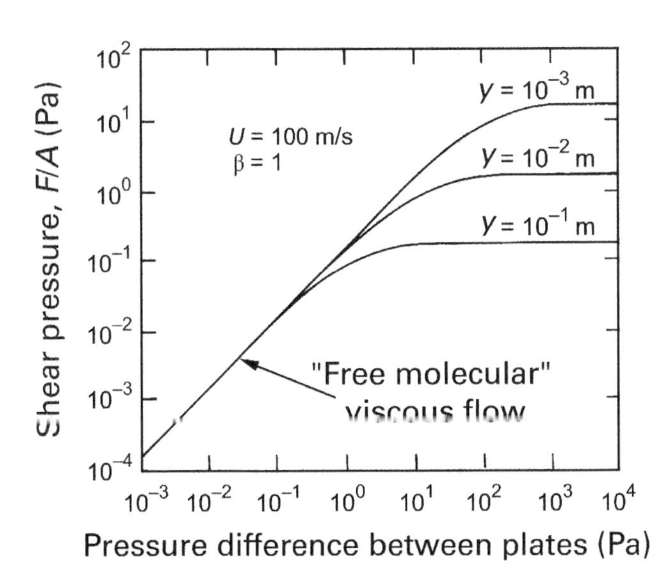

Fig. 2.6 Viscous shear force between two plates at 22°C.

the mean free path is larger than the spacing. This idea was used by Langmuir [6] to construct a viscosity gauge in which viscous damping was proportional to pressure.

2.3.2 Thermal Conductivity

Heat conductivity between two infinite parallel plates is explained by kinetic theory in a manner analogous to that used to explain viscosity. The diagram in Fig. 2.5 could be relabeled to make the top plate stationary at temperature T_2, and the lower plate a stationary plate at a temperature T_1, where $T_1 < T_2$. Heat conduction can be modeled by noting that the molecules moving across the plane toward the hotter plate carry less energy than those moving across the plane toward the cooler surface. The heat flow can be expressed as:

$$H = AK\frac{dT}{dy} \tag{2.24}$$

where H is the heat flow and K is the heat conductivity. The simple theory predicts that the heat conductivity K is expressed by $K = \eta c_v$, where η is the viscosity and c_v is the specific heat at constant volume. This simple theory is correct only to an order of magnitude. A more detailed analysis, which accounts for molecular rotational and vibrational energy, yields:

$$K = \frac{1}{4}(9\gamma - 5)\eta c_v \tag{2.25}$$

where γ is the ratio of specific heats c_v/c_p. When η has the units of Pa-s and c_v has units (J/kg)/K, then K will have units of (W/m)/K. At room temperature, the heat conductivity and viscosity both increase with temperature and decrease with particle diameter. For infinite parallel plates, K does not depend on pressure, as long as the mean free path is smaller than the plate spacing. Heat transfer between two parallel plates in the low-pressure region is given in Eq. (2.26) [7]. Here, Λ is the free-molecular heat conductivity and α_1, α_2, and α_3 are the respective accommodation coefficients of the cold surface-1, the hot surface-2, and the system.

$$E_o = \alpha\Lambda P(T_2 - T_1)$$
$$\alpha = \frac{\alpha_1\alpha_2}{\alpha_1 + \alpha_2 - \alpha_1\alpha_2} \qquad \Lambda = \frac{1}{8}\frac{(\gamma+1)v_1}{(\gamma-1)T_1} \tag{2.26}$$

This equation has the same general form as Eq. (2.23) for free-molecular viscosity. If molecules can thermally equilibrate with the surface, say, by making many small collisions on a rough surface, α will have a value approaching 1. If the same surface is smooth and the molecule recoils without gaining or losing energy, α will approach zero.

The kinetic picture of heat conductivity is rather like viscosity except that *energy* transfer determines the thermal conductivity, and *momentum* transfer determines the viscous drag. Even so, Fig. 2.6 can be sketched for the thermal conductivity of a gas between two parallel plates, where the vertical axis has dimensions of heat flow. In SI Λ has units of W-m^{-2}-K^{-1}-Pa^{-1}, whereas E_o has units of W/m^2. Tables of accommodation coefficient are given elsewhere [3,8]. Accommodation coefficients dependent on the material, its cleanliness, surface roughness, and also on gas adsorption.

When the heated parallel plate is replaced by a heated fine wire, the situation changes. In the case of a heated fine wire, the upper "knee" of the curve is not dependent on the ratio of λ to plate separation, but rather on the ratio of λ to hot wire diameter [9,10]. Both thermocouple and Pirani gauges operate in a region in which the heat conduction from the heated wire is linearly dependent on pressure. However, the "knee" in their linear range begins at a mean free path equal to a few multiples of the wire diameter and not at a mean free path related to the wire-envelope distance.

2.3.3 Diffusion

Diffusion is a complex phenomenon. This discussion has been simplified by restricting it to the situation in a vessel that contains two gases whose compositions vary slowly throughout the vessel but whose total number density is the same everywhere. The coefficient of diffusion D, of two gases is defined in terms of their particle fluxes Γ_1 and Γ_2:

$$\Gamma_1 = -D\frac{dn_1}{dx} \quad \Gamma_2 = -D\frac{dn_2}{dx} \tag{2.27}$$

These fluxes result from the partial pressure gradient of the two gases. The result from kinetic theory, when corrected for the Maxwellian distribution of velocities and for velocity persistence [10], is:

$$D_{12} = \frac{8\left(\frac{2kT}{\pi}\right)^{1/2}\left(\frac{1}{m_1}+\frac{1}{m_2}\right)^{1/2}}{3\pi(n_1+n_2)(d_{o1}+d_{o2})^2} \tag{2.28}$$

where D_{12} is the constant of inter-diffusion of the two gases 1 and 2. In SI, the diffusion constant, has units of m^2/s. For the case of self-diffusion (gas 2 = gas 1), the coefficient is:

$$D_{11} = \frac{4}{3\pi n d_o^2}\left(\frac{kT}{\pi m}\right)^{1/2} \tag{2.29}$$

If the density n is replaced by P/kT, it becomes apparent that the diffusion constant is approximately proportional to $T^{3/2}$ and P^{-1}. The diffusion equation, Eq. (2.30),

$$\left(\frac{dC}{dt}\right) = -D\frac{d^2C}{dt^2} \tag{2.30}$$

whose solutions we do not describe here, contain the term $(Dt)^{1/2}$. This term has the dimensions of length and is called "diffusion length." For long times the diffusion "front" moves through the gas in proportion to $(Dt)^{1/2}$. Diffusion constants for several gases in air are given in Appendix B.2.

Examination of Eq. (2.29) shows the diffusion coefficient will become infinitely large as the density of molecules goes to zero. This does not happen. When the pressure becomes low enough so that the mean free path is much larger than the dimensions of the container, say the diameter of a pipe, gas diffusion is limited by molecules recoiling from walls, rather than from each other. In this regime, $\lambda \gg d$, and the diffusion coefficient is given by:

$$D = \frac{2}{3}rv \tag{2.31}$$

where r is the radius of the pipe and v is the thermal velocity. This is called the Knudsen diffusion coefficient for a long capillary [11].

2.3.4 Thermal Transpiration

When a tube or orifice connects two chambers of different temperatures, their relative pressures depend on the nature (rarefied gas or viscous gas) of the gas in the connecting tubing. The nature of the gas in the tubing or orifice is characterized by (λ/d), where d is the diameter of the connecting tube or orifice. For $\lambda \ll d$ the pressure is everywhere the same in both chambers, $P_1 = P_2$. The densities in the two chambers are related by:

$$\frac{n_2}{n_1} = \frac{T_1}{T_2} \tag{2.32}$$

When the orifice diameter is much smaller than the mean free path $\lambda \gg d$, the flux of gas through the orifice from chamber 1 to chamber 2 is given by Eq. (2.8).

$$\Gamma_{1,2} = \frac{n_{1,2}}{4}\left(\frac{8kT_{1,2}}{\pi m}\right)^{1/2} = \frac{P_{1,2}}{(2kT\pi m)^{1/2}} \tag{2.33}$$

In steady state, the net flux between the two chambers must be zero, and for this example, $\Gamma_1{\rightarrow}_2 = \Gamma_2{\rightarrow}_1$. This results in different chamber pressures.

$$\frac{P_1}{P_2} = \left(\frac{T_1}{T_2}\right)^{1/2} \tag{2.34}$$

Equations 2.32 and 2.34 can be used to calculate the pressures within furnaces or cryogenic enclosures when the pressure gauge is located outside the enclosure at a different temperature. Equation 2.32 is used at high pressure, where $\lambda < d/10$, and Eq. (2.34) is used at low pressure, where $\lambda > 10d$.

Thermal transpiration was discovered by Neumann [12] and studied by Maxwell [13], who predicted the square root dependence given in Eq. (2.34). The geometry and reflectivity from walls of the connecting tubing introduce deviations from the theory. Siu [14] has studied these effects and has predicted that Eq. (2.34) is valid in short tubes only for specular reflection, and in long tubes, only for diffuse reflection.

References

1 Moore, R.W., Jr., in *Transactions of the 8th National Vacuum Symposium of the American Vacuum Society and 2nd International Vacuum Congress, 1961*, L. Preuss, Ed., Vol. **1**, Pergamon, New York, 1962, p. 426.
2 Kennard, E.H., *Kinetic Theory of Gases*, McGraw-Hill, New York, 1938, p. 9.
3 Reference 2, pp. 135–205 and 291–337.
4 Hinds, W.C., *Aerosol Science and Technology*, 2nd ed., Wiley, New York, 1999, p. 49.
5 S. Dushman, "Kinetic Theory of Gases", Chapter 1 in *Scientific Foundations of Vacuum Technique*, 2nd ed., J. M. Lafferty, Ed., Wiley, New York, 1962, p. 35.
6 Langmuir, I., *Phys. Rev.* **1**, 337 (1913).
7 Reference 5, p. 6.
8 Reference 5, p. 68.
9 von Ubisch, H., *Vak. Tech.* **6**, 175 (1957).
10 Pirani, M. and Yarwood, J., *Principles of Vacuum Engineering*, Reinhold, New York, 1961, p. 100.
11 For example, see Lund, L.M. and Berman, A.S., *J. Appl. Phys.* **37**, 2489 (1966).
12 Neumann, C., *Math Phys. K.* **24**, 49 (1872).
13 Maxwell, J.C., *Philos. Trans. R. Soc. London* **170**, 231 (1879).
14 Siu, M.C.I., *J. Vac. Sci. Technol.* **10**, 368 (1973).

3

Gas Flow

In this chapter we discuss gas flow at reduced pressures. The mathematical descriptions of gas flow are complex, and the nature of their solutions depends on pressure, temperature, flow rate, gas properties, duct geometry, and surface properties. We begin by defining flow regimes and introducing the concepts of throughput, speed, conductance, and mass flow. We characterize gas throughput and conductance for several types of flow. We describe approximation techniques and probabilistic methods for solving complex problems, such as flow in ducts containing entrance and exit orifices, apertures, elbows, diameter changes, and practical shapes.

3.1 Flow Regimes

Gas flow regimes are characterized by the *nature* and by *relative quantity* of gas flow. The nature of the gas is determined by comparing the average spacing between molecules to a pipe dimension. A viscous gas is depicted in Chapter 1, Fig. 1.1 (left); its flow is called continuum flow, or viscous flow. A molecular gas is also depicted in Fig. 1.1 (right); its flow is called molecular flow.

Continuum, or viscous, flow can be further divided into turbulent, laminar viscous, and choked. Turbulent flow is chaotic, like the flow behind a moving vehicle or rising smoke from a campfire. Laminar viscous, also called stream flow, occurs when the velocity is small, and gas moves in laminar streamlines analogous to water flow in a garden hose. Choked flow occurs when the velocity of a gas reaches its maximum speed—the speed of sound. In continuum flow, the diameter of the pipe is much greater than the mean free path and the nature of the gas flow is determined by *gas–gas* collisions.

A Users Guide to Vacuum Technology, Fourth Edition. John F. O'Hanlon
and Timothy A. Gessert.
© 2024 John Wiley & Sons, Inc. Published 2024 by John Wiley & Sons, Inc.

A molecular gas has a mean free path much longer than a pipe diameter, and its flow is entirely determined by *gas–wall* collisions. The flow in this region is called molecular flow. The transition from continuum to molecular flow is not abrupt but occurs over a range of densities. In the transition region, gas molecules collide with each other *and* with walls.

Knudsen's number is used to characterize the *state* of a gas. Knudsen's number Kn, is a dimensionless ratio of the mean free path to a characteristic dimension of a component or a system, say, the diameter of a pipe:

$$\mathrm{Kn} = \frac{\lambda}{d} \tag{3.1}$$

Continuum (viscous) gas flow is characterized by $\mathrm{Kn} < 0.01$, and a molecular gas by $\mathrm{Kn} > 1$.

Continuum, or viscous flow can be either turbulent, Fig. 3.1*a*; laminar viscous, Fig. 3.1*b*; or choked viscous flow, Fig. 3.1*c*. The boundary between these regions can be expressed in terms of Reynolds' dimensionless number, **R**, for round pipes:

$$\mathbf{R} = \frac{U\rho d}{\eta} \tag{3.2}$$

where ρ is the mass density (kg/m^3), of the gas of viscosity η flowing with stream velocity U in a pipe of diameter d. Reynolds' number is a ratio of the shear stress due to turbulence to the shear stress due to viscosity. Alternatively, it tells something about the forces necessary to drive a gas system in relation to the forces of

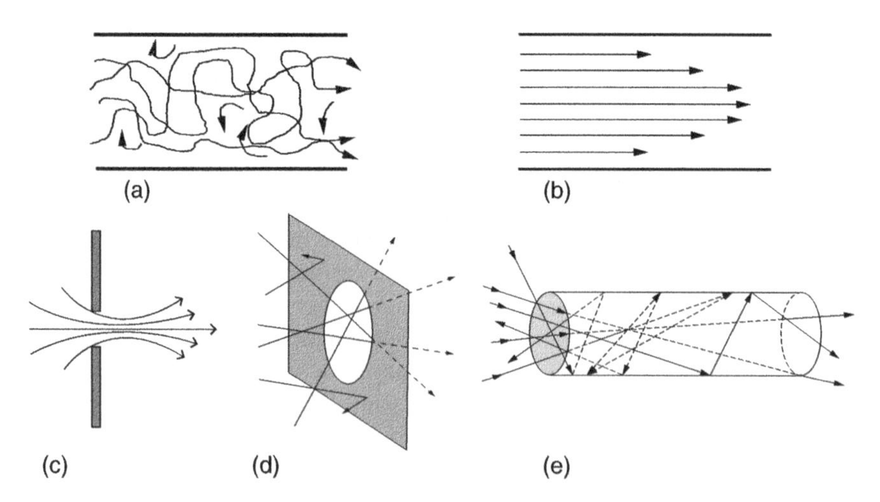

Fig. 3.1 Examples of flow regimes: (a) turbulent flow in a pipe, (b) laminar viscous flow in a pipe, (c) choked viscous flow through an orifice, (d) molecular flow through an orifice, and (e) molecular flow in a pipe.

dissipation due to viscosity. Reynolds [1] found distinct flow situations dynamically similar when this dimensionless number was the same. We will use Reynolds' number and Kundsen's number, together, to describe individual types of gas flow.

When **R** >2200, the flow is turbulent, and when **R** <1200 the flow is laminar viscous [2]. In the region $1200<$ **R** <2200 the flow is either laminar viscous or turbulent, depending on the geometry of the inlet and outlet and on the nature of piping irregularities. During laminar viscous flow, Fig. 3.1*b*, the ordered flow of gas in streamlines, occurs in the region bounded by **R** <1200 Kn <0.01.

We define a rarified gas as having a Kn >1; its flow is called molecular flow. See Fig. 3.1*d,e*. The nature of molecular flow is very different from viscous flow because gas–wall collisions predominate. We do not use Reynolds' number to characterize molecular flow, because classical viscosity cannot be defined in a rarefied gas. In the molecular flow region, the definition of viscosity involves gas–wall collisions, not gas–gas collisions. For most surfaces, diffuse reflection at a wall is a good approximation, i.e., each particle arrives, sticks, rattles around, and is reemitted in a direction independent of its incident path. Thus, it is likely that a particle entering a pipe in which $\lambda \geq d$ may not be transmitted but may be returned to its entrance. In molecular flow, gas molecules do not collide with one another—only with surfaces—and they can flow in opposite directions without interaction.

In the intermediate region, $1>$ Kn >0.01, gas is neither viscous nor molecular, and its flow is difficult to treat theoretically. This region, the slip flow region, is several mean free paths wide, the velocity at the wall is not zero, and reflection is not diffuse, as in free molecular flow. Let us begin by defining throughput, speed, conductance, and mass flow, and develop some useful gas flow formulas.

3.2 Flow Concepts

Throughput, conductance, speed, mass flow, molar flow, and molecular flow are important concepts that are sometimes confused and warrant discussion.

Throughput is the quantity of gas ($P \times V$) that passes a plane in unit time. In mathematical terms, $Q = d/dt(PV) = VdP/dt$, provided the volume does not change with time. If the volume is also changing, as in a piston, then $Q = VdP/dt + PdV/dt$. Its SI unit is the Pa-m^3/s. Since 1 Pa $=$ 1 N/m^2, and 1 J $=$ 1 N-m, throughput could be expressed as J/s or Watts (1 Pa-m^3/s $=$ 1 W). Conceptually, throughput is the energy per unit time (power) crossing a plane. The energy in question is not internal energy of the gas, but the energy of motion. If the gas is not moving, this energy is zero. Expressing flow in units of Watts is awkward and not used, but it helps to explain the concept that throughput is the flow of energy. Throughput is a *volumetric* dimension (*volume* of gas/unit time).

Speed and conductance are two additional volumetric variables of major importance. Speed is defined as the *volume* of gas passing a plane in unit time divided by the pressure *at the location* where the pressure is measured.

$$S = \frac{Q}{P} \tag{3.3}$$

Its SI unit is the m^3/s. Speed measurements at different locations in a pipe with a constant flow, will yield different numerical values, because pressure changes with location. This definition is valid in all flow regimes. Speed has no meaning unless it is referenced to a location.

The ease with which gas flows in a pipe or object of any shape, is dependent on the geometry and construction of the object. The ratio of throughput to the *pressure drop across the object* yields a property known as the conductance of the object.

$$C = \frac{Q}{P_1 - P_2} \tag{3.4}$$

Like speed, the SI unit of conductance is the m^3/s; however, related throughput units of L/s are frequently used for both conductance and speed. Also, units of Pa-L/s are acceptable for flow. Conversion factors are found in Appendix A.3. Unless stated immediately adjacent, all formulas in this chapter use the cubic meter as the volume unit.

The definitions of speed and conductance are often confused. Speed is the property of a *location*. Conductance is the property of an *object*. Figure 3.2 graphically describes this distinction. The pressures P_1 and P_2 in the figure and in Eq. (3.4), refer to pressures measured at each end of the object. For those whose first introduction to flow was with electricity, Q is analogous to electrical current, P is analogous to an electrical potential, and conductance is analogous to susceptance (1/Resistance). As with electrical charge flow, there are situations in which the

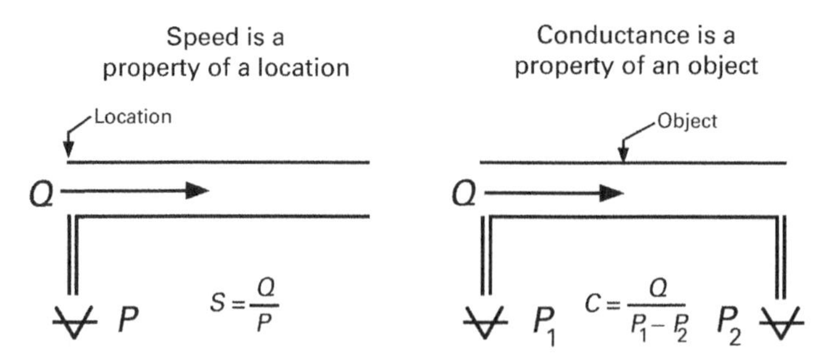

Fig. 3.2 Pictorial description of speed and conductance. Speed is defined at the location of the pressure gauge. Conductance is a property of an object.

gas conductance is nonlinear, that is, a function of the pressure in the pipe. Unlike electrical charge flow, there are many situations like viscous flow in which conductance varies with pressure, and in molecular flow where conductance depends not only on the shape of the object but on the shape of adjacent objects. We will explore this last issue in detail in Section 3.4.5. We will find the electrical analogy does not apply in these cases.

Mass flow, molar flow, and molecular flow are, respectively, the quantity of substance in units of either kg, kg-mole, and molecular number, that pass a plane in unit time. Equation 3.5 describes the relationship between molar flow and throughput.

$$\mathrm{N}'(\text{kg-mol/s}) = \frac{Q}{N_o kT} = \frac{Q}{RT} \tag{3.5}$$

In a similar fashion, mass flow is related to throughput by:

$$\mathrm{N}'(\mathrm{kg/s}) = MQ/N_o kT \tag{3.6}$$

Equations 3.5 and 3.6 show that throughput Q, *cannot* be converted to mass or molar flow unless the temperature is known. It is in many ways unfortunate that vacuum technologists have chosen to use a volumetric unit, because it is an incomplete description of flow. Volumetric flow does not conserve mass. A spatial change in the temperature can alter throughput without altering mass flow. We discuss mass flow in Chapter 6 (flow meters), and in Chapter 14 (cryopumps). Mass flow must be known to calculate supply gas flow to reactive ion etching, "plasma ashing" (removal of photoresist), and reactive sputtering systems.

3.3 Continuum Flow

A gas is called a viscous gas when $\mathrm{Kn} < 0.01$. The flow in a viscous gas can be either turbulent, if $\mathbf{R} > 2200$, or laminar viscous, if $\mathbf{R} < 1200$. Equation 3.2 can be put in a more useful form by replacing the stream velocity *U*, with:

$$U = \frac{Q}{AP} \tag{3.7}$$

In Eq. (3.7), *Q* is the gas flow in Pa-m^3/s, *A* is the area (m^2) and *P* is the pressure (Pa). If we replace mass density, using the ideal gas law, Eq. (3.2) becomes:

$$\mathbf{R} = \frac{4m}{\pi kT\eta}\frac{Q}{d} \quad \text{and for air at 22°C,} \quad \mathbf{R} = 8.41\times10^{-4}\frac{Q(\text{Pa-L/s})}{d(\text{m})} \tag{3.8}$$

In routine use, turbulent flow occurs infrequently. Reynolds' number can reach high values in the piping of a large roughing pump during initial phase of evacuation from atmosphere. A Reynolds' number as high as $\mathbf{R} = 16{,}000$ is possible in a 250-mm-diameter roughing line connecting a chamber to a 47-L/s (100 cfm) pump. When pumping begins under these conditions, expansion cooling creates particles. This is an important topic that is discussed in Section 18.1.2 and in Section 21.3. However, in industries where economics dictates rapid pumping, e.g., coating large sheets of architectural glass, initial roughing is accompanied by unwanted adiabatic cooling.

In the aforementioned systems and in flow restrictors, the gas velocity reaches the velocity of sound, and it is flow is called "choked flow." Further reduction of the downstream pressure cannot be communicated to the high-pressure side and the flow is limited to a maximum flow. The limiting value depends on element geometry, upstream pressure, temperature, molecular weight, and specific heat ratio, γ, of the gas. Chokes can take the form of orifices or short tubes. Shapiro [3] discusses this subject in detail.

Rather than dividing the discussion of continuum flow into viscous, turbulent, and choked, it is convenient to discuss the flow in terms of the pipe geometry. We choose to divide this discussion into flow through orifices, long tubes, and the more complex case—short tubes. We give exact solutions for calculating flow in orifices and long tubes, and a model for short tubes.

3.3.1 Orifice

For a tube of zero length, an orifice, flow is a rather complicated function of pressure. Consider a fixed high pressure, say atmospheric pressure, on one side of the orifice with a variable pressure between atmosphere and zero on the downstream side such as depicted in Fig. 3.1c. (*Note:* upstream pressure does not have to be atmosphere, it can be any pressure in which the gas is in its viscous *state*—we will show pressure *ratios* govern orifice behavior.)

As the downstream pressure is reduced, gas flow through the orifice increases until it reaches a maximum. Gas flowing through an orifice expands adiabatically (i.e., no heat exchange with surroundings). Its flow is described in a complex way through the ratio (P_2/P_1), where P_2 is the downstream pressure and P_1 is the upstream pressure. Turbulent flow is further divided into two regimes: choked, when (P_2/P_1) is less than a critical pressure *ratio*, and unchoked, when (P_2/P_1) is greater than a critical pressure ratio. The *critical pressure ratio* is $P_2/P_1 = (2/(\gamma+1))\gamma^{/(\gamma-1)}$. Values of specific heat ratio, γ, for monatomic, diatomic, and triatomic gases are given in Appendix B.4. The critical pressure ratio is 0.5282 for diatomic gases like dry air, and 0.4871 for monatomic gases like helium and argon.

Adiabatic flow in the unchoked regime is described by:

$$Q = AP_1C'\left(\frac{2\gamma}{\gamma-1}\frac{kT}{m}\right)^{1/2}\left(\frac{P_2}{P_1}\right)^{1/\gamma}\left[1-\left(\frac{P_2}{P_1}\right)^{(\gamma-1)/\gamma}\right]^{1/2}$$
$$\text{for } 1 \geq (P_2/P_1) > \left(2/(\gamma+1)\right)^{\gamma/(\gamma-1)} \tag{3.9}$$

The factor C' accounts for the reduction of cross-sectional area, A, because the high-speed gas stream continues to decrease in diameter after it passes through the orifice. See Fig. 3.1c. This phenomenon is known as the *vena contracta*. For thin, circular orifices, C' is ~0.85.

Speed is defined as flow/pressure Eq. (3.3). We can normalize this by dividing by area to obtain speed per unit area (S/m^2). This normalized *unchoked* speed, $Q/(AP_1)$, is plotted for air and helium at 25°C in Fig. 3.3, on the right-hand side of the graph.

If the downstream pressure P_2 is further reduced, the gas flow will not increase, because the gas in the orifice is traveling at the speed of sound and cannot

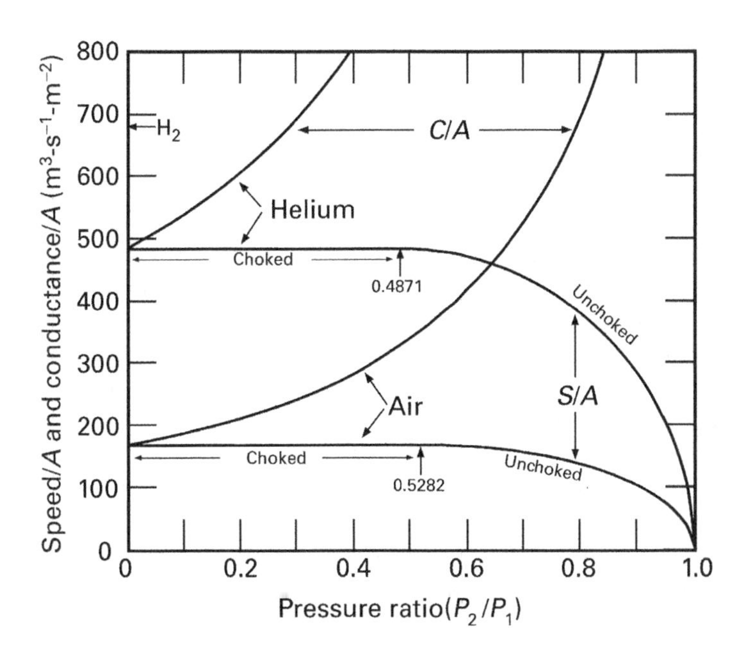

Fig. 3.3 Normalized speed (S/A) and conductance (C/A) of a thin orifice in continuum flow for air and helium at 25°C. Below the critical flow pressure ratio, the flow is choked and constant. Above this ratio, the flow is unchoked and decreases to zero as the ratio →1. Orifice conductance increases with pressure ratio because ΔP rapidly becomes zero as its ratio →1.

communicate with the high-pressure side of the orifice. In this region, P_2 cannot influence the flow so long as $P_2/P_1 < (2/(\gamma+1))\gamma^{/(\gamma-1)}$, the critical pressure ratio. The choked limit is important in describing flow restrictors—devices that control gas flow. In this limit, orifice flow is described by:

$$Q = AP_1C'\left(\frac{kT}{m}\frac{2\gamma}{\gamma+1}\right)^{1/2}\left(\frac{2}{\gamma+1}\right)^{1/(\gamma-1)} \quad \text{for } P_2/P_1 \leq \left(2/(\gamma+1)\right)^{\gamma/(\gamma-1)} \tag{3.10}$$

For air at 25°C, the choked orifice flow limit becomes:

$$\begin{gathered} Q\left(\text{Pa-m}^3/\text{s}\right) = 200P_1AC' = 170P_1\text{A} \\ Q\left(\text{Pa-L/s}\right) = 2\times10^5 P_1\left(\text{Pa}\right)A\left(\text{m}^2\right)C' = 1.7\times10^5 P_1\left(\text{Pa}\right)A\left(\text{m}^2\right) \\ \left(\text{air}, 25°\text{C, and } P_2/P_1 \leq 0.525\right) \end{gathered} \tag{3.11}$$

Normalized speed in the choked region, $Q/(P_1A)$, is also plotted in Fig. 3.3 for air and helium on the left-hand side of the graph.

Conductance is an extremely useful quantity commonly used to characterize components. It is the property of a component, as illustrated in Fig. 3.2. For many objects in the molecular flow region, it has a fixed value, analogous to a fixed resistor (1/conductance) used in electrical circuits. However, in some devices such as orifices in viscous flow, conductance values are flow dependent. In these cases, we need to use care when interpretating conductance values.

For example, examine the conductance values for helium and air chokes are plotted in Fig. 3.3. Clearly, they change with pressure drop. In the choked regime, gas flow is constant. It is the flow through the choke that is important, not its upstream pressure. In the choked regime, an orifice is analogous to an electrical "constant-current source." Constant-current sources provide a constant flow of charge—or in our case—gas. In the electrical analogue, one does not write loop equations containing voltage drops across a current source, rather node equations containing their constant current; it is the same in gas flow. This varying value of conductance has little utility. In the unchoked region, numerical values of conductance approach infinity as $P_2/P_1 \rightarrow 1$. In this limit, $P_2 = P_1$, and $Q \rightarrow 0$. In this limit, adiabatic conditions no longer exist, and conductance has no meaning.

3.3.2 Long Round Tube

The Navier–Stokes equations provide a general description of viscous flow, and their solutions are complex. The simplest and most familiar solution for long straight tubes was independently derived by Poiseuille and Hagen; it is known as the Hagen–Poiseuille equation:

$$Q = \frac{\pi d^4}{128\eta l}\frac{(P_1 + P_2)}{2}(P_1 - P_2) \tag{3.12}$$

For air at room temperature, Eq. (3.12) becomes:

$$Q(\text{Pa-m}^3/\text{s}) = 718.5\frac{d^4(P_1 + P_2)}{l}(P_1 - P_2) \tag{3.13}$$

We define conductance in laminar flow using Eq.'s (3.4) and (3.12) and observe that it is not constant; it is proportional to the *average* pipe pressure.

$$C = \frac{\pi d^4}{128\eta l}\frac{(P_1 + P_2)}{2} = \frac{\pi d^4}{128\eta l}P_{\text{ave}} \tag{3.14}$$

Equation 3.13 illustrates two important properties of laminar viscous flow: first, C is proportional to the fourth power of diameter—a 20% increase in pipe diameter will double its conductance; second, conductance is inversely proportional to length. This is why rough pump piping is smaller in diameter than that used in high vacuum plumbing.

The Poiseuille equation is valid when four conditions are met: (1) fully developed flow—the velocity profile is not position-dependent, (2) laminar viscous flow, (3) zero wall velocity, and (4) incompressible gas. Condition 1 holds for long tubes in which the flow lines are laminar. The criterion for fully developed flow was determined by Langhaar [4] who showed that a distance of $l_e = 0.0568d\mathbf{R}$ into the pipe was required before the flow streamlines developed into their parallel, steady-state profile. For air at 22°C this reduces to l_e (m) = 0.0503Q (Pa-m^3/s). Assumptions 2 and 3 are satisfied when $\mathbf{R} < 1200$ and Kn < 0.01. The condition of incompressibility holds true, if the Mach number U, the ratio of gas-to-sound velocity, is <0.3.

$$\mathbf{U} = \frac{U}{U_{\text{sound}}} = \frac{4Q}{\pi d^2 P U_{\text{sound}}} < \frac{1}{3} \tag{3.15}$$

For the special case of air at 25°C that translates to:

$$Q(\text{Pa-L/s}) < 9.0 \times 10^5 d^2 P \tag{3.16}$$

This value of flow may be exceeded in many cases and would render the Poiseuille equation incorrect. These underlying assumptions of the Poiseuille equation limit its usefulness in many practical situations.

Relationships for viscous flow between long, coaxial cylinders and long tubes of elliptical, triangular, and rectangular cross-section have been tabulated by

Table 3.1 Dimensionless Constant *Y* Used in Rectangular Duct Eqs. (3.17) and (3.18)

h/b	*Y*
1.0	0.4217
0.8	0.41
0.6	0.31
0.4	0.30
0.2	0.175
0.1	0.0937
0.05	0.0484
0.02	0.0197
0.01	0.0099

Holland et al. [5]. Williams et al. [6] give the relation for flow in a long rectangular duct for air at 20°C, as:

$$Q(\text{Pa-L/s}) = 4.6Y \frac{b^2(\text{cm}^2)h^2(\text{cm}^2)}{l(\text{cm})} \frac{(P_1 + P_2)}{2}(P_1 - P_2) \tag{3.17}$$

where the duct cross-section dimensions b and h and the length l are given in cm. The function $Y(h/b)$ is obtained from Table 3.1. In the limit $h \ll b$, the airflow reduces to one-dimensional solution [7]:

$$Q(\text{Pa-L/s}) = 4.6Y \frac{h^3(\text{cm}^3)b(\text{cm})}{l(\text{cm})} \frac{(P_1 + P_2)}{2}(P_1 - P_2) \tag{3.18}$$

Note that h, b, and l are also given in units of cm in Eq.'s (3.17) and (3.18). The flow in Eq.'s (3.17) and (3.18), like that in Eq. (3.12), is inversely proportional to viscosity and may be scaled for other gases. The conductance of these structures is also nonlinear and proportional to their average pressure.

The Poiseuille equation is valid for many situations, such as chambers connected to remotely located pumps, long gas supply lines, mass flow meter tubes, and controlled leaks fabricated from capillary tubes. It does not work in short tubes, where alternatives are needed.

3.3.3 Short Round Tube

In short tubes, flow can change from viscous to critically choked without there being any pressure region in which the Poiseuille equation is valid. Short tubes

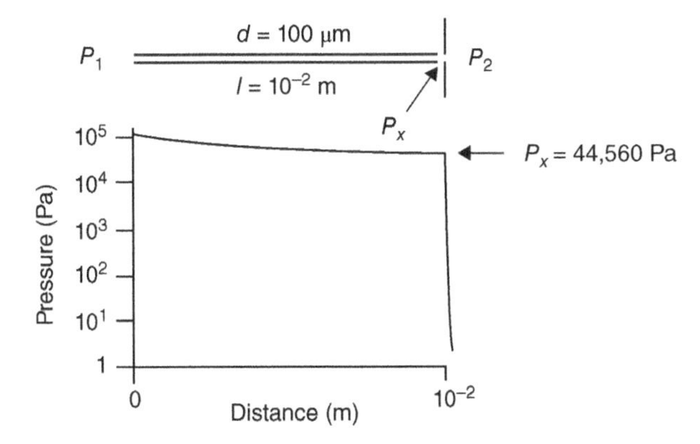

Fig. 3.4 Pressure profile through a fine leak in a vacuum wall, as calculated using Santeler's model. His model assumes Poiseuille flow through the tube with a drop in pressure at the vacuum vessel entrance described by orifice flow at the leak exit.

have been treated in several ways. Dushman [8] devised a nonlinear relation for flow in short round tubes in unchoked flow. Santeler [9] developed and verified a model for a short tube as an aperture in series with a tube of length l'. The solution is formulated by assuming an unknown pressure P_k between the exit of the long tube and the entrance to the "aperture." This is the pressure one would measure with a gauge just inside the end of the tube's exit that pointed upstream. Figure 3.4 illustrates an example using his method—calculating the pressure drop and leak rate through a 100-μm-diameter leak in a 1-cm-thick vacuum wall (or under an O-ring gasket). Santeler's model uses Eq. (3.12) with P_2 replaced by P_x for the flow in the tube. The flow through the aperture was modeled using Eq. (3.10) with P_x as its inlet pressure, and "high vacuum" (set $P = 0$) as its outlet pressure. Since the two flows are in series, they are equal; this the solution yielded $P_x = 44{,}560$ Pa. The answer can be checked to ensure that the assumption of choked flow in the aperture is valid; if not, then Eq. (3.9) must be used in place of Eq. (3.10). This model predicts Poiseuille flow in the leak with significant gas expansion at the vacuum interface.

3.4 Molecular Flow

A gas is called a molecular gas when $\text{Kn} > 1.0$, and its flow, molecular flow. For air at 22°C this becomes $Pd < .67$ Pa-cm, or 0.005 Torr-cm as described in Eq. (2.5). The molecular flow region is well understood. This discussion focuses on molecular flow through orifices, very long tubes, short tubes, complex objects, and series combinations of objects, all of which are of practical use.

3.4.1 Orifice

If two large vessels are connected by an orifice of area A, and the diameter of the orifice is such that Kn > 1, then the gas flow from one vessel (P_1, n_1) to the second vessel (P_2, n_2) is molecular and is given by:

$$Q = \frac{kT}{4} vA\left(n_1 - n_2\right) = \frac{v}{4} A\left(P_1 - P_2\right) \tag{3.19}$$

and the orifice conductance is:

$$C = \frac{Q}{P_1 - P_2} = \frac{v}{4} A \tag{3.20}$$

for air at 22°C, its value is:

$$\begin{aligned} C\left(\text{m}^3/\text{s}\right) &= 116A\left(\text{m}^2\right) \\ C\left(\text{L/s}\right) &= 11.6A\left(\text{cm}^2\right) \end{aligned} \tag{3.21}$$

Note that converting C to units of L/s involves more than changing the units of A, the area, the thermal velocity must also be changed to units of cm/s.

From Eq. (3.19), we note an interesting property of the molecular flow regime. Gas can flow from vessel 2 to vessel 1; at the same time, gas is flowing from vessel 1 to vessel 2 without either of the gases colliding with gas that originated in the other vessel, like ships passing in the night.

Speed was defined in Eq. (3.3). If we allow $P_2 \rightarrow 0$, and define Q as in Eq. (3.19), the speed becomes $S = (vA/4)$, and for the case of air at 22 °C, this reduces:

$$S\left(\text{L/s}\right) = 11.6A\left(\text{cm}^2\right) \tag{3.22}$$

Equation 3.22 is very important and useful. It states that the *maximum* pumping speed of an "ideal hole" in the molecular flow regime is proportional to its inlet area and for air, cannot be larger than 11.6 L/s for each square centimeter area of an ideal pump. We sketch this pictorially in Fig. 3.5. Here we show all the entering particles to be "pumped," that is, they never return to the pump inlet. As we shall see, no pump is ideal, all have measured speeds less than this ideal value. That said, it is an extremely useful concept, as it gives an upper bound, and reinforces the importance of area in determining the speed of any pump in the rarefied gas region.

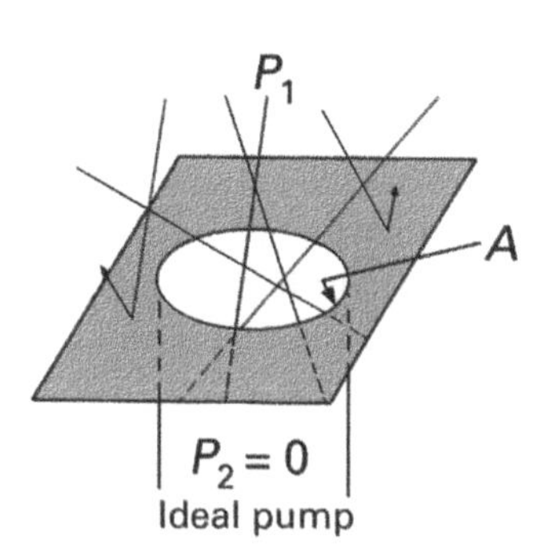

Fig. 3.5 Conceptual view of an ideal pump operating in a rarefied gas. All the gas that enters vanishes; speed is directly proportional to the entrance area.

3.4.2 Long Round Tube

The diffusion method of Smoluchowski [10] and the momentum transfer methods of Knudsen [11] and Loeb [12] were the first to describe gas flow through very long tubes in the free molecular flow region. For tubes of circular cross-section, both derivations yield a conductance value of:

$$C_{\text{tube}} = \frac{\pi}{12} v \frac{d^3}{l}$$
$$C_{\text{tube}}\left(\text{m}^3/\text{s}\right) = 121\frac{d^3}{l} \quad (\text{air}, 22°\text{C}) \tag{3.23}$$
$$C_{\text{tube}}\left(\text{L/s}\right) = 12.1\frac{d^3}{l} \quad (d, l, \text{in cm}; \text{air}, 22°\text{C})$$

Conductance relations have been calculated for some long tubes of noncircular cross-section [13].

3.4.3 Short Round Tube

Equation 3.23 indicates the conductance becomes infinite as the tube length tends toward zero, whereas (3.20) shows the conductance to be $vA/4$. Dushman [14] proposed an empirical solution to the short tube problem by considering the total conductance to be the sum of the reciprocals of component and an aperture of same diameter.

$$\frac{1}{C_{\text{total}}} = \frac{1}{C_{\text{tube}}} + \frac{1}{C_{\text{aperture}}} \tag{3.24}$$

As $l/d \to 0$, Eq. (3.24) reduces to Eq. (3.20), and as $l/d \to \infty$ it reduces to Eq. (3.23). Although Eq. (3.24) gives the correct solution for the extremes, it is not correct for the intermediate. It can be in error by as much as 12–15%.

The difficulty in performing calculations for short tubes lies in the nature of the gas–wall scattering. Lorentz [15] assumed the walls of a pipe to be molecularly rough, i.e., molecules scatter according to the cosine law of diffuse reflection. Molecules hit a wall, oscillate in potential wells, and recoil in a direction that is independent of their arrival angle. In diffuse reflection, scattered molecules have the greatest probability of recoiling at an angle of 90° from the surface. Those not scattering at 90° have equal probability of going forward toward the exit as going backward toward the entrance. See Fig.'s 3.1*e* and 3.6. Clausing [16] solved this problem by calculating the probability that a molecule entering the pipe at one end will escape at the other end after making diffuse collisions with the walls. Clausing's integral is difficult to evaluate. For round pipes, Clausing and others have developed approximate solutions that are

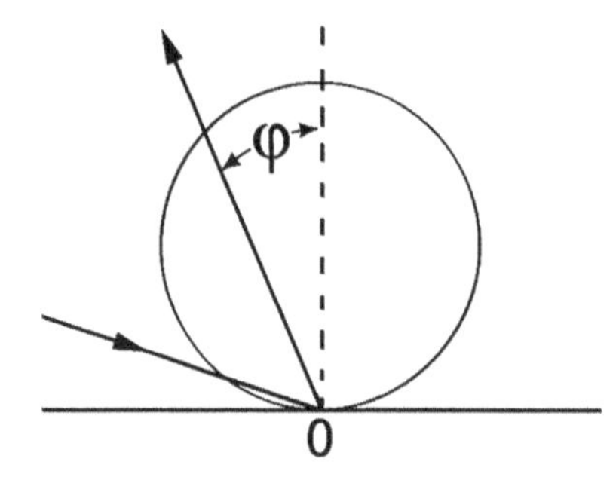

Fig. 3.6 A molecule making a diffuse collision with a wall is scattered in a direction independent of its original path. A molecule forgets its original direction and is emitted with a probability proportional to cosφ from the normal. The most probable angle is φ = 0°. A molecule has an equal probability of going forward as it does backward.

tabulated in many standard texts. Their solutions are given in the form of a transmission probability, a, that an entering molecule will exit. Using this a, the conductance of a pipe is then found from Eq. (3.26), where A is the cross-sectional area of the pipe and ν is the thermal velocity of the gas.

$$C = a\frac{v}{4}A \tag{3.25}$$

For the special case of very long round tubes ($l \gg d$), the expression given in (3.20) can be written in the form of Eq. (3.22) to reveal that the transmission probability a of a long tube is $(4d/3l)$ as shown in Table 3.2.

$$C = \left(\frac{4d}{3l}\right)\frac{v}{4}\left(\frac{\pi d^2}{4}\right) \tag{3.26}$$

For air at 22°C the conductance of a long pipe can be simplified to read:

$$\begin{aligned} C(\text{L/s}) &= 1.16\times 10^5 aA(\text{m}^2) \\ C(\text{L/s}) &= 11.6aA(\text{cm}^2) \end{aligned} \tag{3.27}$$

Equation 3.27 describes the molecular conductance per unit area of any structure in molecular flow. It has a maximum value of 11.6 L/(s-cm^2) for air at 22°C, corresponding to $a = 1$ for a pipe of zero length—a thin aperture. Any pipe longer than a thin aperture will have a conductance less than this value.

DeMarcus [17] used a variational principle to solve the Clausing integral with improved accuracy. Berman [18] made a polynomial fit to DeMarcus' solution and extended it to larger l/d values. Values of the transmission probability for round pipes obtained from his equations are given in Table 3.2 for a range of l/d values. The DeMarcus–Berman results agree with very precise calculations done by Cole [19] to between 4 and 5 decimal places. These values of a can be used in Eq.'s (3.25) and (3.27). "Exit effects," which we shall discuss shortly, are included in these tabulations of probability values.

Table 3.2 Transmission Probability a, for Round Pipes

l/d	a	l/d	a
0.00	1.00000	1.6	0.40548
0.05	0.95240	1.7	0.39195
0.10	0.90922	1.8	0.37935
0.15	0.86993	1.9	0.36759
0.20	0.83408	2.0	0.35658
0.25	0.80127	2.5	0.31054
0.30	0.77115	3.0	0.27546
0.35	0.74341	3.5	0.24776
0.40	0.71779	4.0	0.22530
0.45	0.69404	4.5	0.20669
0.50	0.67198	5.0	0.19099
0.55	0.65143	6.0	0.16596
0.60	0.63223	7.0	0.14684
0.65	0.61425	8.0	0.13175
0.70	0.59737	9.0	0.11951
0.75	0.58148	10.0	0.10938
0.80	0.56655	15.0	0.07699
0.85	0.55236	20.0	0.05949
0.90	0.53898	25.0	0.04851
0.95	0.52625	30.0	0.04097
1.0	0.51423	35.0	0.03546
1.1	0.49185	40.0	0.03127
1.2	0.47149	50.0	0.02529
1.3	0.45289	500.0	0.26479×10^{-2}
1.4	0.43581	5000.0	0.26643×0^{-3}
1.5	0.42006	∞	$4d/3l$

3.4.4 Irregular Structures

The calculation of molecular conductance of any irregular structure is not possible in closed form. It has been solved analytically for noncircular cross-sections in only a few cases, and as a result, probabilistic methods are widely used. The effects of varying accommodation coefficient and sticking coefficient on probability have been simulated by Tanigouchi et al. [20].

3.4.4.1 Analytical Solutions

Analytical solutions exist only for the annular cylindrical pipe [21], the rectangular pipe [22], the elliptical tube, the triangular tube [5], and the thin rectangular slit. This last geometry is encountered in differentially pumped feedthroughs and apertures between differentially pumped chambers. Its rectangular entrance area has thin dimension h, wide dimension b, with the condition $h \ll b$; the length of the duct is l. Berman [18] developed a polynomial fit to solutions for the transmission coefficient and values calculated with the use of his formula are given in Table 3.3. His results agree with those of Neudachin et al. [23]. The conductance of a slit can also be calculated from Eq. (3.25) using the transmission probability from Table 3.3 and an inlet area of $A = bh$. "Exit conductance" effects are also included in these transmission probabilities. The

Table 3.3 Transmission Probability a, for Thin, Rectangular, Slit-like Ducts

l/h	a	l/h	a
0.0	1.00000	15	0.18664
0.1	0.95245	20	0.15425
0.2	0.90958	30	0.11648
0.3	0.87097	40	0.09471
0.4	0.83617	50	0.08035
0.5	0.80473	60	0.07008
0.6	0.77620	70	0.06234
0.7	0.75021	80	0.05627
0.8	0.72643	90	0.05136
0.9	0.70457	100	0.04731
1.0	0.68438	200	0.02722
2.0	0.54206	500	0.01276
3.0	0.45716	1000	0.70829×10^{-2}
4.0	0.39919	2000	0.38914×10^{-2}
5.0	0.35648	5000	0.17409×10^{-2}
6.0	0.32339	10,000	0.94000×10^{-3}
7.0	0.29684	20,000	0.50472×10^{-3}
8.0	0.27496	50,000	0.22023×10^{-3}
9.0	0.25655	100,000	0.11705×10^{-3}
10	0.24080	200,000	0.61994×10^{-4}

transmission probability has been experimentally shown to decrease with increased surface roughness [24]. Other short tube cross-sections and complex structures must be treated with statistical techniques such as the Monte Carlo technique described below.

3.4.4.2 Statistical Solutions

The Monte Carlo statistical methods developed for the calculation of molecular flow conductance by Davis [25] and by Levenson et al. [26] were a major breakthrough in the calculation of the conductance of complex, practical, objects, such as elbows, traps, and baffles. The Monte Carlo technique simulates the individual trajectories of a large number of randomly chosen molecules. Figure 3.7 is a computer-generated graphical model of the trajectories of 15 random molecules entering an elbow. In this example, three molecules exited the structure, yielding a transmission probability of $a = 3/15 = 0.2$. When modeled with a larger number of particles, a transmission probability of 0.31 was calculated. This points to one difficulty of the Monte Carlo technique; its accuracy depends on the number of molecular trajectories. Modern desktop computing has simplified the computational effort for complex structure calculations [27]. Figures 3.8–3.15 contain examples of Monte Carlo solutions for many structures of interest [25,26,28]. The molecular conductance is the product of the probability, a, and conductance of the structure's entrance aperture. The computational effort needed to perform Monte Carlo calculations has driven others to develop approximate solutions by combining cylindrical tubes, orifices, and baffle plates. Several methods for calculating conductance of complex geometries have been developed [29,30,31,32,33,34], but geometries of most interest require the Monte Carlo method.

3.4.5 Components in Parallel and Series

Implicit in the definition of conductance, Eq. (3.4), is the understanding that molecules will arrive at the entrance to a component in a *random distribution*. This definition implies that the component is connected to chambers of *infinite dimension* at its entrance and its exit. In this case, flow through the component is calculated on the assumption that molecules arrive randomly. This does not happen in practice, because it is practical to interconnect components with the shortest flow path, thus molecules can arrive in a focused or beamed distribution. Connecting objects in parallel is straightforward. The total conductance of components connected in parallel between two vessels is the simple sum of their individual values.

$$C_T = C_1 + C_2 + C_3 + \dots \tag{3.28}$$

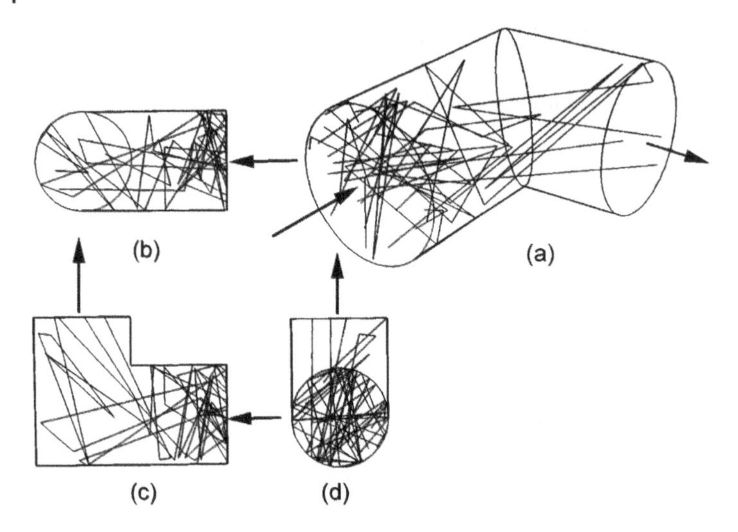

Fig. 3.7 A computer graphical display of the trajectories of 15 molecules entering an elbow in free molecular flow. (a) Isometric view, (b–d) *x*-, *y*-, and *z*-axis views. Courtesy of A. Appel, IBM T. J. Watson Research Center, Yorktown Heights, New York.

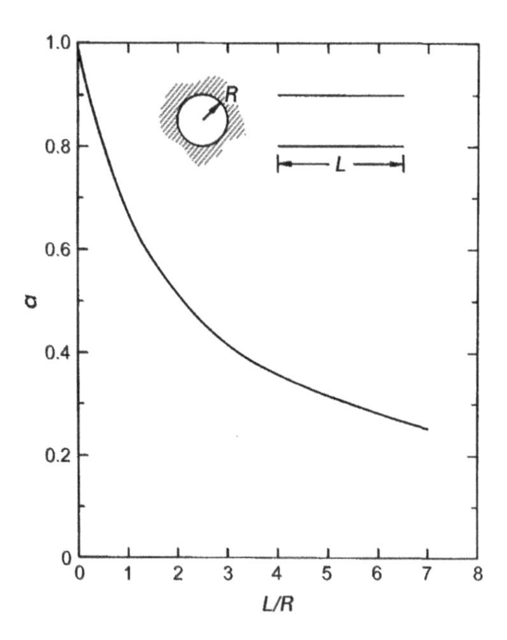

Fig. 3.8 Molecular transmission probability of a round pipe. L. L. Levenson et al. [26]/ Reproduced with permission from Société Française du Vide.

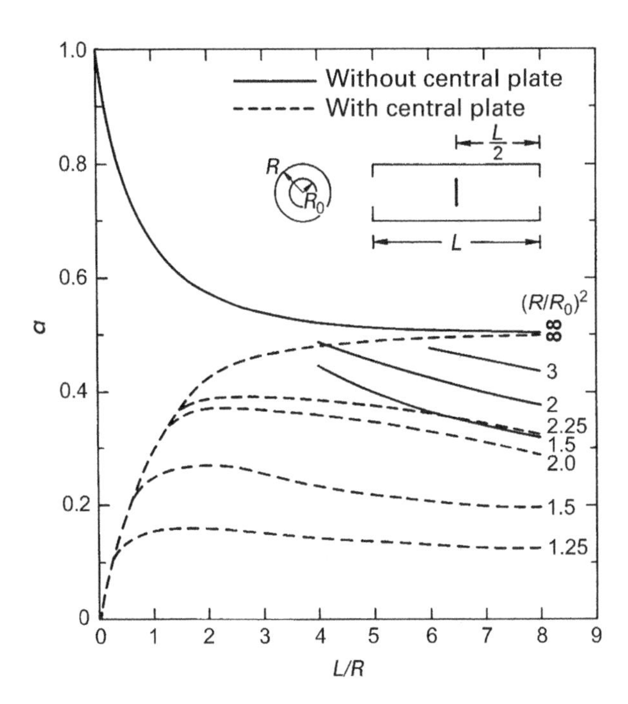

Fig. 3.9 Molecular transmission probability of a round pipe with entrance and exit apertures. D.H. Davis [25]/Reproduced with permission from AIP Publishing.

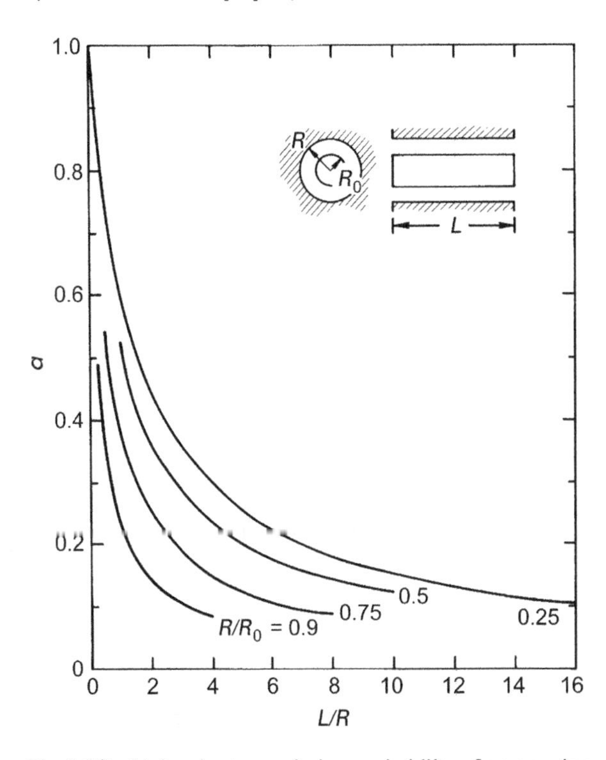

Fig. 3.10 Molecular transmission probability of an annular cylindrical pipe. L. L. Levenson et al. [26]/Reproduced with permission from Société Française du Vide.

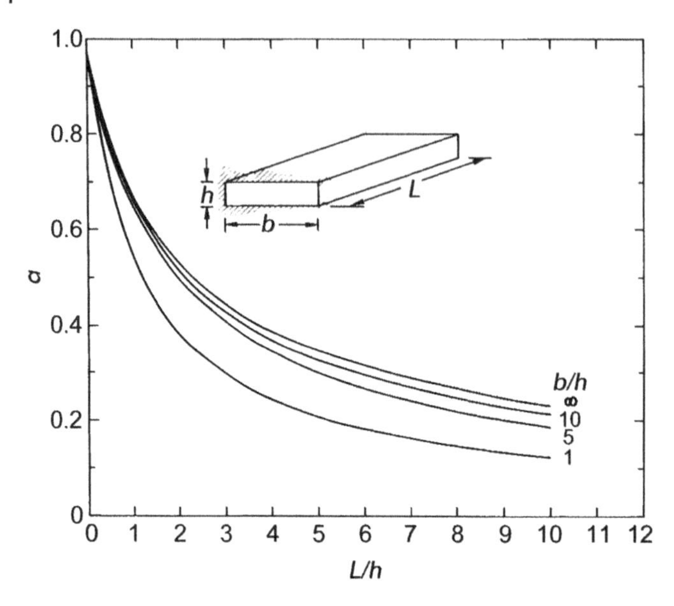

Fig. 3.11 Molecular transmission probability of a rectangular duct. L.L. Levenson et al. [26]/Reproduced with permission from Société Française du Vide.

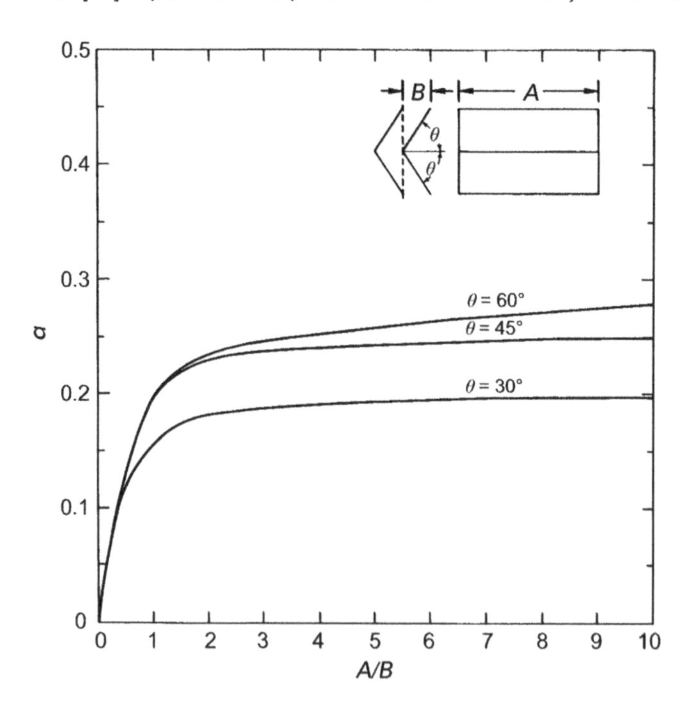

Fig. 3.12 Molecular transmission probability of a chevron baffle. L.L. Levenson et al. [26]/Reproduced with permission from Société Française du Vide.

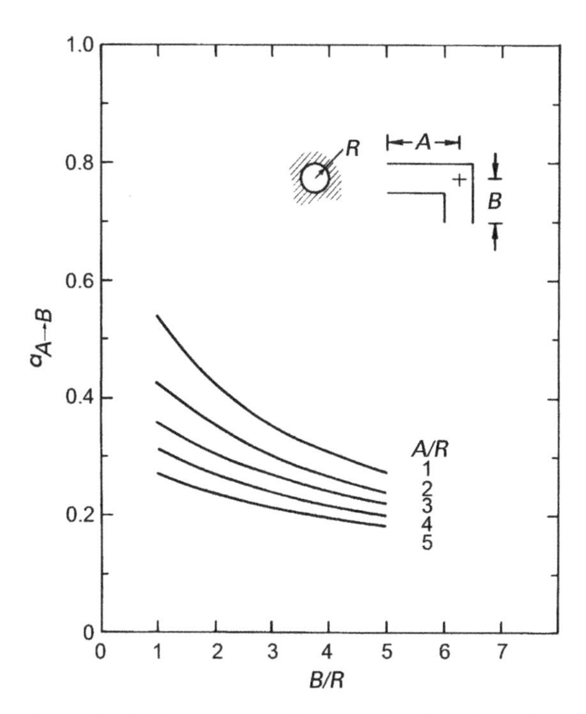

Fig. 3.13 Molecular transmission probability of an elbow. D.H. Davis [25]/Reproduced with permission from AIP Publishing.

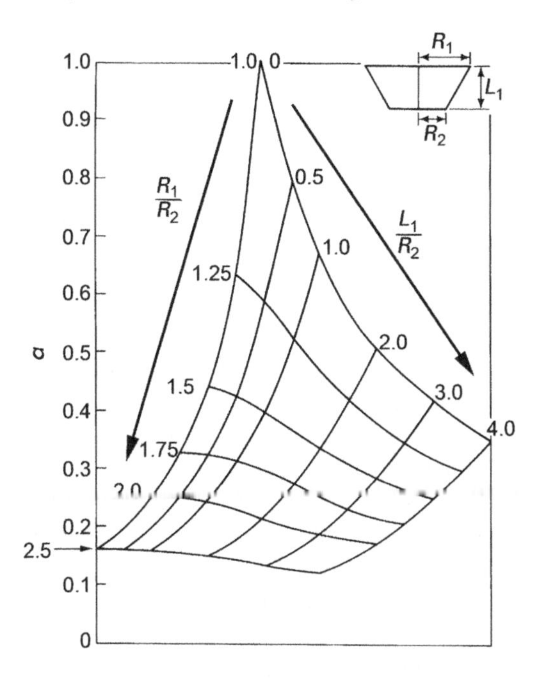

Fig. 3.14 Molecular transmission probability of a chevron baffle. J.D. Pinson and A.W. Peck [28]/Reproduced with permission from Macmillan.

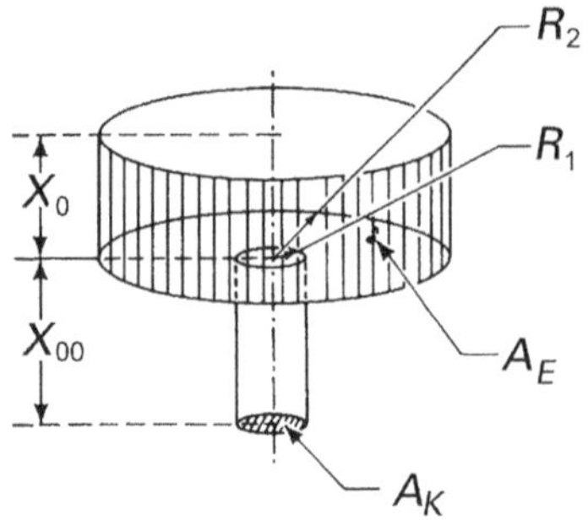

$\frac{R_2}{R_1}$	$\frac{X_0}{R_1}$	$\frac{X_{00}}{R_1}$	a_M	$\frac{P_1}{P_2}$
1	2	0	0.241	0.966
2	2	0	0.120	0.957
3	2	0	0.078	0.928
4	2	0	0.060	0.955
5	2	0	0.045	0.896
3	2	1	0.054	0.650
3	2	2	0.041	0.494
3	2	3	0.034	0.407
3	2	4	0.029	0.353
5	1	0	0.065	0.647
5	3	0	0.032	0.966
2	1	0	0.186	0.744
2	3	0	0.080	0.964

Fig. 3.15 Molecular transmission probability of a parallel plate model. J.D. Pinson and A.W. Peck [28]/Reproduced with permission from Macmillan.

Series conductance of truly independent components in molecular flow, i.e., joined by large chambers, will yield a total conductance of:

$$\frac{1}{C_T} = \frac{1}{C_1} + \frac{1}{C_2} + \frac{1}{C_3} + \ldots \tag{3.29}$$

Equation (3.29) gives the value of conductance we would measure if the elements were isolated from each other by large volumes as sketched in Fig. 3.16*a*. A large volume allows molecules exiting the first component to equilibrate and randomize before entering the second component.

The simple reciprocal rule, Eq. (3.29), does not work when two adjacent components are combined directly with no intervening large volume to randomize the directions of arriving molecules. See Fig. 3.16*b*. As an example, consider two tubes each with length-to-diameter ratio $l/d = 1$. From Table 3.1 we find $a = 0.51423$ for each tube. If we combine them according to the reciprocal rule in Eq. (3.29) we will obtain a value of $a = 0.25712$. We know this is incorrect, because this same table gives the transmission probability of a tube with $l/d = 2$ as $a = 0.35685$. The error in this example is 27.9%. Why do we have this large error? Because molecules that did

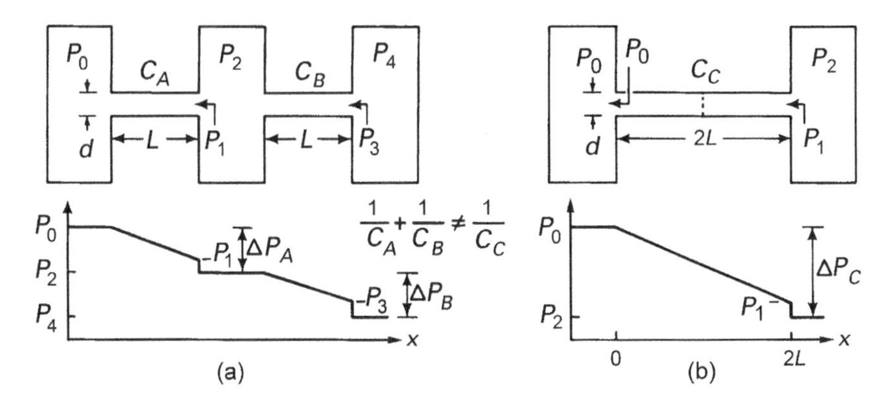

Fig. 3.16 Series conductance of two elements: (a) the pipes are isolated by a large volume and (b) the pipes are connected directly together. The pressure readings are those measured by a gauge in the gas stream pointing upstream and parallel to the flow direction.

not transmit the second pipe returned to the first with a skewed angular distribution. The reciprocal rule assumed molecules returned from the second pipe arrived at random angles, but they did not, when the second pipe was connected directly to the first. This difference in arrival direction is called an "end effect."

Figure 3.17 sketches the angular distribution of molecules exiting the first pipe as its length-to-diameter ratio changes—an effect called "beaming." This sketch shows molecules become beamed as they traverse a long tube toward its exit. Beaming results in a larger fraction being transmitted through the next pipe. As a mental exercise, think of a bullet shot along the axis of a straight pipe. It will exit ($a=1$) independent of the pipe's length! All real objects, e.g., cold traps, baffles, elbows, valves, etc, have unique molecular distributions, and varying beaming effects.

Several researchers [30,31,32,33] have used the concept of a probability factor a, to calculate the conductance of a series combination of vacuum components in free molecular flow. First, we describe the method developed by Oatley [32]. Figure 3.18 illustrates the concept with a single component; Γ molecules per second enter at the left-hand side, $a\Gamma$ molecules per second exit at the right-hand side, and $(1-a)\Gamma$ molecules per second are returned to the source vessel. The conductance is expressed by:

$$C = a\frac{v}{4}A \tag{3.30}$$

Figure 3.19 illustrates Oatley's method for two tubes in series. From the Γ molecules per second that enter the first tube, $a_1\Gamma$ enter the second; $\Gamma(1-a_2)a_1$ of these are returned to the first tube and $\Gamma a_1 a_2$ enter the second. From the

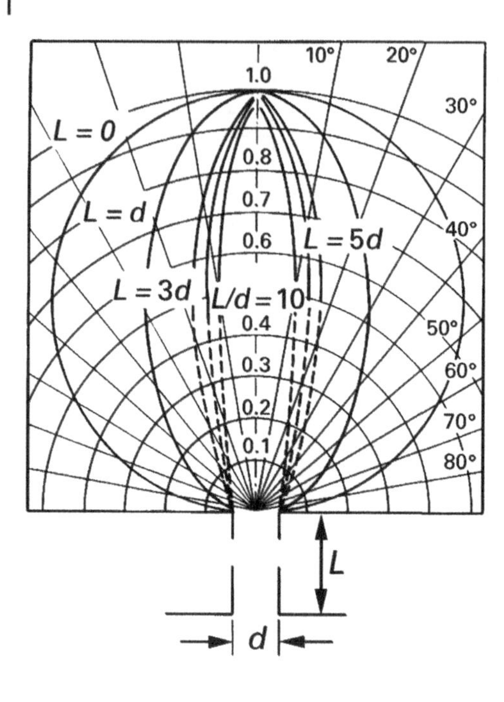

Fig. 3.17 Angular distribution of particles exiting tubes of various ratios of length to diameter. L. Vályi [35]/Reproduced with permission from John Wiley & Sons.

Fig. 3.18 Model for calculating the transmission probability of a single element.

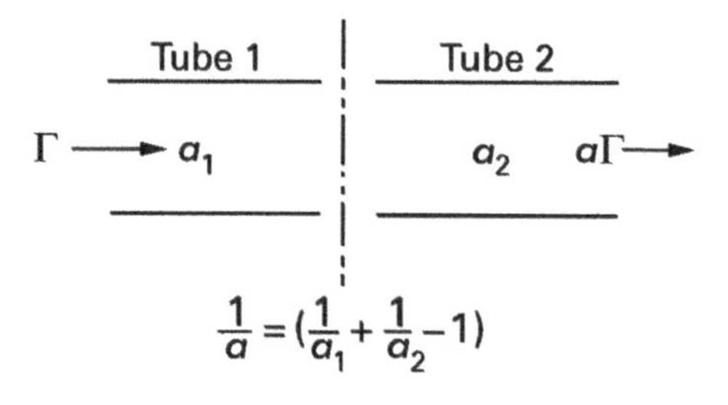

Fig. 3.19 Model for calculating the transmission probability of two elements in series. C.W. Oatley [32]/ Reproduced with permission from IOP Publishing.

group $\Gamma(1-a_2)a_1$ molecules returned to the first tube $\Gamma a_1(1-a_2)(1-a_1)$ are returned to the second, and so on, until an infinite series expression was developed that simplified to:

$$\frac{1}{a} = \frac{1}{a_1} + \frac{1}{a_2} - 1 \tag{3.31}$$

The last term represents the effects of gas scattering and beaming. When generalized to several elements in series, this becomes:

$$\frac{1-a}{a}=\frac{1-a_1}{a_1}+\frac{1-a_2}{a_2}+\frac{1-a_3}{a_3}+\ldots \tag{3.32}$$

Now return to the previous example of two short pipes in series and use Eq. (3.32) to calculate the series conductance of the two pipes of $l/d=1$. Using Oatley's method, we obtain a value $a=0.3460$. This is much closer to Clausing's value of $a=0.35685$ obtained from Table 3.1. It is in error by only 2.93%.

Second, we consider the method developed by Haefer [33]. Oatley's formula, as given in Eq. (3.32) or (3.32), applies to objects of the same diameters. Haefer's addition theorem extends it to differing diameters.

His theorem relates the total transmission probability of n elements a_{1n}, to the transmission probability a_i and the inlet area A_i of each component. Extra terms are included in the equation whenever a cross-sectional area *decreases* upon entering the next element but not when the area *increases*. It is given here without proof.

$$\frac{1}{A_1}\left(\frac{1-a_{1\to n}}{a_{1\to n}}\right)=\sum_1^n\frac{1}{A_i}\left(\frac{1-a_i}{a_i}\right)+\sum_1^{n-1}\left(\frac{1}{A_{i+1}}-\frac{1}{A_i}\right)\delta_{i,i+1} \tag{3.33}$$
$$\text{where } \delta_{i,i+1}=1 \text{ for } A_{i+1}<A_i, \text{ and } \delta_{i,i+1}=0 \text{ for } A_{i+1}\geq A_i$$

The use of this formula is demonstrated with the example depicted in Fig. 3.20. Shown is a combination of three pipe sections of inlet areas A_1, A_2, and A_3. The pipes have corresponding transmission probabilities a_1, a_2, and a_3. Using Eq. (3.33), one obtains:

$$\begin{aligned}\frac{1}{A_1}\left(\frac{1-a_{1\to 3}}{a_{1\to 3}}\right)&=\frac{1}{A_1}\left(\frac{1-a_1}{a_1}\right)+\frac{1}{A_2}\left(\frac{1-a_2}{a_2}\right)+\frac{1}{A_3}\left(\frac{1-a_3}{a_3}\right)+\left(\frac{1}{A_3}-\frac{1}{A_2}\right)\\ \frac{1}{a_{1\to 3}}&=\frac{1}{a_1}+\left(\frac{A_1}{A_2}\right)\frac{1}{a_2}+\left(\frac{A_1}{A_3}\right)\frac{1}{a_3}-\left(\frac{A_1}{A_2}\right)2\end{aligned} \tag{3.34}$$

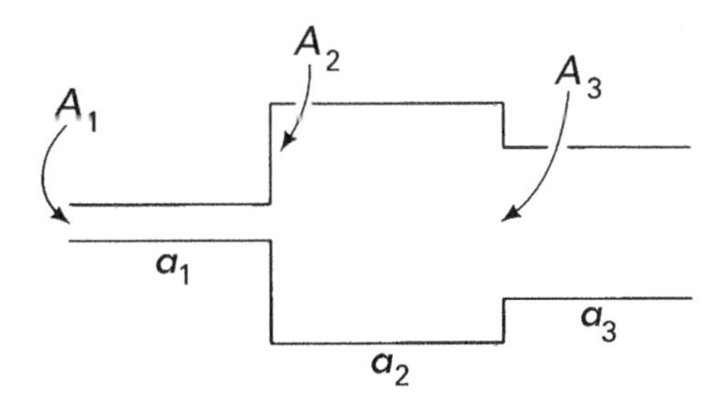

Fig. 3.20 Example conductance evaluated using Haefer's addition theorem

This answer reduces to Eq. (3.32) when the pipe areas are all the same. Haefer's method must be applied with consistency because the total transmission coefficient one calculates is a function of which end of the structure is chosen to be the origin. If the transmission probability were calculated from right to left, a_{n-1}, we would have to use an inlet area of A_n in calculating the conductance. However, this does not affect the total conductance of a structure because:

$$C = \left(a_{1\to n}\right)\frac{v}{4}A_1 = \left(a_{n\to 1}\right)\frac{v}{4}A_n \tag{3.35}$$

We can also see that the answer would have been different if the order of the second and third pipes were interchanged. Interchanging the order of components in a series configuration *will* affect the total transmission probability. As a rule, the conductance of the actual configuration should be calculated, because errors enter when components are mathematically rearranged [36]. It is generally true that a complex structure made up of several series objects has the maximum conductance when the elements are arranged in increasing sizes because the exit losses are the smallest and the beaming is the greatest. Arranging objects in alternating large and small diameters introduces wall scattering, which makes the input to the following sections approach a random distribution.

Problems arise in conductance calculations because Eq. (3.29), which is valid for independent elements, is indiscriminately applied to series elements that are not isolated. The methods developed by Oatley and Haefer account for the largest errors, but neither method corrects for entrance effects, that is, a beamed entrance flux.

Pinson and Peck [28] discuss beaming errors for pipe sections with and without baffles and show the difference between a calculation and Monte Carlo technique is ≤10% with the greatest differences seen when a baffle is not used. Beaming corrections for tube combinations have been developed by Santeler [36]. Components like chevron baffles and elbows tend to scatter the gas. For example, the conductance of a nondegenerate elbow can be calculated by using the conductance values of the individual arms, obtained from Fig. 3.8, and summing these values with Eq. (3.31). Saksaganski [37] discusses efficient methods for analyzing complex systems and shows the angular coefficient and integral kinetic methods as alternatives to the Monte Carlo method.

The formulas developed by Haefer and Oatley may be used to calculate the pumping speed at the inlet of pipes connected to a pump. In this case the pump is characterized by its inlet area and Ho coefficient (defined in Chapter 7). The pump is effectively a conductance of entrance area A and transmission probability a_p equal to its Ho coefficient.

3.5 Models Spanning Molecular and Viscous Flow

The theory of gas flow in the transition region between molecular and viscous flow is not well developed. Thomson and Owens [38], and Loyalka et al. [39], have reviewed the state of the theory. DeMuth and Watson [40] have done additional work on the transition between molecular and isentropic flow in orifices. The simplest treatment of this region, due to Knudsen, discussed in many texts, states:

$$Q = Q_{\text{viscous}} + Z'Q_{\text{molecular}} \tag{3.36}$$

for long circular tubes Z' is given by:

$$Z' = \frac{1+2.507(d/2\lambda)}{1+3.095(d/2\lambda)} \tag{3.37}$$

In addition to Dushman's model for the transition region discussed above, we discussed Santeler's model for flow in a short tube in the viscous and choke regions. Santeler's [36] model for a short tube separates the tube component from the exit effect. It can be applied to a series of tubes by modeling the system as a series of tubes with one exit loss after the last tube section. Tison [41] developed an empirical fit with a form similar to Knudsen's that described the flow through a metal capillary from the molecular to the viscous region.

Other relations have been developed for specialized geometrical shapes. One is a relation developed by Kieser and Grundner [42] for the thin, slit-like tube. The thin, rectangular slit-like tube with one side in a rarefied gas and the other at atmospheric pressure is encountered in atmosphere-to-vacuum continuous feed systems and reel-to-reel coating systems known as web, or roll, coaters. This relation, which is valid in the molecular, transition, and viscous flow regions, combines the ideas of Dushman and Knudsen. It is valid for any inlet pressure but only for low exhaust pressures ($P_o < 0.528P_i$). Kieser and Grundner begin with Dushman's relationship, which assumed a duct to be composed of a pipe and an entrance aperture. The conductance of the series combination is:

$$\frac{1}{C_{\text{total}}} = \frac{1}{C_{\text{pipe}}} + \frac{1}{C_0} \tag{3.38}$$

for air at 20°C, this reduces to:

$$\frac{1}{C_{\text{total}}} = \frac{1}{(0.1106eP_i + Z')C_M} + \frac{1}{C_0} \tag{3.39}$$

where C_0 and C_M in Eq. (3.39) are given by:

$$C_0\left(\text{L/s}\right)=11.6ew\left[\frac{10+0.5\left(e/\lambda\right)^{3/2}}{10+0.3412\left(e/\lambda\right)^{3/2}}\right]$$
$$C_M\left(\text{L/s}\right)=11.6ew\left(\frac{a}{1-a}\right) \tag{3.40}$$

In the above relationships, e is the channel thickness, and w is the channel width. The form of Eq. (3.40) allows the (air) conductance of the aperture to vary from 11.6 L/s in molecular flow to 17 L/s in the choked limit. Equation 3.40 is an experimental fit to the transition region for air [42].

The pipe conductance given in Eq. (3.38), is due to Knudsen and is a superposition of continuum and molecular flow. Since the aperture is included in the C_0 term, we have removed it from Eq. (3.40) [43]. The pipe conductance C_M, calculated by Berman already contains an end effect C_0. The premise on which Eq.'s (3.36) and (3.37) are based, i.e., the representation of the total conductance as a series combination of a "pipe" and an "aperture" conductance, is known to be in error by 10–15% in the region $l/e=1$–5. For longer pipes there is another small error. Equation 3.39 implicitly lets the average pressure in the "pipe" portion of the conductance be $P_i/2$. This forces the pressure across the viscous conductance to an incorrect value. However, the pipe conductance is in series with a choked orifice, so any error in the "pipe" conductance is greatly attenuated by the large pressure drop of the choke. This is the analog of a series combination of a large and small resistor; the series resistance is not greatly affected by the value of the small resistor.

Few publications have provided flow relations that are valid over several regions. For example, Schumacher [44] summarized the flow through small round tubes in graphical form, while Levina [45] developed nomographs for the same problem.

The values of gas flow, pressure, and pipe size each extend over a wide range. We can summarize this discussion by sketching a (normalized) plot of flow divided by pipe size (Q/d) versus pressure times distance (Pd). Figure 3.21 depicts the various regions discussed in this chapter. Molecular flow occurs in the region $\mathbf{R}<1200$, and $\text{Kn}>1$. The flow is proportional to the first power of the pressure (slope = 1). When $\text{Kn}<0.01$ the gas is viscous, and the flow is either turbulent, fully developed (Poiseuille), undeveloped, or choked. Observe that fully developed flow is proportional to the square of the pressure (slope = 2). The boundary between turbulent and laminar viscous is determined by Reynolds' number. Langhaar's number and the Mach number determine the boundary between fully developed and undeveloped flow. The region between molecular and viscous flow is the transition region. We see the transition from completely free molecular flow to completely viscous flow can take place over a two-decade pressure range.

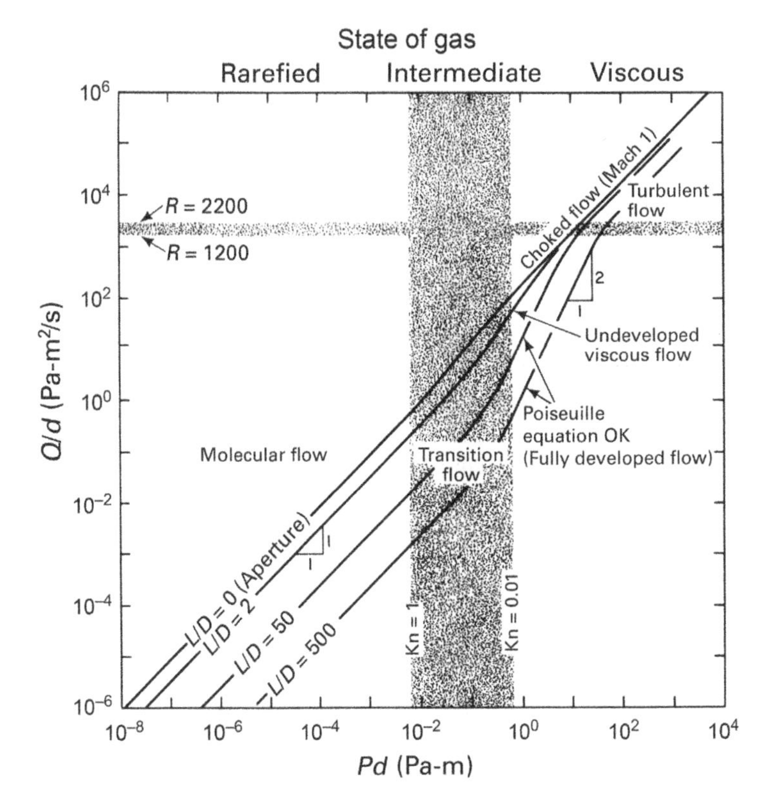

Fig. 3.21 Normalized gas flow–pressure regimes.

In this chapter we have considered the equations of flow in each distinct flow region. The equations developed here related flow to pressure drop in several pressure regions and geometric cross sections. When combined with the dynamical equations of gas flow from a chamber, the time required to reach a particular pressure can be calculated.

References

1 Reynolds, O., *Philos. Trans. R. Soc. London* **174** (1883).

2 Guthrie, A.O. and Wakerling, R.K., *Vacuum Equipment and Techniques*, McGraw-Hill, New York, 1949, p. 25.

3 Shapiro, A.H., *Dynamics and Thermodynamics of Compressible Fluid Flow*, Ronald, New York, 1953.

4 Langhaar, H.L., *J. Appl. Mech.* **9**, A-55 (1942).

5 Holland, L., Steckelmacher, W., and Yarwood, J., *Vacuum Manual*, E. & F. Spoon, London, 1974, p. 26.

6 Williams, B.J., Fletcher, B., and Emery, J.A.A., *Proceedings of the 4th International Vacuum Congress, 1968*, Institute of Physics and the Physical Society, London, 1969, p. 753.

7 Sasaki, S. and Yasunaga, S., *J. Vac. Soc. Jpn.* **25**, 157 (1982).

8 S. Dushman, "Kinetic Theory of Gasses", Chapter 1 in *Scientific Foundations of Vacuum Technique*, 2nd ed., J. M. Lafferty, Ed., Wiley, New York, 1962, p. 35.

9 Santeler, D.J., *J. Vac. Sci. Technol., A* **4**, 348 (1986).

10 von Smoluchowski, M., *Ann. Phys.* **33**, 1559 (1910).

11 Knudsen, M., *Ann. Physik* **28**, 75 (1909); **35**, 389 (1911).

12 Loeb, L.B., *The Kinetic Theory of Gases*, 2nd ed., McGraw-Hill, New York, 1934 Chapter 7.

13 Steckelmacher, W., *J. Phys. D: Appl. Phys.* **11**, 473 (1978).

14 Reference 8, p. 91.

15 Lorentz, H.A., *Lectures on Theoretical Physics*, Vol. **1**, Macmillan, London, 1927 Chapter 3.

16 Clausing, P., *Ann. Phys.* **12**, 961 (1932), English Translation in *J. Vac. Sci. Technol.*, **8**, 636 (1971).

17 W.C. DeMarcus, "*The Problem of Knudsen Flow: Addendum, Solution for One-Dimensional Systems*," Vol. 1302, K-Series, Part 3, Union Carbide Corporation, Oak Ridge, TN, 1957.

18 Berman, A.S., *J. Appl. Phys.* **36**, 3365 (1965), and erratum, *ibid*, **37**, 4598 (1966).

19 Cole, R.J., in *Rarefied Gas Dynamics*, J.L. Potter, Ed., (*10th International Symposium Rarefied Gas Dynamics*), *Progress in Astronautics and Aeronautics*, Part 1, Vol. **51**, American Institute of Aeronautics and Astronautics, 1976, p. 261.

20 Tanigouchi, H., Ota, M., and Aritomi, M., *Vacuum* **47**, 787 (1996).

21 Berman, A.S., *J. Appl. Phys.* **40**, 4991 (1969).

22 Santeler, D.J. and Boeckmann, M.D., *J. Vac. Sci. Technol., A* **9**, 2378 (1992).

23 Neudachin, I.G., Porodnov, B.T., and Suetin, P.E., *Sov. Phys.- Tech. Phys.* **17**, 1036 (1972).

24 Sugiyama, W., Sawada, T., and Nakamori, K., *Vacuum* **47**, 791 (1996).

25 Davis, D.H., *J. Appl. Phys.* **31**, 1169 (1960).

26 Levenson, L.L., Milleron, N., and Davis, D.H., *Le Vide* **103**, 42 (1963).

27 Pace, A. and Poncet, A., *Vacuum* **41**, 1910 (1990).

28 Pinson, J.D. and Peck, A.W., in *Transactions of the 9th National Vacuum Symposium*, G.H. Bancroft, Ed., Macmillan, New York, 1962, p. 407.

29 Füstöss, L. and Töth, G., *J. Vac. Sci. Technol.* **9**, 1214 (1972).

30 Harries, W., *Z. Angew. Phys.* **3**, 296 (1951).

31 W. Steckelmacher, Proceedings of the 6th International Vacuum Congress, Kyoto, Japan, *Jpn. J. Appl. Phys.*, Vol. 13, Suppl. 2-1, H. Kumagai, Ed., p. 117 1974.

32 Oatley, C.W., *Br. J. Appl. Phys.* **8**, 15 (1957).
33 Haefer, R., *Vacuum* **30**, 217 (1980).
34 Ballance, J.O., *Transactions of the 3rd. International Vacuum Congress*, Vol. **2**, Pergamon, Oxford, 1967, p. 85.
35 Vályi, L., *Atom and Ion Sources*, 86 (1977).
36 Santeler, D.J., *J. Vac. Sci. Technol., A* **4**, 338 (1986).
37 Saksaganski, G.L., *Molecular Flow in Complex Systems*, Gordon and Breach, New York, 1988.
38 Thomson, S.L. and Owens, W.R., *Vacuum* **25**, 151 (1975).
39 Loyalka, S.K., Storvick, T.S., and Park, H.S., *J. Vac. Sci. Technol.* **13**, 1188 (1976).
40 DeMuth, S.F. and Watson, J.S., *J. Vac. Sci. Technol., A* **4**, 344 (1986).
41 Tison, S.A., *Vacuum* **44**, 1171 (1993).
42 Kieser, J. and Grundner, M., *Proceedings of the 8th International Vacuum Congress, 1980, Cannes, France, Le Vide*, Suppl. **201**, F. Abeles, Ed., 1978, p. 376.
43 O'Hanlon, J.F., *J. Vac. Sci. Technol. A* **5**, 98 (1987).
44 Schumacher, B.W., *Proceedings of the 8th National Vacuum Symposium 1961*, Vol. 2, Pergamon, New York, 1962, p. 1192.
45 Levina, L.E., *Sov. J. Nondestruct. Test.* **16**, 67 (1980).

4

Gas Release from Solids

All the gas could be pumped from a vacuum chamber in a very short time if it were located in only the volume of the chamber. Consider a 100-L chamber previously roughed to 10 Pa and just connected to a 1000 L/s high vacuum pump. Equation 18.1 states that the pressure would drop from 10 Pa, when the high vacuum valve is opened, to 4.5×10^{-4} Pa in 1 s. In practice this will never happen, because gases and vapors residing on and in the interior walls desorb slowly. The gas to be removed by the pump consists of gas released from the surface and gas within the volume. For this example, 15–60 min will be required to reach the mid-10^{-4}-Pa (10^{-2}-Torr) range in an unbaked, but clean system.

This chapter discusses the mechanisms of gas evolution from solid surfaces and explains how they affect the pumping rate and ultimate pressure in vacuum chambers. Gas is dissolved in and adsorbed on solids. This gas release, collectively referred to as outgassing, is actually a result of several processes. Figure 4.1 shows the possible sources of gas in addition to the gas located in the volume of the chamber. Gases and vapors released from the surface are a result of vaporization, thermal desorption, diffusion, permeation, and stimulated desorption.

4.1 Vaporization

A vapor is a gas above its condensation temperature and vaporization is the thermally stimulated entry of molecules into the vapor phase. In dynamic equilibrium the rate at which molecules leave the surface of a solid or liquid equals the rate at which they arrive at the surface. The pressure of the vapor over the surface in dynamic equilibrium is the vapor pressure of the solid or liquid, provided that the solid or liquid and the vapor are at the same temperature. In Chapter 2 we stated the molecular flux of vapor crossing a plane was $nv/4$. In equilibrium, this is

A Users Guide to Vacuum Technology, Fourth Edition. John F. O'Hanlon and Timothy A. Gessert.
© 2024 John Wiley & Sons, Inc. Published 2024 by John Wiley & Sons, Inc.

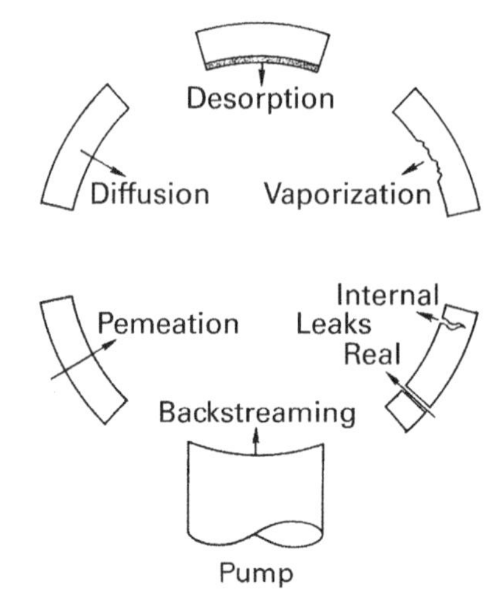

Fig. 4.1 Potential sources of gases and vapors in a vacuum system.

therefore the rate of molecular release from the surface. For the case of free evaporation of a solid from a heated source, Eq. (2.9) given here in a different form, may be used to calculate the maximum rate of evaporation of a solid based on knowledge of its temperature T, vapor pressure P, surface area A, and molecular weight M:

$$\Gamma(\text{molecules/s}) = 2.63 \times 10^{24} \frac{PA}{(MT)^{1/2}} \tag{4.1}$$

The vapor pressure–temperature curves for many gases given in Appendixes B.5 and C.7 provide vapor pressures of the solid and liquid elements.

4.2 Diffusion

Diffusion is the transport of one material through another. Gas diffusion to the interior wall of a vacuum system followed by desorption into the chamber contributes to the system outgassing. The gas pressure in the solid establishes a concentration gradient that drives molecules or atoms to the surface, where they desorb. Because diffusion is often a much slower process than desorption, the rate of transport through the bulk to the surface will usually govern the rate of release into the vacuum. The outgassing rate from a solid wall containing gas at an initial concentration, $C_{o,}$ is obtained from the classic diffusion equation [1]. One solution for a uniform initial concentration of dissolved gas is:

$$q = C_o \left(\frac{D}{\pi t}\right)^{1/2} \left[1 + 2\sum_{1}^{\infty} (-1)^n \exp\left(\frac{-n^2 d^2}{Dt}\right)\right] \tag{4.2}$$

where D is the diffusion constant in m^2/s and $2d$ is the thickness of the material. C_0, the internal pressure of the gas dissolved in the solid, is given in units of Pa, q has units of (Pa-m^3)/(m^2-s) or in less physically recognizable terms, W/m^2. We do not have to derive or solve this equation in order to understand some basic ideas about diffusion. We only need to know its value for short and long times. Let us examine the solutions for these two limits. In the limit as we approach $t = 0$, the terms in the sum become zero and the rate of gas release from the surface is given by:

$$q = C_o\left(\frac{D}{\pi t}\right)^{1/2} \tag{4.3}$$

Equation 4.3 describes the slow decrease in outdiffusion that is experimentally observed. The initial outdiffusion from a solid containing a uniform gas concentration varies as $t^{-1/2}$. This gas release rate is much slower than first-order desorption.

For long times, the infinite series in Eq. (4.2) does not converge rapidly. By the maneuver of placing the mathematical origin at one surface, instead at the center of the solid, the diffusion equation can be solved again to yield:

$$q = \frac{2DC_o}{d}\sum_{o}^{\infty}\exp\left[-\frac{(2n+1)^2\pi^2 Dt}{4d^2}\right] \tag{4.4}$$

Equations 4.2 and 4.4 are equivalent but Eq. (4.4) converges rapidly for large values of t. Only the first term is significant for long times, and the solution reduces to:

$$q = \frac{2DC_o}{d}\exp\left(-\frac{\pi^2 Dt}{4d^2}\right) \tag{4.5}$$

Equation 4.5 states that the outdiffusion rate decreases as $\exp(-\alpha Dt)$ at long times. This rapid exponential decrease in outdiffusion is experimentally observed and corresponds to the near depletion of the dissolved gas in the solid. Figure 4.2 describes

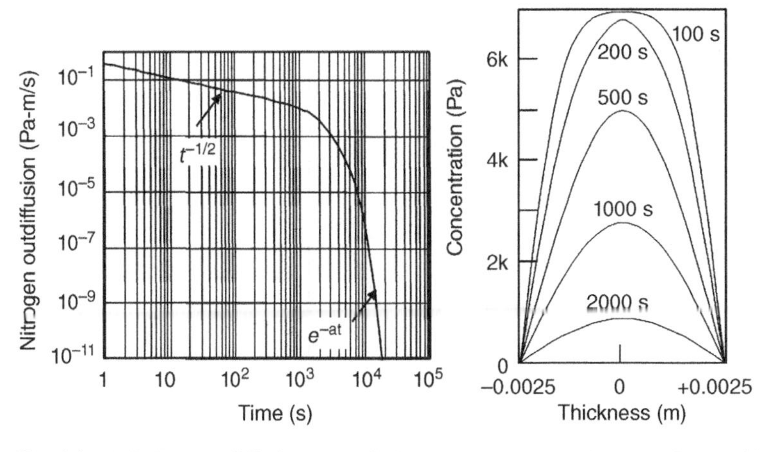

Fig. 4.2 Left: The outdiffusion rate of nitrogen from one surface of a 5-mm-thick silicone rubber sheet, $T = 25\,°C$. Right: the internal nitrogen concentration profile.

the rate at which nitrogen diffuses from one side of a two-dimensional silicone rubber surface that is immersed in vacuum. We first observe the rate to decrease slowly as $t^{-1/2}$, and after some time decrease as $\exp(-\alpha Dt)$, when the solid is nearly exhausted of its gas supply. The transition from slow decay to rapid decay occurs at $t \sim d^2/6D$.

4.2.1 Reduction of Outdiffusion by Vacuum Baking

The diffusion constant is a function of the thermal activation energy of the diffusing gas in the solid and is given by:

$$D = D_o e^{-E_D/kT} \tag{4.6}$$

Because of this exponential dependence on temperature, a rather modest increase in temperature will sharply increase the initial outdiffusion rate and reduce the time necessary to unload the total quantity of gas dissolved in the solid.

Figure 4.3 illustrates how vacuum baking reduces the final outdiffusion rate to a level far below that which is possible in the same time without baking. A solid exposed to vacuum at ambient temperature T_1 outdiffuses along the initial portion of the lower curve with a slope of $t^{-1/2}$, as given by Eq. (4.3). At time corresponding to point A, the temperature is increased to temperature T_2. Because the gas concentration in the solid cannot change instantaneously, the outdiffusion rate increases to the value given by point B on the high temperature diffusion curve but shifted in time to point C. The dotted line connecting A and B is a line of constant concentration. It has a slope of −1 and the value of q_B is given by $q_A t_A = q_B t_B$. Outdiffusion continues along the high temperature curve but is

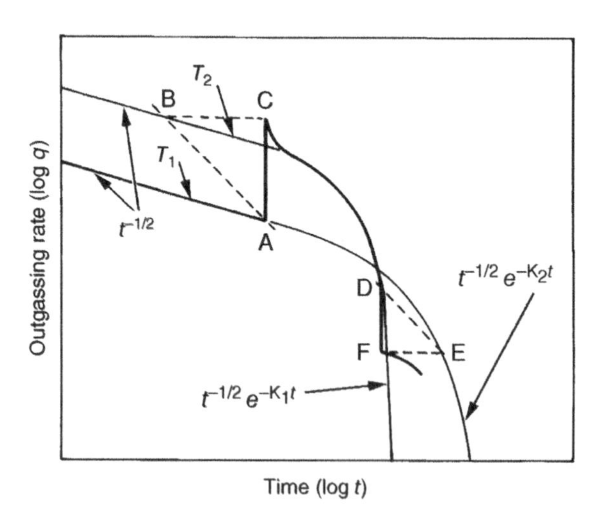

Fig. 4.3 Change in outdiffusion rate for an increase in temperature from T_1 to T_2 for a diffusion process. D.J. Santeler et al. [2]/NASA/Public Domain.

displaced in time. The plot looks curved near point C, because of the logarithmic axis. At time corresponding to point D the baking operation is terminated, and the temperature is reduced to T_1. The outdiffusion rate is reduced to a value corresponding to point E on the low temperature curve but at an earlier time given by point F. Outdiffusion continues at a rate given by the low temperature curve but shifted to an earlier point in time. The new pressure at point F is given by [3]:

$$P_f = \left(D_1 / D_2\right) P_d \tag{4.7}$$

where $D_{1,2}$ are the diffusion coefficients at temperatures $T_{1,2}$, respectively. The net effects of baking are the reduction in the outgassing rate and the reduction of the time required to remove the initial concentration of gas dissolved in the solid.

4.3 Thermal Desorption

Thermal desorption is the heat-stimulated release of gases or vapors previously adsorbed on the interior walls of the system. They may have been adsorbed on the chamber surface, while it was exposed to the environment and then slowly released as the pump removed gas from the chamber. Alternatively, they may have reached the inner surface by diffusion from within, or permeation from the outside. The rate of desorption is a function of the molecular binding energy, the temperature of the surface, and the surface coverage. Gas is bound on surfaces by physisorption and chemisorption. Physisorbed molecules are bonded to the surface by weak van der Waals forces of energy <40 MJ/(kg-mol). Adsorption at energies greater than this value is known as chemisorption. Weakly physisorbed molecules are removed quickly from solid surfaces at ambient temperature and do not hinder pumping. Chemisorbed particles desorb slowly unless external energy is provided in the form of phonons, photons, or charged particles. The effects of readsorption must be included in real vacuum systems, when determining the net or measured thermal desorption rate. If readsorption is significant, the true outgassing rate will be much higher than the measured outgassing rate.

In this section we will first consider desorption without readsorption. Our discussion of readsorption will focus on the practical issue of pumping adsorbed water vapor, a most important concern in the performance of both quick-cycle, unbaked vacuum systems, and well baked accelerator systems.

4.3.1 Zero Order

Surfaces covered with multilayers of molecules, such as water vapor will desorb at a constant rate [4].

$$\frac{dn}{dt} = \alpha e^{E_v/kT} \tag{4.8}$$

This process is called zero-order desorption; n is molecular surface density, E_v is the latent heat of vaporization, and α is a constant. It describes the rate at which molecules depart from a surface saturated with large quantities of vapor. E_v ~40,600 MJ/(kg-mol) or 9.7 kcal/g-mol for water.

4.3.2 First Order

When the surface coverage becomes less than one monolayer, molecules that do not dissociate on adsorption depart without return at a rate proportional to their surface concentration. This can be expressed mathematically as:

$$\frac{dC(t)}{dt} = -K_1 C(t) = \frac{e^{-E_d/(N_o kT)}}{\tau_o} C(t) \tag{4.9}$$

Desorption with a rate $dC(t)/dt$, which is proportional to the concentration of atoms or molecules on the surface C, is called first-order desorption. This is a description of how monoenergetic atoms of, for example, helium or argon desorb from a metal or glass, or how water vapor desorbs from itself or a glass. The rate constant K_1 is strongly dependent on desorption energy E_d and the temperature. It can be described by:

$$\frac{1}{K_1} = \tau_r = \tau_o e^{E_d/(N_o kT)} \tag{4.10}$$

where τ_o is the vibrational frequency of a molecule or atom in an adsorption site and is typically 10^{-13} s. K_1 is also the reciprocal of the average residence time τ_r that a molecule or atom spends on the surface. The desorption rate will also decrease with time as the surface layer becomes depleted. By integrating Eq. (4.3) we can show:

$$\frac{dC(t)}{dt} = C_o K_1 e^{-K_1 t} = C_o K_1 e^{-t/\tau_r} \tag{4.11}$$

Equation 4.11 describes the manner in which desorption rate decreases with time. It predicts rapid (exponential) decay of the desorption rate in a few τ_r.

Table 4.1 shows the strong dependence of the residence time on the temperature and desorption energy for three representative gas–metal systems: water weakly bonded to itself, hydrogen strongly bonded to molybdenum, and the intermediate case was water bonded to a metal. These residence times assume, of course, that the molecule accommodated to the surface after its arrival. Not all molecules do so; or said another way, the sticking coefficients may be less than 1. Reliable sticking coefficients are lacking. These residence times were calculated using desorption energies given by Ehrlich [5], and Moraw and

Table 4.1 Average Residence Time of Chemisorbed Molecules

System	Desorption Energy (10^6 J/(kg-mol))	Residence Time at		
		77 K (s)	22°C (s)	450°C (s)
H_2O/H_2O	40.6	10^{15}	10^{-5}	10^{-9}
H_2O/metal	96	—	10^{5}	10^{-5}
H_2/Mo	160	—	10^{17}	1

Prasol [6]. The room temperature residence time for water adsorbed on itself is less than a microsecond and baking is not necessary to remove the bulk of this physisorbed water—it is removed quickly beginning at about 2330 Pa (17.5 Torr), the vapor pressure of water at 20°C. Hydrogen is strongly chemisorbed at energies of 160 MJ/(kg-mol) or 40,000 cal/(g-mol). It cannot be removed without a high temperature bake or the use of stimulated desorption. One monolayer of water is chemisorbed on a metal with energies in the range of 92–100 MJ/(kg-mol) or 22,000–24,000 cal/(g-mol) [6]. It is a problem because it is bound just strongly enough to stick at room temperature. Although it is hard to remove at room temperature, it is easily desorbed by baking at temperatures over 250°C. Cooling the surface has a dramatic effect on the residence time of all molecules—physisorbed and chemisorbed. Residence times become very long as cooled surfaces become traps.

Water is the most troublesome vapor to remove without baking because the last monolayer desorbs over a long time period, ranging from hours to days at room temperature. Desorption of the remaining monolayer is difficult without some form of baking or nonthermal radiation. Hydrogen is the dominant gas that desorbs from stainless steel surfaces.

4.3.3 Second Order

First-order desorption does not describe the situation where a gas dissociates on adsorption and recombines before desorption. Diatomic gases on metals, for example, hydrogen on steel or molybdenum, dissociate on adsorption. The atoms must recombine on the surface before desorbing ($H + H \rightarrow H_2$). A reaction that is proportional to the concentration of each of two species is called a second-order reaction; the rate equation is given by:

$$\frac{dC(t)}{dt} = -K_2 C^2(t) \tag{4.12}$$

This can be solved to yield:

$$\frac{dC(t)}{dt} = \frac{-K_2 C_o^2}{\left(1 + C_o K_2 t\right)^2} \tag{4.13}$$

We see from Eq. (4.13) that the time to clean up a surface is longer than for first order because the rate decays as $1/t^2$. Remember that K_2 also contains the energy and temperature dependent term $\exp(-E_d/kT)$ so that second-order desorption, like first-order desorption, can proceed rapidly at high temperatures and weak binding energies.

Figure 4.4 shows a sketch of desorption rate versus time for a first and a second-order process. Knize and Cecchi [7] performed an experiment that nicely illustrates the two processes. Their vacuum system contained Zr–Al getters whose surfaces were contaminated with a small amount of deuterium. Since two deuterium atoms need to recombine and form D_2 before desorption can occur, surfaces with a low concentration of deuterium desorbed as a second-order process according to Eq. (4.13). By flooding the chamber with a high, constant pressure of hydrogen, they created a surface in which the small number of deuterium atoms was flooded with H atoms. Since an H atom was frequently hitting each D atom, H–D was formed and desorbed at a rate proportional to the concentration of D atoms. Their data showed the deuterium desorption rate decreased as $1/t^2$ without the addition of hydrogen gas. With the addition of hydrogen, the HD desorption rate

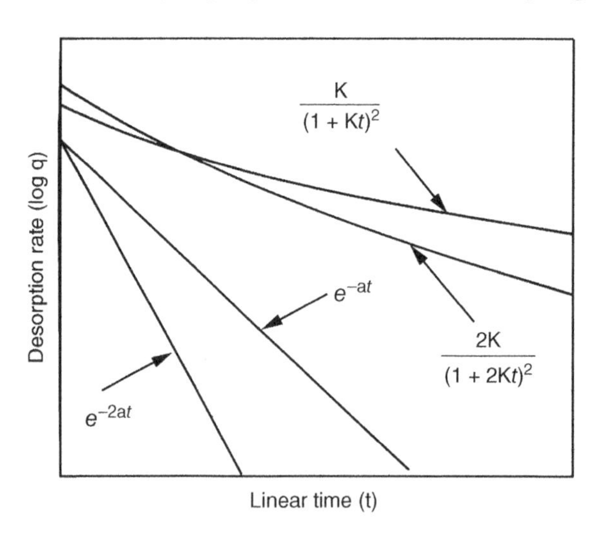

Fig. 4.4 Desorption rate (log) versus time (linear) for first-order desorption K_1, and second-order desorption K_2. Two activation energies are shown for each type of desorption. In both cases, the rate constants are energy dependent, but after long times second-order desorption is always slower than first-order.

decreased exponentially; a second-order process was converted into a first-order process described by Eq. (4.11).

4.3.4 Desorption from Real Surfaces

If atoms and molecules simply desorbed from a surface and exited to the pump in a single bounce—never to return again—the relationships described in Section 4.3.1 would accurately characterize their rate of removal. Unfortunately, real vacuum systems cannot be characterized in this manner. In the molecular flow regime, molecules leaving a surface make myriad collisions with other surfaces before encountering the pump entrance. During some of those collisions, a molecule will re-adsorb or stick, whereas the molecule will be merely scattered, usually diffusely, after the remainder of the collisions. The sticking coefficient s represents the fraction of collisions in which the molecules re-adsorbs at an available surface site. It remains at the surface for a residence time τ_r. The picture is further complicated by the variation of the sticking coefficient with the number of available adsorption sites, and by the ability of molecules to diffuse to the surface from complex, porous, near-surface regions. Here we discuss experimental outgassing, practical baking of vacuum chambers and models for thermal desorption.

4.3.5 Outgassing Measurements

Outgassing may be measured by several methods [8]. The throughput method, where the gas is pumped through a known conductance, is one commonly used method. Gas accumulation or rate of rise, where the pressure rise is measured after the chamber has been isolated, is another common method. Typical results are displayed in Fig. 4.5 for the outgassing of 304L stainless steel with differing initial water vapor exposures [9].

Room temperature outgassing data for most gases sorbed on metals, including water vapor, show the outgassing rate to vary inversely with time, at least for the first 10 h of pumping [9,10,11]. This can be expressed as:

$$q_n = \frac{q}{t^{\alpha_n}} \tag{4.14}$$

where the subscript n denotes the time in hours for which the data apply. See Fig. 4.6. The exponent α will range from 0.7–2, with 1 the most common value. Equation 4.14 is often misinterpreted to be an equation by which outgassing data, such as given in Appendixes C.1–4, C.7, can be determined for all time. That is not its intent. The values given in the appendixes and other publications are points and slopes at discrete times. Tabulated experimental data are usually given for one or two representative times: for example, 1 and 4 h or 1 and 10 h. To further

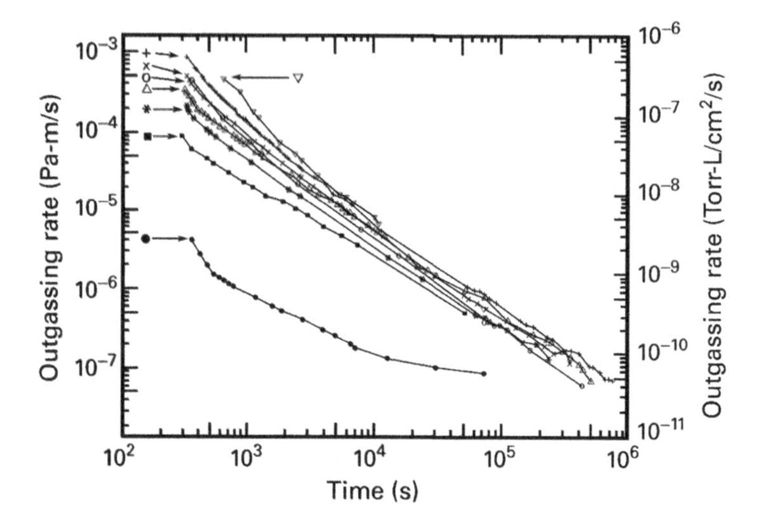

Fig. 4.5 Outgassing measurements for different H_2O exposures during venting of a 304 stainless steel chamber of inner surface area 0.4747 m^2: ○ Ambient air-exposed, 7.8 mL absorbed; ▽ 600 mL exposed, 16.8 mL absorbed; + 400 mL exposed, 9.2 mL absorbed; × 200 mL exposed, 7.2 mL absorbed; Δ 100 mL exposed, 3.6 mL absorbed; ★ 10 mL exposed, 2.3 mL absorbed; ■ N_2 gas with <10 ppm H_2O exposed, 0.7 mL absorbed; • dry N_2 gas exposed, 0.017 mL absorbed; M. Li and H.F. Dylla [9]/Reproduced with permission from AIP Publishing.

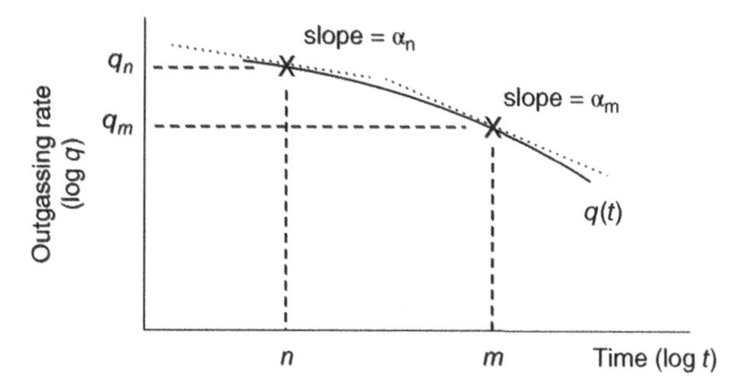

Fig. 4.6 Outgassing data notation, as used in Appendixes C2, C3, C4, and C7.

confuse the situation, published studies may not have all chosen the same $t = 0$. This starting time is best chosen as the time at which the chamber pressure equals the saturation vapor pressure of water. Outgassing data for engineering surfaces typically show wide variation, representative of both the methods used to accumulate data, and by which samples were prepared. We will discuss this in detail in Chapter 15.

4.3.6 Outgassing Models

The mathematical modeling of outgassing from stainless steel has been studied by numerous investigators. There are two distinct models. Gas can be assumed to originate in the near-surface region and be released after diffusing to the surface [9,10,11]. In this model, readsorption is not included. Alternatively, gas can be assumed to originate in an adsorbed layer on the surface [4,6,12,13]. As Redhead pointed out, both adsorption and diffusion occur, and the two models differ in their assumption of which process is the rate-limiting step [4]. Both models are able to predict the $1/t$ dependence experimentally observed for water on type 304 stainless steel.

Dayton [10] modeled this $1/t$ behavior as the sum of the successive outdiffusion of molecules of a range of energies. He proposed a uniform distribution of activation energies like that found in a surface oxide containing pores of varying diameters will result in a sum of individual outgassing curves. It will appear as one curve with a slope of about −1 on a log–log plot of outgassing rate versus pumping time, as sketched in Fig. 4.7. Each one of the curves in Fig. 4.7 is an outgassing curve whose shape we discussed previously. Redhead [4] assumed the short-term sub-monolayer outgassing of water from stainless steel was dominated by desorption from a reversible phase. He used an extended Temkin isotherm to describe the surface coverage–pressure relation, and with it, he closely modeled experimental data using a sticking coefficient of $s=0.1$.

Experimental measurements of the sticking coefficient of water on stainless steel at less than one monolayer coverage are not in agreement but do show the same trend—the sticking coefficient increases as the number of available sites increases. Tuzi et al. [13], showed the sticking coefficient to increase from $s<0.001$ at 5×10^{-3} Pa to $s=0.04$ at $P\leq10^{-8}$ Pa. Moraw and Prasol [6] found s varied from 0.0018 at $P=10^{-4}$ Pa to 0.003 at $P=10^{-6}$ Pa. A pulsed laser desorption experiment [13] yielded $s=0.2$ for clean surfaces and $s=0.01$ at a coverage of 2×10^{17} molecules/m^2.

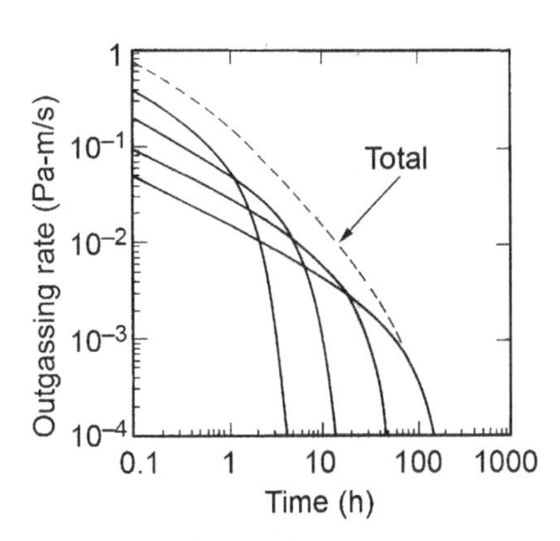

Fig. 4.7 Total outgassing rate as a sum of four rates, each resulting from a single outgassing time constant whose value depends on the shape of the surface oxide pores and the activation energy for desorption. B.B. Dayton [10]/Reproduced with permission from Elsevier.

4.3.7 Reduction by Baking

According to Dayton's model [10], baking will reduce the outgassing time in the same manner as outdiffusion time. The rate of gas release from the surface of a solid containing dissolved gas is slow because the rate is determined by the mobility of the gas in the solid. When the solid is nearly exhausted, the slope becomes much greater (exponential).

Readsorption of molecules leaving a surface will lengthen the time a molecule remains in the chamber. The average number of collisions a molecule makes with a surface before departing the chamber depends on the pump speed S, the internal surface area A, and the average velocity [6].

$$f = \frac{v}{4}\left(\frac{1}{S/A}\right) \tag{4.15}$$

This results in an increased residence time τ_c in a chamber of:

$$\frac{t_c}{t_r} = \left(1 + \frac{v}{4}\frac{s}{S/A}\right) \tag{4.16}$$

When baking vacuum systems, it is imperative that all surfaces be baked, as any area not baked will contribute an exceedingly large gas flux, relative to the baked area. For example, if 10% of the area were not baked, and its outgassing rate was 1000× that of the baked area, then the unbaked area would contribute 90% of the desorbed gas flux.

Alternatively, one can draw the same conclusion using a residence time model. Consider two surfaces, as shown on the left side of Fig. 4.8. In this image, a water molecule is shown bouncing from two stainless steel surfaces on its sojourn to the pump. One surface is unheated at 295 K (22°C) whereas the other surface is heated to 80°C. This image would be representative of a vacuum system in which only half the wall area was heated. Half the molecular bounces would be from heated surfaces and half would be from unheated surfaces. In constructing this model, we have assumed that the sticking coefficients were the same for both surfaces. Using Eq. (4.10) and the sorption energy for H_2O on stainless steel given in Table 4.1, residence times of 9800 s and 16 s were calculated for the cold and warm surfaces, respectively, near one monolayer water coverage. The right-hand illustration in Fig. 4.8 shows that a

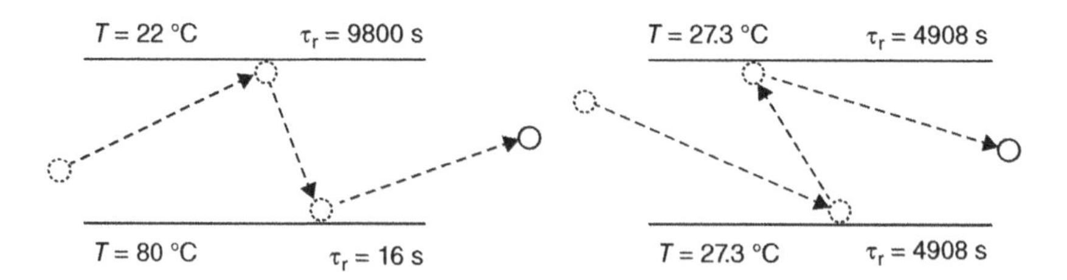

Fig. 4.8 The total residence time for a water molecule after two bounces from a metal surface is shown to be the same for two sets of surface temperatures; a sticking coefficient of one was assumed. This example illustrates the necessity of baking all surfaces within a vacuum chamber. Unbaked surfaces dominate the behavior of the system.

uniform temperature of 27.3°C will result in the same net residence time. Were the water molecules to have bounced from a system with 1/3 of the wall area at ambient temperature and 2/3 of wall area heated to 80°C, the equivalent uniform baking temperature would have been 30.5°C. We see that baking a fraction of the chamber (1/2–2/3 in this example) to a high temperature is equivalent to baking the entire system to just a few degrees above ambient, 5–8°C in this example!

4.4 Stimulated Desorption

Electrons, atoms, molecules, ions, or photons incident on solid surfaces can release adsorbed gases and generate vapors in quantities large enough to limit the ultimate pressure in a vacuum chamber. Many reactions are possible when energetic particles collide with a surface [14]. Among them are electron-stimulated desorption, ion-stimulated desorption, electron- or ion-induced chemical reactions, and photodesorption.

4.4.1 Electron-Stimulated Desorption

An energetic electron incident on the surface gas layer excites a bonding electron in an adsorbate atom to a nonbonding level. This produces a repulsive potential between the surface and atom, which allows the atom to desorb as a neutral or an ion [15,16,17,18]. The electron-stimulated desorbed neutral gas flux can be as high as 10^{-1} atom per electron, whereas the desorbed ion flux is of order 10^{-5} ion/e. This desorption process is specific; it depends on the manner in which a molecule is bonded to the surface. These gas release phenomena have been shown to cause serious errors in pressure measurement with the Bayard-Alpert gauge [19,20] and residual gas analyzer [21] and to increase the background pressure in systems that use high-energy electron or ion beams.

4.4.2 Ion-Stimulated Desorption

Ion-stimulated desorption has been studied by Winters and Sigmund [22] and Taglauer and Heiland [23]. Winters and Sigmund studied desorption of previously chemisorbed nitrogen on tungsten by noble gas ions in the range up to 500 V. They also showed that the adsorbed atoms were removed by a sputtering process as a result of direct knock-on collisions with impinging and reflecting noble gas ions. Ion-stimulated desorption is responsible for part of the gas release observed in sputtering systems, Bayard-Alpert gauges, and glow discharge cleaning.

Edwards [24] has measured the electron- and ion-stimulated desorption yields of type 304 stainless steel and aluminum as a function of cleaning technique.

Table 4.2 Desorption Yields for Stainless Steel and Aluminum

	100 eV Ar^+ Ions		500 eV electrons	
Gas	**304 SS**	**Aluminum**	**304 SS**	**Aluminum**
He	2.13	2.38	0.15	0.18
CO	3.22	3.00	0.06	0.05
CO_2	1.55	1.35	0.21	0.16
CH_4	0.075	0.07	0.0025	0.003

Note: Cleaned by a soap wash, water rinse, acetone rinse, methanol rinse, and air dried followed by 200°C, 60-h bake, and 2-day equilibration.
Source: D. Edwards [24]/Reproduced with permission from AIP Publishing.

Table 4.2 shows his results for degreased stainless and aluminum samples. Methane desorption is most likely caused by an electron- or ion-induced chemical reaction.

4.4.3 Stimulated Chemical Reactions

Ion- and electron-stimulated chemical reactions may also occur at solid surfaces. For example, hydrogen and oxygen can react with solid carbon to produce methane and carbon monoxide. An ion-stimulated chemical reaction is responsible for rapid etch rates observed in reactive-ion etching. In reactive ion etching of silicon, high-energy ions in a collision cascade greatly enhance the reaction of neutral F with Si and produce a much greater yield of volatile SiF_4 than is possible without ion bombardment.

4.4.4 Photo Desorption

Lichtman et al. [25] observed greater than band gap desorption. The effect occurs in semiconductors (CrO_2 on stainless steel). However, one should be aware that the fraction of the energy greater than band gap, available from a mercury arc lamp, is insufficient to clean a vacuum chamber. Most desorption results from optically induced surface heating. Gröbner et al. [26], Mathewson [27] and Ueda et al. [28] measured the energetic photon-induced desorption of H_2, CO_2, CO, and CH_4 from aluminum, copper, and stainless steel used in high-energy accelerators. They measured desorption efficiencies of 0.1–0.001 molecules per photon.

4.5 Permeation

Permeation is a three-step process. Gas first adsorbs on the outer wall of a vacuum vessel, diffuses through the bulk, and lastly desorbs from the interior wall. Permeation through glass, ceramic, and polymeric materials is molecular. Molecules do not dissociate on adsorption. Hydrogen does dissociate on metal surfaces and diffuses as atoms that recombine before desorption on the vacuum wall.

4.5.1 Atomic and Molecular Permeation

The movement of atoms and molecules that do not dissociate when moving through a solid is described by [29]:

$$q(\text{Pa-m/s}) = \frac{DS'P}{d} \sum_{m=1}^{\infty} 2(-1)^m \exp\left(\frac{-m^2\pi^2 Dt}{d^2}\right) \tag{4.17}$$

where D is the diffusion constant, S' is the gas solubility, and d is the thickness of the material through which the gas permeates. Equation 4.17 converges slowly for short times. The alternative relationship for short times, given in Eq. (4.15), can be obtained by the same method as used for Eq. (4.18).

$$q(\text{Pa-m/s}) = 2S'P\left(\frac{D}{\pi t}\right)^{1/2} \sum_{0}^{\infty} \exp\left(\frac{d^2}{4dt}(2m+1)^2\right) \tag{4.18}$$

Figure 4.9 illustrates the permeation of helium through type 7740 glass at 300 K for glass that is initially devoid of helium. In time, equilibrium will be established, and helium will permeate the glass at a constant rate. Steady-state permeation therefore behaves like a constant leak. The product of the diffusion constant and the solubility is the permeation constant K_p:

$$K_p = DS' \tag{4.19}$$

In SI, both permeation constant and diffusion constant have units of m^2/s. Solubility is the dimensionless ratio of the gas-to-solid volume at the temperature of the measurement. For example, the experimentally determined solubility of He in type 7740 glass described in Fig. 4.9 is 0.0065. The concentration inside the glass surface C, the solubility S', and the external gas pressure P are related through Henry's law, $C = S'P$. The solubility of many materials is not strongly temperature dependent. Because of this, the activation energies for permeation and diffusion are similar for many engineering materials. Consequently, both quantities increase rapidly with temperature.

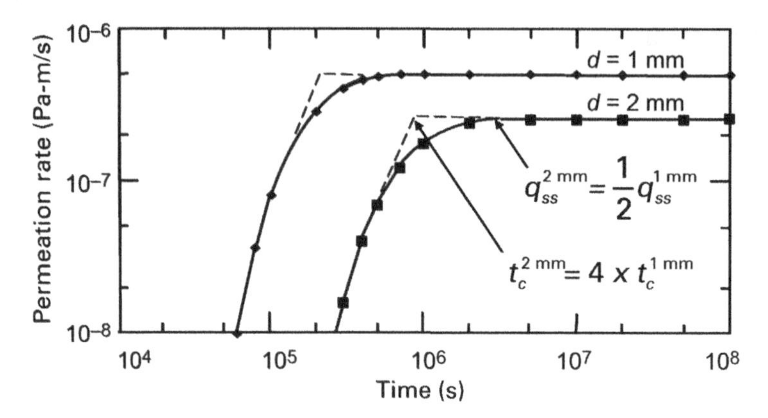

Fig. 4.9 Permeation of helium through d=1-mm- and 2-mm-thick 7740 glass at 27°C that was initially free of helium. The steady-state permeation *rate* varies inversely with the glass thickness. The critical *time* increases with the square of the glass thickness.

The steady-state permeation flux of an atom or molecule that does not dissociate on adsorption can be expressed as:

$$Q_k = q_k A = \frac{K_p P A}{d} \tag{4.20}$$

where q_k is the flux in units of (Pa-m^3)/(s-m^2) and P is the pressure drop across the solid of thickness d. This is the characteristic steady-state flux for all gases that permeate glass, ceramic, and polymeric materials.

Since the initial gas load must traverse a solid (devoid of gas) before it reaches the vacuum wall, a period of time will pass before it is first observed in the vacuum chamber. The time to reach the steady state is known as the critical time t_c and is given by [29]:

$$t_c = \frac{d^2}{6D} \tag{4.21}$$

Solubility S' and diffusion D play distinct roles in gas permeation. The product of $S'D$ in Eq. (4.19) determines the steady-state flux, whereas the time to reach steady state is only dependent on the diffusion constant Eq. (4.21). See Fig. 4.9. Doubling the glass thickness reduced the steady-state flux by 2× but increased the critical time to reach steady state by 4×.

4.5.2 Dissociative Permeation

Reactive molecules dissociate when they adsorb on metal surfaces, and then diffuse through a solid as atoms or smaller compounds. Their behavior differs from that of nonreactive atoms or molecules.

The steady-state permeation rate of hydrogen through metal is an example of this process. First, molecular hydrogen dissociates on the surface with equilibrium constant k_1, then each atom is incorporated into the solid with equilibrium constant k_2. The total concentration within the metal is found to be $[H]_{solid} = k_2(k_1[H_2]_{gas})^{1/2}$. Thus, the driving force for diffusion of hydrogen in metals is $P^{1/2}$. This is fundamentally different from atomic diffusion described in Eq. (4.20). Here the permeation flux is:

$$Q_k = q_k A = \frac{K_p\left(P_2^{1/2} - P_1^{1/2}\right)A}{d} \tag{4.22}$$

The permeability K_p for gases that dissociate has units of $Pa^{1/2}m^2/s$.

4.5.3 Permeation and Outgassing Units

Permeation units recorded in the literature are inconsistent. They are not standardized in gas quantity, surface area, thickness, or pressure drop. Some researchers used a differential pressure of 10 Torr. Others used 1 Torr, or 1 cm Hg, or 1 atm; some researchers normalized data to 1-mm-thick samples; others reported data per cm thickness. The "scc" and "Standard Atmosphere" of gas quantity are traditionally referenced to 0°C, whereas the volumetric flow units for permeation flux are referenced to the measurement temperature, thus permeation measurements recorded at high temperature should be corrected for density change $273/T_{meas}$ [29]. Noncoherent units are as cumbersome to convert between each other, as they are to SI units. Converting older units directly to m^2/s facilitates comparison, and is an important reason to work in SI.

Permeability is a measure of gas transport through a solid by diffusion. It is the product of the dimensionless number (solubility) and the diffusion constant. We routinely use units of cm^2/s or m^2/s (SI) for diffusion constant, so it is hard to explain the origin of such cumbersome permeability units. If the permeability and the diffusion constant are stated in SI units of m^2/s, the distinct roles of solubility and diffusion in determining steady-state flux q and critical time t_c are obvious. We will revisit this topic when we discuss the very important topic of gas permeation through elastomer O-rings in Chapter 16.

Total flux Q permeating *through an object* is given in units of Pa-m^3/s or sometimes mbar-L/s. The older English unit is the Torr-L/s. One Pa-m^3/s = 1 W. Measuring permeation flux in Watts is not useful, but it illustrates that gas release *from a surface* has the dimensions of energy flow. The permeation per unit surface area is given in units of Pa-m/s, or mbar-m/s. Conversion factors between older units and SI units are given in Appendix A.3.

4.6 Pressure Limitations During Pumping

In the introduction to this chapter, we alluded to the fact that outgassing, not volume gas removal determined the final pressure in the high vacuum region. We know this to be true from another viewpoint. One monolayer of gas on the interior of a 1-L sphere, if desorbed instantly, would result in a pressure of 2.5 Pa (18 mTorr). This is a pressure far above high vacuum. Diffusion, permeation, and stimulated desorption are also important. In the previous section we made no attempt to numerically relate these various rates to a real vacuum system. The relative roles of surface desorption, diffusion, and permeation are a function of the materials used for construction (steel, aluminum, ceramic, or glass), the seals (metal gaskets, or single- or differentially pumped elastomer gaskets), and the system history (newly fabricated, unbaked, chemically cleaned, or baked). The mathematical description of this pumping problem can be solved easily with certain approximations as:

$$P = P_o e^{-St/V} + \frac{Q_o}{S} + \frac{Q_D}{S} + \frac{Q_k}{S} \tag{4.23}$$

The first term in Eq. (4.23) represents the time-dependence of the pressure that is due to the gas in the chamber volume. The remaining terms represent the contribution of other gas sources. These terms represent outgassing, diffusion, and permeation. They are slowly varying functions and become dominant after the initial pumping period has passed. After some time, the first term on the right-hand side is zero and the pressure is determined by outgassing, diffusion, desorption, and permeation. The first term (outgassing) decays slowly, often as t^{-1}. The second term is the diffusion term. It decays initially as $t^{-1/2}$ and at long times as $\exp(-\alpha Dt)$. The last term in Eq. (4.23) is due to permeation and is constant. From this equation we can construct a composite pump-down curve that will illustrate the relative roles of these phenomena.

Figure 4.10 shows the high vacuum pumping portion of an unbaked system sealed with metal gaskets. In the initial stages the pressure is reduced exponentially with time as the volume of gas is removed. This portion of the pumping curve takes only a short time because a typical system time constant is only seconds. It is expanded here for clarity. In the next phase surface desorption controls the rate of pressure decrease. In a typical unbaked system, most of the gas load is water vapor; however, nitrogen, oxygen, carbon oxides, and hydrocarbons are also present. The material and its history determine the total quantity of gas released. Glass or steel that has been exposed to room ambient for extended periods may contain up to 100 monolayers of water vapor, whereas a carefully vented chamber may contain little water vapor. A slow decrease in

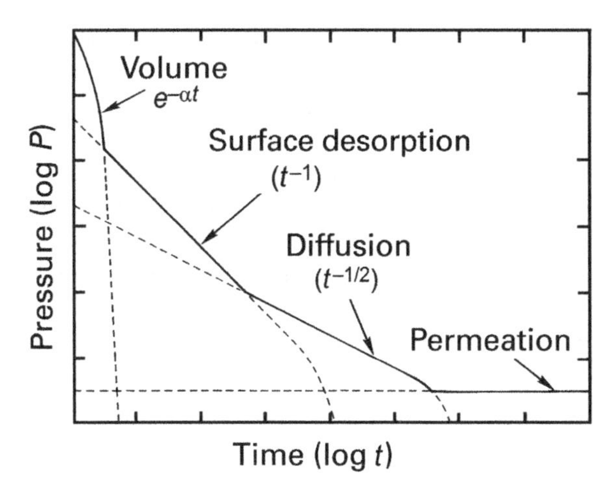

Fig. 4.10 Rate limiting steps during the pumping of a vacuum chamber.

pressure is to be expected, if the system has a number of large elastomer O-rings, or a large interior surface area. Unbaked, rapidly cycled systems are never pumped below the outgassing-limited range.

If the system is allowed to continue pumping without baking, the surface gas load will ultimately be removed and the outdiffusion of gases in solution with the solid walls will be observed. The slope of the curve will change from t^{-1} to $t^{-1/2}$. For example, hydrogen that diffuses into steel in a short time at high fabrication temperatures outdiffuses from near-surface regions exceedingly slowly at room temperatures. The system would now be at its ultimate pressure given by $P = Q_k/S$. Particular gases are known to permeate materials. Hydrogen perhaps from water vapor is a potential permeation source through stainless steel. Helium permeates glass walls, and water and air permeate elastomers. In a real system, the time required to achieve the ultimate pressure depends strongly on choice of materials and cleaning procedures.

The order of importance of the processes is not always as shown in Fig. 4.10. Elastomer gaskets have a high permeability for atmospheric gases and if the system contained a significant number of elastomer gaskets, it would be the dominant permeation source.

The level of outgassing resulting from various electron- and ion-stimulated desorption processes is not shown in Fig. 4.10. These processes play a variable but important role in determining the ultimate pressure in many instances, because they are desorbed atoms that were not removed by baking. The desorption yield from energetic electrons and ions is a function of incident particle's energy, the material and its cleaning treatment, and the time under vacuum.

Electron-stimulated desorption in an ion gauge is easily observable during gauge outgassing. A 10 mA electron flux can desorb 6×10^{14} neutrals/s. Because a gas flux of 2×10^{17} atoms/s is equivalent to 1 Pa-L/s at 22°C, this electron flux can initially provide a desorption flux of 2×10^{-3} Pa-L/s. If the system is pumping on the ion gauge tube at a rate of 10 L/s, this desorption peak can reach a pressure of an order of 10^{-4} Pa in the gauge tube. The scattered electrons from Auger and LEED systems can stimulate desorption in a similar manner, with an increase in background pressure when the beam is operating. Electron-induced desorption is a first-order process that should produce a simple exponential decay of the desorbed species. If readsorption is included, the initial rate will decay exponentially to a steady-state level [30]. A 20 mA ion beam can desorb surface gas at an initial rate of 10^{-3} Pa-L/s, assuming a desorption efficiency of two atoms per incident ion.

References

1 Carlslaw, H.S. and Jaeger, J.C., *Conduction of Heat in Solids*, 2nd ed., Oxford, New York, 1959, p. 97.
2 D.J. Santeler, D.J., Jones, D.W., Holkeboer, D.H., and Pagano, F., *Vacuum Technology and Space Simulation*, NASA SP-105, National Aeronautics and Space Administration, Washington, DC, 1966, p. 191.
3 Bills, D.G., *J. Vac. Sci. Technol.* **6**, 166 (1969).
4 Redhead, P., *J. Vac. Sci. Technol.* **13**, 2791 (1995).
5 Ehrlich, G., *Transactions of the 8th National Vacuum Symposium of the American Vacuum Society and 2nd International Vacuum Congress*, L. Preuss, Ed., Pergamon, New York, 1962, p. 126.
6 Moraw, M. and Prasol, H., *Vacuum* **49**, 353 (1998).
7 Knize, R.J. and Cecchi, J.L., *J. Vac. Sci. Technol., A* **1**, 1273 (1993).
8 Redhead, P.A., *J. Vac. Sci. Technol., A* **20**, 1667 (2002).
9 Li, M. and Dylla, H.F., *J. Vac. Sci. Technol., A* **11**, 1702 (1993).
10 Dayton, B.B., *1961 Trans. 8th Nat. Vac. Symp. and Proc. 2nd Intl. Congr. on Vac. Sc. Technol.*, **1**, Pergamon, New York, 1962, p. 42.
11 Dayton, B.B., *1960 Trans. 7th Vac. Symp*, Pergamon, New York, 1961, p. 101.
12 Shiokawa, Y. and Ichikawa, M., *J. Vac. Sci. Technol., A* **16**, 1131 (1998).
13 Tuzi, Y., Kurokawa, K., and Takeuchi, K., *Vacuum* **44**, 477 (1993).
14 See Redhead, P.A., Hobson, J.P., and Kornelsen, E.V., *The Physical Basis of Ultrahigh Vacuum*, Chapman and Hall, London, 1968, Chapter 4.
15 Drinkwine, M.J. and Lichtman, D., *Progress in Surface Science*, Vol. 8, Pergamon, New York, 1977, p. 123.
16 Menzel, D. and Gomer, R., *J. Chem. Phys.* **41**, 3311 (1964).
17 Redhead, P., *Can. J. Phys.* **42**, 886 (1964).

18 Knotek, M.L. and Feibelman, P.J., *Phys. Rev. Lett.* **40**, 964 (1978).
19 Redhead, P.A., *Vacuum* **12**, 267 (1962).
20 Hartman, T.E., *Rev. Sci. Instrum.* **34**, 1190 (1963).
21 Marmet, P. and Morrison, J.D., *J. Chem. Phys.* **36**, 1238 (1962).
22 Winters, H.F. and Sigmund, P., *J. Appl. Phys.* **45**, 4760 (1974).
23 Taglauer, E. and Heiland, W., *J. Appl. Phys.* **9**, 261 (1976).
24 Edwards, D., Jr., *J. Vac. Sci. Technol.* **16**, 758 (1979).
25 Fabel, G.W., Cox, S.M., and Lichtman, D., *Surface Sci.* **40**, 571 (1973).
26 Gröbner, O., Mathewson, A.G., and Marin, P.C., *J. Vac. Sci. Technol., A* **12**, 846 (1994).
27 Mathewson, A.G., *Vacuum* **44**, 479 (1993).
28 Ueda, S., Matsumoto, M., Kobari, T., Ikeguichi, T., Kobayashi, M., and Hori, Y., *Vacuum* **41**, 1928 (1990).
29 Rogers, W.A., Buritz, R.S., and Alpert, D., *J. Appl. Phys.* **25** (7), 868 (1954).
30 Drinkwine, M.J. and Lichtman, D., *J. Vac. Sci. Technol.* **15**, 74 (1978).

Part II

Measurement

In these five chapters, we discuss the instruments used to characterize gases in vacuum systems: pressure gauges, flow meters, and residual gas analyzers (RGAs). Pressure gauges are used to monitor and control process performance and identify problems. Flow meters are now an integral part of any deposition or etch system. Residual gas analyzers add a considerable degree of sophistication to our analytical capability. RGAs can single out a gas or vapor that is limiting the system pressure or causing a process problem. They greatly reduce the difficulty in troubleshooting large and complex systems. Like gauges and flow controllers, they are now an integral part of process control in high technology manufacturing.

Chapter 5 is devoted to a discussion of commonly used pressure gauges. Chapter 6 describes the operation of flow meters, especially the thermal mass flow meter. Chapter 7 describes pumping speed measurement techniques. Pumping speed is a quantity derived from pressure and flow, so both instruments are needed. Chapter 8 discusses the operation and installation of RGAs. Chapter 9 concludes this section with a description of qualitative and quantitative methods for interpreting the information contained in an RGA spectrum.

The illustration on the following page describes both the variety and operational ranges of instruments and techniques used for measuring pressure and gas flow.

A Users Guide to Vacuum Technology, Fourth Edition. John F. O'Hanlon
and Timothy A. Gessert.
© 2024 John Wiley & Sons, Inc. Published 2024 by John Wiley & Sons, Inc.

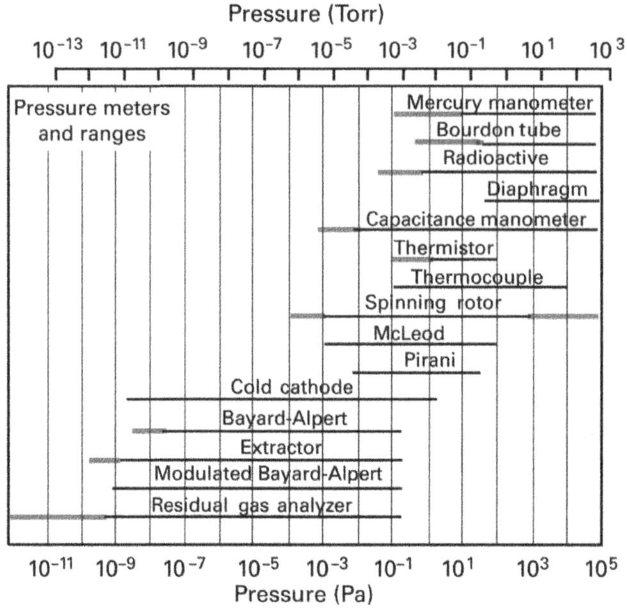

Source (above) Pressure regimes for various gauges. Adapted with permission from *Scientific Foundations of Vacuum Technique*, 2nd ed., J. M. Lafferty, Ed., p. 350. Copyright 1962, John Wiley and Sons.

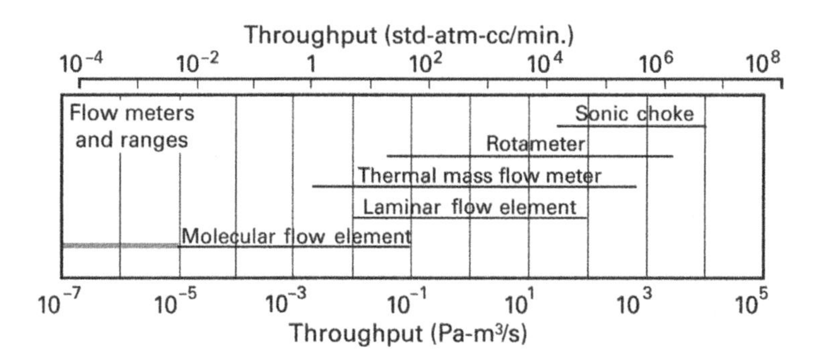

5

Pressure Gauges

In this chapter we discuss common pressure gauges. To discuss in detail each existing pressure gauge would result in a reduced presentation of those gauges that are the most important. Descriptions of the less frequently used gauges are contained in other texts [1,2,3]. Descriptions of ultrahigh vacuum gauges have been added to this discussion, as they are needed in production systems with ultrahigh vacuum base pressures.

Many techniques have been developed for the measurement of reduced pressures. A number of these designs are classified in Fig. 5.1. Gauges are either direct- or indirect-reading. Those that measure pressure by calculating the force exerted on the surface by incident particle flux are called direct reading gauges. Indirect gauges record the pressure by measuring a gas property that changes in a predictable manner with gas density. Examples of many gauge designs and their ranges are illustrated in the figure accompanying the Measurement Section introduction on the preceding page.

5.1 Direct Reading Gauges

Diaphragm, Bourdon, and capacitance manometers are the most commonly used direct reading gauges. Two rather well-known gauges which have a necessary place in pressure measurement, the U-tube manometer and the McLeod gauge, are not described in detail here, because they are no longer in routine use. In its simplest, form a manometer consists of a U-tube that contains a low vapor pressure fluid such as mercury or oil. One arm is evacuated and sealed, the other is connected to the unknown pressure. The unknown pressure is read as the difference in the two liquid levels. The McLeod gauge is a mercury manometer in which a volume of gas is compressed before measurement; for example, compressing a

A Users Guide to Vacuum Technology, Fourth Edition. John F. O'Hanlon
and Timothy A. Gessert.
© 2024 John Wiley & Sons, Inc. Published 2024 by John Wiley & Sons, Inc.

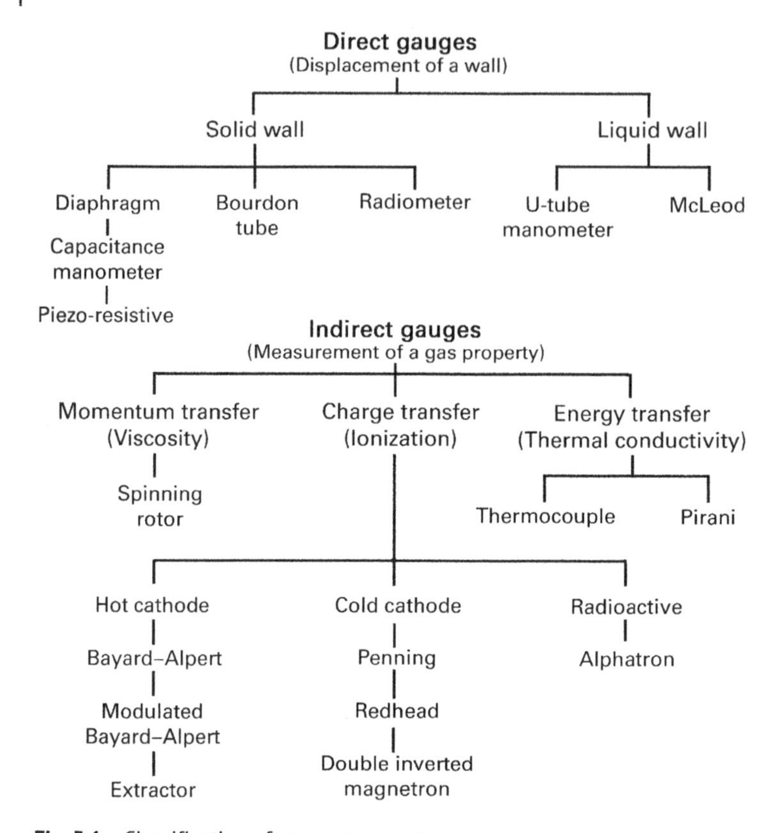

Fig. 5.1 Classification of pressure gauges.

small volume of gas at 10^{-2} Pa by a factor of 10,000 results in a measurable pressure of 100 Pa. The U-tube manometer, which is used in the 10^{2}–10^{5} Pa range, and the McLeod gauge, which is a primary standard in the 10^{-3}–10^{2} Pa pressure range, are described elsewhere [1,2,3].

5.1.1 Diaphragm and Bourdon Gauges

The simplest mechanical gauges are diaphragm and Bourdon gauges. Both use gears and levers to transmit the deflection of a solid wall to a pointer. The Bourdon tube is a coiled tube of elliptical cross-section, fixed at one end and connected to the pointer mechanism at the other end. Evacuation of the gas in the tube causes rotation of the pointer. The diaphragm gauge contains a pressure-sensitive element from which the gas has been evacuated. By removing gas from the region surrounding the element, the wall is caused to deflect, and in a manner similar to the Bourdon tube, the linear deflection of the wall is converted to angular deflection of the pointer.

Simple Bourdon or diaphragm gauges—for example, those of the 50-mm-diameter variety—will read from atmospheric pressure to a minimum pressure of about 10^3–5×10^3 Pa (10–40 Torr). They are inaccurate and used only as a rough indication of pressure.

Diaphragm and Bourdon gauges, which are more accurate than those described above, are available in a variety of ranges from 10^3 to 2×10^5 Pa (10–760 Torr) and with sensitivities of order 25/Pa (0.2/Torr). The differential Bourdon gauge is quite suitable for rough pressure measurement of clean systems when fabricated from 316 stainless steel. When made from Inconel, it can be designed to measure corrosive gases.

5.1.2 Capacitance Manometer

A capacitance manometer is simply a diaphragm gauge in which the deflection of the diaphragm is measured by observing the change in capacitance between it and a fixed counter electrode. Alpert, Matland, and McCoubrey [4] described the first gauge in 1951. They used a differential gauge head as a null reading instrument between the vessel of unknown pressure and another whose pressure was independently adjustable and monitored by a U-tube manometer.

The capacitance of the diaphragm–counter-electrode structure is proportional to geometry (area/gap) and the dielectric constant of the measured gas relative to air. For most gases the pressure may be calculated from the geometry and the observed capacitance change; this is a true, absolute-pressure measurement. A few gases have relative dielectric constants significantly different from air, for example, certain conductive and heavy organic vapors, or gases ionized by radioactivity. For these gases, a 1% difference in dielectric constant will result in a 0.5% error. A single-sided structure is not dependent on the dielectric constant of the measuring gas, because both electrodes are on the vacuum (reference) side.

Capacitance manometers consist of two components, a transducer and an electronic sense unit that converts the membrane position to a signal linearly proportional to the pressure. A common transducer design is shown in Fig. 5.2. The flexible metal diaphragm, which has been stretched and welded in place, is located between two fixed electrodes. The differential transducer shown in Fig. 5.2 may be a null detector or a direct reading gauge. When used as a null detector, the pressure at the reference side P_r is adjusted until the diaphragm deflection is zero. In this mode a second gauge is necessary to read pressure. To use as a direct reading gauge the reference side must be pumped to about 10^{-5} Pa. After calibration the instrument may be used directly over the pressure range for which it was designed. Transducers are available with the reference side open for evacuation or

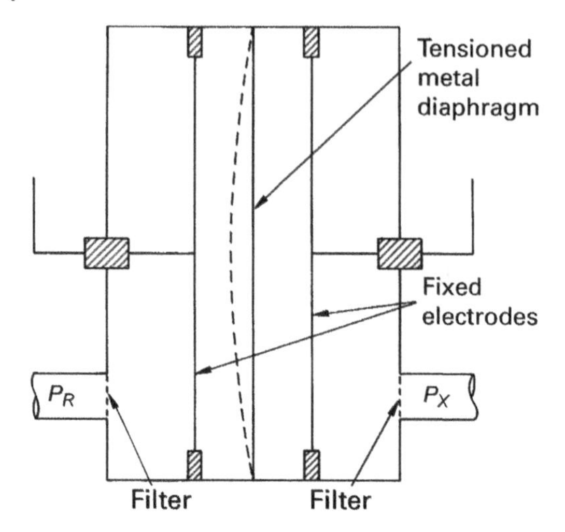

Fig. 5.2 Double-sided capacitance manometer head assembly. Industrial Research/Development, 1976/Reproduced with permission from Technical Publishing Co.

evacuated and permanently sealed with a copper pinch seal; a gettering surface is activated inside the tube at the time of manufacture.

Care should be taken in attaching the transducer to a system. It is generally advisable to have a bellows section in series with one tube, if the transducer is to be permanently welded to a system with short tube extensions. Even though some transducers contain filters to prevent particles from entering the space between the diaphragm and the electrodes, it is advisable to force argon through a small diameter tube placed inside the tube or bellows extension during welding. The end of this small tube should be pushed beyond the weld location to allow the argon flow to flush particles out of the tube and away from the sensor during welding. This procedure stops oxide formation on the tube's interior walls.

Because the transducer contains ceramic insulators, cleanliness is in order; a contaminated head is difficult to clean. Cleaning solvents are difficult to remove from the ceramic and may cause contamination of the system at a later time. To avoid this problem one transducer has been designed with a single-sided sensor. Both electrodes are on the reference side. See Fig. 5.3. One electrode is placed at the center of the diaphragm and the second is an annular ring located around the center electrode. For zero deflection of the membrane the circuit is adjusted for zero output signal. The deflection or bowing of the diaphragm causes a capacitance imbalance, which is converted to a voltage proportional to pressure. A proper choice of materials results in a transducer suitable for service in corrosive environments without head damage or in extremely clean environments without contamination by the head.

The electronic sensing unit applies an ac signal to the electrodes. The changes in signal strength produced by the diaphragm are amplified and demodulated in

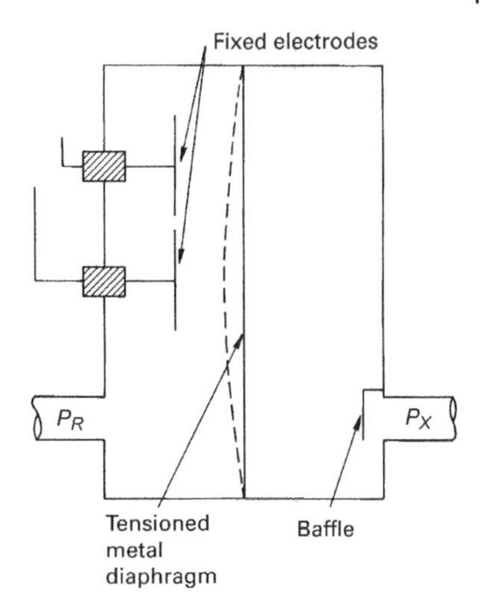

Fig. 5.3 Single-sided capacitance manometer head assembly. The outer electrode is an annular ring. Industrial Research/Development, 1976/Reproduced with permission from Technical Publishing Co.

phase to minimize the noise level. The dc output is then used to drive an analog or digital readout. Because the resolution of the instrument is limited by system noise, the system bandwidth must be stated when specifying resolution; noise is proportional to the half-power of the bandwidth. Typical capacitance manometers have a resolution of 1 part per million (ppm) full scale, at a bandwidth of 2 Hz.

Just as low electronic noise is of prime importance in obtaining high resolution, thermal stability of the head is necessary for stable, accurate, and drift-free operation. The diaphragm deflection in the transducer can be as low as 10^{-9} cm; therefore, motion of parts due to temperature change becomes a large source of error. Transducers are available with heaters that maintain the ambient temperature at about 50°C and avoid some of the problems of ambient temperature change. Many transducers can be operated at temperatures as high as 250°C. At high temperatures, the readings must be corrected for thermal transpiration. See Section 2.3.4. Stable operation of a transducer requires that the thermal expansion coefficients of the diaphragm and electrode assemblies be well-matched. Practical designs must make a trade-off between expansion coefficient and corrosion resistance. Without proper temperature regulation a transducer may have zero and span coefficients of 5–50 ppm full scale and 0.004–0.04% of reading per degree Celsius, respectively, at ambient temperature [5]. Proper temperature regulation can result in an order of magnitude improvement in the zero and span coefficients.

Capacitance manometers can be operated over a large dynamic range, a factor of 10^4–10^5 for most instruments. The overall system accuracy deteriorates at small fractions of full head range. Transducers with a full-scale deflection of 130 Pa (1 Torr) have been checked in the 2.5×10^{-2}–6.5×10^{-4} Pa (1.5×10^{-4}–5×10^{-6} Torr) pressure range by volumetric division [6]. Methods for measurement and calibration of capacitance manometers are described in an ISO Standard [7].

5.2 Indirect Reading Gauges

In this section, we describe the most useful indirect reading gauges. Indirect gauges report pressure by measuring a pressure-dependent property of the gas. In the pressure range above 0.1 Pa, energy and momentum transfer techniques can be used for pressure measurement. Thermal conductivity gauges incorporate the principle of energy transfer between a hot wire and a room temperature gauge wall. A Pirani [8] or a thermocouple [9] gauge is found on every vacuum system for measuring pressure in the medium vacuum region. The spinning rotor gauge [10,11] operates on the principle of momentum transfer. Ionization gauges, which measure gas density, have found wide acceptance. Hot and cold cathode gauges are based on the principle of ionization. The Bayard-Alpert [12] and extractor [13] hot cathode gauges and the cold cathode [14] gauge span the range $0.1–10^{-10}$ Pa.

5.2.1 Thermal Conductivity Gauges

Thermal conductivity gauges are a class of pressure-measuring instruments that operates by measuring the rate of heat transfer between a hot wire and its surroundings. In the low-pressure range, free molecular heat flow increases linearly with pressure. It is first observable when the pressure is large enough for the heat flow from the wire to be larger than the end and radiation losses. It is not related to the tube diameter. In the high-pressure limit von Ubisch [15,16] found heat flow from the wire to become nonlinear at a $\lambda \sim 1$ wire diameter and saturate at 9–12 wire diameters [17]. Molecules leaving the heated wire collided with nearby gas molecules and returned heat to the wire. von Ubisch varied the tube diameter from 0.25 to 10 cm with little significant change in the characteristic response [18]. Ultimately, at very high pressures, the heat flow from wire to tube would become constant, were it not for convection currents within the gauge causing increased heat flow to the wall. See Fig. 5.4. Thermocouple gauges using serpentine heaters fabricated in a planar thin-film structure [19] would begin to saturate at a pressure corresponding to a value where the distance between the heater and thermal sink is a few mean free paths. In the lowest pressure region radiation and conduction to the supports account for most of the heat flow losses.

$$H = A\sigma\varepsilon_1\left(T_2^4 - T_1^4\right) + \text{end losses} \tag{5.1}$$

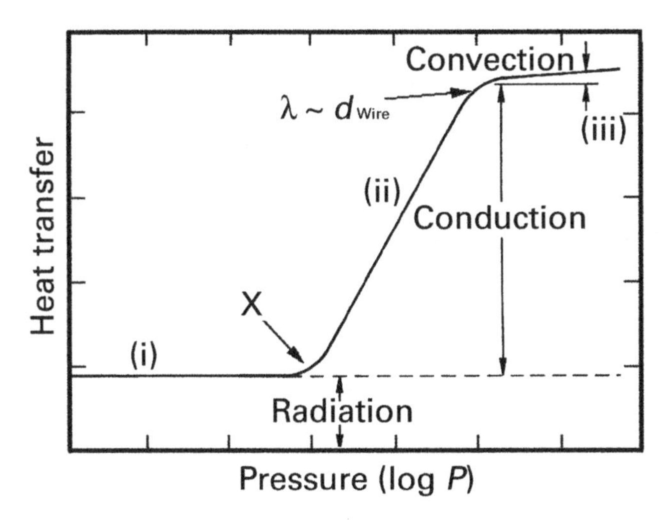

Fig. 5.4 Heat transfer regimes in thermal conductivity gauges, such as thermocouple gauges or Pirani gauges constructed from fine, heated wires. Three regions are illustrated: (i) $\lambda \gg d_{\text{Wire}}$, (ii) $\lambda \ll d_{\text{Wire}}$, and (iii) intermediate. The location of the upper knee will differ if the heated element is fabricated in a planar or coiled geometry. X = low-pressure limit; it is reached when the heat transfer equals the radiation and end conduction losses.

To extend the range of a gauge to its lowest possible pressure limit it is necessary to reduce both radiation losses and conduction losses. End losses dominate when the wire is short, (in thermocouple tubes), whereas radiation losses (and sensitivity) increase with wire temperature. Two commonly manufactured sensors have upper pressure limits of 15 and 150 Pa (0.1 and 1 Torr). Sensors reading to 10^5 Pa (760 Torr), which take advantage of the pressure-dependent convection losses, are commercially available [20].

The sensitivity of the gauge is determined by tube construction and gas as well as by the technique for sensing the change in heat flow with pressure. Tungsten is commonly used for the heater wire because it has a large thermal resistance coefficient. When a semiconductor is used as the heat-sensitive element, the device is referred to as a thermistor gauge, even though it is configured identically to a Pirani gauge. The accommodation coefficient α for clean materials can be of an order of 0.1, but for contaminated surfaces it can be as high as unity. For most cases, α is stable but not known. With all other factors well controlled, changes in emissivity and accommodation coefficient are large enough to allow thermal conductivity gauges to be used as only rough indicators of vacuum.

The change in wire temperature can also be detected by monitoring the resistance of the heated wire. When a Wheatstone bridge circuit is used to measure the resistance change, the device is termed a Pirani gauge.

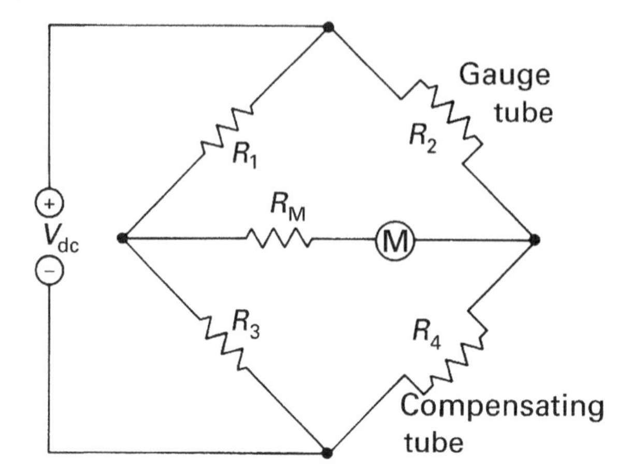

Fig. 5.5 Basic Pirani gauge circuit. A. Guthrie [21]/With permission from John Wiley & Sons.

Alternatively, the temperature change can be measured directly with a thermocouple, in which case it is called a thermocouple gauge.

5.2.1.1 Pirani Gauge

The term Pirani gauge is given to any type of thermal conductivity gauge in which the heated wire forms one arm of a Wheatstone bridge. A simple form of this circuit is shown in Fig. 5.5. The gauge tube is first evacuated to a suitably low pressure, say 10^{-4} Pa and R_1 is adjusted for balance. A pressure increase in the gauge tube will unbalance the bridge because the increased heat loss lowers the resistance of the hot wire. By increasing the voltage, more power is dissipated in the hot wire, which causes it to heat, increase its resistance, and move the bridge toward balance. In this method of gauge operation, called the constant temperature method and the most sensitive and accurate technique for operating the bridge, each pressure reading is taken at a constant wire temperature. To correct for temperature-induced changes on the zero adjustment, an evacuated and sealed compensating gauge tube is used in the balance arm of the bridge. Bridges with a compensating tube can be used to 10^{-3} Pa.

The constant voltage and constant current techniques were devised to simplify the operation of the Pirani gauge. In each case the total bridge voltage or current is kept constant. The constant voltage method is widely used in modern instruments because no additional adjustments need to be made after the bridge is nulled at lower pressures. The out-of-balance current meter is simply calibrated to read the pressure.

The constant temperature method is the most sensitive and accurate because at constant temperature the radiation and end losses are constant. Because the wire

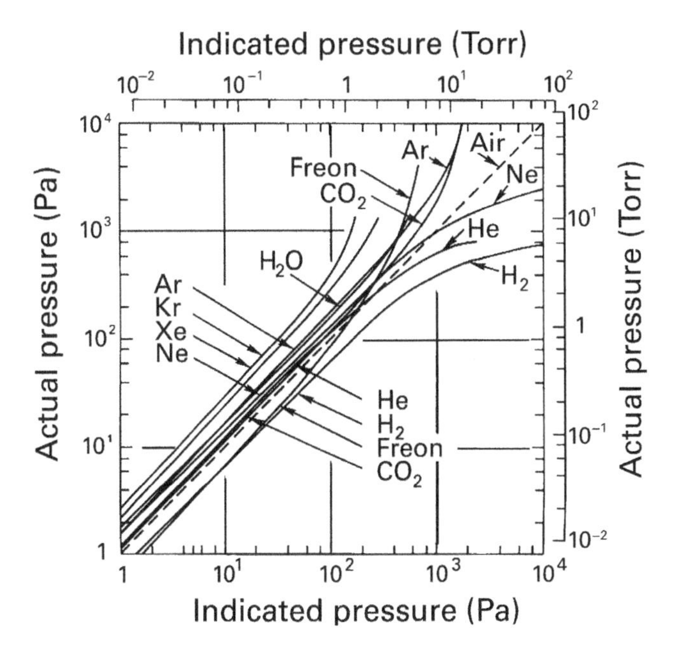

Fig. 5.6 Gas dependence curves for the Oerlikon Leybold Vacuum TTR91 Pirani gauge tube. Reproduced with permission from Leybold.

temperature is constant, the sensitivity is not diminished in the high-pressure region. This method does not lend itself to easy operation; balancing is required before each measurement. A sudden drop in pressure can also cause overheating of the wire if the bridge is not immediately rebalanced. Direct reading constant-temperature bridges that need only a zero adjustment are now commercially available, although at somewhat greater expense than a constant voltage or current bridge. Modern circuitry has eliminated tedious bridge balancing. Because the heat conductivity varies considerably among gases and vapors, the calibration of the gauge is dependent on the nature of the gas. Most instruments are calibrated for air; therefore, a chart like the one shown in Fig. 5.6 is needed for each specific gauge, when the pressure of other gases is to be measured. ISO Standard 19685:2017 describes specification and measurement methods for Pirani gauges [22].

5.2.1.2 Thermocouple Gauge

The thermocouple gauge measures pressure-dependent heat flow. Constant current is delivered to the heated wire, and a tiny Chromel–Alumel thermocouple is carefully spot welded to its midpoint. As the pressure increases heat flows to the walls and the temperature of the wire decreases. A low-resistance dc microammeter is connected to the thermocouple, and its scale is calibrated in pressure units.

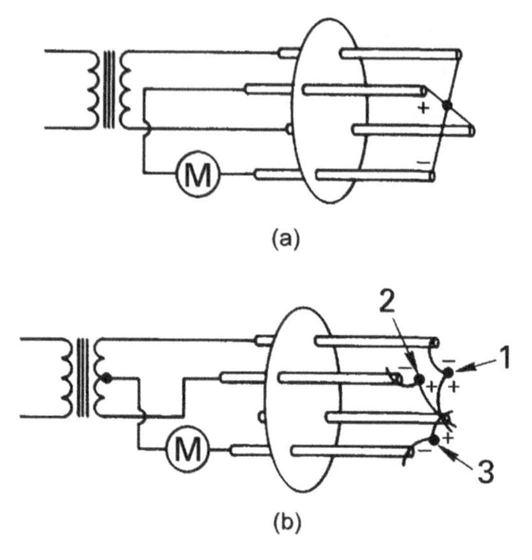

Fig. 5.7 Thermocouple gauge tube commonly used in the 0–100 Pa (0–1000 Torr) range. (a) Uncompensated gauge tube and (b) compensated gauge tube (thermocouple No. 3 is the compensating couple).

Figure 5.7 shows the four-wire and three-wire versions of the gauge tubes. The four-wire gauge tube uses a dc meter to read the temperature of the thermocouple, whereas the power supply is regulated to deliver a constant current to the wire. The current can be ac or dc. The three-wire gauge circuit reduces the number of leads between the gauge tube and controller. It reduces the number of vacuum feedthroughs by using ac to heat the wires, and a dc microammeter to read the voltage between one thermocouple wire and the center tap of the transformer, which is a dc connection to the other junction. In both tubes the power delivered is not constant; instead, the wire current is constant. Because the resistance of the wire is temperature-dependent, the power delivered decreases slightly at high pressures. Both gauge forms are rugged and reliable but inaccurate. Calibration curves for one thermocouple gauge are given in Fig. 5.8.

5.2.1.3 Stability and Calibration

A detailed review of the properties of thermal conductivity gauges determined its critical parameters for stability [23]. It concluded that the critical parameter in the low-pressure portion of their operating range was the wire emissivity. Midrange accuracy was found strongly dependent on the thermal accommodation coefficient—a function of surface contamination. The full-scale region of the thermal conductivity gauge had no critical parameter; good stability was predicted and observed.

Thermal conductivity gauges can be calibrated by direct comparison to standard reference gauges, such as capacitance manometers or spinning rotor gauges.

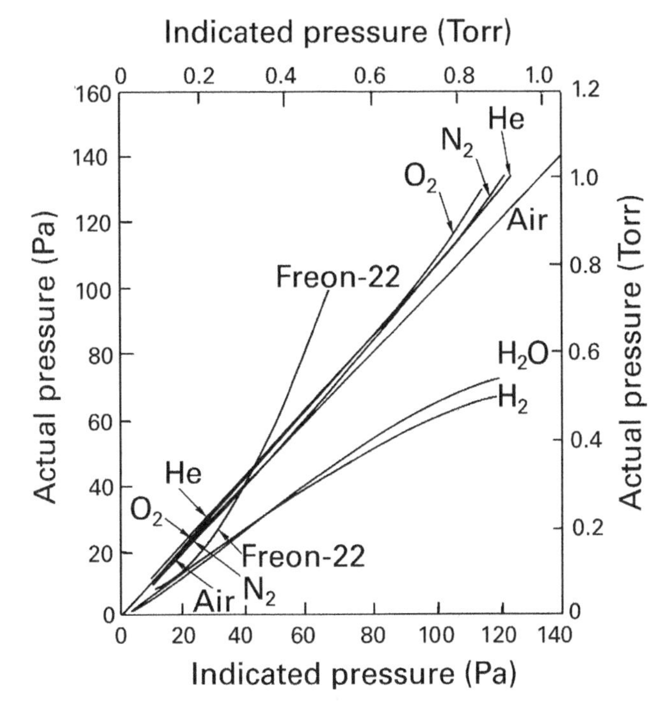

Fig. 5.8 Calibration curves for the Hastings DV-6 M thermocouple gauge tube. Reproduced with permission from Hastings Instruments Co.

Detailed recommended practices and procedures for calibrating thermal conductivity vacuum gauges have been published by the American Vacuum Society [24].

5.2.2 Spinning Rotor Gauge

The spinning rotor gauge [10,11] measures pressure by measuring the rate at which a rotating ball slows due to the viscosity of the surrounding gas. Gas molecules collide with the ball, causing its angular momentum to decrease. At high pressure, the rotating ball slows rapidly, at low pressures, slowly. The spinning rotor gauge is an indirect gauge—it does not measure pressure on a wall. However, it has been proven so accurate and stable that it has found use as a secondary or transfer standard [25,26].

The rate at which a rotating ball loses angular velocity is given by:

$$-\frac{\omega}{\omega_0} = \sigma\left(\frac{10}{\pi}\right)\left(\frac{1}{rd}\right)\frac{P}{v_{\text{ave}}} \tag{5.2}$$

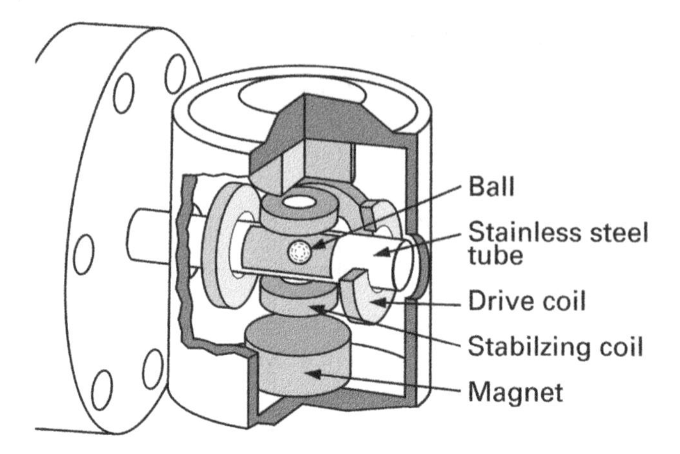

Fig. 5.9 Spinning rotor gauge sensor. Reproduced with permission from MKS Instruments.

where σ is the coefficient of tangential momentum transfer, ω is the rotational velocity, ω_0 is the rotational velocity at time zero—typically 400 revolutions per second, r is the ball radius, and d is the ball density. An illustration of a commercial sensing head is shown in Fig. 5.9. Only the rotating ball is located within the vacuum system; it is levitated inside a small stainless steel closed-end tube that is welded to a 70-mm-diameter metal flange. Equation 5.2 shows that the rate of decay is proportional to the $\sqrt{m}$ of the measured gas. The user is required to know the gas being measured and its calibration factor. Hinkle and Jacobs [27] have shown that the effective mass of a gas mixture can be entered into the spinning rotor gauge software as:

$$m_{\text{eff}} = \left(\sum_{i=1}^{n} a_i \sqrt{m_i} \right)^2 \tag{5.3}$$

They note that simultaneous use of a capacitance manometer and spinning rotor gauge allows the user to observe changes in gas mixtures.

The fundamental property that makes this instrument so stable is the constancy of the quantity σ— the coefficient of tangential momentum transfer. Long-term studies have shown that this variable remains constant for long periods of time and shows little variation between similar rotor balls. In one sequence of measurements on eight instruments in three national laboratories, σ ranged from 0.9927 to 1.0075, with standard deviations between 0.0003 and 0.0012 in the values reported for individual instruments [28].

As the pressure tends to zero, there is a fixed "offset" rate at which the ball slows. This rate is due to eddy currents induced in the ball by the rotor fields and by coulomb forces developed by the magnets. Thus, to read the lowest pressure possible, this offset rate must be subtracted. It is claimed that commercial instruments can read to 10^{-5} Pa (10^{-7} Torr) [29]; however, care is required. The ball requires spinning for approximately 5 h at a pressure one decade lower than the lowest pressure to be measured in order to achieve equilibrium and obtain a precise measurement of the residual drag [30]. Even after that measurement, the residual drag must be measured frequently during use. At high pressures, the rate at which the ball velocity decreases is a constant and is determined by the properties of the (viscous) gas in which it is immersed. This rate has been calculated to be 50 Pa, but the real device becomes nonlinear above 0.1 Pa. Algorithms contained within commercial instruments attempt to correct for this nonlinearity and thermal changes introduced at high pressure. With refined algorithms the gauge can measure pressures of 100 Pa (1 Torr) to $\pm 1\%$ [31]; it can measure pressures of ~1000 Pa (10 Torr) with less accuracy.

The spinning rotor gauge is an excellent instrument when used by a trained operator in a laboratory setting. It is the instrument of choice for calibrating other indirect gauges within its pressure range; it can be used by a skilled operator to identify gas mixture problems.

5.2.3 Ionization Gauges

In the high and ultrahigh vacuum region, where the particle density is extremely small it is not possible, except in specialized laboratory situations, to detect the minute forces that result from the direct transfer of momentum or energy between the gas and a solid wall. For example, at a pressure of 10^{-8} Pa the particle density is only $2.4 \times 10^{12}/m^3$. This may be compared with a density of $3 \times 10^{22}/m^3$ at 300 K, which is required to raise a column of mercury 1 mm. Even a capacitance manometer cannot detect pressures lower than 10^{-4} Pa.

In the region below 10^{-3} Pa, pressure is measured by ionizing gases, then counting and amplifying the ion signal. Each ionization gauge has its own lower pressure limit at which the ionized particle current is equal to a residual or background current. Ionization gauges normally used in the high vacuum region have a background limit of ~10^{-8} Pa (10^{-10} Torr). Both hot and cold cathode gauges have been developed for pressure measurement. Each technique has its own operating range, advantages, and disadvantages.

5.2.3.1 Hot Cathode Gauges

The operation of the ion gauge is based on electron impact ionization. The ionized molecular current to the collector electrode is proportional to pressure, provided that all other parameters, including temperature, are held constant. The number of positive ions formed is actually proportional to the number density, not the pressure; the ion gauge is not a true pressure measuring instrument, but rather it is a particle-density gauge. Its reading is proportional to pressure only when the temperature is constant.

The earliest form of ion gauge, the triode gauge, consisted of a filament surrounded by a grid wire helix and a large-diameter, solid cylindrical ion collector. This gauge, which is not illustrated here, looked a lot like a triode vacuum tube. Electrons emitted by the heated filament were accelerated toward the grid wire, which was held at a positive potential of about 150 V. The large area external collector was biased about −30 V with respect to the filament; it collected the positive ions generated in the space between the filament and the ion collector. This gauge recorded pressures as low as 10^{-6} Pa; it would not give a lower reading, even if indirect experimental evidence indicated the existence of lower pressures. Further progress was not made until after 1947, when Nottingham [32] suggested that the cause of this effect was X-ray-generated photocurrent. Nottingham proposed that soft X-rays generated by the electrons striking the grid wire collided with the ion collector cylinder and caused photoelectrons to flow from the collector to the grid. Some photo emission is also caused by ultraviolet radiation from the heated filament. As they leave the collector, these photoelectrons produce a current in the external circuit, which is not distinguishable from the positive ion flow toward the ion collector and mask the measurement of reduced pressures.

In 1950, Bayard and Alpert [12] designed a gauge in which the large area collector was replaced with a fine wire located in the center of the grid (Fig. 5.10*a*). Because of its smaller area of interception of X-rays, this gauge could measure pressures as low as 10^{-8} Pa. Today this gauge is the most popular design for the measurement of high vacuum pressures.

The proportionality between the collector current and pressure is given by:

$$i_c = S' i_e P \quad \text{or,} \quad P = \frac{1}{S'}\frac{i_c}{i_e} \tag{5.4}$$

where i_c and i_e are the collector and emission currents, respectively, and S′ is the sensitivity of the gauge tube. This sensitivity has units of reciprocal pressure, which in SI is 1/Pa. The sensitivity is dependent on the tube geometry, grid and collector voltages, the type of control circuitry, and the nature of the gas being measured. For the standard Bayard-Alpert tube with external control circuitry, a grid voltage of +180 V, and filament bias voltage of +30 V. See Fig. 5.11. The sensitivity for nitrogen is typically ~0.07/Pa. Variations in tube design, voltage, and control circuitry can cause it to range from 0.05 to 0.15/Pa. The tube's sensitivity for other gases varies with ionization probability. Alpert [33] suggested that the relative sensitivity (the ratio of the absolute sensitivity of an unknown gas to nitrogen) should be independent of structural and electronic variations and therefore be more meaningful to tabulate than absolute sensitivity.

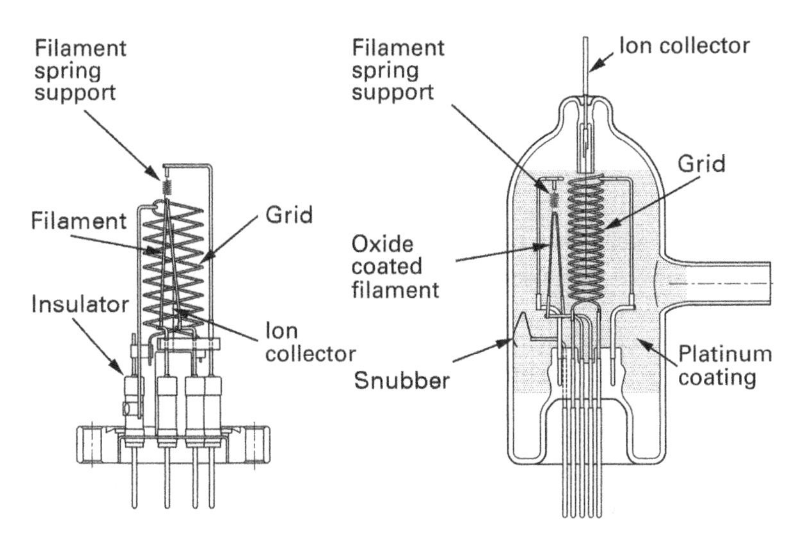

Fig. 5.10 Cross-sectional views of two Bayard–Alpert sensors. Left: Bakeable, flange-mounted nude gauge. Right: Glass-encapsulated gauge. Reproduced with Permission from The Fredericks Company.

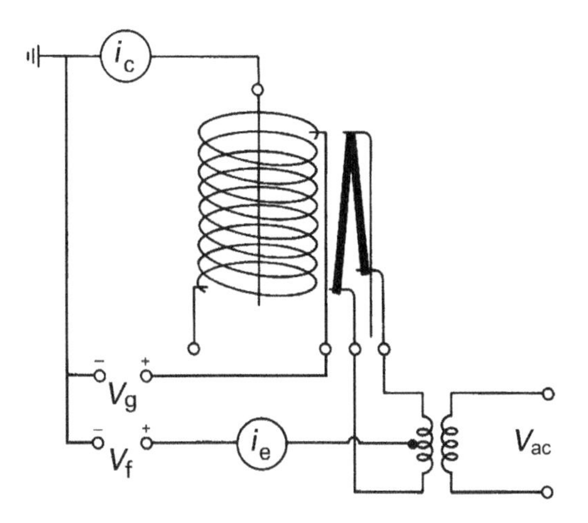

Fig. 5.11 Control circuit for a Bayard–Alpert ionization gauge tube.

The relationship between the pressure of an unknown gas $P(x)$ and the meter reading is:

$$P(x) = \frac{S(N_2)}{S(x)} P(\text{meter reading}) \tag{5.5}$$

or after normalizing the gas sensitivities, by dividing each sensitivity by that of nitrogen: $S(x)_{\text{relative}} = S(x)/S(N_2)$, we obtain:

$$P(x) = \frac{P(\text{meter reading})}{S(x)_{\text{relative}}} \tag{5.6}$$

By use of Eq. (5.6) and Table 5.1 [34], the pressure of gases other than nitrogen can be measured with an ion gauge, even though all ion gauges are calibrated for nitrogen.

Gauge sensitivity is often given in units of microamperes of collector current per unit of pressure per manufacturer's specified emission current; for example, a typical nitrogen sensitivity is (100/μA/mTorr)/10 mA. This is a confusing way of saying the sensitivity is 10/Torr, but it does illustrate an important point; not all gauge controllers operate with the same emission current and not all gauge tubes have the same sensitivity. Checking the instruction manual can avoid potential embarrassment.

The classical control circuit is designed to stabilize the potentials and emission current while measuring the plate current. The plate current meter is then calibrated in appropriate ranges and units of pressure. The accuracy of the gauge is dependent in part on moderately costly, high-quality, emission current regulation.

Table 5.1 Approximate Relative Sensitivity of Bayard–Alpert Gauge Tubes for Several Gases

Gas	Relative Sensitivity
H_2	0.42–0.53
He	0.18
H_2O	0.9
Ne	0.25
N_2	1.00
CO	1.05–1.1
O_2	0.8–0.9
Ar	1.2
Hg	3.5
Acetone	5

The pressure of an individual gas is found by dividing the ion gauge reading by its relative sensitivity.
Source: T.A. Flaim and P.D. Ownby [34]/Reproduced with permission from AIP Publishing.

Tungsten and thoriated iridium (ThO_2 on iridium) are two commonly used filament materials. Thoriated iridium filaments are not destroyed when accidentally subjected to high pressures—an impossible feat with fine tungsten wires—but they do poison in the presence of some hydrocarbon vapors. The remarks in Section 8.2 about filament reactivity with gases in the residual gas analyzer ionizer also apply to the ion gauge.

Ion gauge grids are outgassed by direct or electron bombardment heating. The grid wire can be directly heated by connecting it to a low-voltage, high-current transformer. Alternatively, the grid and plate wire can be connected to a high-voltage transformer and heated by electron bombardment. It is best to wait until the pressure is on a suitably low scale (~10^{-4} Pa) before outgassing. An unbaked glass encapsulated gauge should be outgassed until the glass has desorbed. (The pressure may be monitored during outgassing of gauges with resistance-heated grids.) The time for this initial outgassing is variable, but 15–20 min is typical. After the initial outgassing the tube should be left on. Subsequently, only short outgassing times, say 15 s, are periodically needed to clean the electrodes. Figure 5.12 illustrates the gas release from a grid during electron bombardment. Notice the large transients at $M/z = 28$, 16, 12, and to a lesser extent, 44. This identifies the desorbed gases as CO, CO_2, and some CH_4. If the outgassing power supply is inadequately designed, desorption will not be complete [35]; see the discussion of electron-stimulated desorption in the following subsection.

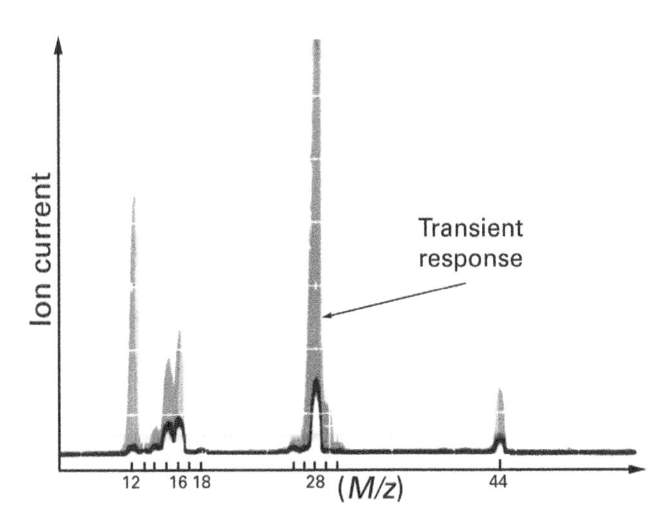

Fig. 5.12 Transient desorption during degassing a glass-encapsulated Bayard–Alpert ion gauge tube as recorded by a storage tube oscilloscope. The gray traces indicate the transient release of carbon monoxide along with some methane and carbon dioxide on grid heating. The solid black line represents the background spectrum.

Extending the operating region of the hot cathode gauge below 10^{-8} Pa (10^{-10} Torr) is not so easy. Few ultrahigh vacuum hot cathode gauges are commercially available. The Extractor gauge [13] and uhv-24 Bayard-Alpert sensor and are two commercially available ultrahigh vacuum gauges. An efficient Bayard-Alpert design increased the sensitivity of the gauge tube by capping the end of the grid to prevent electron escape [36]. It reduced the X-ray limit by use of a fine, 125-μm-diameter collector wire. This sensor design has a sensitivity of 0.15/Pa (20/Torr) [37]. See Fig. 5.10 (left). An Extractor gauge design is illustrated in Fig. 5.13. The hemispherical ion reflector, maintained at grid potential, reflects ions to the collector wire. A modern extractor gauge, e.g., the Leybold IE-514 has a low-pressure limit of 1×10^{-10} Pa (0.75×10^{-12} Torr), and a sensitivity of 0.66/Pa (8.9/Torr) [38].

The modulated Bayard-Alpert gauge was constructed like a standard Bayard-Alpert gauge but had a modulator wire located within the grid region [39]. When a potential was applied to the modulator wire located that equaled that of the collector, some of the gas phase ion flux was diverted to the modulator and the collector current decreased. However, the X-ray-generated charge flow remained constant. By measuring this fractional decrease at pressures far above the X-ray limit, the modulation factor was determined, and the gauge calibrated. Its sensitivity was measured to be 0.9/Pa (12/Torr) [40].

5.2.3.2 Hot Cathode Gauge Errors

Despite its deceptively simple operating principle, measurement of pressure in the ultrahigh vacuum region with a hot cathode gauge must be done with considerable care. Numerous effects can alter the indicated ion current [35,41,42].

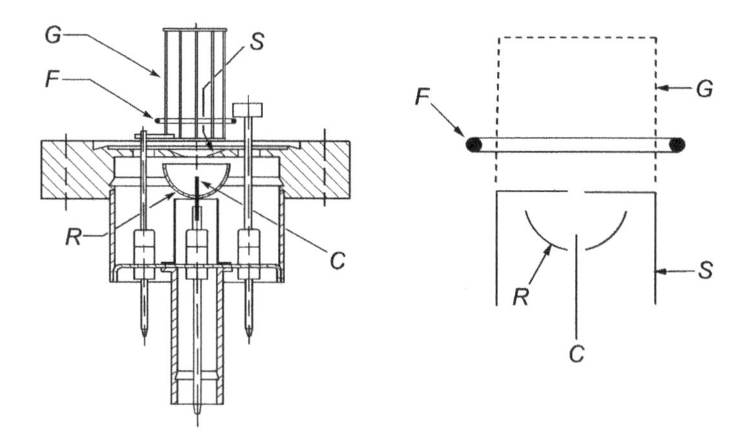

Fig. 5.13 IE514 Extractor gauge. Left: Sensing head. Right: Expanded detail of the electrode structure. G, grid; F, filament; S, suppressor; R, reflector; C, collector. Reproduced with permission from INFICON AG.

The previously described X-ray generated photocurrent, electron-stimulated grid (and ion collector) desorption, wall outgassing caused by thermal heating from the hot cathode, chemical effects on heated cathodes, controller errors, wall diameter, and cathode evaporation can each introduce measurement error.

Historically, the most attention has been paid to reducing the X-ray limit; it was the most significant limiting factor; indeed, it remains a concern. The uhv-24 sensor design was reported to have an X-ray limit of 4×10^{-9} Pa (2.8×10^{-11} Torr) [35]. The modulated Bayard-Alpert gauge was shown to have an X-ray limiting pressure of 9×10^{-9} Pa (7×10^{-12} Torr) [38]. Careful attention to materials, structural design, and manufacturing details are required to produce stable ultra high vacuum (UHV) gauge tubes with very low X-ray limits.

Electron-stimulated desorption (ESD) is a significant, but a variable and an unpredictable source of error; in many cases ESD errors can be significantly larger than those caused by X-rays. Electrons striking the grid and ion collector release previously adsorbed gases. Prior exposures to oxygen results in desorption of CO formed by a reaction of dissociated oxygen and carbon impurities in metals. Other studies point to enhanced ESD in systems containing small amounts of water vapor or other oxygen-containing gases. Using optical metrology (with no heated filaments), Looney [43] demonstrated ESD of carbon monoxide from Bayard-Alpert gauges and RGA sensors. Figure 5.14 illustrates the decrease and increase of the CO signal as RGA and Bayard-Alpert gauge hot filaments were first turned off, then on. The instantaneous change in signal levels indicated that CO released

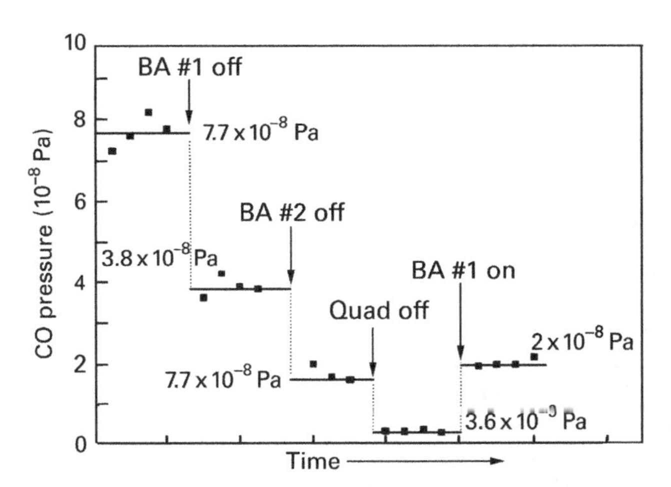

Fig. 5.14 Carbon monoxide concentration in an ultrahigh vacuum system as measured optically (no hot filament). The CO partial pressure is seen to decrease as hot filament devices are turned off. J.P. Looney et al. [43]/Reproduced with permission from AIP Publishing.

from surfaces was not due to the slow heating or cooling of nearby surfaces. For decades, many researchers have inferred CO to be an artifice of hot filaments and not a true ultrahigh vacuum background gas. This data unambiguously confirms indirect evidence. Kendall [35] observed that ESD disappeared when a Bayard-Alpert grid was heated above ~800 K. He concluded that gases could not adsorb on grids at this temperature. Watanabe's careful ESD measurements agree [42]. Figure 5.15 shows the magnitude of the ESD signal in an experimental gauge capable of distinguishing gas phase ions from their energetically different ESD counterparts. Since many new ionization gauge controllers operate at emission currents of 4 mA or less, they may not heat the grid to a sufficiently high temperature to prevent ESD.

Wall outgassing caused by heating of the gauge walls, or surfaces near nude gauges, will introduce measurement errors. Watanabe noted that stainless steel, surrounding a typical ultrahigh vacuum gauge, had two undesirable characteristics: high emissivity and low thermal conductivity. The nearby walls are hot because they absorb heat easily and dissipate it poorly. Watanabe used these concepts to construct an experimental gauge using a gold-plated copper fixture, which operated with reduced heater power and reduced wall outgassing [44].

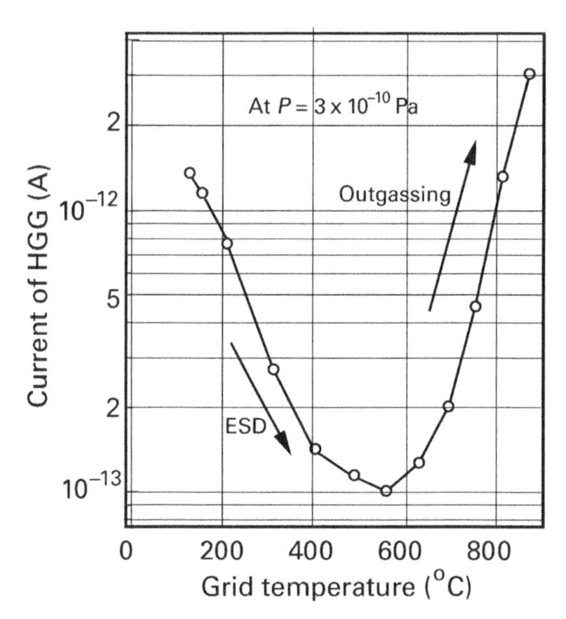

Fig. 5.15 Ion current versus grid temperature in a hot cathode UHV ion gauge. Electron-stimulated desorption of the grid decreases with temperature up to a temperature of ~800 K. The increased signal at high grid temperatures is believed to be caused by the outdiffusion of hydrogen from the grid wire. F. Watanabe [42]/Reproduced with permission from Elsevier.

Kendall observed that mounting flange virtual leaks and dimensional manufacturing tolerances, as well as inaccurate electrometer and voltage divider resistors contributed to gauge error [35]. Surface chemical effects can synthesize ions not part of the system background. The gauge envelope is also part of the measurement circuit. Glass charges to its floating potential. Redhead demonstrated that the sensitivity factor for a Bayard-Alpert gauge was strongly dependent on the filament distance and the electric fields near the filament [45]. The diameter of the glass or metal envelope affects electron orbits [46]. Filippelli demonstrated that envelope diameter changes could produce sensitivity changes as large as 50% with some Bayard-Alpert sensors [47]. He found that extractor gauges were less susceptible to wall diameter changes, because of the sensor's low profile—much of the gauge was located within its mounting flange.

Measurement of pressure in the ultrahigh vacuum range requires care; it is essential that those who are serious about this subject begin by reading papers by Kendall [35], Redhead [41], and Watanabe [42]. The Stabil-Ion gauge [48] design focuses on reducing many errors, with a resulting stability of about ±4% [49].

5.2.3.3 Cold Cathode Gauge

The cold cathode gauge, developed by Penning in 1937 [14] provides an alternative to the hot cathode gauge, which in many respects is superior to a Bayard-Alpert gauge. A commercial sensor tube is illustrated in Fig. 5.16. This gauge is based on the inverted magnetron geometry [51]. The arrangement of the electric and magnetic fields causes electrons to travel long distances in spiral paths before finally colliding with the anode. Long electron trajectories enhance the ionization

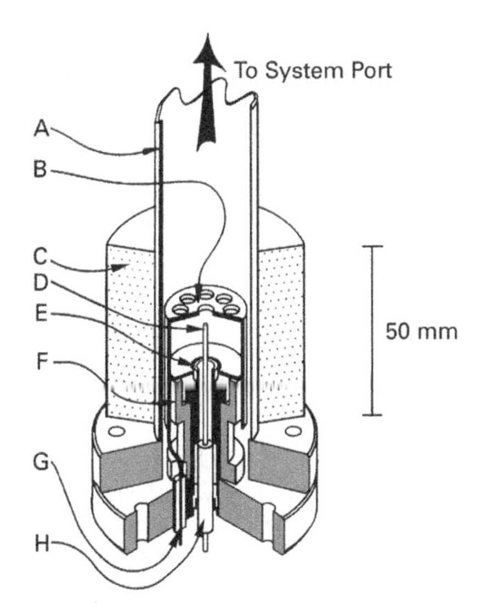

Fig. 5.16 Cold cathode gauge: A, envelope; B, cathode; C, magnet; D, anode; E, guard electrode; F, ceramic support; G, cathode current feedthrough; H, high voltage feedthrough. R.N. Peacock et al. [50]/ Reproduced with permission from AIP Publishing.

probability and improve sensitivity. The discharge begins when one electron or ion gains sufficient energy to ionize a gas molecule. The electron density increases until it is space charge limited. The time required for a gauge to reach steady state decreases as pressure increases. This can be described by a starting parameter, typically of order 50–500 μPa-s (0.5–5 μTorr-s) for gauges without auxiliary starting sources [52]. Starting the gauge at low pressures can be difficult unless the gauge contains an auxiliary source; radioactive sources, photo emitters, hot filaments, and field emitters have been used to provide additional starting electrons. The range of operation of the cold cathode gauge is $1–10^{-9}$ Pa ($10^{-2}–10^{-11}$ Torr). It is a viable option to the hot cathode gauge for systems with 10^{-9}-Pa-range base pressures if it is not mounted immediately adjacent to a hot cathode gauge or residual gas analyzer. Cold and hot cathode gauges have sensitivities that vary with gas species and in a similar manner. Like hot cathode gauges, cold cathode gauges measure gas density. Unlike hot cathode gauges, they have no X-ray limit and have little ESD or thermally induced wall outgassing; they do not have filaments to change. Gas release from a cold cathode gauge was observed after the gauge had been contaminated [53]. The magnetic field surrounding a cold cathode gauge is a concern in some applications. Fringing fields have been reduced with alternative magnetic configurations [52]. Figure 5.17 illustrates a conventional and two “double magnetron” designs; one uses opposing radial magnets whereas the other uses opposing longitudinal magnets.

Cold cathode gauges can pump; however, their pumping speeds have been found similar to those of a hot cathode gauge [50]. They should not be connected

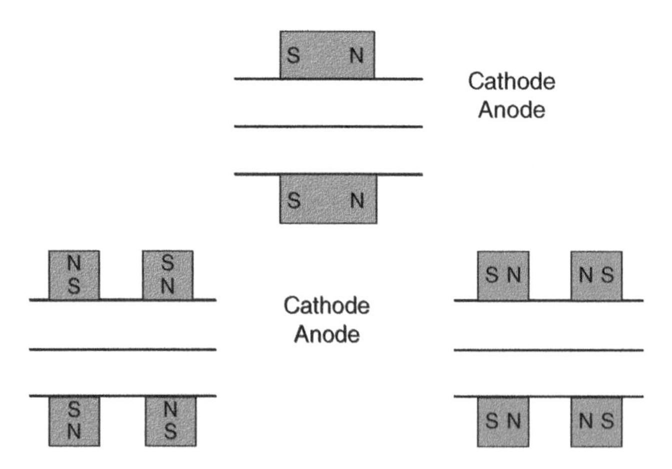

Fig. 5.17 Simplified view of the magnet configurations in modern inverted magnetron cold cathode gauges. Top: Conventional single inverted magnetron. Lower left: Radial double magnets. Lower right: Axial double magnets. B.R.F. Kendall and E. Drubetsky [52]/ Reproduced with permission AIP Publishing.

to a system with tube of a smaller diameter, or a pressure drop will result. Cold cathode gauges should be mounted in a way that will not allow metal particles to fall inside the tube.

5.2.3.4 Gauge Calibration

Direct comparison, static expansion, and continuous expansion are techniques used to calibrate high vacuum gauges against primary standards. Direct comparison is performed by comparing standard and unknown gauges on the same chamber. The static expansion method uses Boyle's law to calculate the known quantity of gas, as it expands from one chamber to a second. By repeating this process, successively lower pressures can be reached. In the continuous expansion method, gas from a vessel of known pressure flows to a calibration chamber through an orifice of known size. From the calibration chamber, the gas flows through another known orifice into the pump. Knowing the dimensions of the calibration chamber, the temperature, and the difference between the two flow rates, one can calculate the pressure of the intermediate chamber. Series expansion is claimed to be the most accurate method [54]. ISO Standard 27 894:2009 describes calibration methods for hot cathode ion gauges [55]. The 2018 AVS Recommended Practice for calibrating ion gauges [56], describes calibration methods suitable for hot cathode and cold cathode ion gauges. The average user does not have the resources to calibrate high vacuum gauges but may wish single point comparison to a secondary standard such as a spinning rotor gauge or a capacitance manometer. Ion gauges have been found to be linear to low pressures [57], so that direct comparison with a standard in their upper range is valid. Cold cathode gauges often exhibit a small "kink" in their current–pressure response at low pressure [50]. Hot cathode gauge tubes tend to be more repeatable than accurate; however, the issues discussed in the previous section can affect accuracy. Reducing errors generated by improper use or mounting can be as significant as calibration.

The future direction of pressure standards and measurement in the ultra- and extreme-high-vacuum region is focused on optical and quantum techniques [58].

References

1 Leck, J.H., *Pressure Measurement in Vacuum Systems*, 2nd ed., Chapman and Hall, London, 1964.
2 Roth, J.P., *Vacuum Technology*, North-Holland, Amsterdam, 1982.
3 Berman, A., *Total Pressure Measurements in Vacuum Technology*, Academic Press, New York, 1985.
4 Alpert, D., Matland, C.G., and McCoubrey, A.C., *Rev. Sci. Instrum.* **22**, 370 (1951).
5 Sullivan, J.J., *Ind. Res. Dev.*, 41 (1976).

6 Loriot, G. and Moran, T., *Rev. Sci. Instrum.* **46**, 140 (1975).

7 ISO Standard 20146:2019, *Specification, Calibration, and Measurement Methods for Capacitance Diaphragm Gauges*, International Standards Organization, Geneva, Switzerland, 2019.

8 Pirani, M., *Verhandl. Dent. Physik. Ges.* **8**, 686 (1906).

9 Voege, W., *Physik Zs.* **7**, 498 (1906).

10 Beams, J.W., Spitzer, D.M., Jr., and Wade, J.P., Jr., *Rev. Sci. Instrum.* **33**, 151 (1962).

11 Fremerey, J.K., *J. Vac. Sci. Technol.* **9**, 108 (1972).

12 Bayard, R.T. and Alpert, D.A., *Rev. Sci. Instrum.* **21**, 571 (1950).

13 Redhead, P.A., *J. Vac. Sci. Technol.* **3**, 173 (1966).

14 Penning, F.M., *Physica* **4**, 71 (1937).

15 von Ubisch, H., *Arch. F. Mat. Astro och fysik* **36A**, 4 (1948); *Nature* [London], **161**, 927 (1948).

16 von Ubisch, H., *Vak. Tech.* **6**, 175 (1957).

17 Pirani, M. and Yarwood, J., *Principles of Vacuum Engineering*, Reinhold, New York, 1961, pp. 100–101.

18 Leck, J.H., *Pressure Measurement in Vacuum Systems*, Chapman and Hall, London, 1957, pp. 33–38.

19 Alvesteffer, W.J., Jacobs, D.C., and Baker, D.H., *J. Vac. Sci. Technol., A* **13**, 2980 (1995).

20 For example: Granville-Phillips Convectron Gauge, Series 275, Granville Phillips Co, Boulder CO. The Leybold TR201, TR211 or TTR90 gauges, (Leybold GmbH, Köln, Germany), as well as the HPS-MKS Series 317 and 907 gauges, (HPS Division of MKS Instruments, 5330 Sterling Drive, Boulder, CO, 80301) can also measure pressure to one atmosphere by use of the convection principle.

21 Guthrie, A., *Vacuum Technology*, 163 (1963).

22 ISO Standard 19685:2017, *Specification, Calibration, and Measurement Methods for Pirani Gauges*. International Standards Organization, Geneva, Switzerland, 2017.

23 Jitschin, W. and Ruschitzka, M., *Vacuum* **44**, 607 (1993).

24 Ellefson, R.E. and Miiller, A.P., *J. Vac. Sci. Technol., A* **18**, 2568 (2000).

25 Comsa, G., Fremerey, J.K., Lindenau, B., Messer, G., and Röhl, P., *J. Vac. Sci. Technol.* **17**, 642 (1980).

26 Reich, G., *J. Vac. Sci. Technol.* **20**, 1148 (1982).

27 Hinkle, L.D. and Jacobs, R.P., *J. Vac. Sci. Technol., A* **11**, 261 (1993).

28 Jousten, K., Filippelli, A.R., and Tilford, C.R., *J. Vac. Sci. Technol., A* **15**, 2395 (1997).

29 Ueda, E., Hirohata, Y., Hino, T., and Yamashina, T., *Vacuum* **44**, 587 (1993).

30 McCulloh, K.E., Wood, S.D., and Tilford, C.R., *J. Vac. Sci. Technol., A* **3**, 1738 (1985).

31 Setina, J. and Looney, J.P., *Vacuum* **44**, 577 (1993).

32 Nottingham, W.B., *7th Annual Conference on Physical Electronics*, MIT, Cambridge, MA, 1947.

33 Alpert, D., *J. Appl. Phys.* **24**, 7 (1953).

34 Flaim, T.A. and Ownby, P.D., *J. Vac. Sci. Technol.* **8**, 661 (1971).

35 Kendall, B.R.F., *J. Vac. Sci. Technol., A* **17**, 2041 (1999).

36 Nottingham, W.B., *1954 Vacuum Symposium, Committee on Vacuum Techniques*, 76, Boston, MA, 1955.

37 Hseuh, H.C. and Lanni, C., *J. Vac. Sci. Technol., A* **5**, 3244 (1987).

38 *Leybold Full Line Catalog*, Leybold, GmbH, Cologne, Germany, 2021.

39 Redhead, P.A., *Rev. Sci. Instrum.* **31**, 343 (1960).

40 Filippelli, A.R., *J. Vac. Sci. Technol., A* **5**, 3234 (1987).

41 Redhead, P.A., *Vacuum* **44**, 559 (1993).

42 Watanabe, F., *Vacuum* **53**, 151 (1999).

43 Looney, J.P., Harrington, J.E., Smyth, K.C., O'Brian, T.R., and Lucatorto, T.B., *J. Vac. Sci. Technol., A.* **11**, 3111 (1993).

44 Watanabe, F., *J. Vac. Sci. Technol., A* **11**, 1620 (1993).

45 Suginuma, S. and Hirata, M., *Vacuum* **53**, 177 (1999).

46 Redhead, P., *J. Vac. Sci. Technol.* **6**, 848 (1969).

47 Filippelli, A.R., *J. Vac. Sci. Technol., A* **14**, 2953 (1996).

48 Arnold, P.C., Bills, D.G., Borenstein, M.D., and Borichevsky, S.C., *J. Vac. Sci. Technol., A* **12**, 580 (1994).

49 Fedchak, J.A. and Defibaugh, D.R., *J. Vac. Sci. Technol., A* **30**, 061601 (2012).

50 Peacock, R.N., Peacock, N.T., and Hauschulz, D.S., *J. Vac. Sci. Technol., A* **9**, 1977 (1991).

51 Hobson, J.P. and Redhead, P.A., *Can. J. Phys.* **36**, 271 (1958).

52 Kendall, B.R.F. and Drubetsky, E., *J. Vac. Sci. Technol., A* **18**, 1724 (2000).

53 Mukugi, K., Tsuchidate, H., and Oishi, N., *Vacuum* **44**, 591 (1993).

54 Jitschin, W., Migwi, J.K., and Grosse, G., *Vacuum* **41**, 1799 (1990).

55 ISO Standard 27894:2009, *Specification, Calibration, and Measurement Methods for Hot Cathode Ion Gauges*, International Standards Organization, Geneva, Switzerland, 2009.

56 Fedchak, J.A., Abbott, P.J., Hendricks, J.H., Arnold, P.C., and Peacock, N.T., *J. Vac. Sci. Technol., A* **36** (3), 030802 (2018).

57 Filippelli, A.R. and Dittmann, S., *J. Vac. Sci. Technol., A* **9**, 2757 (1991).

58 Scherschligt, J., Fedchak, J.A., Ahmed, Z., Barker, D.S., Douglass, K., Eckel, S., Hanson, E., Hendricks, J., Klimov, N., Purdy, T., and Ricker, J., *J. Vac. Sci. Technol., A* **36** (4), 040801 (2018).

6

Flow Meters

Flow measurements are performed to characterize components and to monitor and control systems. Pump manufacturers and some users measure the speed of pumps; they require measurement of gas flow to determine pumping speed. Process engineers control the gas flow in systems for plasma deposition or etching, chemical vapor deposition, reactive sputtering, or ion milling. Scientists measure the gas flow into a chamber to calibrate systems used for studying gas desorption kinetics. For some applications, accuracy is important; however, repeatability is important for other applications.

At atmospheric pressure a moderately large 50-L/s (100-cfm) mechanical pump has a gas throughput of 5×10^6 Pa-L/s (3.75×10^4 Torr-L/s). A small 100-L/s ion pump operating at 10^{-5} Pa pumps 10^{-3} Pa-L/s (7.5×10^{-6} Torr-L/s)—a range of more than 9 orders of magnitude. The figure in Part II, preceding Chapter 5, describes the ranges of several flow meters. No one flow meter covers the entire range. Old techniques, such as moving oil or mercury pellets and time to exhaust a reservoir [1,2,3], are still used to measure the very low flows needed to make pumping speed measurements.

In this chapter, we define molar flow and mass flow, relate them to throughput, and review several methods and devices for flow measurement.

6.1 Molar Flow, Mass Flow, and Throughput

Gas flow can be expressed in two ways. It is frequently expressed in non-conservable units of throughput, such as Pa-m^3/s or Torr-L/s. It may also be expressed in terms of the conservable quantities kg-mol/s or kg/s. Confusion arises because the two ways of describing flow are not dimensionally the same. In SI, throughput has units of Pa-m^3/s. Although we do not express it in these

A Users Guide to Vacuum Technology, Fourth Edition. John F. O'Hanlon and Timothy A. Gessert.
© 2024 John Wiley & Sons, Inc. Published 2024 by John Wiley & Sons, Inc.

dimensions, throughput has the dimensions of power and $1\,\text{Pa-m}^3/\text{s} = 1\,\text{J/s} = 1\,\text{W}$. This is the power required to *transport* the gas. One Pa-m^3 is the quantity of gas contained in $1\,\text{m}^3$ at a pressure of 1 Pa. Molar flow and mass flow have dimensions of kg-mol/time or mass/time, respectively. Mass/time and energy/time are not the same. These can be related to throughput, only if the temperature of the gas is known.

Mass conservation is an important distinction between mass flow and throughput. Throughput is not a conservable quantity. The volumetric unit of pressure × time unit such as Pa-m^3 or Torr-L, does not define the number of molecules. One Pa-m^3 of air could contain 2.45×10^{20} molecules/m^3 at 300 K, or it could contain 1.225×10^{20} molecules/m^3 at 600 K. Moles and mass are conservable quantities. Knowledge of the molar flow rate (mol/s), or mass flow rate (kg/s), allows us to perform calculations when the system temperature is nonuniform. We should know when it is appropriate to use molar flow and when to use throughput, and how to convert.

The molar flow rate N' has SI units of (kg-mol)/s and represents the total number of kg-moles of gas passing a plane in one second. Molar flow and throughput are related through the ideal gas law. If we replace n in Eq. (2.13) with N/V, we get:

$$PV = NkT = \frac{N}{N_o}\left(N_o k\right)T = N'RT \tag{6.1}$$

where N' is the molar quantity (kg-mol) of gas, and $R = N_o k = 8314.3\,\text{kJ/(K-kg mol)}$. The molar flow rate at constant temperature is obtained by taking the time derivative of Eq. (6.1):

$$\begin{aligned} \frac{d}{dt}\left(PV\right) &= Q = RT\frac{dN'}{\text{dt}} \\ \frac{dN'}{dt}\left(\text{kg-mol/s}\right) &= \frac{Q}{RT} = 1.21 \times 10^{-4}\left(\frac{Q}{T}\right) \end{aligned} \tag{6.2}$$

Sometimes we wish to express the flow as mass flow. Its units will be kg/s in SI. Each kg-mol has a mass of M kg:

$$\frac{dm}{dt}\left(\text{kg/s}\right) = \left(\frac{MQ}{RT}\right) = 1.21 \times 10^{-4}\left(\frac{MQ}{T}\right) \tag{6.3}$$

M is the molecular weight and Q has units of $\text{Pa-m}^3/\text{s}$ at temperature T. The flow may be expressed alternatively as the number of molecules per second passing a plane: $\Gamma(\text{molecules/s}) = N_o dN'/dt$. Using these relationships, we can convert from throughput Q to molar flow rate dN'/dt, mass flow rate dm/dt, or molecular flow rate Γ. Ehrlich [4] notes that it is customary to label standard leaks with units of "atm-cc/s at T." He reminds us that the value given is numerically equal to Q and not to kg-mol/s. The statement "at T" is included to allow conversion from throughput to molar flow. Some of the equations given here have flow given with units of throughput and others with units of molar or mass flow.

Throughput, mass flow, or molar flow can each be used in calculations. Throughput is a convenient term to use when the system is at a constant temperature for measurements such as pumping speed calculations. Molar flow is best used for studying reaction kinetics and for calculations, which would otherwise have to be referred to several temperatures. For example, a calibrated leak labeled in "Pa-m^3/s at 23°C" is connected to a system whose chamber is at 35°C and pump is at 50°C. In this example, it would be much easier to have the leak labeled in kg-mol/s. The important concept to remember is the fundamental difference between throughput Q and molar flow dN'/dt. Throughput is dimensionally distinct from molar flow and is not a quantity that is conserved. A kg-mol of gas is a fixed number of molecules that do not change with temperature.

6.2 Rotameters and Chokes

A rotameter [5] is a flow-measuring device constructed from a precision tapered bore that contains a ball of accurately ground diameter and known mass. See Fig. 6.1*a*. The gas flow through the tube raises the ball to a height proportional to the throughput or mass flow. The general equation for continuum flow in an orifice was given earlier in (3.7), and its choked flow limit in Eq. (3.8). The equation for continuum flow in a rotameter with a small pressure drop can be obtained from Eq. (3.8) by letting $P_2 = (P_1 - \Delta P)$. When ΔP is small compared to P_1 or P_2, we get:

$$Q = \left(\frac{2kT\Delta P P_1}{m} \right)^{1/2} A \tag{6.4}$$

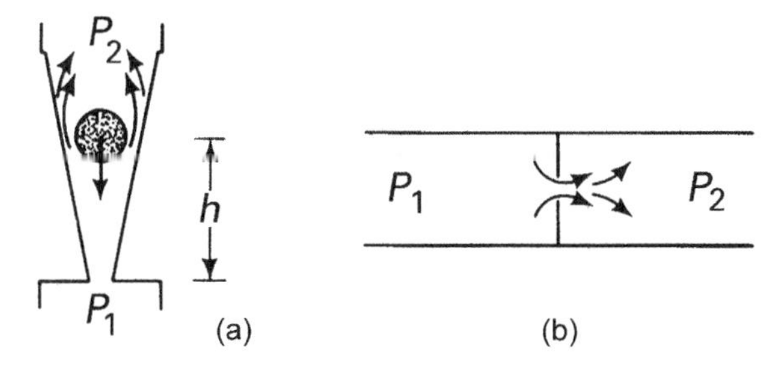

Fig. 6.1 (a) Rotameter and (b) choke.

Except for a geometrical factor, which accounts for the nonzero thickness of the orifice, Eq. (6.4) describes gas flow through a rotameter. The gas flow is a function of the inlet pressure P_1, gas temperature, molecular weight, height h, and mass of the ball, m. The mass of the ball is constant and creates a constant pressure difference ΔP:

$$Q \propto \left(\frac{T}{M} \Delta P P_1 \right)^{1/2} f(h) \tag{6.5}$$

The mass flow rate m' can be expressed as:

$$m' \propto \left(\frac{M}{T} \Delta P P_1 \right)^{1/2} f(h) \tag{6.6}$$

From (Eq.'s (6.5) and (6.6)), we see that both inlet pressure and temperature must be known to calibrate the throughput and mass flow. Rotameters are initially calibrated for one gas and must be recalibrated for use with any other gas. Rotameters are made for flows ranging from 5×10^3 to 5×10^6 Pa-L/s. The accuracy of these instruments is of order 10–20% full scale, whereas repeatability is about 2–3%.

Flow limiting chokes [6,7], see Fig. 6.1*b*, are used to set a predetermined throughput, and are used in the range 5×10^4–10^7 Pa-L/s. When the flow through an orifice reaches its sonic limit ($P_2/P_1 < 0.52$), it is practically independent of outlet pressure, and its behavior was described by Eq. (3.8). The mass flow for air at 22°C, in units of kg/s is:

$$\frac{dm}{dt}(\mathrm{kg/s}) = 2.58 \times 10^{-3} P_1 C' A(\mathrm{m}^2) \tag{6.7}$$

Throughput is dependent on the area, inlet pressure, temperature, and gas species. These devices are not accurate in small sizes (<1-mm diameter), because the nature of the choke is critically dependent on the length of the hole as well as the radius and shape of the entrance edge. Van Atta [5] discusses large radius orifices, which are designed to make the flow more uniform and repeatable. Chokes are useful as flow-restricting devices where accuracy and repeatability are not necessary. Chokes are used to limit flow from gas cylinders, to make calibrated leaks, to prevent aerosol formation during the initial phase of rough pumping, and, along with diffuse plugs, to prevent turbulence during chamber venting.

6.3 Differential Pressure Devices

Rotameters and chokes measure gas flow at a known inlet pressure and an essentially constant pressure drop. They do not operate in the flow ranges below 5×10^3 Pa-L/s (40 Torr-L/s). Low flow rates are easily measured in the molecular

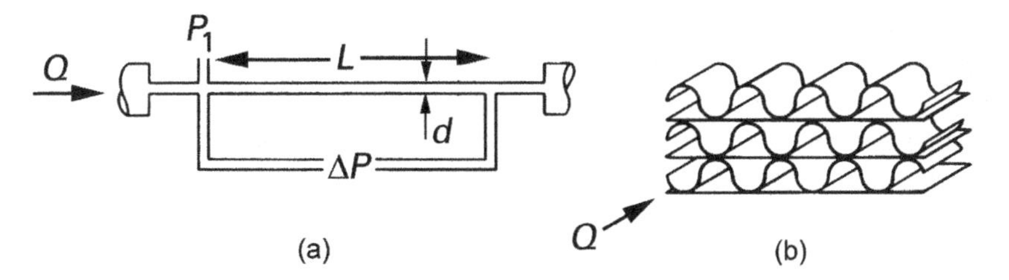

Fig. 6.2 Differential pressure flow elements, (a) laminar and (b) molecular.

or laminar viscous flow region by measuring the pressure drop across a known conductance. The concept is the same for low, medium, and high vacuum. Only the form of the conductance and the pressure gauge differ. Measurements of extremely small flows in the high- and ultrahigh-vacuum region are usually limited to pumping speed measurements. These are treated separately in Chapter 7.

A molecular or laminar viscous element is used in combination with a capacitance manometer to measure the gas flow in the low and medium vacuum range. A laminar flow element [8] is incorporated into a flow meter as sketched in Fig. 6.2*a*. It is simply a capillary tube long enough to satisfy the Poiseuille equation. See Eq. (3.10) rewritten here as:

$$Q = \frac{\pi d^4}{128\eta l} P_{ave} \Delta P \tag{6.8}$$

The flow is proportional to ΔP and the average pressure in the tube. Flow measurement with a long capillary requires two pressure gauges and knowledge of the temperature as well as the gas species. This is an accurate technique, but not the most convenient. It is most often used for calibration of thermal mass flow meters.

Molecular-flow elements [9] are constructed from a parallel bundle of capillaries, v-grooves, or similar shapes. See Fig. 6.2*b*. The diameter of each channel must be kept small for the flow to remain molecular at usably high pressures, say 100 Pa. Typically, a single channel will have a diameter of a fraction of a millimeter. The conductance of one channel is of order 10^{-4} L/s so that a large number of parallel channels (>10,000) are necessary to achieve a practical device. The flow through such an element can be expressed as:

$$Q = C\Delta P = xa' \frac{v}{4} \Delta P \tag{6.9}$$

where x is the number of channels and a' is the transmission probability of one channel. We see the flow to be proportional to $(T/M)^{1/2}$. As long as the line pressure is less than 100 Pa, it is not necessary to know its value. These devices are available commercially for use in the range 0.01–100 Pa-L/s, attitude-insensitive, stable, and easy to use. However, the holes can become clogged. They have a slow response time (5–60 s) and can have a high pressure drop; they cannot be used at inlet pressures greater than 100 Pa. The output of the capacitance manometer that measures ΔP can be used to control the opening and closing of an adjacent valve and achieve closed loop flow control.

6.4 Thermal Mass Flow Technique

Mass flow can be determined from the quantity of heat per unit time required to raise the temperature of a gas stream a known amount. Practical flow measuring devices are sensitive to either thermal conductivity [10,11,12] or heat capacity [13,14]. Devices sensitive to heat capacity have become widely accepted, because of their accuracy and ability to measure large gas flows with low power input. When combined with a rapidly acting valve, they become mass flow controllers (MFCs).

6.4.1 Mass Flow Meter

The concept of the Thomas thermal mass flow meter [15] is illustrated in Fig. 6.3. We can measure the gas flow by applying constant power to the uniformly spaced grid and observing the temperature rise of the gas on the downstream side of the grid. The amount of heat that is required to warm the gas stream is linearly dependent on the mass flow and the specific heat

$$\frac{dm}{dt} = \frac{H}{C_p\left(T_2 - T_1\right)} \tag{6.10}$$

For example, if we apply a nitrogen mass flow of 0.001 kg/s to this device, we will observe a temperature rise of 10°C for each 8.1 J/s of heat input to the grid.

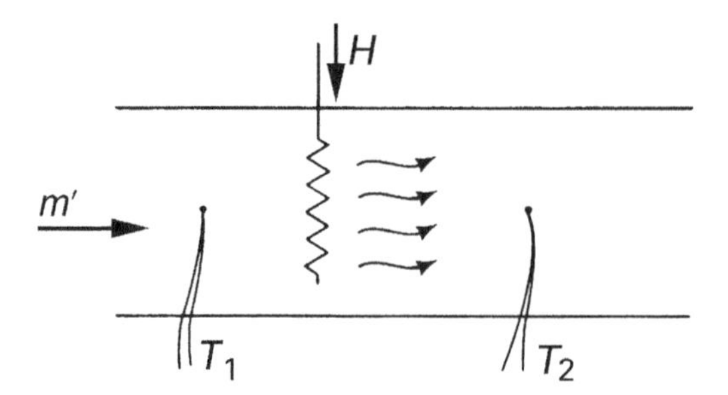

Fig. 6.3 Principle of thermal mass flow measurement.

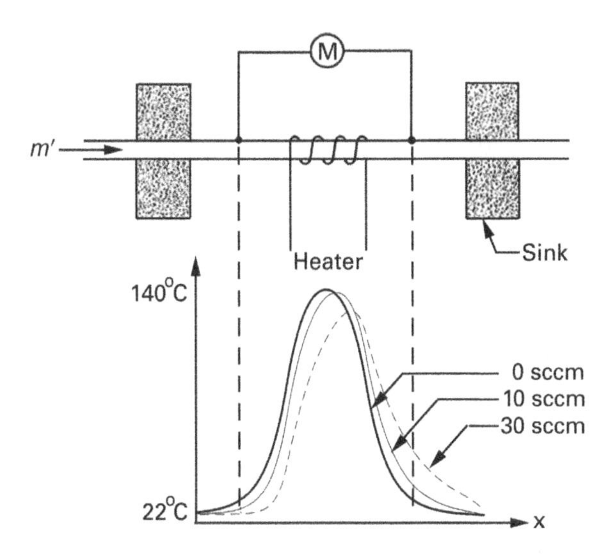

Fig. 6.4 Thermal mass flow sensor and temperature profile. Upper: Cross-sectional view of the sensor tube, resistance thermometers, and heat sinks. Lower: Temperature profile. Temperature profile reproduced with permission from William J. Alvesteffer [16]. Copyright 2000, William J. Alvesteffer.

The change in thermal capacity with temperature is small and has only a slight effect on the measurement of mass flow. Typical thermal coefficients due to heat capacity variations range from +0.075%/°C (CO_2) to +0.0025%/°C (Ar).

Operation of a thermal mass flow meter is based on Eq. (6.10). One early form of the device is sketched in Fig. 6.4. Heat is externally applied at the center of the tube. While there is no gas flow, the temperature profile is symmetrical. With gas flow, the thermal profile is skewed. Two thermocouples (or resistance thermometers) measure the change in temperature profile between the no-flow and flow condition. The meter must be mounted in the position shown, as the heat distribution is attitude sensitive. Another form of the device is illustrated in Fig. 6.5. It uses a bridge circuit to keep the temperatures constant. With this technique, the mass flow is proportional to the amount of power required to maintain constant temperatures. Sampling tubes of 0.2–0.8-mm internal diameter are usually made from stainless steel; however, Inconel or Monel can be used for corrosive gases. Flow meter manufacturers prefer units of standard cubic centimeters per minute (sccm) for display of small flows and standard liters per minute for large gas flows. Flow meters are manufactured with full-scale deflections ranging from 1 sccm (1.6 Pa-L/s) to 500 standard liters per minute (~10^6 Pa-L/s). Devices that measure flows greater than 10–200 sccm (~20–3500 Pa-L/s) use a laminar flow bypass, which diverts a fixed percentage of the flow to the small tube. Hinkle and Marino have shown that the linearity of the flow bypass is design-dependent; the bypass ratio was affected by end effects. The thermal properties of the

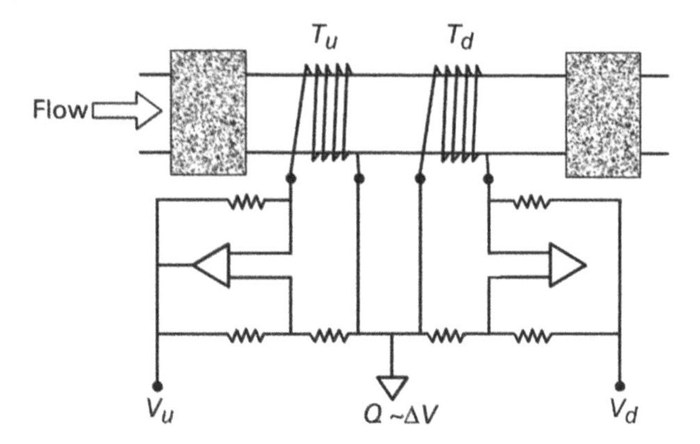

Fig. 6.5 Constant power mass flow controller circuit. Reproduced with permission of Luke Hinkle, Copyright 2000, Luke Hinkle.

system were dependent on the Nusselt number, which varied with flow. If boundary conditions vary, it will also be a function of position [17]. These effects can be addressed by software algorithms.

A thermal mass flow meter directly measures the amount of heat absorbed by the gas stream, and therefore indirectly measures the mass flow. However, the gauge scales are normally calibrated in units of throughput—usually air. We can convert throughput to mass flow with the aid of Eq. (6.3). A gauge calibrated in units of throughput has a different temperature coefficient than one calibrated in units of mass flow, because throughput is density dependent. According to Eq. (6.3), a mass flow of 10^{-6} kg/s (air) at a temperature of 20°C corresponds to a throughput of 83.64 Pa-L/s. If the gas were heated to 30°C, a mass flow of 10^{-6} kg/s would correspond to a throughput of 86.5 Pa-L/s. Heated air is less dense than room temperature air. Near room temperature the temperature coefficient due to density changes is about −0.33%/°C. The temperature coefficient for throughput is therefore slightly less than this because of the small positive temperature coefficient of the heat capacity previously discussed. Temperature stability is improved by use of an insulation layer. Adding insulation increases the response time from 1–2 s to 6–10 s.

The mass flow sensor will have to be readjusted for gases other than air. We can purchase a sensor especially calibrated for one gas, or we can multiply the meter reading by a correction factor. To first order, the meter deflection is proportional to the gas density and heat capacity, so we can make an approximate correction with the aid of Eq. (6.11):

$$Q_x = \left(\frac{\rho(\text{air}) C_p(\text{air})}{\rho(x) C_p(x)} \right) Q_{\text{meter}} = f_x Q_{\text{meter}} \tag{6.11}$$

The factor in parentheses is known as the meter correction factor f. This factor is approximate. The actual correction may differ due to small changes in gas viscosity, specific heat with temperature, or the rate at which a gas transfers heat to and from the tube wall. Example correction factors are given in Table 6.1. Hinkle and Marino [17] have shown that a function, not a factor is required for some gases. The factors were found to be correct for small gas flows. For high flows, they demonstrated that significant error could be introduced when measuring gases with different thermal diffusion constants. Figure 6.6 illustrates this error for two gases, helium, and hydrogen, when measured in a flow sensor calibrated for nitrogen. A flow of 100 sccm introduced a 7% error in the helium correction factor, whereas the same flow of hydrogen introduced a 12% error in the correction factor.

Thermal mass flow sensors have the advantages of convenience, stability, accuracy (<1%), and moderately short response time. These sensors are attitude-sensitive and have a high temperature coefficient. Designs using very-small-diameter tubes can clog, especially when reactive gases contact minute impurities of water vapor.

6.4.2 Mass Flow Controller

A thermal MFC consists of a thermal mass flow sensor, a rapidly acting valve, and an electronic control system. In its basic form, a MFC can maintain a constant, operator-set flow. Complex instruments can maintain ratios of gas flows or be integrated into process control systems. Flow valves may be controlled by a solenoid, a piezoelectric stack. For ultraclean applications, a piezoelectric controller with a metal diaphragm will produce the fewest particles.

The maximum flow and the pressure drop across the flow sensor and valve must be known to choose the proper size combination. We first determine the equivalent air throughput from $Q_{meter} = Q_x/f_x$. We then choose a flow meter with the next largest full-scale meter deflection. From the manufacturer's data, we determine the pressure drop ΔP across the meter and valve combination at maximum flow. The value of ΔP is of little concern, if the delivery pressure is above atmosphere. Vapors such as CCl_4, which are liquid at room temperature, have vapor pressures below atmospheric pressure. If the gas or vapor source is at a reduced pressure, we must size the meter and valve so that the pressure drop is less than the difference between P_1 and P_2, the delivery and chamber pressures, respectively, $\Delta P < (P_1 - P_2)$. The pressure drop can be reduced by choosing a low conductance valve or, if necessary, a somewhat larger flow meter than might otherwise be desired.

Table 6.1 Thermal Mass Flow Meter Correction Factors

Gas	Heat Capacity J/(kg-°C)	Density (kg/m³) at 0°C	Correction Factor f[a]
Air	1004.2	1.293	1.00
NH_3	2058.5	0.760	0.73
Ar	520.5	1.782	1.45
AsH_3	488.3	3.478	0.67
BCl_3	535.1	5.227	0.41
CCl_4	692.5	6.86	0.31
Cl_2	478.7	3.163	0.86
CF_4	692.0	3.926	0.42
B_2H_6	2125.5	1.235	0.44
SiH_2Cl_2	627.6	4.506	0.40
C_2H_6	1714.2	1.342	0.50
He	5192.3	0.1786	1.45
H_2	1430.0	0.0899	1.01
HBr	360.0	3.61	1.00
HF	1455.6	0.893	1.00
Kr	2481.5	3.793	1.54
CH_4	2229.2	0.715	0.72
Ne	1029.3	0.9	1.46
NO	974.0	1.339	0.99
N_2	1039.7	1.250	1.00
NO_2	808.8	2.052	0.74
NF_3	751.9	3.168	0.48
O_2	917.6	1.427	1.00
PH_3	993.3	1.517	0.76
SiH_4	1334.3	1.433	0.60
$SiCl_4$	531.4	7.58	0.60
SF_6	666.1	6.516	0.26
CCl_2FCClF_2	673.6	8.36	0.20[b]
WF_6	338.9	13.28	0.25
Xe	158.2	5.856	1.32

[a] $Q_x = f\,Q_{meter}$.
[b] At 60°C.
Source: Reproduced with permission from MKS Instruments, 6 Shattuck Road, Andover, MA 01810.

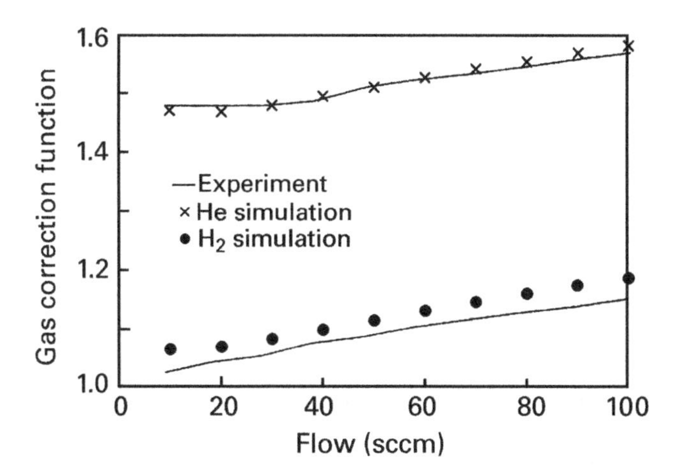

Fig. 6.6 Simulation and experimental gas correction functions for He and H_2 relative to N_2. L. D. Hinkle and C. F. Marino [17]/Reproduced with permission from AIP Publishing.

6.4.3 Mass Flow Meter Calibration

Calibration of a mass flow meter can be done by comparison with primary or secondary standards. Primary standards are most often maintained at National laboratories. They determine flow by comparison with independently calibrated fundamental measurements. Secondary standards are instruments that have been calibrated against primary standards. Secondary standards are found in the laboratories of large research institutions or manufacturing firms. Volumetric piston and rate of rise techniques are used as primary standards, whereas the pressure drop across a laminar flow element, or another mass flow meter can be used as secondary standards. Methods for low throughput measurement are found in the AVS Recommended procedure for pumping speed measurement [3]. Hinkle and Uttaro have reviewed primary and secondary standards and presented data on long-term mass flow meter stability [18]. No AVS Recommended Practices have been published for calibration of mass flow meters; however, standards have been published by the ISO [19].

References

1 Normand, C.E., in *Transactions of the 8th National Vacuum Symposium of the American Vacuum Society and 2nd International Vacuum Congress*, L.E. Preuss, Ed., Pergamon Press, Elmsford, NY, 1962, p. 534.

2 Stevenson, D.J., in *1961 Trans. 8th. Vac. Symp.*, L.E. Preuss, Ed., Pergamon Press, Elmsford, NY, 1962, p. 555.

3 Hablanian, M., *J. Vac. Sci. Technol., A* **5**, 2552 (1987) (American Vacuum Society Recommended Procedure for Measuring Pumping Speeds; see its Appendix A).
4 Ehrlich, C., *J. Vac. Sci. Technol., A* **4**, 2384 (1986).
5 Van Atta, C.M., *Vacuum Science and Engineering*, McGraw-Hill, New York, 1965, Chapter 7.
6 Kuzara, R.W., in *Flow–Its Measurement and Control in Science and Industry*, W.W. Durgin, Ed., Vol. 2, Instrument Society of America, Research Triangle Park, NC, 1981, p. 741.
7 ASME, *Flow Measurement*, PTC 19.5.4 American Society of Mechanical Engineers, New York, 1959 Chapter 4.
8 Todd, D.A., Jr., in *Flow–Its Measurement and Control in Science and Industry*, W.W. Durgin, Ed., Vol. 2, Instrument Society of America, Research Triangle Park, NC, 1981, p. 695.
9 Kiesling, R.M., Sullivan, J.J., and Santeler, D.J., *J. Vac. Sci. Technol.* **15**, 771 (1978).
10 C.E. Hastings and C.R. Wcislo, *AIEE*, March 1951.
11 Mac Donald, F., *Instrum. Control Syst.* (1969).
12 Laub, J.H., *Electr. Eng.* **66**, 1216 (1947).
13 Benson, J.M., Baker, W.C., and Easter, E., *Instrum. Control Syst.* **43**, 85 (1970).
14 Hawk, C.E. and Baker, W.C., *J. Vac. Sci. Technol.* **6**, 255 (1969).
15 Thomas, C.C., *J. Franklin Institute* **152**, 411 (1911).
16 William J. Alvesteffer, An Improved Thermal Mass Flow Controller for Hazardous and Precision Applications, M.S. Thesis, Old Dominion University, 2000.
17 Hinkle, L.D. and Marino, C.F., *J. Vac. Sci. Technol., A* **9**, 2043 (1992).
18 Hinkle, L.D. and Uttaro, F.L., *Vacuum* **47**, 523 (1996).
19 ISO 14511:2019 (2019). *Measurement of Fluid Flow in Closed Conduits—Thermal Mass Flow Meters*, International Standards Organization, Geneva, Switzerland.

7

Pumping Speed

Pumping speed is a quantity few of us would ever measure. We may be content with published data, but we should know how it is measured, so that we can interpret data meaningfully. Pumping speed is measured by established techniques. Recommended practices have been improved; however, much data taken in old test fixtures pervades the literature. We should understand the accuracy of available data.

There are situations where we wish to measure pumping speed, e.g., to check for proper operation, or to find speed for a specific gas, pump temperature, or fluid. If we know the speed at the pump inlet and connecting conductance, we can calculate speed at chamber entrance with reasonable accuracy.

We begin by reviewing pumping speed and then describe its measurement in low, medium and high vacuum pumps. We include a simple technique for checking the speed of an operating system and discuss measurement of water speed in high vacuum pumps. We conclude with a discussion of the errors inherent in standard test fixtures.

7.1 Definition

Pumping speed was defined in Eq. (3.3) and accompanying Fig. 3.2 (left), as the volumetric rate at which gas is transported across a plane, i.e., the ratio of gas flow to pressure at point of measurement. Like conductance, it has dimensions of volume per unit time, and in SI it is expressed in units of m^3/s. Units of L/s or m^3/h are acceptable. Unlike conductance, pumping speed is not a property of a passive component like a length of pipe or a trap. See Eq. (3.4).

Measurement of speed implies independent measurements for both throughput Q, and pressure P. Pressure gauges are described in Chapter 5. For measuring

A Users Guide to Vacuum Technology, Fourth Edition. John F. O'Hanlon and Timothy A. Gessert.
© 2024 John Wiley & Sons, Inc. Published 2024 by John Wiley & Sons, Inc.

permanent gas pressures, liquid nitrogen cold traps may be used between the gauge and test fixture to avoid errors due to condensable gas impurities. Throughput measurement is described in Chapter 6, with the exception of the conductance method that is described in Section 7.3.1.

7.2 Mechanical Pump Speed Measurements

The test fixture used for measuring mechanical pump speeds is illustrated in Fig. 7.1 (right). The fixture and pump diameters are equal, unless the pump diameter is less than 50 mm. For pumps whose inlet diameter is less than 50 mm, a conical adapter flange is used to make the transition from test fixture to pump. This flange is described in the AVS Recommended procedures for measuring the performance of positive displacement mechanical pumps [2]. This fixture is rather similar to that described in the more recent ISO Standard [3]. Gauge accuracy must be within ±5% above 0.1 Pa, and ±10% in the range 0.001–0.1 Pa. The capacitance manometer, spinning rotor, and ion gauge (≤0.01 Pa) may be used. The recommended procedures require flow-measuring devices with 2% accuracy. Thermal mass flow meters may be used in the range 10–10,000 Pa-L/s, whereas the

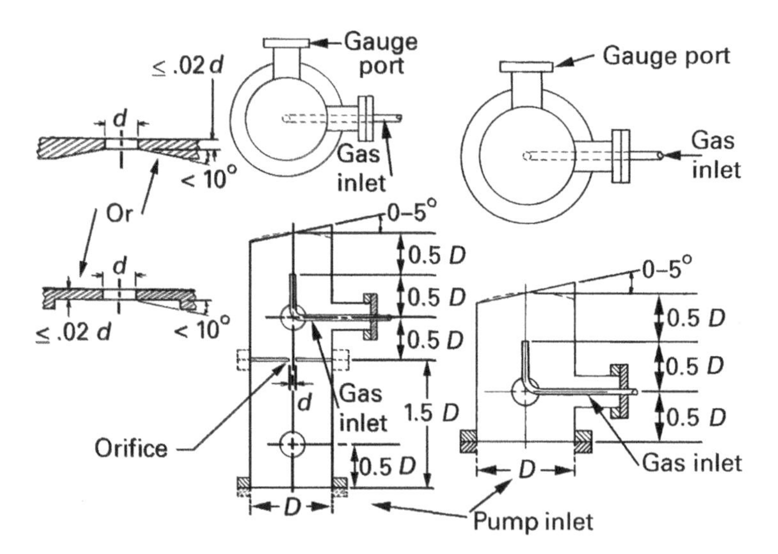

Fig. 7.1 Test domes for the measurement of mechanical and high vacuum pumps. Right: Flowmeter method test dome. This dome is used for speed measurement in pumps with a diameter greater than 50 mm. Left: Conductance (orifice) test dome. This dome is used for speed measurement at low gas flows. M. Hablanian [1]/Reproduced with permission from AIP Publishing.

conductance method is appropriate for flows less than 0.1 Pa-L/s. The intermediate range may be covered by a differential capacitance flow meter. After the gas has been flowing into the test fixture for at least 3 min, the equilibrium pressure is recorded and the speed is calculated from Eq. (7.2), where P is the measured pressure and P_b is the base pressure with zero gas flow.

$$S = \frac{Q}{\left(P - P_b\right)} \tag{7.1}$$

Three pumping speed measurements per decade are made over the pump's operating range. The ambient temperature, barometric pressure, rotation speed, and type of oil in the pump should be recorded. Pumping speed measurements are typically performed with nitrogen; liquid-nitrogen-cooled cold traps are used between the gauge sensor and system to eliminate the effects of pump oil vapors.

7.3 High Vacuum Pump Speed Measurements

Measurement of high vacuum pumping speed requires flow measurements considerably below the range of the techniques discussed in Chapter 6. Speed measurements are further complicated by the anisotropic gas flow patterns in the test fixture, and a flowing gas stream. Here we examine two techniques for measuring the speed of a pump and examine one technique for estimating speed at the entrance to the working chamber.

7.3.1 Methods

Volumetric pumping speed is measured with a standard metering fixture. Myriad early fixture designs have been replaced by a design similar to that described in the AVS Recommended procedure for measuring pumping speeds [1] and described in Fig. 7.1.

The flowmeter method uses a fixture described in Fig. 7.1 (right). Pumping speed is calculated in the same manner as given in Eq. (7.2). For very low flows, the orifice method described in Fig. 7.1 (bottom center) is used. Use of this technique requires a flow in the molecular regime. The numerical value of the flow is determined by the pressure drop across a known conductance $C(P_1 - P_2)$; the resulting speed is calculated from:

$$S = C\left(\frac{P_1 - P_{01}}{P_2 - P_{02}} - 1\right) \tag{7.2}$$

7.3.2 Gas and Pump Dependence

The ISO standard specifies the use of nitrogen; however, speed can be and is measured for many other gases. Turbo, ion, getter, and cryopumps behave differently from diffusion pumps, as their speed versus mass dependence's are not the same. Diffusion pumps pump all gases; ideally, their speed should vary as $1/m^{1/2}$. In Chapter 12 we show the light gas pumping speeds to be somewhat higher than the speeds for air, or nitrogen, but not as great as predicted $1/m^{1/2}$. Ion, getter, and cryogenic pumps do not pump all gases. Consequently, capture pumps are characterized with specific gases because their speed is a function of gas composition as well as prior history.

Some time may be required for a capture pump to reach equilibrium after admitting gas at a fixed flow rate [1]. This is particularly true for ion pumps at low pressure [4]. If pumping-speed data are recorded too quickly, data taken in order of decreasing pressure will yield an incorrectly small value of speed, while those taken in order of increasing pressure will yield an incorrectly large value of speed. It may also be necessary to erase the pump's memory for one or more gases pumped before measurement. This can be accomplished by pumping for 1 h at a pressure of 10^{-3} Pa with the gas under study [1].

The operating characteristics of each pump should be recorded with the speed measurements. The type of fluid, the size and type of forepump, the foreline pressure in compression pumps, diffusion pump boiler power, turbo pump rotational frequency, and the refrigeration capacity of cryogenic pumps are some of the factors that need to be reported when measuring speed.

Few measurements of water vapor pumping speed have been made because of the experimental problems. Water is difficult to degas: it can freeze on evaporation, boil at room temperature, and plug valves. It will sorb on the test fixture at very low pressures. Landfors et al. [5] have measured the water vapor pumping speed of a diffusion pump with and without a liquid nitrogen cold trap and the speed of a cryogenic pump. They constructed a chamber for admitting known amounts of degassed water vapor at constant pressure into an ISO test fixture. Speed was measured at pressures greater than 10^{-2} Pa to avoid sorption effects. They found the water pumping speed to be essentially the same as air for a diffusion pump without an LN_2 trap. The water pumping speed for an LN_2-trapped diffusion pump was approximately the same as that for a cryopump of the same throat area. They concluded the pumping speed for water should be given by the projected area of the cold surface at the inlet of the trap as reduced by the conductance of any intervening pipe. AVS Recommended Practices for measuring performance of closed-loop helium cryopumps describe methods for measuring speed of cryopumps [6]. Although cryopumps have a higher Ho coefficient than other pumps, the variations allowed in the recommended practice have little impact on the pumping speed results [7].

7.3.3 Approximate Speed Measurements

Only pump manufacturers and large projects will have the necessary test fixtures for measuring pumping speed, so it is necessary to have an approximate method anyone can use on an existing system. The approximate pumping speed at the entrance to the chamber can be measured without the trouble of attaching an elaborate fixture and gas metering system. The speed can be deduced approximately, if it is assumed to be independent of pressure in the region of interest. If this is true, then it follows that:

$$S = \left(\frac{Q_2 - Q_1}{P_2 - P_1} \right) \tag{7.3}$$

where Q_1 is the flow that results in P_1, and so on. It can be shown that this flow is equivalent to:

$$S = V \frac{\left(\frac{dP_2}{dt} \right) - \left(\frac{dP_1}{dt} \right)}{\left(P_2 - P_1 \right)} \tag{7.4}$$

To measure the pumping speed, the system is first pumped to its base pressure P_1 and the high vacuum valve is closed. At this time the pressure rise dP_1/dt is plotted over at least a one-decade pressure increase. The high vacuum valve is then opened, and the system is pumped to its original base pressure. A test gas is then admitted through a leak valve until the pressure rises to a value P_2, which is several times that of the base pressure. The high vacuum valve is closed again, and dP_2/dt is recorded. The system volume is estimated. and the speed is calculated by use of Eq. (7.5). This method, called the rate of rise or constant volume method, is only approximate, because gas flow at the base pressure Q_1 is, in general, background desorption and not the same gas species as admitted through the leak. Fixing the starting pressure P_2 at a value of at least 10 times P_1 will ensure reasonable accuracy.

The pressure at the pump throat can be estimated by subtracting the impedance drop of the conducting pipe and trap located between the gate valve and the pump.

7.3.4 Errors

High vacuum pumping speed measurement errors arise from three sources. First, the definition of speed given by Eq. (7.1) requires Q and P to be measured at the same surface. This is not true for the fixture described in Fig. 7.1. Second, the definitions of Q and P assume random molecular arrival and that is not the case when gas enters from a pipe. Third, the gas flow is sometimes measured inaccurately. All three sources of measurement error occur, because we do not completely compensate for the nonuniform gas distribution in the system. The nonuniform gas distribution affects both the capture coefficient (also known as the Ho coefficient of the pump) and the measurement of pressure. The Ho coefficient is the ratio of actual pumping speed to the maximum pumping speed of an aperture of the same size in which no molecules are reflected backward toward the source [8].

In Chapter 3 we defined the intrinsic conductance of a tube for randomly distributed incoming molecules and noted this definition did not apply to beamed flow. The same effect occurs at the inlet of a pump. The capture probability is a function of molecular arrival angle. When we place a pipe at the inlet of a pump, the capture coefficient will change, and the pump no longer has its "intrinsic" speed—that is, the speed it would have, if appended to an extremely large chamber. Its speed may increase because molecules shot straight into the pump will have a greater probability of being captured than will those arriving randomly [9], and because molecules bounce around—the "maze" effect [10]. Test fixtures constructed from pipes of the same diameter as the pump are criticized because the inlet flux is not random, and the resultant speed is not what is measured on a large chamber. Proponents of equal-diameter test fixtures argue that the measured speed is meaningful because pumps are usually connected to a pipe of the same diameter.

Pressure measurement is the second consideration. The pressure in the fixture is not isotropic or uniform. Mathematically, the pressure in the fixture can be described by a tensor [11]. The pressure is a function of gauge location and orientation, and surface temperatures may not be uniform.

Examine the pump and pipe sketched in Fig. 7.2. The pump is an ideal pump from which no molecules return ($a = 1$). Gas is flowing to the pump inlet. Gauge 1 reads the pressure at the pump entrance, whereas gauge 2 reads a pressure corresponding to the flux incident on its opening. One-half of the gas arriving at the entrance to gauge 2 will come from the top, and one-half from the bottom. Since there is no flow from the bottom, gauge 2 will read $P_1/2$. The speed we would calculate from the side-mounted gauge located in the plane of the pump inlet would be two times the actual speed. Side reading gauge error has been understood for quite some time and has been discussed by many authors [9,11,12,13,14]. Projecting parts and low sticking coefficients will usually prevent a pump from capturing all molecules. A fraction $(1 - a)$ of them will be returned and gauge 2 will read $P_2 = P(1 - a/2)$, where a is the transmission coefficient. The error resulting from the use of a side reading

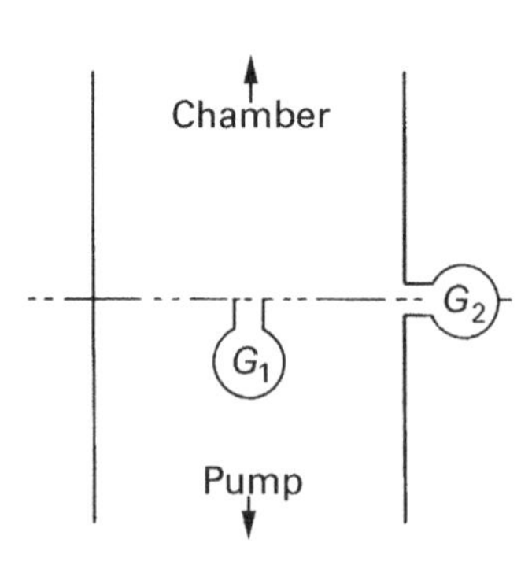

Fig. 7.2 The effect of orientation on pressure gauge readings.

gauge is a function of the Ho coefficient. The relation between the actual Ho coefficient a, and the apparent Ho coefficient a', due to the side reading gauge is:

$$a = \left(\frac{a'}{1 + \frac{a'}{2}} \right) \tag{7.5}$$

The apparent speed of a perfect pump ($a = 1$) is $2S$. The error will become less as the Ho coefficient goes to zero. Equation 7.5 is valid provided that the reflected gas is randomly distributed. In general, this is not true, because objects such as inlet screens and chevrons may distort the flow.

Feng and Xu [15] calculated the pumping speed measurement error of the 1980 AVS/ISO test fixture to be within 4% for Ho coefficients in the range of 0.2–1.0. They found the error in the original AVS fixture [16] to be greater than 10%. Measurement error was the reason for changing the gauge location in the new AVS Recommended Practice. This is sufficiently accurate for the applications envisioned. It is not necessary to know the pumping speed to a fraction of a percent, because we will use it in formulas whose transmission coefficients have not been corrected for beaming when joining series components.

References

1. Hablanian, M.H., *J. Vac. Sci. Technol., A* **5**, 2552 (1987).
2. Kendall, B.R.F., *J. Vac. Sci. Technol., A* **7**, 2403 (1989).
3. ISO-21360–x:2020, *Vacuum technology—Standard methods for measuring vacuum-pump performance*. Six parts: –1. *Part 1: General description*; –2, Positive Displacement Pumps; –3, Mechanical Booster Pumps; –4, Turbomolecular Pumps; –5, NEG Pumps; and –6, Cryopumps. International Standards Organization, Geneva, Switzerland, 2020.
4. Andrew, D., *Vacuum* **16**, 653 (1966).
5. Landfors, A.A., Hablanian, M.H., Herirck, R.F., and Vaccarello, D.M., *J. Vac. Sci. Technol., A* **1**, 150 (1983).
6. Welch, K., Andeen, B., de Rijke, J.E., Foster, C.A., Hablanian, M.H., Longsworth, R.C., Millikin, W.E., Jr., Sasaki, Y.T., and Tzemos, C., *J. Vac. Sci. Technol., A* **17**, 3081 (1999).
7. Nezterov, S.B., Vassiliev, Y.K., and Longsworth, R.C., *J. Vac. Sci. Technol. A* **19**, 2287 (2001).
8. Ho, T.L., *Physics* **2**, 386 (1932).
9. Steckelmacher, W., *Vacuum* **15**, 249 and 503 (1965).
10. Buhl, R. and Trendelenburg, E.A., *Vacuum* **15**, 231 (1965).

11 Dayton, B.B., *Ind. Eng. Chem.* **40**, 795 (1948).

12 Santeler, D.J., Jones, D.W., Holkeboer, D.H., and Pagano, F., *Vacuum Technology and Space Simulation*, NASA SP-105, National Aeronautics and Space Administration, Washington, DC, 1966, p. 119.

13 Dayton, B.B., *Vacuum* **15**, 53 (1965).

14 Denison, D.R., *J. Vac. Sci. Technol.* **12**, 548 (1975).

15 Feng, Y. and Xu, T., *Vacuum* **30**, 377 (1980).

16 Hablanian, M.H. "Apparatus of AVS Tentative Standard 4.1, 4.7, 4.8", *J. Vac. Sci. Technol.* **8**, 664 (1971).

8

Residual Gas Analyzers

The first 180° magnetic sector mass spectrometer was developed in 1918 by Dempster [1]. By 1943, Nier [2] designed a metal version of his glass instrument used to measure uranium isotope ratios [3] and to leak test mass separation equipment at Oak Ridge Laboratory [4]. In 1960 Caswell used the RGA to study the residual gases in vapor deposition systems, and he demonstrated that the performance of a vacuum system could be improved with Viton gaskets, Meissner traps, and getters [5]. Caswell also used the RGA to characterize the effects of residual gases on the properties of tin and indium films [6,7].

Since then, RGAs have proved their value in the solution of many vacuum-related manufacturing problems. They have been used extensively to diagnose problems and verify environments. RGAs are now part of sophisticated process control systems where they are used to measure gas purity, gas loads evolved from films and sputtering targets, vapor fluxes from physical vapor deposition sources, and background gas composition during sputtering and other plasma processes, as well as from ultraclean processes such as molecular beam epitaxy.

This chapter is concerned with the theory of the operation of rf quadrupoles, as well as installation methods and data collection on high vacuum and plasma processing chambers. The interpretation of RGA spectra is discussed in Chapter 9.

8.1 Instrument Description

Mass spectrometers are used to measure the ratio of mass-to-electric charge (M/z) of a molecule or atom. Mass measurement is based on the ^{12}C scale—that is, the weight of carbon is 12.0000 atomic mass units (AMU). The charge of the ion is ze, where z is the number of electrons removed from the outer shell of the atom, and e is the charge of the electron. Figure 8.1 displays a mass spectrum recorded on an

A Users Guide to Vacuum Technology, Fourth Edition. John F. O'Hanlon and Timothy A. Gessert.
© 2024 John Wiley & Sons, Inc. Published 2024 by John Wiley & Sons, Inc.

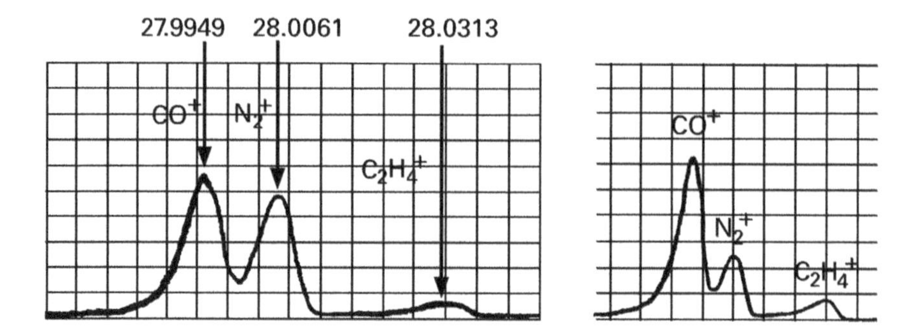

Fig. 8.1 Left: Expanded sweep over the $M/z = 28$ triplet showing resolution of CO, N_2, and C_2H_4 at a time shortly after pump down. Right: Expanded sweep over the $M/z = 28$ triplet after about 2 h of pumping showing the diminution of N_2 relative to CO and C_2H_4. W.L. Fite and P. Irving [8]/Reproduced with permission from AIP Publishing.

analytical quadrupole mass spectrometer [8]. Examination of this spectrum shows three well-resolved peaks. The resolving power of this instrument permitted identification of the triplet CO^+ ($M/z = 27.9949$), N_2^+ ($M/z = 28.0061$), and $C_2H_4^+$ ($M/z = 28.0313$). The spectrum shown in Fig. 8.1 illustrates the capability of a precision analytical laboratory instrument. The small, portable, inexpensive instruments we describe here are called residual gas analyzers (RGAs). These instruments are capable of resolving approximately single mass units ($\Delta M/z \sim 1$), and they scan an M/z range of 1–50 or perhaps 1–400. In an RGA spectrum, the triplets illustrated in Fig. 8.1 would be displayed as one peak. Figure 8.1 (right) showed that the N_2 peak decreased after pumping, whereas the CO peak increased; we now know that CO is generated in the hot filaments used in RGAs and ion gauges.

All mass spectrometers have three parts: an ionizer, a mass filter, and a detector. See Fig. 8.2. The simplicity and reduced cost of an RGA results in an instrument with far less resolving power and sensitivity than is possible in an analytical

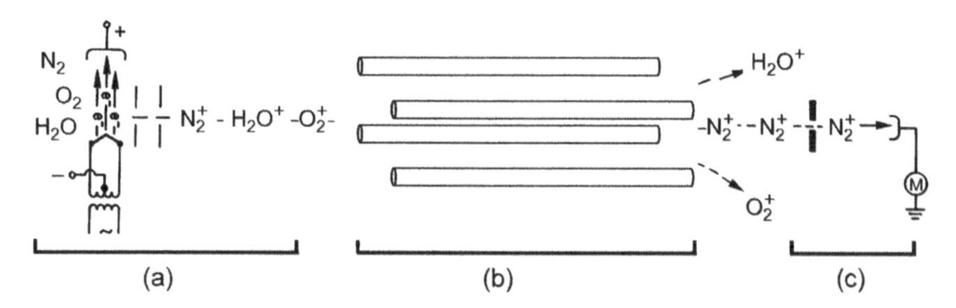

Fig. 8.2 Three stages of partial pressure analysis: (a) Ionizer: hot filament illustrated; (b) mass filter: quadrupole filter illustrated; and (c) detector: Faraday cup illustrated.

laboratory instrument. Here we review two commonly used ion sources, two mass filters, and two detectors used in commercial RGAs.

8.1.1 Ion Sources

The only technique applicable to the production of positive ions in commercial residual gas analyzers is electron impact ionization. Other techniques such as field ionization and chemical ionization are useful in research instruments, such as the triple-quadrupole atmospheric pressure ionization mass spectrometer (APIMS) or tandem magnetic sector analytical mass spectrometer. In a modern RGA, the ion source is either immersed in the vacuum ambient or locally isolated by nearby walls from the source gas. These ion sources are referred to, respectively, as open, and closed ion sources.

8.1.1.1 Open Ion Sources

Figure 8.3 sketches an open ion source that might be used in a residual gas analyzer. The electrons from the filament are drawn across the chamber to the anode. Some of the electrons collide with gas molecules, strip off one or more outer electrons, and create positive ions. Not all ionization chambers are geometrically similar to the one sketched in Fig. 8.3. One instrument looks very much like a Bayard-Alpert ionization gauge except for the absence of the wire collector and the addition of an electron reflector. These and other ionizers were designed to maximize ion production and sensitivity.

As in the ion gauge, positive ion production is not equal for all gases. The RGA differs from the ion gauge in that it sorts ions by their mass-to-charge ratio (M/z) and counts each ratio separately. For example, when measuring nitrogen, an ion gauge makes no distinction between current due to $^{14}N^+$ ($M/z = 14$), $^{15}N^+$

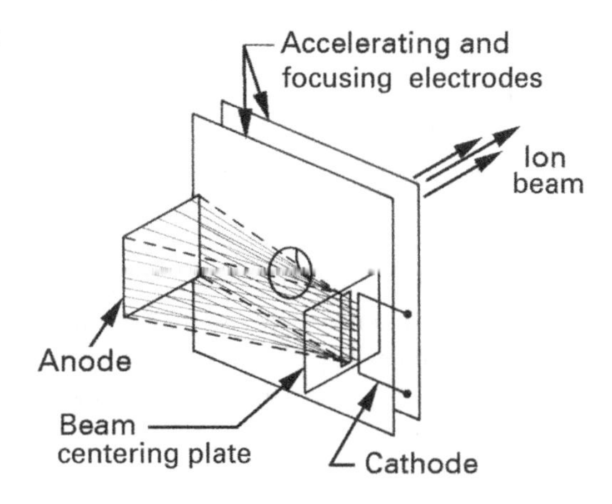

Fig. 8.3 One form of an open ion source.

Table 8.1 Experimental Total Ionization Cross Sections (70 V) for Selected Gases Normalized to Nitrogen

Gas	Relative Cross-Section
H_2	0.42
He	0.14
CH_4	1.57
Ne	0.22
N_2	1.00
CO	1.07
C_2H_4	2.44
NO	1.25
O_2	1.02
Ar	1.19
CO_2	1.36
N_2O	1.48
Kr	1.81
Xe	2.20
SF_6	2.42

Source: D. Rapp and P. Englander-Golden [9]/Reproduced with permission from AIP Publishing.

($M/z = 15$), or $^{14}N_2^+$ ($M/z = 28$), whereas the RGA distinguishes each ion current. Table 8.1 gives total positive ion cross-sections, relative to N_2, for several common gases at an ionizing energy of 70 V [10]. Although the ionization cross section does not peak at the same energy for all gases, it is generally the largest for most gases in the range of 50–150 V. For this reason, many ionizers operate at an electron energy of 70 V. Some instruments make provision for adjustment of the electron energy because it is sometimes desirable to reduce the potential in order to reduce the dissociation of complex molecules. This is essential in qualitative analysis.

The ion production of each species is proportional to its density or partial pressure. Consider a sample of a gas mixture containing only equal portions of nitrogen, oxygen, and hydrogen, whose total pressure is 3×10^{-5} Pa. A mass scan of this mixture would show three main peaks of unequal amplitudes. Since the sensitivity and gain are not constant for all masses, the peak ratios will not be in direct proportion to their partial pressures. If however, the total pressure of the gas mixture were increased to 6×10^{-5} Pa, the amplitudes of each of the three dominant peaks would double. In other words, the instrument is linear with pressure. Ionizers are linear to a pressure of ~10^{-3} Pa (10^{-5} Torr). At higher pressures the effects of space charge and gas collisions become important. The ions produced in the space between the filament and anode are drawn out of that region, focused, and accelerated toward the mass filter. The acceleration energy depends on the type of mass analyzer; in a magnetic sector, ion acceleration is part of the mass filtering process.

8.1.1.2 Closed Ion Sources

Electron- and ion-stimulated desorption from an open-source limits an RGA's sensitivity, when sampling gases through pressure reducing components. Consider a process chamber operating at 1 Pa (10^{-2} Torr) connected through a pressure reducer to a differentially pumped RGA, whose ionizer and analyzer are operating at 10^{-4} Pa (10^{-6} Torr). In this example, imagine 10^{-7} Pa of water vapor was desorbed from the analyzer walls by stimulated desorption and thermal heating of nearby surfaces. This water vapor signal would be indistinguishable from 10^{-3} Pa of water vapor in the chamber, whose pressure is reduced to 10^{-7} Pa by the pressure reducer. The smallest partial pressure of water vapor from the process chamber that could be detected within the analyzer would therefore be 10^{-3} Pa or 0.1% of the process gas concentration. The closed ion source was developed to overcome the effects of stimulated desorption from walls when sampling from high-pressure chambers.

The sketch in Fig. 8.4 illustrates how gas flows through the small sample tube of one closed ion source. An external filament emits electrons, some of which pass through the hole in the sample tube and ionize the gas within. The only electron-stimulated wall desorption is generated by the small local area inside the sample

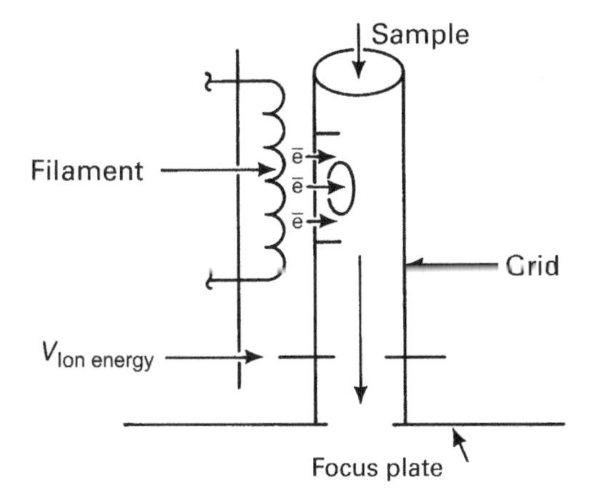

Fig. 8.4 Schematic view of a closed ion source. Gas enters the upper end of sample tube, which is at a higher pressure than the analytical chamber, before flowing into the analyzer. The filament is located outside this tube and is at the same pressure as the analyzer. B.S. Brownstein et al. [11]/ Reproduced with permission from AIP Publishing.

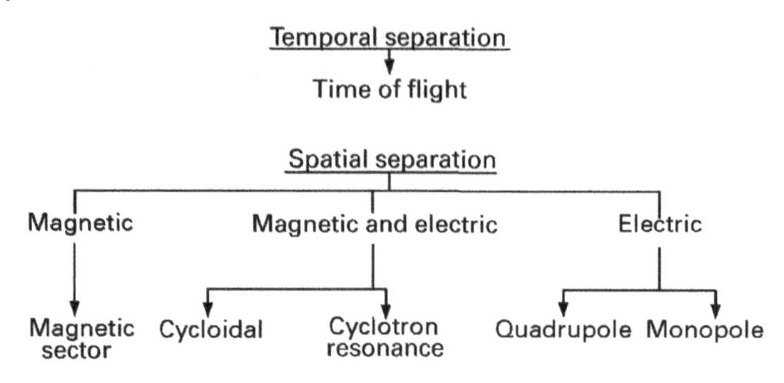

Fig. 8.5 Mass separation methods.

tube. The ionizer efficiency of a closed source will be less than that of an open source, because less than a quarter of the electron flux enters the hole in the sample tube. The closed ion source is particularly useful in sampling chambers operating at high pressures, which require differential pumping. We will discuss the use of this source in Section 8.2.2.

8.1.2 Mass Filters

Numerous techniques have been developed for filtering mass and these are outlined in Fig. 8.5. For many years, magnetic sector filters were widely used; however, the rf quadrupole developed by Paul [10] and coworkers is now used almost exclusively, except in some reduced cost leak detectors. We will limit our discussion to these two types. Many mass filtering techniques are given in the paper by Rapp and Englander-Golden [9].

8.1.2.1 Magnetic Sector

The magnetic sector mass filter separates ions of different mass-to-charge ratios by first accelerating the ions through a potential V_a and then directing them into a uniform magnetic field perpendicular to the direction of ion motion. While under the influence of this magnetic field the ions are deflected in circular orbits of radii r:

$$r = \frac{1}{B}\left(\frac{2mV_a}{ze}\right)^{1/2} \tag{8.1}$$

A practical 60° magnetic sector mass filter is shown in Fig. 8.6. In principle, any angle will work, but 60° is common for leak detectors and older RGAs. It provides sufficient mass separation with a minimum amount of magnetic material.

Differentiation of Eq. (8.1) reveals the mass dispersion Δx to be mass dependent. For ions of equal energy traversing a uniform magnetic sector the mass

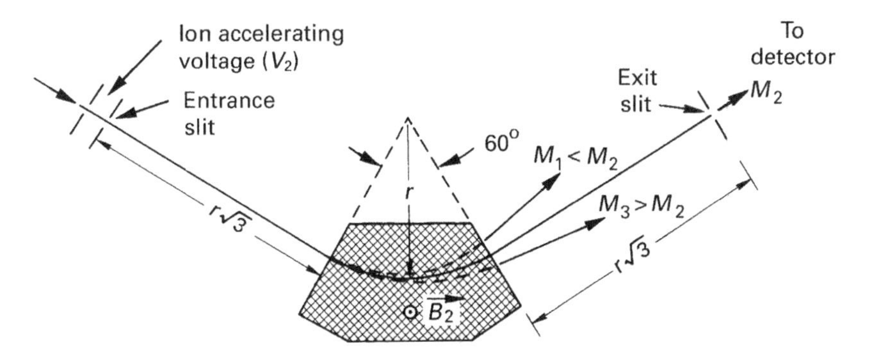

Fig. 8.6 A Magnetic sector mass separator (60°) with symmetrical entrance and exit slits. F.A. White [3]/Reproduced with permission from John Wiley & Sons.

dispersion or spatial separation between adjacent peaks of mass m and $m+1$, was found to be [3]:

$$\Delta x \propto \frac{r}{m} \tag{8.2}$$

Spatial separation is called "resolving power," and the mass-dependent resolving power of the magnetic sector makes it extremely useful for separating light atoms and molecules. High resolving power and low cost, make magnetic sector mass filters highly suitable for leak detectors that can be tuned to ^3He, ^4He, or H_2. These devices will be discussed in Chapter 24.

8.1.2.2 RF Quadrupole

The wide acceptance of the quadrupole was due to the development of the required stable, high-power quadrupole power supplies that are now available at low cost. Figure 8.7 illustrates the mass filter geometry and the path of a filtered ion. The ideal electrodes are hyperbolic in crosssection. In practice, they are realized by four rods of cylindrical cross-section located to provide the optimum approximation to the hyperbolic fields. Each of the rods is spaced a distance r_o from the central axis. Mosharrafa [12] provided a nonmathematical explanation of quadrupole operation. The two rods with positive dc potential, $+U$ in Fig. 8.8, create a potential valley near the axis in which positive ions are conditionally stable. The potential is zero along the axis of symmetry, shown in the dashed curve of Fig. 8.8. This field is zero only if the potential $-V$ is simultaneously applied to the other pair of quadrupole rods. It is a property of a quadrupole, not a dipole, field. The addition of an rf field of magnitude greater than the dc field ($U + V\cos(\omega t)$) creates a situation in which positive ions are on a potential "hill" for a small portion of the cycle. Heavy ions have too much inertia to be affected by this short

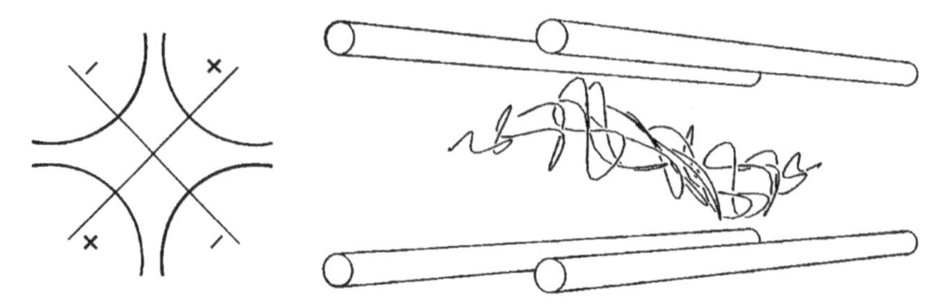

Fig. 8.7 Quadrupole mass filter. Left: Idealized hyperbolic electrode cross section. Right: Three-dimensional computer-generated representation of a stable ion path. Courtesy of A. Appel, IBM T. J. Watson Research Center, Yorktown Heights, NY.

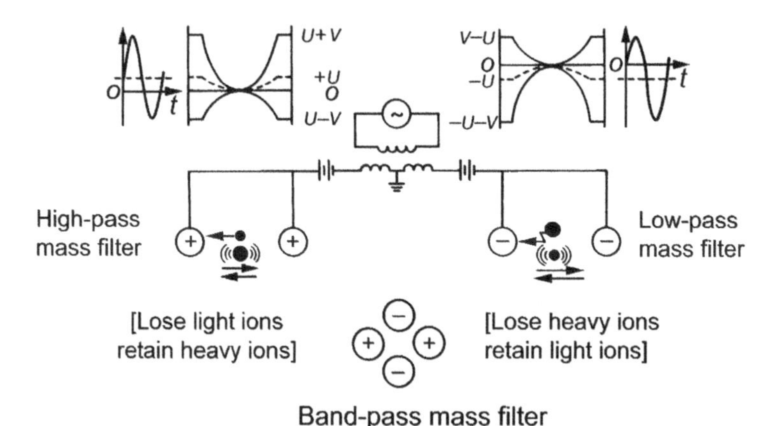

Fig. 8.8 Electric fields in a quadrupole mass filter. M. Mosharrafa [12]/Reproduced with permission from Technical Publishing Co.

period of instability, but light ions are quickly collected by the rods after a few cycles. The lighter the ion, the fewer number of cycles required before ejection from the stable region. This rod pair acts as a "high pass" filter.

The rod pair with the negative dc potential $-U$ creates a potential "hill" that is unstable for positive ions. However, the addition of the rf field creates a field $-(U + V\cos(\omega t))$, which allows a potential "valley" to exist along the axis of the quadrupole for a small portion of the cycle, provided that $V > U$. In this field, light ions are conditionally stable, and heavy ions drift toward the electrodes, because the potential "hill" exists for most of the cycle. This half of the quadrupole forms a "low pass" filter.

Together the high and low pass filters form a band-pass filter that allows ions of a particular mass range to go through a large number of stable, periodic oscillations while traveling in the z direction. The width of the pass band is a function of the ratio of dc to rf potential amplitudes U/V. The "sharpness" of the pass band is determined by the electrode uniformity, electrical stability, and ion entrance velocity and angle. A detector is mounted on the z axis at the filter's exit to count the transmitted ions. Ions of all other M/z ratios will follow unstable orbits. The rods will collect them before exiting the filter. The stability limits for a particular M/z ratio are determined from the solutions of the equations of motion of an ion through the combined rf and dc fields. The solutions involve ratios of ω, M/z, and r_0^2 and the potentials U and V. A thorough discussion of the rf quadrupole has been given by Dawson [13]. By sweeping the rf and dc potentials linearly in time the analyzer can scan a mass range. Scan times as slow as 10–20 min and as fast as 80 ms are typically attainable in commercial analyzers with M/z ranges of 1–300. One noticeable distinguishing feature of the quadrupole is that no additional restriction other than linear sweeping of the rf and dc potentials is needed to obtain a graphical display that is linear in mass scan.

Although the stability of the trajectory of an ion may be calculated without consideration of the z component of the ion velocity or the beam divergence, experimentally the situation is more complicated. There is a reasonable range of velocities and entrance angles that yields stable trajectories. In the magnetic sector both the ion energy and the magnetic field determine the focus point of an ion. One of the advantages of the quadrupole is that ions with a range of energies or entrance velocities will focus, although not with the same resolution. The slow ions are resident in the filter for a longer time and therefore are subjected to a greater number of oscillations in the rf field than are those ions with larger z components of velocity. As a result, the slow ions are more finely resolved, but suffer more transmission losses than do the light ions. For this reason, quadrupole transmission usually decreases with increasing mass. In a typical instrument that can resolve mass peaks 1 AMU apart, the gain is constant to about $20 < M/z < 50$, after which it decays at the rate of approximately a decade per 150 AMU. See Fig. 8.9. This is only typical; there is considerable instrumental and manufacturer variation. By proper choice of the potentials U and V, the mass dependence of the transmission can be considerably improved at the expense of resolution. If accurate knowledge of the transmission versus mass is desired, it must be measured for the particular filter and potentials in question. A typical mass scan taken on a small oil

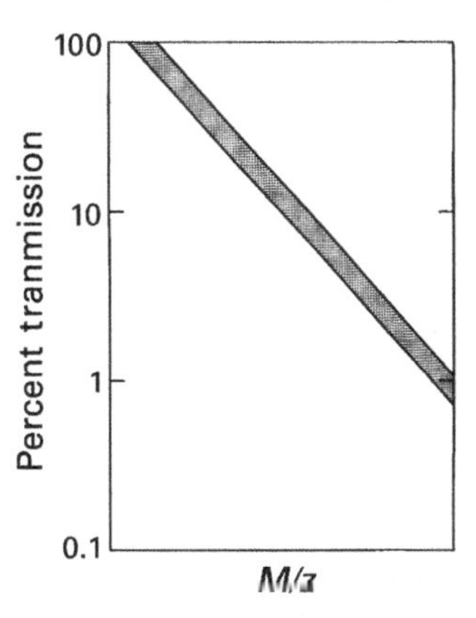

Fig. 8.9 Relative transmission of a typical rf quadrupole as a function of M/z when adjusted for unity absolute resolution. This may be varied by changing the sensitivity.

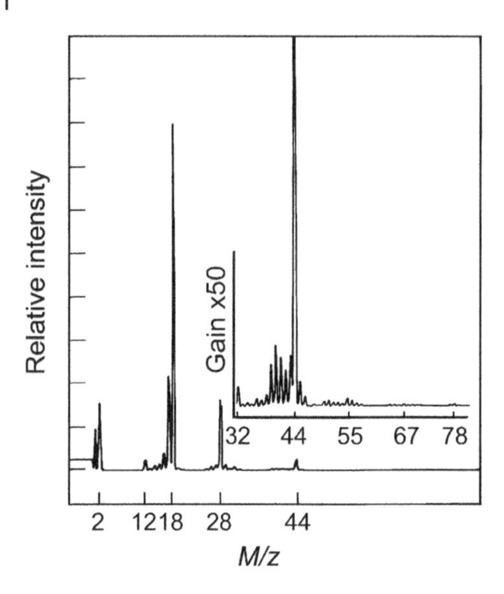

Fig. 8.10 Mass scan taken on a small oil diffusion pumped chamber by an rf quadrupole instrument.

diffusion pumped system with a quadrupole adjusted for constant absolute resolution is displayed in Fig. 8.10.

8.1.2.3 Resolving Power

The resolving power of an RGA is a measure of the ion-separating ability of the instrument of a given mass-to-charge ratio M/z. The width of the peak at 10% of the peak height $\Delta(M/z)_{10}$ is one way to measure resolving power. The resolving power of an RGA is the dimensionless ratio of peak height to the peak width at 10% peak height, $(M/z)/\Delta(M/z)_{10}$. See the AVS Recommended Practice for the Calibration of Mass Spectrometers and Partial Pressure Analyzers [14]. Analytical spectrometry requires the discrimination of mass peaks separated by extremely small fractional mass units. A resolving power of 2000 is needed to distinguish $^{32}S+$ ($M/z = 31.9720$) from $^{16}O_2+$ ($M/z = 31.9898$). RGAs or PPAs easily resolve peaks separated by about 1 AMU.

8.1.3 Detectors

The ion current detector, located at the exit of the mass filter, must be sensitive to small ion fluxes. The ion current at mass n is related to the pressure in the linear region by:

$$i_n = S'_n \, P_n \tag{8.3}$$

where i_n, S'_n, and P_n are, respectively, the ion current, sensitivity of the ionizer and filter, and partial pressure of the nth gas. A typical sensitivity for nitrogen is 5×10^{-6}

to 2×10^{-5} A/Pa. We might ask why the sensitivity is defined with dimensions of current per unit pressure, instead of reciprocal pressure, as in ion gauge tubes. The answer is that the ion sources used in RGAs are sometimes space charge controlled, and their ion current is not linearly proportional to their emission current. Some instruments with high sensitivity use high emission currents, up to 50 mA, but a typical ionizer with a nitrogen sensitivity of 7×10^{-6} A/Pa will have an emission current of 1–5 mA. For an emission current of 1 mA the sensitivity, defined as an ion gauge, would be 7×10^{-3} Pa^{-1}, which is an order of magnitude smaller than that of an ion gauge. The design of the mass analyzer is responsible for the low "ion gauge" sensitivity of the mass analyzer. Not all the ions generated in the ionizer are extracted through the focus electrode. Not all of these extracted ions traverse the mass filter. If we assume an average sensitivity of 10^{-5} A/Pa and a dynamic pressure range of 10^{-1}–10^{-12} Pa, the ion current at the entrance to the detector can range from 10^{-6} to 10^{-17} A. For the upper half of this range a simple Faraday cup detector, followed by a stable, low-noise, high-gain FET amplifier will suffice. Below 10^{-12} A, an electron multiplier is required.

8.1.3.1 Discrete Dynode Electron Multiplier

Figure 8.11 illustrates a typical installation in which high gain is required—a combination Faraday cup–electron multiplier. When the Faraday cup is in operation, the first dynode is grounded to avoid interference. When the electron multiplier is used, the Faraday cup is grounded or connected to a small negative potential to improve the focus of the ions as they make a 90° bend toward the first dynode. In quadrupole analyzers the first dynode is generally located off-axis to avoid X-ray and photon bombardment.

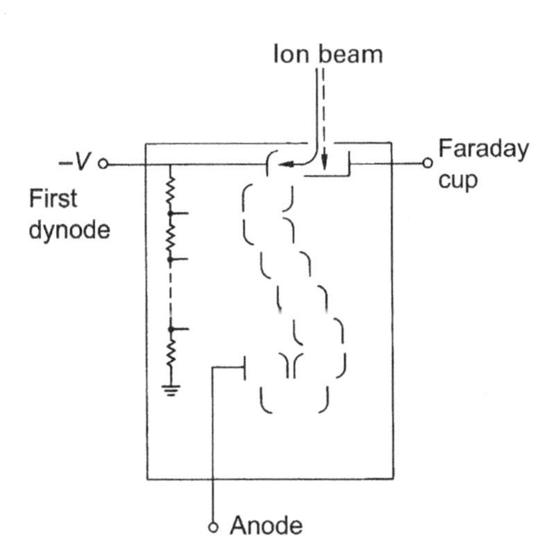

Fig. 8.11 Combination Faraday cup–electron multiplier detector. Reproduced with permission from Uthe Technology Inc.

Amplification in an electron multiplier is achieved when positive ions incident on the first dynode, generate secondary electrons. Secondary electrons are amplified as they collide with each succeeding dynode. Multipliers are operated with a large negative voltage (−1000 to −3000 V) on the first dynode. The gain of the multiplier is given by:

$$G = G_1 G_2^n \tag{8.4}$$

where G_1 is the number of secondary electrons generated on the first dynode per incident ion and G_2 is the number of secondary electrons per incident electron generated on each of the n succeeding dynodes. The values of G_1 and G_2 depend on the material, energy, and nature of the incident ion. For a 16-stage multiplier, whose overall gain $G = 10^6$, $G_2 = 2.37$; for $G = 10^5$, G_2 would be 2.05.

The multiplier sketched in Fig. 8.11 typically uses a Cu–Be alloy, containing 2–4% Be, as the dynode material. When suitably heat-treated to form a beryllium oxide surface, it will have an initial gain as high as 5×10^5. These tubes should be stored under vacuum at all times because a continued accumulation of contamination will cause the gain to decrease slowly. If the tube has been contaminated with vapors, such as halogens or silicone-based pump fluids, the gain will drop below 10^3; at this point, the multiplier must be cleaned or replaced. If the multiplier has had only occasional exposure to air or water vapor and has not been operated at high output currents near saturation for prolonged times, its gain can usually be restored. A typical treatment uses successive ultrasonic cleanings in toluene and acetone, followed by an ethyl alcohol rinse, air drying, and baking in air or oxygen for 30 min at 300°C. Tubes contaminated with silicones cannot be reactivated. Tubes contaminated with chemisorbed hydrocarbons or fluorocarbons often form polymer films on the final stages in which the electron current is great. Such contamination can be quickly removed by plasma ashing.

8.1.3.2 Continuous Dynode Electron Multiplier

A channel electron multiplier is illustrated in Fig. 8.12. The structure consists of a finely drawn PbO–Bi_2O_3 [15] tube, typically 5 mm OD and 1 mm ID. A high voltage is applied between the ends of the tube. The high resistivity of the glass makes it act like a resistor chain, which causes secondary electrons generated by incoming ions to be deflected in the direction of the voltage gradient. The tubes are curved to prevent positive ions generated near the end of the tube from traveling long distances in the reverse direction, gaining a large energy, colliding with the wall, and releasing spurious, out-of-phase secondary electrons [16,17]. If the tube is curved, the ions will collide with a wall before becoming energetic enough to release unwanted secondary electrons. An entrance horn can be provided if a larger entrance aperture is needed. A channel electron multiplier can be operated at pressures up to about 10^{-2} Pa and at temperatures up to 150°C; it has a distinct advantage over the

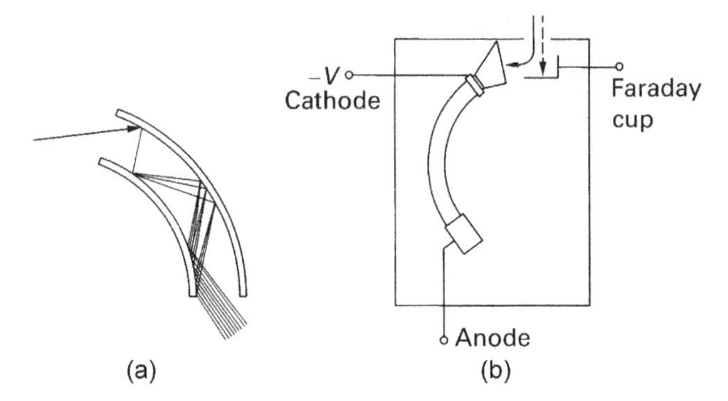

Fig. 8.12 Channeltron® electron multiplier: (a) Schematic detail of capillary; (b) Incorporation into mass analyzer. Reproduced with permission from Galileo Electro-Optics Corp.

Be–Cu-type multiplier in that it is relatively unaffected by long or repeated exposure to atmosphere. Because the secondary emission material is a sub-oxide, its emission is adversely affected by prolonged operation at high oxygen pressures. It seems to be less affected by some contaminants than a Be–Cu multiplier. Under normal operating conditions the gain of either multiplier will degrade in or 2 years to a point at which replacement or rejuvenation is necessary. Little increase in gain will be realized by increasing the voltage on a multiplier whose gain is less than 1000.

Although channel electron multipliers can be exposed to air, they saturate at a lower current than Be–Cu multipliers and cannot be operated with a linear output at high pressures, unless the operating voltage or the emission current is reduced. The manufacturer's literature should be consulted for the precise values of saturation current and range of linearity before any electron multiplier is used. The gain of a multiplier is not a single-valued function. Ions illuminate the first dynode and electrons collide with each succeeding dynode. The gain of the latter dynodes G_2 is dependent on the inter-stage voltage and the dynode material, whereas the gain of the first dynode G_1 is complex. The gain of the first stage is primarily dependent on the material, mass, and the energy of the impinging ion. Ideally, the gain is proportional to $m^{1/2}$, but this is valid only for heavy, low energy ions. The gain of any multiplier is dependent on other factors. Stray magnetic fields can distort the path of the electrons and complex molecules may dissociate on impact, to produce more electrons than a simple compound or element of the same mass. Isotopes and doubly ionized molecules also react somewhat differently, because of the variations in outer electron binding energy [18].

If the gain of a specific electron multiplier needs to be accurately known, a calibration curve of gain versus mass must be experimentally measured for all gases and vapors of interest. The gain of an electron multiplier is never constant. It may be measured by taking the ratio of the currents from the Faraday cup and the electron multiplier output, a tedious process, done only when semiquantitative analysis is required. Even then the gain of the multiplier must be periodically checked with one major gas to account for day-to-day aging of the tube. Only periodic checking of the gain at mass 28 with N_2 is necessary for routine gas analysis.

8.2 Installation and Operation

RGAs may be mounted within high and ultrahigh vacuum chambers as well as process chambers operating below the 10^{-3} Pa range. Standard RGAs may be used to sample the gas in a medium vacuum chamber by extracting gas through an orifice or capillary tube into a low-pressure sample chamber. Alternatively, miniature quadrupoles can be installed directly in chambers that operate at pressures up to 1 Pa (10 mTorr). We first discuss RGA mounting and analysis within high vacuum chambers, and then we discuss gas sampling and miniature instruments for analyzing high-pressure process chambers.

8.2.1 Operation at High Vacuum

RGAs are attached to chambers operating in the molecular flow region for leak detection and monitoring preprocess chamber conditions. They are extensively used for monitoring the partial pressures of background impurities and source vapor fluxes. Mounting and stability are two issues the user needs to understand in order to generate meaningful data.

8.2.1.1 Sensor Mounting

For routine leak detection or background gas analysis, the RGA head is mounted on a port immediately adjacent to the chamber. A valve is usually placed between the head and the process chamber to keep the electron multiplier clean, especially when the chamber is frequently vented to air. This is important with a Be–Cu electron multiplier. The head should be positioned to achieve maximum sensitivity in the volume being monitored. If the instrument is to monitor beams, a line-of-sight view is necessary. It is often worthwhile to mount the head inside the chamber and shield its entrance so that the beam impinges on a small portion of the ionizer. This shielding will reduce the contamination of the ionizer and associated ceramic insulators. It is counterproductive to mount the RGA in remote locations where it samples the chamber ambient through convoluted pathways.

An automated data collection system and printer simplify data analysis. Online graphics are useful for observing the gas "fingerprint" and for leak hunting. The electron multiplier may saturate at high gain, if the pressure is too high, say above

10^{-3} Pa; in this case, one may reduce the electron multiplier voltage or switch to the Faraday cup. High-gain observation of very small signals should be done with low-speed scans, say 1 AMU/s, or the amplifier's long time constant will mask the signal. The ground connection of a turbomolecular pump and other noise sources should be connected to the ground of the most sensitive input of the highest gain amplifier. That point should be connected to the nearest quiet ground.

8.2.1.2 Stability

Stable operation of an RGA in a high vacuum or ultrahigh vacuum chamber is a function of many variables. Some of these variables are related to the instrument design, whereas others are operator- or system-dependent. Stability is affected by materials of filament and ionizer construction, exposure to contaminants during prior use, ion currents and energies, electron-stimulated desorption (ESD), and operation of nearby hot filaments. ESD is as much of a problem in the operation of RGAs as it is in Bayard-Alpert ion gauges. However, the ability to vary the ion energy allows the RGA operator to discern its presence, as the relative sensitivity is a function of ion energy [19]. If the ion energy is reduced and the spectrum changes, then ESD is a concern. The use of silicone-based pump fluids will result in the deposition of insulating films on nearby surfaces. Insulating films and polymer surfaces can store surface charge and alter the energy of the ions entering the mass filter and alter electron orbits. The absence of fluid fragments from the spectrum does not imply the surface is polymer free. Polyphenylether or perfluoropolyether diffusion pump fluids will dissociate as vapors under electron bombardment; they are recommended for use in RGAs and leak detectors.

Ceramic insulators will often exhibit a "memory effect" in which they continuously evolve fragments of hydrocarbons, fluorides, or chlorides after having been exposed to their vapors. This memory effect often confuses analysis when a mass head is moved from one system to another. Ion sources can be cleaned by vacuum firing at 1100°C.

Filaments made from tungsten, thoriated iridium, and rhenium are used in RGAs. Tungsten filaments are recommended for general-purpose work, although they generate copious amounts of CO and CO_2. Rhenium filaments do not consume hydrocarbons. Thoriated iridium filaments are used in oxygen environments, although their emission characteristics are easily changed after contamination by hydrocarbons or halocarbons. As in the ion gauge, the thoriated iridium or "non-burnout" filaments are advantageous when a momentary vacuum loss occurs. Thoriated iridium or rhenium filaments also operate at reduced temperature with less adjacent wall outgassing [20]. Preconditioning the filament by heating it above its normal operating temperature should eliminate less volatile materials [21]. Table 8.2 gives some of the properties of three filament materials as tabulated by Raby [22]. Molecular hydrogen dissociates at a temperature of 1100 K [23]. This can be avoided

Table 8.2 Properties of RGA Filament Materials

Property	ThO_2	W/3% Re	Re
CO production	Unknown	High	High
CO_2 production	High	Moderate	Moderate
O_2 consumption	Low	High	High[a]
H-C consumption	Unknown	High	Low
Water entrapment	Maybe high	Low	Low
Volatility in O_2	Low	High	Very high[b]
Good filament for . . .	Nitrogen oxides	Hydrogen	Hydrogen
	Oxygen	Hydrogen halides	Hydrocarbons
	Sulfur oxides	Halogens	Hydrogen halides
		Halocarbons[c]	Halocarbons[c]
			Halogens

[a] Loss of one filament caused O_2/N_2 ratio to increase 17.6%.
[b] Exposure to air at 10^{-4} Pa caused failures of the filament pair at 30 and 95 h.
[c] Freons, etc.
Source: Reproduced with permission from Uthe Technology Inc.

by the use of a lanthanum hexaboride filament operating at 1000 K [24]. Outgassing of nearby walls can be reduced by the use of EX-processed aluminum [25], or gold-coated copper [26], because they are excellent thermal conductors.

Bayard-Alpert ion gauge tubes should be turned off when the RGA is operated. They can produce ion- and electron-stimulated desorption and increase thermally stimulated outgassing of nearby walls. In the ultrahigh vacuum region, inert gas ions may be pumped, whereas hydrogen and $M/z = 16$, 28 will be desorbed by electron bombardment [27]. Ion-stimulated desorption from the extraction plate has been observed; it was shown that the amount of desorption is gas- pressure- and ion-energy dependent [28]. Maintaining a positive filament potential with respect to ground has been shown to reduce outgassing rates of RGAs [29].

Mass heads and pressure gauges should be baked at the same or higher temperature than the chamber, and they should be cooled slowly to prevent the condensation of vapors. The maximum baking temperature will affect electron multiplier choice. Detectors with channel electron multipliers can be baked to 320°C and operated at temperatures up to 150°C. Detectors using a Be–Cu multiplier can usually be baked to about 400°C and operated at temperatures up to 185°C. These temperature limitations are due to the materials used in channel electron structures and the glass-encapsulated resistors in Be–Cu multiplier chains.

8.2.2 Operation at Medium and Low Vacuum

At pressures greater than ~10^{-3} Pa (10^{-5} Torr), the standard RGA design will not function. An APIMS [30], in which the gas is field ionized at high pressures, can sample impurities in process gas streams to the level of parts per trillion of one atmosphere (~10^{-7} Pa). This analytical laboratory technique is only suitable for specialized applications. For routine analysis of plasma systems at pressures as high as 1 Pa (10^{-2} Torr), one may sample the gas stream by means of a pressure reduction scheme or use a miniature quadrupole. Miniature quadrupoles have dimensions of order of the mean free path. Differentially pumped gas analysis and miniature quadrupoles are useful techniques for determining the time constants of contaminant decay during preprocessing and for sampling the background gases during processing.

8.2.2.1 Differentially Pumped Analysis

Analysis of gas in a plasma etching or deposition chamber can be done by differentially pumping the source gas. Honig [31] has given three primary conditions for gas analysis with a differentially pumped RGA: (1) The beam intensity should be directly proportional to the pressure of the gas in the sample chamber but should be independent of its molecular weight. (2) In a gas mixture the presence of one component should not affect the peaks due to another component. (3) The gas flow through the orifice should be constant during the scan. These conditions are satisfied if there is good mixing in the chamber, the leaks are molecular, and the pump is throttled.

Unfortunately, it is not always possible to meet these conditions in all applications. Many references state that the superior method of sampling from high pressures is a two-stage reduction scheme, where the gas is first drawn through a long capillary tube to an intermediate chamber. A mechanical pump attached to this chamber removes most of the gas. A sample of that gas is then further pressure divided in the molecular flow regime. See Fig. 8.13. Fractionation can occur in

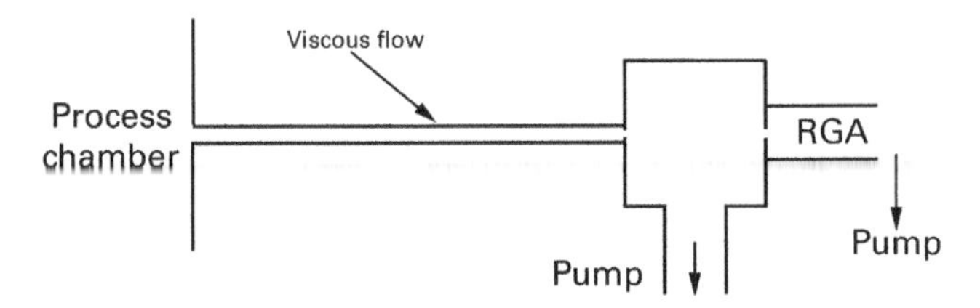

Figure 8.13 Analysis of a differentially pumped residual gas analyzer using a long viscous flow capillary sampling tube, followed by a molecular flow pressure divider. G.B. Bunyard [32]/Reproduced with permission from Elsevier.

either of these pressure dividers. Consider first the case of a short capillary not meeting the requirements of Poiseuille flow. Such a case was illustrated in Fig. 3.2. In that case, we showed the net flow was determined by equating Poiseuille flow in the tube to choked flow in the orifice. Poiseuille flow depends on viscosity and molecular diameter, whereas choked flow depends on the nonlinear ratio of specific heats. Contrary to popular belief, gas mixtures will fractionate. Consider next a long capillary tube connecting the high-pressure chamber to the intermediate chamber, as described in Fig. 8.13; flow in this tube is governed by (3.12). Fractionation of a gas mixture can result at the exit during expansion. Gas mixtures will not fractionate in the limit of small Knudsen numbers. If the flow is in the transition region, the mixture will be significantly dependent on the ratio of gases; baro-diffusion and the diffusion baro effect are small but they do depend on the Knudsen number in viscous flow [33].

When sampling in molecular flow, the relative size of the pump and second orifice must be considered. Examine the model of the differentially pumped system sketched in Fig. 8.14 that operates in molecular flow. In this model, P_c is the pressure in the sputtering chamber, P_s is the pressure in the spectrometer chamber, and P_p is the pressure in the auxiliary pump whose speed is S_p. Connecting areas with conductance values C_1 and C_2, which connect these chambers, are schematically shown as capillary tubes, but in practice, C_1 may be a closed ion source or myriad small holes milled in a thin plate. The gas flow through the auxiliary system, which is everywhere the same, leads to the following equation:

$$C_1\left(P_c-P_s\right)=C_2\left(P_s-P_p\right)=S_pP_p \tag{8.5}$$

When P_p is eliminated, we obtain:

$$P_s=\frac{P_c}{1+\left(\frac{C_2}{C_1}\right)\left(\frac{S_p}{S_p+C_2}\right)} \tag{8.6}$$

Now consider two cases: Case 1: Let the conductance C_2 be much larger than the speed of the pump. The auxiliary pump is located in or immediately adjacent

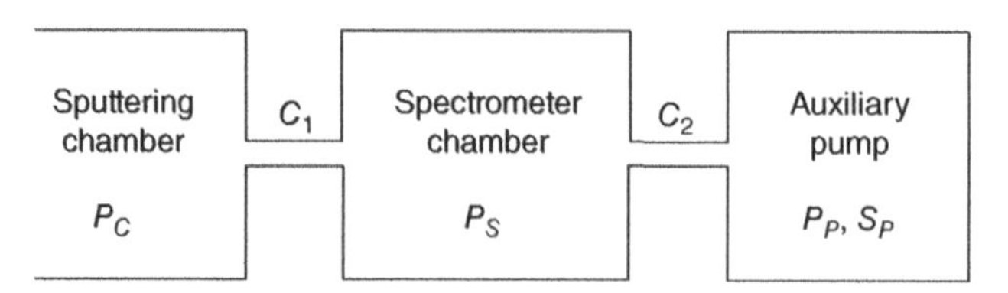

Fig. 8.14 Analysis of a differentially pumped residual gas analyzer using a molecular flow pressure divider.

to the spectrometer chamber with no interconnecting conductance. Equation 8.6 then reduces to:

$$P_s = \frac{P_c}{1+(S_p/C_1)} \tag{8.7}$$

The conductance C_1 has a mass dependence that varies as $1/m^{1/2}$, whereas the pumping speed has a mass dependence that is a function of the pump type. For example, an ion pump has different speeds for noble and reactive gases. Diffusion pump speeds increase for light gases, but not as rapidly as $1/m^{1/2}$. The important result is that the ratio of the gas pressure in the spectrometer to the pressure in the chamber P_s/P_c is mass-dependent when the auxiliary pump is appended directly to the chamber. Gases will not exit the spectrometer chamber in the same proportion as they exit the sputtering chamber. Honig's first criterion will not be satisfied.

Case 2 considers the situation in which a small conductance C_2 is placed between the spectrometer and pump, such that $C_2 \ll S_p$. Then Eq. (8.6) becomes:

$$P_s = \frac{P_c}{1+(C_2/C_1)} \tag{8.8}$$

Light gases will still pass from the sputtering chamber to the spectrometer chamber more rapidly than heavy gases, but also, they will exit to the pump. Stated another way, the mass dependence of C_1 and C_2 are the same and cancel each other so that Eq. (8.8) is mass independent. Therefore, to sample the ratios of gases in the sputter chamber accurately, the auxiliary pump should be throttled to about 1/10 of its speed. If this is done, the pressure ratio of two gases in the chamber, P_{ac}/P_{bc} will be the same as their ratio in the spectrometer chamber P_{as}/P_{bs}.

The flow ratio of two gases passing through an orifice in molecular flow is given by Eq. (2.20); this implies that the gas composition in the sputtering chamber would eventually change with time. However, the flow to the sampling system is so small that it cannot affect the process flow. Sullivan and Busser [34] noted that the time constant of the spectrometer chamber (V_s/C_2) must be equal to or less than the reaction time of the process to obtain accurate time-dependent information concerning the process.

When C_1 in Fig. 8.14 is a closed ion source, the gas pressure in the ionizer can be as high as 100 times that of the spectrometer chamber and this further increases the sensitivity for impurity detection. Using a closed ion source, one can detect ~100 ppm for H_2O and 200 ppb for other gases.

The background gases in the chamber should first be analyzed while the chamber is at its ultimate pressure, in order to ensure there are no leaks. One should save a background spectrum for later comparison. After initiating the process gas flow to the chamber, a process flow background is recorded. Next a process spectrum is taken with the discharge operating. The large pressure reduction factor and ever-present background gases reduce the detection limit of the RGA, unless a closed ion source is used.

8.2.2.2 Miniature Quadrupoles

As early as 1973, Visser [35] demonstrated that miniature quadrupoles designed to operate at short mean free path could be operated directly in a process chamber at pressures up to 0.5 Pa. Commercial instruments have been constructed by scaling standard quadrupoles to operate at pressures as high as 1 Pa (10 mTorr) without any pressure reduction stages [36]. Gas scattering at pressures above 0.1 Pa required internal correction for the (nonlinear) sensitivity. Ion–molecule reactions occur in all instruments in which ions can collide with molecules and the collision probability increases linearly with pressure. Such reactions are observable downstream from the corona discharge ionizer in an APIMS, as well as in miniature quadrupoles when operated at pressures above 0.1 Pa. For example, N_3^+ can be formed from a collision between N_2^+ and N_2. In turn, the N_3^+ can react with O_2 to form NO^+ or NO_2^+. See Section 9.1.5.

8.3 Calibration

Calibration of an RGA is done for the purpose of checking the linearity over a wide pressure range, measuring the sensitivity at important M/z values, and verifying the electron multiplier gain. The gain of an electron multiplier can be checked quite simply by taking the ratio of an ion current at the electron multiplier exit to the electron current at the Faraday cup. The instrument sensitivity and linearity can be verified by direct comparison to a transfer standard, by use of a pressure dividing calibration chamber, or by use of an orifice flow system. Detailed descriptions of these techniques and test chambers are given in the AVS Recommended Practice for Calibration of Mass Spectrometers and Partial Pressure Analyzers [14].

Many users wish to have a quick, *in situ* method for verifying the RGA sensitivity and pumping speed. The pulse injection method was developed by Kendall [37]. In its original form, a small quantity of gas was isolated and injected into the vacuum system by means of a piston with two closely spaced O-ring gaskets. Alternatively, a fuel injector was used to inject a series of extremely short pulses. The gas pulse (P_iV_i) can be approximated by isolating a small quantity of gas in a small pipe between two valves. This requires knowledge of the small volume V_i, and its pressure P_i. Knowing the chamber volume V_c one can calculate the instantaneous pressure rise of the chamber $P_c = (P_iV_i/V_c)$. The RGA sensitivity is found by dividing I_p the peak ion current at $t = 0$ by P_c. The sensitivity is $S' = I_pV_c/(P_IV_I)$. Figure 8.15 illustrates the instantaneous peaks and decay rates of four separate

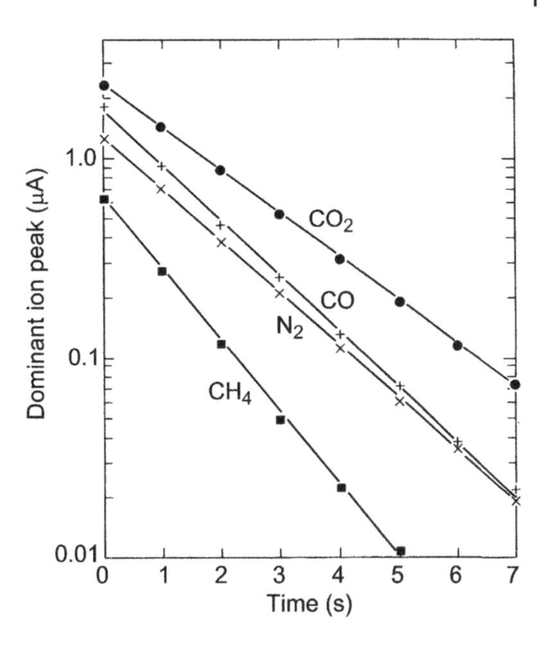

Fig. 8.15 Measurement of the sensitivity and pumping speeds of four gases using the pulsed injection technique. J.F. O'Hanlon [38]/Reproduced with permission from AIP Publishing.

gases. In addition to determining the RGA sensitivity, one can determine the pumping speed for each test gas at the location of the RGA sensor. Since the ion current decays to a value of $1/e$ in time τ, the system time constant can be determined graphically from a plot like that given in Fig. 8.15. The time required for the ion current to decay to a value of I_p/e can be determined from the semilog plot of log I versus linear t. Pumping speeds for individual gases are given by $S_i = V_c/\tau_i$. The advantage of this method is that the data are taken on the system and are representative of the system.

8.4 Choosing an Instrument

The instrument chosen for residual gas analysis work must be simple, reliable and have adequate resolution, adequate sensitivity, and mass range. Numerous commercial instruments are available, ranging from elementary sector or quadrupole, with an M/z range of 1–50, a Faraday cup detector, and the capability to resolve 1 AMU to those with M/z extending to 800 AMU. They differ in complexity and cost. Today, quadrupole instruments are most commonly used.

For simple monitoring of background gases in a cycled, unbaked high vacuum chamber, the simplest of instruments with a range of $M/z = 1$–50 is adequate. With such an instrument, the dominant fixed gases up to $M/z = 44$ and hydrocarbons at $M/z = 39$, 41, and 43 may be monitored. Units with Faraday cup detectors and partial pressure limits of 10^{-8} Pa are commercially available. Most have provisions for alarms to alert the operator when the ion current at a particular M/z value has been exceeded.

Detailed residual gas analysis with a sector or quadrupole requires increased sensitivity and mass range. Partial pressures of 10^{-12} Pa (10^{-14} Torr) are detectable with an electron multiplier. An M/z range of at least 1–80 is necessary to distinguish pump oils, but some solvents such as xylene have major peaks in the M/z range of 104–106. An M/z range of 1–200 permits identification of many heavy solvents. The ability to detect partial pressures in the 10^{-10}–10^{-12} Pa (10^{-12}–10^{-14} Torr) range is needed to leak check, analyze the background of ultra-high vacuum systems, and to do serious semiquantitative analysis. An electron multiplier is also necessary for differentially pumped systems. Instruments for these purposes need both Faraday cup and electron multiplier to measure the gain and operate over a wide pressure range.

Analysis of gases in medium and low vacuum pressures can be done with either a closed-source ionizer or a miniature quadrupole. The spectrum from a closed source instrument can be interpreted in a straightforward manner, whereas miniature quadrupoles operated at high pressure will always produce peaks generated by ion–molecule reactions.

Microprocessor-based instruments offer convenience and ease of storing and manipulating large quantities of data and interpretive displays. The operator convenience and databases are quite useful, but they cannot overcome the influences of instrument design, instrument history, cleanliness, and operating parameter adjustments, in particular, they cannot compensate for instabilities caused by ESD. Ignorance of these effects can seduce the user into believing that the data are representative. Partial pressure equivalents derived from ion currents are not meaningful unless the instrument has been recently calibrated.

References

1 Dempster, A.J., *Phys. Rev.* **11**, 316 (1918).
2 Nier, A.O.C., *Rev. Sci. Instrum.* **11**, 212 (1940).
3 White, F.A., *Mass Spectrometry in Science and Technology*, Wiley, New York, 1968, p. 11.
4 Nerkin, A., *J. Vac. Sci. Technol., A* **9**, 2036 (1992).
5 Caswell, H.L., *J. Appl. Phys.* **32**, 105 (1961).
6 Caswell, H.L., *J. Appl. Phys.* **32**, 2641 (1961).
7 Caswell, H.L., *IBM J. Res. Dev.* **4**, 130 (1960).
8 Fite, W.L. and Irving, P., *J. Vac. Sci. Technol.* **11**, 351 (1974).
9 Rapp, D. and Englander-Golden, P., *J. Chem. Phys.* **43**, 1464 (1965).

10 Paul, W. and Steinwedel, H., *Zeitschrift für Naturforsch* **8A**, 448 (1953).
11 Brownstein, B.S., Fraser, D.B., and O'Hanlon, J.F., *J. Vac. Sci. Technol., A* **11**, 694 (1993).
12 M. Mosharrafa, *Industrial Research/Development*, March 1970, p. 24.
13 Dawson, P.H., Ed., *Quadrupole Mass Spectrometry and Its Applications*, Elsevier, New York, 1976.
14 Basford, J.A., Boeckman, M.D., Ellefson, R.E., Filippelli, A.R., Holkeboer, D.H., Lieszkovszky, L., and Stupak, C.M., *J. Vac. Sci. Technol., A* **11**, A22 (1993).
15 H.M. Smith, R.R. Thompson, and B. Deragoorian, U.S. Pat. #3,492,523., assigned to Bendix Corp. (issued January 27, 1970).
16 Timothy, J.G. and Bybee, R.L., *Rev. Sci. Instrum.* **48**, 292 (1977).
17 Kurz, E.A., *Am. Lab.* **11**, 67 (1979).
18 M.G. Inghram and R.J. Hayden, *A Handbook on Mass Spectroscopy: Nuclear Science Series Report No. 14*, National Research Council Publication No. 311, National Academy of Science, Washington, DC, 1954.
19 Lieszkovszky, L. and Filippelli, A., *J. Vac. Sci. Technol., A* **8**, 3838 (1990).
20 Kurokouchi, S., Watanabe, S., and Kato, S., *Vacuum* **47**, 763 (1996).
21 Batey, J.H., *Vacuum* **43**, 15 (1992).
22 B. Raby, UTI Technical Note 7301, May 18, 1977.
23 Hickmott, T.W., *J. Vac. Sci. Technol.* **2**, 257 (1965).
24 Hickmott, T.W., *J. Chem. Phys.* **32**, 810 (1960).
25 Ishimaru, H., *J. Vac. Sci. Technol., A* **7**, 2439 (1989).
26 Watanabe, F., *J. Vac. Sci. Technol., A* **11**, 1620 (1993).
27 Rozgonyi, G.A. and Sosniak, J., *Vacuum* **18**, 1 (1968).
28 Müller, N., *Vacuum* **44**, 623 (1993).
29 Kurokouchi, S. and Kato, S., *J. Vac. Sci. Technol., A* **19**, 2820 (2001).
30 Siefering, K., Berger, H., and Whitlock, W., *J. Vac. Sci. Technol., A* **11**, 1593 (1993).
31 Honig, R.E., *J. Appl. Phys.* **16**, 646 (1945).
32 Bunyard, G.B., *Vacuum* **44**, 633–638 (1993).
33 Sharipov, F. and Kalempa, D., *J. Vac. Sci. Technol., A* **20**, 814 (2002).
34 Sullivan, J.J. and Busser, R.G., *J. Vac. Sci. Technol.* **6**, 103 (1969).
35 Visser, J., *J. Vac. Sci. Technol.* **10**, 464 (1973).
36 Holkeboer, D.H., Karndy, T.L., Currier, F.C., Frees, L.C., and Ellefson, R.E., *J. Vac. Sci. Technol., A* **16**, 1157 (1998).
37 Kendall, B.R.F., *J. Vac. Sci. Technol., A* **5**, 143 (1987).
38 O'Hanlon, J.F., *J. Vac. Sci. Technol., A* **9**, 1 (1991).

9

Interpretation of RGA Data

The method of interpreting mass scans like that shown in Fig. 8.10, is based on detailed knowledge of the cracking or fragmentation patterns of gases and vapors found in the system. After becoming familiar with an RGA, qualitative analysis of many major constituents is rather straightforward, whereas precise quantitative analysis requires careful calibration and complex analysis techniques. This section discusses cracking patterns, some rules of qualitative analysis, and methods of determining the approximate partial pressures of the gases and vapors in the spectrum.

9.1 Cracking Patterns

When molecules of a gas or vapor are struck by electrons whose energy can cause ionization, fragments of several mass-to-charge ratios are created. The mass-to-charge values are unique for each gas species, whereas the peak amplitudes are dependent on the gas and instrumental conditions. This pattern of fragments, called a cracking pattern, forms a fingerprint that may be used for absolute identification of a gas or vapor. The various peaks are primarily created by dissociative ionization, isotopic mass differences, and multiple ionization. In an ionizer operating at pressures greater than 0.1 Pa (10^{-3} Torr), ions and molecules may react to form new ionized radicals.

9.1.1 Dissociative Ionization

The cracking pattern of methane CH_4, illustrated in Fig. 9.1, shows, in addition to the parent ion CH_4^+, the electron dissociation of the molecule into lighter fragments, CH_3^+, CH_2^+, CH^+, C^+, H_2^+, and H^+. Fragments containing ^{13}C are not shown. Except for H_2^+, fragments shown in this cracking pattern were produced by dissociative ionization.

A Users Guide to Vacuum Technology, Fourth Edition. John F. O'Hanlon and Timothy A. Gessert.
© 2024 John Wiley & Sons, Inc. Published 2024 by John Wiley & Sons, Inc.

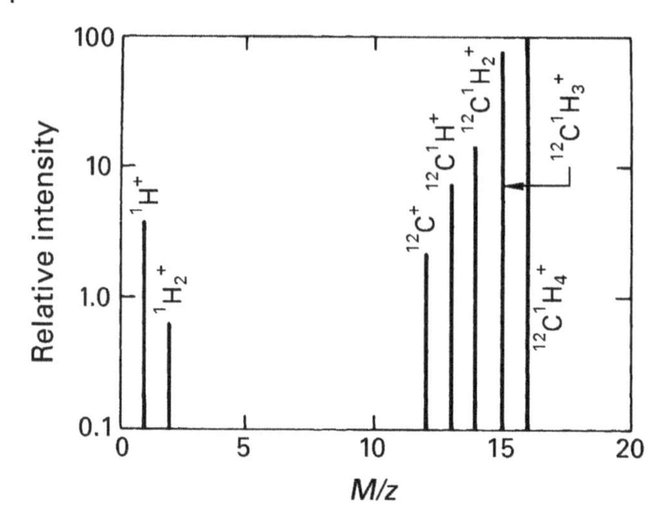

Fig. 9.1 This cracking pattern of methane illustrates the five largest dissociative ionization peaks.

9.1.2 Isotopes

Figure 9.2 illustrates several isotopic peaks in the mass spectrum of singly ionized argon. By comparing the peak heights of the isotopes with the relative isotopic abundance given in Appendix D, we see that this is a spectrum of naturally occurring argon. The relative isotopic peak heights observed on the RGA will generally mirror those given in Appendix D unless the source was enriched, or the sensitivity of the RGA was not constant over the isotopic mass range. Some compressed gas cylinders may contain gas in which a rare component has been selectively removed.

9.1.3 Multiple Ionization

Higher degrees of ionization are also visible in the argon spectrum. Argon has three isotopes of masses 36, 38, and 40; the doubly ionized peaks Ar^{++} show up at $M/z = 18$, 19, and 20. The cracking patterns of heavy metals used in filaments (W, Re, Mo, and Ir) may show triply ionized states.

9.1.4 Combined Effects

The cracking pattern of a somewhat more complex molecule CO, given in Appendix E.2, is illustrated in Fig. 9.3 to show the combined effects of isotopes, dissociation, and double ionization. The amplitude of the largest line $^{12}C^{16}O^+$ has been normalized to 100, while the amplitude of the weakest line shown, $^{16}O^{++}$, is 10^6 times smaller. The spectrum, as sketched in Fig. 9.3, can be observed only

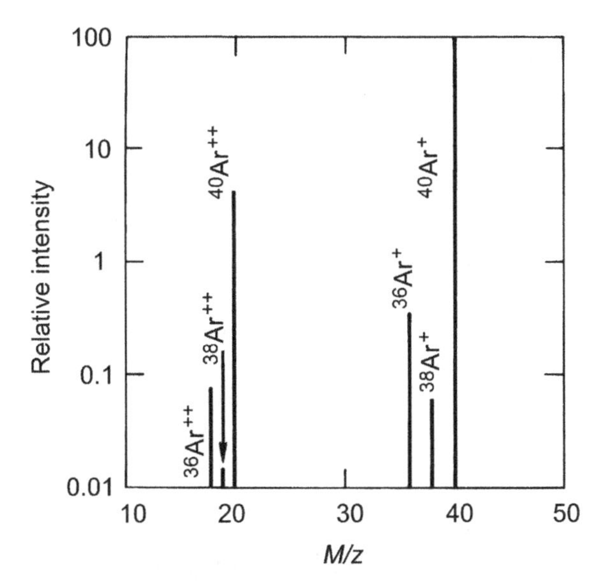

Fig. 9.2 The argon cracking pattern illustrates isotopic mass differences (three isotopes) and two degrees of ionization (single and double).

Fig. 9.3 The cracking pattern of CO contains peaks due to dissociative ionization, isotopic ionization, and double ionization.

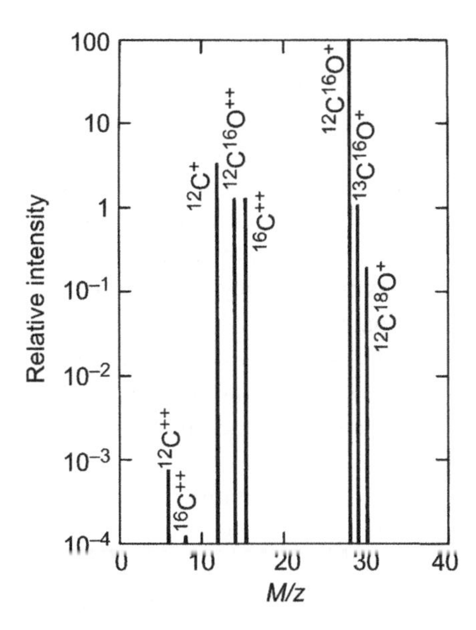

under carefully controlled laboratory conditions. During normal operation of an RGA the spectrum will be cluttered with other gases and only the four or five most intense peaks will be identifiable as a part of the CO spectrum. Even then, some of the peaks will overlap the cracking patterns of other gases. Under the assumption that this clean spectrum has been obtained, it is instructive to classify the eight lines according to their origin. The main peak at $M/z=28$ is due to the single ionization of the dominant isotope $^{12}C^{16}O$. Energetic electrons decompose some of these molecules into two other fragments $^{12}C^{+}$ and $^{16}O^{+}$. Because carbon has two isotopes and oxygen three, several isotopic combinations are possible. The two most intense lines are $^{13}C^{16}O^{+}$ and $^{12}C^{18}O^{+}$. Isotopic fragments such as $^{13}C^{+}$ and $^{18}O^{+}$ are too weak to be seen here. Other complexes—for example, $^{12}C^{16}O^{++}$ ($M/z=14$), $^{16}O^{++}$ ($M/z=8$), and $^{12}C^{++}$ ($M/z=6$)—have high enough concentrations to be seen.

The relative amplitudes of the eight lines of CO are determined by the source gas and instrument. If isotopically pure source gas were used (^{12}C and ^{16}O), only six peaks would be found. Changes in the operating conditions of the instrument also affect the relative peak heights. The ion temperature and electron energy affect the probability of dissociation and the formation of higher ionization states. Dissociation of a molecule into fragments also changes the kinetic energy of the ion fragment. This can be a serious problem in a magnetic sector instrument because the kinetic energy of the ion directly affects the focusing and dispersion of the instrument. Quadrupoles can focus ions with a greater range of initial energies and therefore are less sensitive to transmission losses of fragment ions than sector instruments.

Electron- and ion-stimulated desorption of atoms from surfaces [1,2] can also add to the complexity of the cracking pattern. Oxygen, fluorine, chlorine, sodium, and potassium are some of the atoms that can be released from surfaces by energetic electron bombardment. In a magnetic sector, instrument energetic oxygen or fluorine will often occur at fractional mass numbers of mass 16-1/3 and 19-1/3, respectively [3]. These peaks represent gas phase oxygen and fluorine that dissociated on adsorption, then desorbed from the walls. They occur at fractional mass numbers because they are formed at energies different from those corresponding to the ionized states of free molecules [1] and leave the surface with some kinetic energy [3]. The generation of gases resulting from the decarburization of tungsten filaments can add other spurious peaks to the cracking pattern.

Representative cracking patterns are given in the appendixes. Appendixes E.2 and E.3 give the cracking patterns of some common gases and vapors, and Appendix E.4 contains the patterns of frequently encountered solvents. Appendix E.5 describes the patterns of gases used in semiconductor processing and partial cracking patterns of six pump fluids are given in Appendix E.1. The patterns in the appendices are intended to be representative of the substances; they are not unique. It cannot be emphasized too strongly that each pattern is quantitatively meaningful only to those who use the same instrument under identical operating conditions. Nonetheless, there are enough similarities to warrant tabulation.

9.1.5 Ion–Molecule Reactions

At pressures greater than 0.1 Pa (10^{-3} Torr), ion–molecule reactions will be large enough to be detectable. Ion–molecule reactions are visible in miniature quadrupoles operating at high pressure because gas–gas collisions are linearly proportional to pressure. Ar_2^+ formed by Ar^+ and Ar gas collisions has an abundance of 500 ppm at 1 Pa and is within the operating region of a physical vapor deposition system. A standard RGA typically operates at pressures <0.01 Pa, so the Ar dimer abundance predicted by the equation in Table 9.1 is only 5 ppm and is not normally observed in the spectrum. Table 9.1 shows representative ion–molecule reactions generated by collisions between of argon, nitrogen, and wet or dry air. Spectra from air will yield NO. If oxygen is present, it dominates the reaction with nitrogen to produce NO; however, if water vapor is present instead of O_2, it will react with nitrogen to produce NO. Charge exchange reactions ($X + Y^+ \rightarrow X^+ + Y$) can cause the apparent or observed ion ratios to differ from their actual concentration ratios. These and other reaction products can be as large as parts-per-thousand at a pressure of 1 Pa. One should understand

Table 9.1 Common Ion–Molecule Reactions

Plasma Constituents Ion–Molecule Reaction(s)	Parent Ion	Reaction Product Ion	Reaction/Parent Ratio Pres. Dependence[a]
Argon:			
$Ar^+ + Ar \rightarrow Ar_2^+$	Ar^+	Ar_2^+	5×10^{-4} *P*(Ar)
Dry N_2 (300 ppm H_2O)			
$N_2^{+*} + N_2 \rightarrow N_3^+ + N$	N_2^+	N_3^+	8×10^{-4} *P*(N_2)
$N_2^{+*} + H_2O \rightarrow NO^+ + H_2 + N$	N_2^+	NO^+	8×10^{-5} *P*(N_2)
Dry Air (400 ppm H_2O):			
$N_2^{+*} + O_2 \rightarrow NO^+ + NO$			
$N_2^{+*} + N_2 \rightarrow N_3^+ + N$			
$N_3^+ + O_2 \rightarrow NO^+ + O + N_2$	N_2^+	NO^+	2.1×10^{-3} *P*(N_2)
$\rightarrow NO_2^+ + N_2$	N_2^+	N_3^+	5.3×10^{-4} *P*(N_2)
$N_2^{+*} + H_2O \rightarrow NO^+ + H_2 + N$			
$N_2^{+*} + H_2O \rightarrow NO^+ + H_2 + N$			
Wet Air (2.7% H_2O):			
(same reactions as for dry air, above)	N_2^+ N_2^+	NO^+ N_3^+	2.2×10^{-3} *P*(N_2) 5.1×10^{-4} *P*(N_2)

[a] As taken on an INFICON Transpector XPR2 miniature quadrupole mass spectrometer.
Source: Reproduced with permission from INFICON AG.

that ion–molecule products would be generated in any ionizer operating at high pressure, e.g., mini-quadrupole, APIMS, standard closed-source ionizer, or micromechanical mass spectrometer.

9.2 Qualitative Analysis

Perhaps the most important aspect of the analysis of spectra for the typical user of an RGA is qualitative analysis; that is, the determination of the types of gas and vapor in the vacuum system. In many cases the existence of a particular molecule points the way to fixing a leak or correcting a process step. The quantitative value or partial pressure of the molecular species in question usually does not need to be known because industrial process control is frequently done empirically. The level of a contaminant—for example, water vapor, which will cause the process to fail—is determined experimentally by monitoring the quality of the product. An inexpensive RGA tuned to the mass of the offending vapor is then used to indicate when the vapor has exceeded a predetermined partial pressure. With experience many gases, vapors, residues of cleaning solvents, and traces of pumping fluids will be recognizable without much difficulty.

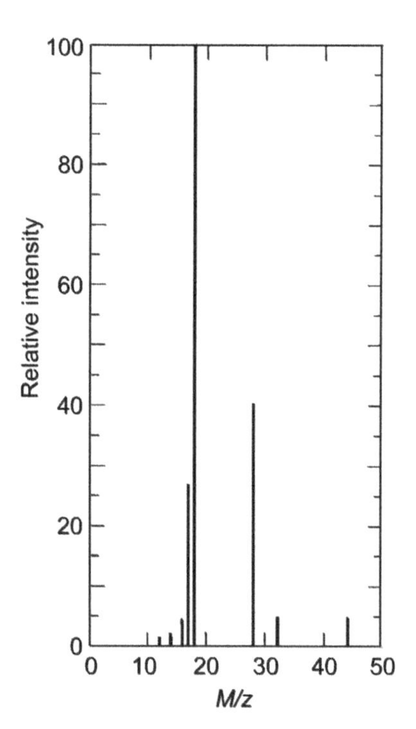

Fig. 9.4 Background spectrum constructed from a (20:4:4:1:1) mixture of H_2O, N_2, CO, O_2, and CO_2.

Mass spectra contain considerable information about the present condition as well as the history of the systems on which they were recorded. To help in their interpretation examine the hypothetical spectrum in Fig 9.4, which was artificially constructed from five gases H_2O, N_2, CO, O_2, and CO_2 in the ratio (20:4:4:1:1) from the cracking patterns in Appendix E. Let us study this pattern under the assumption that its origin and composition are not known. Examination of the cracking patterns of common gases, like those tabulated in Appendix E, quickly verifies the presence of carbon dioxide, oxygen, and water vapor. Notice that the mass 32 peak is not due to the dissociation of carbon dioxide. The presence of oxygen at

32 AMU usually indicates an air leak unless it is being intentionally introduced. Analysis of the mass 28 peak is not so clear. Some of this peak is certainly due to the dissociation of CO_2; however, if we assume that the sensitivity of the instrument is reasonably constant over the mass range in question, that contribution cannot be very great. The majority of the peak amplitude would then be attributable to N_2 or CO or both. To distinguish these gases further the amplitudes of the peaks at $M/z=16$ (O^+), $M/z=14$ (N^+, N_2^{++}, CO^{++}), and $M/z=12$ (C^+) are examined. In practice it is difficult to conclude much from the presence of carbon because it originates from so many sources, both organic and inorganic. The $M/z=14$ peak is largely due to N^+. Therefore, nitrogen is present. Analysis of the $M/z=16$ peak is complicated by the fact that there are other sources of atomic oxygen beside CO, namely O_2 and CO_2, as well as electron-stimulated desorption of O^+ from the walls. Oxygen desorption is a common phenomenon. Referring to the cracking pattern tables, we see that the $M/z=16$ peak looks too large to be accounted for totally by the dissociation of CO_2 and O_2. The $M/z=14$ peak looks too small to be only a fragment of nitrogen. We then conclude that both nitrogen and carbon monoxide are present but in undetermined amounts.

After some familiarity with the combined effects of these common background gases, they should be easily identifiable in the spectrum shown in Fig. 8.10. The large peak is located at 18 AMU, and is due to water vapor, which is typically the dominant peak in an unbaked vacuum system. This spectrum also shows organic fragments. One way to determine the nature of the organic in the system qualitatively is to become familiar with the cracking patterns of commonly used solvents, pump fluids, and elastomers.

The cracking patterns of several solvents are listed in Appendix E.4. A common characteristic of organic molecules is their high probability of fragmentation in a 70-eV ionizer. It is so great that the parent peak is rarely the most intense peak in the spectrum. The lighter solvents such as ethanol, isopropyl, and acetone have fragment peaks bracketing nitrogen and carbon monoxide at $M/z=27$ and 29, but each has a prominent peak at a different mass number. Methanol and ethanol each have a major fragment at $M/z=31$; they can be distinguished by the methanol's absence of a fragment at $M/z=27$. Solvents that contain fluorine or chlorine have characteristic fragments at $M/z=19$, 20, 35, 36, 37, and 38, respectively. The extra fragments at $M/z=20$ (HF) and 35–38 (HCl) seem to be present, whether or not the solvent contains hydrogen. Fragments due to CF, CCl, CF_2, and CCl_2 are also characteristic of these compounds. As with all fragments, their relative amplitudes will vary with instrumental conditions. The fragments at $M/z=27$, 29, 31, 41, and 43 are prominent in these solvents. They are also common fragments of many pump fluids, which further complicates the interpretation of a spectrum.

Appendix E.1 lists the partial cracking patterns of six common pump fluids. All the fragment peaks up to mass number 135 are tabulated. For the sector instrument the largest peak (100%) often occurs at a higher mass number and therefore is not shown. The complete spectra of the four fluids, which were taken on the sector instrument, were tabulated by Wood and Roenigk [4]. Most organic pump fluid molecules are quite heavy, and the parent peaks are not often seen in the system because only the lighter fragments backstream through a properly operated trap. Saturated straight-chain hydrocarbon oils are characterized by groups of fragment peaks centered 14 mass units apart and coincide with the number of carbon atoms in the chain. Figure 9.1 shows the fragment peaks for group C_1 in the mass range 12–16. Higher carbon groups have similar characteristic arrangements; C_2 (M/z=24–30), C_3 (M/z=36–44), C_4 (M/z=48–58), C_5 (M/z=60–72), and so on. The spectrum of Apiezon BW diffusion pump oil taken by Craig and Harden [5], shown in Fig. 9.5, illustrates these fragment clusters. In most hydrocarbon oils the fragments at M/z=39, 41, 43, 55, 57, 67, and 69 are notably stable, and their presence is a guarantee of hydrocarbon contamination. These odd-numbered peaks, which are more intense than the even-numbered peaks in straight-chain hydrocarbons, are visible in the scans shown in Figs. 8.10 and 9.5. Hydrocarbon oils are used in most rotary mechanical pumps, except those for pumping oxygen or corrosive gases, and in many diffusion pumps. Hydrocarbon diffusion pump oils are not resistant to oxidation and will decompose when exposed to air while heated. Their continued popularity for certain applications is due to their low cost.

Silicones are an important class of diffusion pump fluids. They have very low vapor pressures and also have extremely high oxidation resistance [6]. Cracking

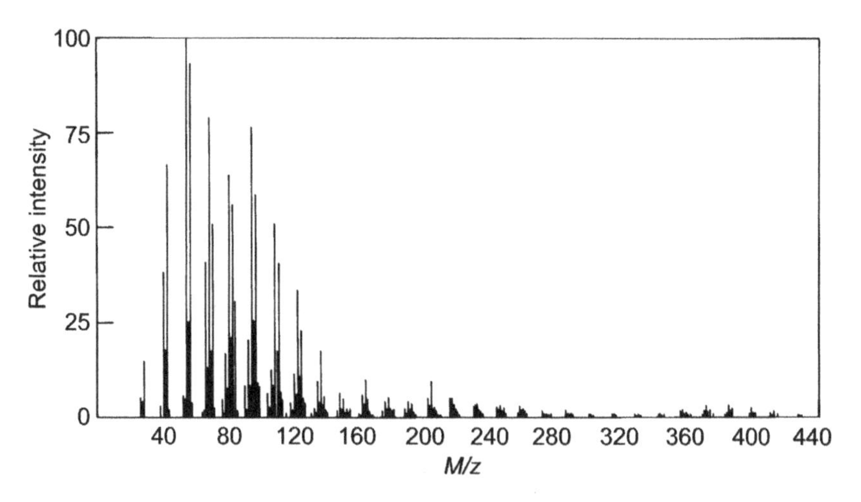

Fig. 9.5 Mass spectrum of Apiezon BW oil (obtained using MS9 sector spectrometer), source temperature 170°C. R.D. Craig and E.H. Harden [5]/Reproduced with permission from Elsevier.

patterns for Dow Corning DC-704 and DC-705, given in Appendix E.1, show many of the characteristic fragments of benzene. By way of illustration, a partial spectrum taken in a contaminated diffusion pumped system using DC-705 fluid is compared with the cracking pattern of benzene in Fig. 9.6. The peaks labeled M are due to mechanical pump oil. Systems that have been contaminated with a silicone pump fluid will always show the characteristic groups at $M/z = 77$ and 78 and usually those at $M/z = 50$, 51, and 52. Notice the lack of these peaks in Fig 8.10, which was taken on a system with a straight chain hydrocarbon oil in the diffusion pump. Wood and Roenigk [4] observed fragments of DC-705 at $M/z = 28$, 32, 40, and 44, all of which could naturally occur in a vacuum system. These peaks are due to dissolved gas.

Esters and polyphenylethers are also widely used pump fluids because they polymerize to form conducting layers. They find use in systems that contain mass analyzers, glow discharges, and electron beams. Octoil-S is characterized by its repeated C_mH_n groupings. Polyphenylether (Santovac 5 and BL-10) also contains the characteristic fragments of the phenyl group which include the fragments at $M/z = 39$, 41, 43, 44, and 64. Cracking patterns will vary greatly from one instrument to another and the inability to match the data exactly to the patterns in any table should not be considered evidence of the absence or existence of a particular pumping fluid in the vacuum system. In fact, the cracking pattern for DC-705 given in Appendix E.1 does not show the same relative intensities at $M/z = 50$, 51, 52, and 78, as shown in Fig. 9.6 or seen by other workers. The spectrum taken from a gently heated liquid pump fluid source contains proportionally more high mass decomposition products than the spectrum of backstreamed vapors from a trapped diffusion pump because the trap is effective in retaining high molecular

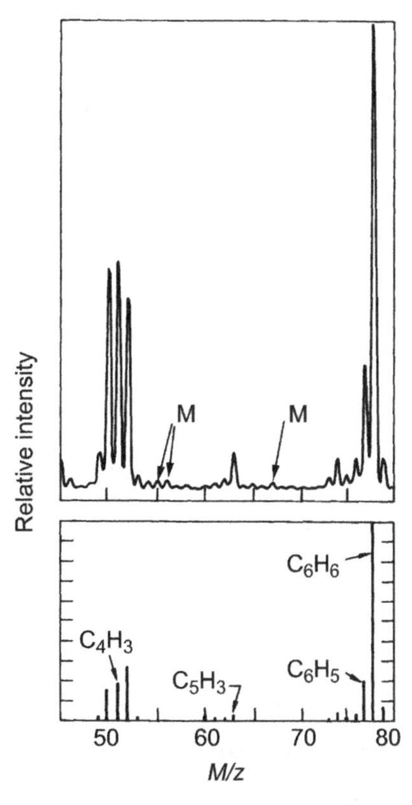

Fig. 9.6 Comparison of residual gas background in a system contaminated with DC-705 fluid (top), with the cracking pattern of benzene (bottom). Peaks labelled M are from mineral oil.

weight fragments. See Section 12.4. A liquid nitrogen trap is a mass-selective filter. These patterns cannot be used to differentiate between backstreamed DC-704 and DC-705. High-mass ion currents are severely attenuated when a quadrupole instrument is operated at constant absolute resolution. This built-in attenuation, sketched in Fig. 8.9, can easily lead to the conclusion that the environment is free of heavy molecules.

Elastomers are found in all systems with demountable joints except those using metal gaskets. The most notable property of all elastomers is their ability to hold gas and release it when heated or squeezed. The mass spectrum of Buna-n is shown in Fig. 9.7 during heating [7]. Also shown is the initial desorption of water followed by the dissociation of the compound at a higher temperature. The decomposition temperature is quite dependent on the material. Mass spectra obtained during the heating of Viton fluoroelastomer, depicted in Fig. 9.8, show the characteristic release of water at low temperatures and the release of carbon monoxide and carbon dioxide at higher temperatures. At a temperature of 300°C the Viton begins to decompose. Silicone rubbers have a polysiloxane structure. They are permeable, and their mass spectra usually show a large evolution of H_2O, CO, and CO_2. At high temperatures they begin to decompose. Their spectra show groups of peaks at $Si(CH_3)_n$, $Si_2O(CH3)_n$, and $Si_3O_2(CH3)_n$; $n=3, 5, 6$, respectively, are the largest [8]. Polytetrafluoroethylene (Teflon) is suitable for use up to 300°C, although it outgasses considerably. A spectrum taken at 360°C shows major fragments at $M/z=31$ (CF), 50 (CF_2), 81 (C_2F_3), and 100 (C_2F_4) [8].

An extensive and extremely useful collection of cracking patterns is found in the CERN Accelerator School (CAS) tutorial on Interpretation of RGA spectra [9].

The potential limits of qualitative analysis become clear after some practice with an RGA. Knowledge of the instrument, the cleaning solvents, and the pumping fluids used in mechanical, diffusion, or turbomolecular pumps combined with periodic background scans will result in the effective use of the RGA in the solution of equipment and process problems.

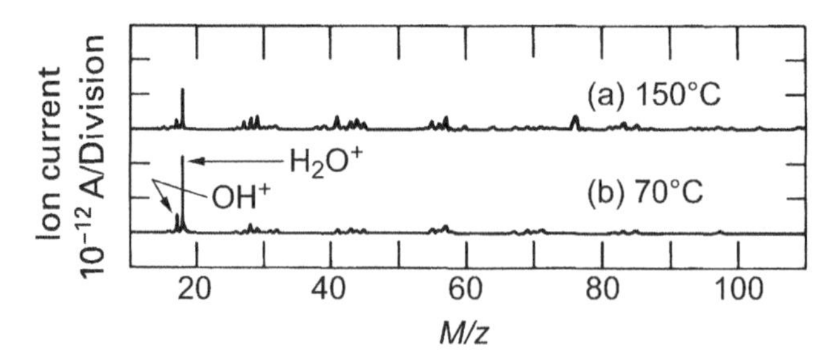

Fig. 9.7 Mass spectra obtained during the heating of Buna-N rubber. (a) $T=150$ °C; (b) $T=70$ °C. R.R. Addis et al. [7]/Reproduced with permission from Elsevier.

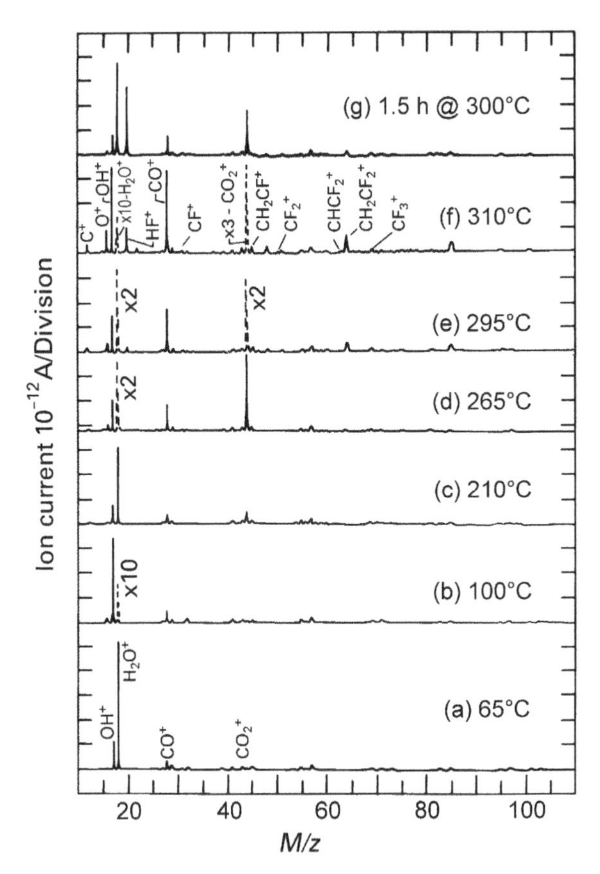

Fig. 9.8 Mass spectra obtained during the heating of Viton fluoro-elastomer. (a) T = 65°C; (b) T = 100°C; (c) T = 210°C; (d) T = 265°C; (e) T = 295°C; (f) T = 310°C; (g) T = 300°C at 1.5 h. R.R. Addis et al. [7]/Reproduced with permission from Elsevier.

9.3 Quantitative Analysis

The RGA is not intended to be an analytical instrument. It cannot eliminate overlapping peaks, nor is it as stable as an analytical spectrometer. Quantitative analysis of a single gas or vapor or combination of gases and vapors with unique cracking patterns is a simpler task than the analysis of combinations that contain overlapping peaks. This section demonstrates approximation techniques for quickly obtaining quantitative partial pressures within a factor of 10 but points out that the acquisition of data accurate to, say, 10% requires careful calibration. Either crude data are obtained quickly, or accurate data are obtained painstakingly; there is little middle ground. Let us consider techniques, both approximate and reasonably precise, for gases with isolated and overlapping cracking patterns.

9.3.1 Isolated Spectra

Approximate analysis techniques for gases with isolated spectra will yield results of rough accuracy with minimal effort. For example, a mixture of Ar, O_2, N_2, and H_2O would be reasonably easy to examine, because of unique peaks at $M/z=40$, 32, 28, and 18. An approximate measure of the partial pressures can be obtained by summing the heights of all peaks of significant amplitude that are caused by these gases, followed by the division of that number into the total pressure. The resulting sensitivity factor, expressed in units of pressure per unit scale division, is then applied to all the peaks without further correction. Some improvement in accuracy can be made if the ion currents are first corrected for the ionizer sensitivity by dividing by the values given in Table 8.1. There is an alternative technique that is equally accurate and does not require the knowledge of the total pressure. It relates the partial pressure of a given gas to the ion current, sensitivity of the mass analyzer, and gain of the electron multiplier according to:

$$P(x)=\frac{\text{total ion current}(x)}{GS(\mathrm{N}_2)} \tag{9.1}$$

The sensitivity is usually provided by the manufacturer for nitrogen and is typically of the order of 10^{-5} A/Pa. By taking the ratio of electron multiplier current to the Faraday cup current we can determine the gain G of the multiplier. From this information the partial pressure of the gas is obtained. These two techniques are accurate to within a factor of five.

A more accurate correction accounts for the gas ionization sensitivity and the mass dependencies of the multiplier gain and mass filter transmission. The mass dependence of the multiplier is often approximated as $M^{1/2}$, but this is not always valid. The transmission of a fixed radius sector with variable magnetic field is independent of mass, but the most common RGA, the quadrupole, has a transmission that is dependent on the energy, focus, and resolution settings. The more accurate correction may turn out to be less accurate unless a significant amount of calibration is done to obtain the mass dependence of the mass filter and electron multiplier accurately. The time spent in applying corrections is probably out of proportion to the information gained.

Accurate measurements of the partial pressures of gases with nonoverlapping spectra are best accomplished by calibrating the system for each gas of interest. The vacuum system must be thoroughly cleaned and baked if possible before the background spectrum is recorded. It is then backfilled with gas to a suitable pressure so that the cracking pattern can be recorded and the gas sensitivity, measured. The values of all the ionizer potentials and currents, the gain of the electron multiplier, and the pressure should be recorded at that time. The system should be thoroughly pumped and cleaned between each successive background scan and gas admission. Calibration procedures are discussed in Section 8.3. Calibration is a laborious process and is only done when precise knowledge of the partial pressure of a particular species is required. Even then a periodic check of the multiplier gain, for example, at $M/z = 28$, remains necessary.

9.3.2 Overlapping Spectra

Analysis of overlapping spectra is made more difficult by the fact that the peak ratios for a given gas may not be stable with time for two reasons: electron multiplier contamination and trace contaminants in the system. To gain an appreciation of the problems involved in determining partial pressures, two simplified numerical examples are solved here.

Consider a mass spectrum taken on a system that contains peaks mainly attributable to N_2 and CO. Peaks due to nitrogen appear at $M/z = 28$ and 14. Peaks due to CO appear at $M/z = 28$, 16, 14, and 12. Trace amounts of carbon present in the system from other sources dictate that the amplitude of the $M/z = 12$ peak cannot be relied on for accurate determination of the CO concentration. In a similar manner, the $M/z = 16$ peak is of questionable value because of the surface desorption of atomic oxygen, methane, or other hydrocarbon contamination. Therefore, the analysis in this simplified example is weighted heavily in favor of using the peaks at $M/z = 14$ and 28. From cleanly determined experimental cracking patterns the relative peak heights for nitrogen were found to be 0.09 and 1.00 for $M/z = 14$ and 28, respectively. Values of 0.0154 and 1.10 were measured at the same locations for CO. (The CO cracking patterns have been corrected for the difference in ionizer sensitivity; see Table 8.1.) The sensitivity and multiplier gain of the instrument were $S = 10^{-5}$ A/Pa and $G = 10^{5}$. From the mass spectrum the ion currents were $i_{14} = 10.54\,\mu A$, and $i_{28} = 210\,\mu A$. The individual partial pressures were then found by solving the following two equations simultaneously:

$$
\begin{aligned}
i_{28} &= SG\left[a_{11}P(N_2) + a_{12}P(CO)\right] \\
i_{14} &= SG\left[a_{21}P(N_2) + a_{22}P(CO)\right] \\
&\text{or,} \\
210\,\mu A &= 1\,A/Pa\left[1.00 \times P(N_2) + 1.10 \times P(CO)\right] \\
10.54\,\mu A &= 1\,A/Pa\left[0.09 \times P(N_2) + 0.0154 \times P(CO)\right]
\end{aligned}
\tag{9.2}
$$

which yielded $P(N_2) = P(CO) = 10^{-4}$ Pa.

Now consider how a change unaccounted for in the cracking pattern would affect the accuracy of this calculation. The actual cracking pattern of nitrogen, and consequently the measured ion currents, could have changed without the knowledge of the operator, because of contamination in the first dynode or a

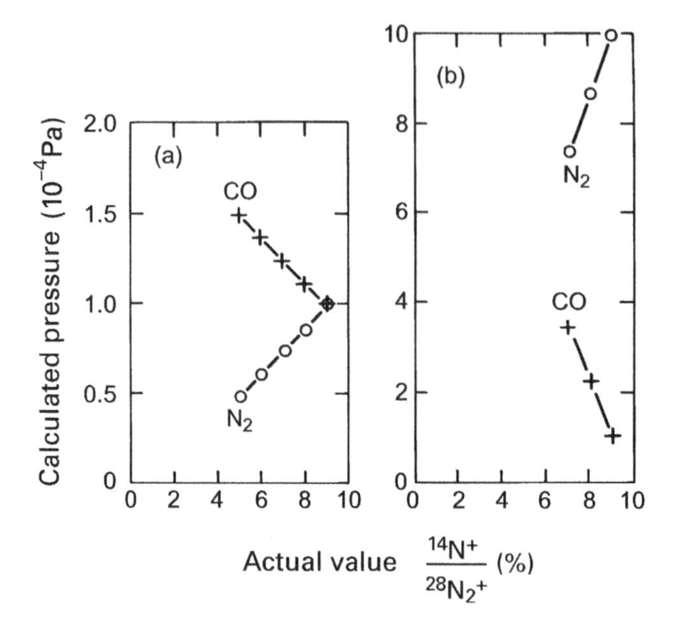

Fig. 9.9 Errors induced in the calculation of pressures of mixtures of N_2 and CO resulting from physical changes in the dissociation of nitrogen (the ratio of ^{14}N to $^{14}N_2$) which were not compensated for by appropriate changes in the coefficient a_{21} of the nitrogen cracking pattern. (a) 1:1 mixture of N_2 to CO, (b) 10:1 mixture of N_2 to CO.

change in the temperature of the ion source. If this had happened, an error would have been introduced into the calculation, because the coefficients a_{mn} in the right-hand side of Eq. (9.2) were not altered simultaneously. Figure 9.9*a* shows the calculated values of $P(N_2)$ and $P(CO)$ that would be obtained for the example in (9.2) if the actual $M/z = 14$ fragment of nitrogen were changed from 9% to 5% of the $M/z = 28$ peak without our knowledge and therefore without our having made the corresponding change in the coefficient a_{21}. It can be seen that even with moderate changes in the cracking pattern the partial pressures of the two gases can still be determined within 25% for this example in which the N_2 and CO are present in equal amounts. If greater accuracy is desired, the cracking pattern of the gases should be taken frequently and the coefficients a_{mn} adjusted to account for these instrumental changes.

For unequal concentrations of the two gases, the errors are far greater than when the gases are present in equal proportions. Figure 9.9*b* illustrates the pressure measurement error as a function of the cracking pattern change for a 10:1 ratio of N2:CO. Again, this represents an actual change in cracking pattern which was not accounted for by a corresponding change in the coefficient a_{21}. This demonstrates that even modest changes in the cracking pattern ratios of the major constituent can cause the error in partial pressure calculation of the minor constituent to be as great as 200–400% when the major and minor constituents have overlapping cracking patterns.

These two illustrations demonstrate that quantitative analysis of overlapping spectra requires accurate and often, frequent measurements of the cracking patterns and that accurate quantitative measurements of trace gases are difficult when the trace gas peaks overlap those of a major constituent.

The effects of certain cracking pattern errors in a residual gas spectrum have been illustrated by Dobrozemsky [10]. These data taken on an Orb–Ion pumped system are presented in Table 9.1. Column 1 shows the correctly analyzed partial pressures and their standard deviations, Column 2 and 3, respectively, show the effects of incorrectly interchanging the peaks at $M/z = 17$ and 18, and the effects of doubling the peak height at $M/z = 14$. Dobrozemsky's analysis demonstrates vividly the effects of cracking pattern errors in the accuracy of calculating trace gas compositions. Column 1, the correct analysis, shows a standard deviation for H_2O of 0.3%, while for a trace gas such as hydrogen the standard deviations is large enough to render the measurement useless. The standard deviation for hydrogen is large, because the signal at $M/z = 2$ arises from many sources. Literally any hydrocarbon that is ionized in the RGA has a fragment at $M/z = 2$ and the standard deviation of each fragment is additive. The result is that the hydrogen concentration, if any, is not known. It demonstrates the ease with which false data can be generated when cracking patterns are not accurately known. Note also that the incorrect analyses, shown in Column 2 and Column 3 of Table 9.2, yield unreasonably large standard deviations for all gases even when only one ion current was in error.

Even though only a limited number of low molecular weight gases are present in a vacuum system, the analysis procedure is complicated by the fact that there are often several gases that produce peaks at the same M/z values. For example, CO, N_2, C_2H_4, and CO_2 produce ion current at $M/z = 28$ and CO_2, O_2, CH_4, and H_2O produce ion currents at $M/z = 16$. For a system containing n gases and m ion current peaks, Eq. (9.2) becomes:

$$\begin{bmatrix} i_1 \\ i_2 \\ \bullet \\ i_m \end{bmatrix} = \begin{bmatrix} a_{11} & \bullet & \bullet & a_{1n} \\ a_{12} & \bullet & \bullet & a_{2n} \\ \bullet & \bullet & \bullet & \bullet \\ a_{m1} & \bullet & \bullet & a_{mn} \end{bmatrix} \begin{bmatrix} P_1 \\ P_2 \\ \bullet \\ P_n \end{bmatrix} \tag{9.3}$$

where i_m is the ion current at mass m, a_{mn} are the components of the cracking pattern matrix, and P_n is the partial pressure of the nth gas. Most gases have more than one peak, so that $m > n$, and the system is over-specified. Least mean squares or some other smoothing criterion is then applied to the data to get the best fit.

Table 9.2 Analysis of Background Gases in an Orb–Ion Pumped System

	Partial Pressures × 10^{-8} Torr		
Gas	**1**	**2**	**3**
H_2	2.21 ± 3.15	2.85 ± 8.29	2.07 ± 2.67
He	15.54 ± 0.62	14.95 ± 22.6	14.94 ± 7.2
CH_4	5.37 ± 0.49	-13.50 ± 5.13	9.64 ± 1.63
NH_3	2.64 ± 0.89	49.56 ± 6.83	5.03 ± 2.17
H_2O	50.95 ± 0.15	15.40 ± 5.29	48.44 ± 1.68
Ne	4.54 ± 0.54	4.39 ± 16.5	4.38 ± 5.24
N_2	-1.99 ± 3.26	34.88 ± 57.6	-13.60 ± 18.3
CO	4.54 ± 2.91	-30.70 ± 51.4	15.38 ± 16.4
C_2H_6	3.42 ± 0.74	8.44 ± 18.4	1.89 ± 5.84
O_2	0.00 ± 0.13	-0.99 ± 4.87	0.30 ± 1.55
Cl	-0.02 ± 0.12	0.04 ± 4.64	0.02 ± 1.48
Ar	1.36 ± 0.09	1.29 ± 3.29	1.31 ± 1.04
CO_2	5.67 ± 0.14	5.30 ± 3.22	5.50 ± 1.02
C_3H_8	2.72 ± 0.56	0.27 ± 12.9	3.32 ± 4.11
Acetone	2.99 ± 0.31	7.94 ± 6.91	1.33 ± 2.2

(1) Correct analysis; (2) Introduce error—interchange $M/e = 17$ and 18; (3) introduce error— double amplitude of $M/e = 14$.
Source: R. Dobrozemsky [10]/Reproduced with permission from AIP Publishing.

If the cracking patterns are carefully taken and if the standard deviations are measured as well, then accuracy of a few percent may be obtained for major constituents [10,11]. The matrix for a real problem would contain about 10×50 elements and would require the assistance of a computer in order to obtain a solution. In this case careful data acquisition, instrument control, and analysis are required [11]. These experiments are expensive and time-consuming and are only performed in situations where such precision is required. Beware of instruments that directly convert ion currents to partial pressures and generate graphical displays. Cleanliness, sensitivity, pressure, and internal power supply voltages affect cracking patterns. Simple algorithms cannot accurately determine pressure. If one wishes to perform quantitative analysis, then frequent *in situ* recalibration is necessary. A full set of masses is not necessary, as relative sensitivities are reasonably stable, that for most work, calibration with a light (H_2) and a heavy (N_2) gas will suffice, and these can be injected via a calibration port or via fixed calibration standards [9].

References

1. Marmet, P. and Morrison, J.D., *J. Chem. Phys.* **36**, 1238 (1962).
2. Redhead, P.A., *Can. J. Phys.* **42**, 886 (1964).
3. Robins, J.L., *Can. J. Phys.* **41**, 1383 (1963).
4. Wood, G.M., Jr. and Roenigk, R.J., Jr., *J. Vac. Sci. Technol.* **6**, 871 (1969).
5. Craig, R.D. and Harden, E.H., *Vacuum* **16**, 67 (1966).
6. Solbrig, C.W. and Jamison, W.E., *J. Vac. Sci. Technol.* **2**, 228 (1965).
7. Addis, R.R., Jr., Pensak, L., and Scott, N.J., *Transactions of the 7th National Symposium of the American Vacuum Society, 1960*, Pergamon, Oxford, 1961, p. 39.
8. Beck, H., *Handbook of Vacuum Physics*, Vol. 3, Macmillan, New York, 1964, p. 243.
9. B. Jenninger and P. Chiggaito, *CERN Accelerator School (CAS) on Vacuum for Particle Accelerators, CAS Tutorial on RGA Interpretation of RGA Spectra*, Glumslöv, Sweden, June 2017, page 72. https://indico.cern.ch/event/565314/contributions/2285748/attachments/1467497/2273709/RGA_tutorial-interpretation.pdf
10. Dobrozemsky, R., *J. Vac. Sci. Technol.* **9**, 220 (1972).
11. Ramondi, D.L., Winters, H.F., Grant, P.M., and Clarke, D.C., *IBM J. Res. Dev.* **15**, 307 (1972).

Part III

Production

Vacuum pumps are classified according to the physical or chemical phenomena responsible for their operation. In practice, this is a bit awkward because some pumps combine two or more principles to pump a wide range of gases or to pump over a wide range of pressures. In this section the discussion of pump operation is described in six chapters. Chapter 10 discusses mechanical vacuum pumps. Mechanical vacuum pumps operate by displacing gas from the work chamber to the pump exhaust. Rotary vane pumps operate in the low vacuum region, whereas single and multistage Roots pumps operate in the medium vacuum region. Scroll, dry piston, and screw pumps are nominally "oil-free" and are used in place of oil-lubricated pumps for clean pumping applications and for pumping particle-laden or chemically reactive gases. The turbomolecular pump is a mechanical high vacuum pump that is the subject of Chapter 11. It transports gas from regions of low pressure to high pressure by momentum transferred from high-speed blades. Molecular drag pumps are combined with magnetically levitated turbomolecular pumps to provide "oil-free" high vacuum pumps for ultraclean applications.

Chapter 12 is devoted to the diffusion pump, which, like the turbomolecular pump, is a momentum transfer pump. The diffusion pump was the mainstay of the vacuum industry for many decades and is now largely supplanted by cryogenic and turbomolecular pumps. However, it remains so commonly used in older systems that a discussion of its operational principles and system use is warranted.

Capture pumps or entrainment pumps bind molecules and atoms to a surface instead of expelling them to the atmosphere. Chapter 13 describes getter and ion pumps. Getter pumps, such as the titanium sublimation pump, remove gases by chemical reactions that form solid compounds; ion pumps ionize gas molecules and imbed them in a wall. The sputter-ion pump combines getter pumping with ion pumping. Other entrainment pumps are based on condensation and sorption.

A Users Guide to Vacuum Technology, Fourth Edition. John F. O'Hanlon and Timothy A. Gessert.
© 2024 John Wiley & Sons, Inc. Published 2024 by John Wiley & Sons, Inc.

Sorption pumps physisorb gas molecules in materials of high surface area. These surfaces are usually cooled to enhance their pumping ability. Another capture pump, the cryogenic pump, is the subject of Chapter 14. A helium gas refrigerator cryogenic pump uses at least two stages of cooling. The warmer stage pumps by condensation or adsorption on a cooled metallic surface. In addition, the colder stage uses both a metal and an adsorbate-coated surface. The adsorbate pumps gases that are not entrained by the other two surfaces. The following figure describes the operating pressure ranges of many pumps and pump combinations.

These chapters do not cover all of the many techniques by which high vacuum may be achieved. Here we review the operation of commonly used pumps in a concise manner and focus on turbo, cryo, and newer dry mechanical pumps.

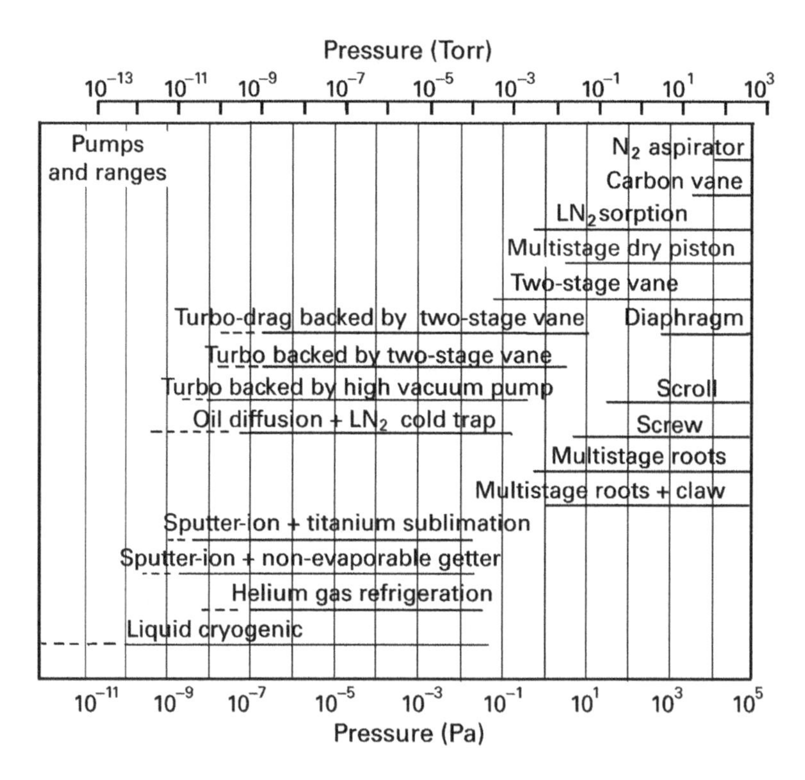

10

Mechanical Pumps

In this chapter we review the operation of five low vacuum pumps: rotary vane, scroll, screw, diaphragm, and dry reciprocating piston; and the lobe and claw medium vacuum pumps. Vane pumps are widely used for backing high vacuum pumps and for initial chamber evacuation. Lobe blowers are used with vane pumps to rough large systems, to back large pumps, and to pump large quantities of gas in plasma processing systems. Lobe pumps incorporating claw stages are particularly well suited for pumping particulate-laden atmospheres. Newly designed multistage lobe pumps have found use in applications formerly reserved for scroll pumps. Screw, scroll, diaphragm, and dry, reciprocating piston pumps are used in oil-free applications. Fluid-sealed piston pumps were formerly used for some situations in place of large vane pumps, but their inherent vibration has rendered them obsolete for all but a few applications.

10.1 Rotary Vane

The rotary vane pump is an oil-sealed pump that has a useful operating range of $1–10^5$ Pa (10^{-2} Torr–1 atm). Vane pumps of 10–200 m^3/h displacement are used for rough pumping and for backing diffusion or turbomolecular pumps.

In this pump, illustrated in Fig. 10.1, gas enters the suction chamber (A), is compressed by the rotor (3) and vane (5) in region B and then expelled to the atmosphere through the discharge valve (8) and fluid reservoir. An airtight seal is made by one or more spring or centrifugally loaded vanes and closely spaced sealing surfaces (10). The vanes and the surfaces between the rotor and housing are sealed by the low vapor pressure fluid, which also serves to lubricate the pump and fill the volume above the discharge valve. Pumps that use a speed-reduction pulley operate in the 400–600 rpm range, whereas direct drive pumps operate at speeds of

A Users Guide to Vacuum Technology, Fourth Edition. John F. O'Hanlon and Timothy A. Gessert.
© 2024 John Wiley & Sons, Inc. Published 2024 by John Wiley & Sons, Inc.

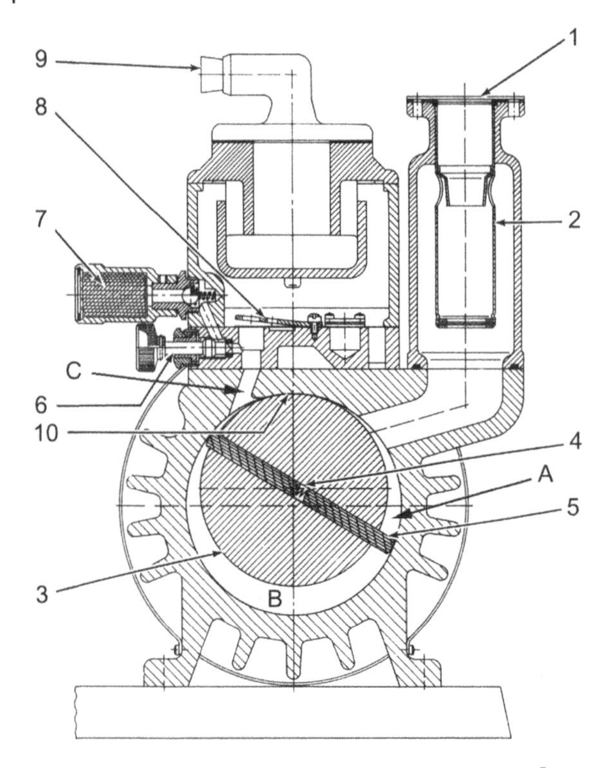

Fig. 10.1 Sectional view of the Pfeiffer DUO-35, 35 m^3/h double-stage, rotary vane pump: (1) intake, (2) filter, (3) rotor, (4) spring, (5) vane, (6) gas ballast valve, (7) filter, (8) discharge valve, (9) exhaust, (10) sealing surface, (A) suction, (B) compression, and (C) exhaust chambers. Reproduced with permission from Pfeiffer Vacuum GmbH.

1500–1725 rpm. The fluid temperature is considerably higher in the direct drive pumps than in the low-speed pumps, typically 80 and 60°C, respectively. These values will vary with the viscosity of the fluid and the quantity of air being pumped.

Single-stage pumps consist of one rotor and stator block (Fig. 10.1). Lower pressures may be reached if a second stage is added as shown schematically in Fig. 10.2, by connecting the exhaust of the first stage to the intake of the second. Physically, the second pumping stage is located adjacent to the first and on the same shaft. The pumping speed characteristics of single-stage and two-stage rotary vane pumps are shown in Fig. 10.3.

The ultimate pressure of a two-stage pump is less than a single-stage pump for two reasons: First, because the stages are in series, and second, because the fluid circulating in the second stage has been degassed by its isolation from the reservoir.

Kendall [1] demonstrated the effect of dissolved gas in the fluid on the ultimate pressure of a two-stage rotary vane pump. He showed a rotary vane pump could

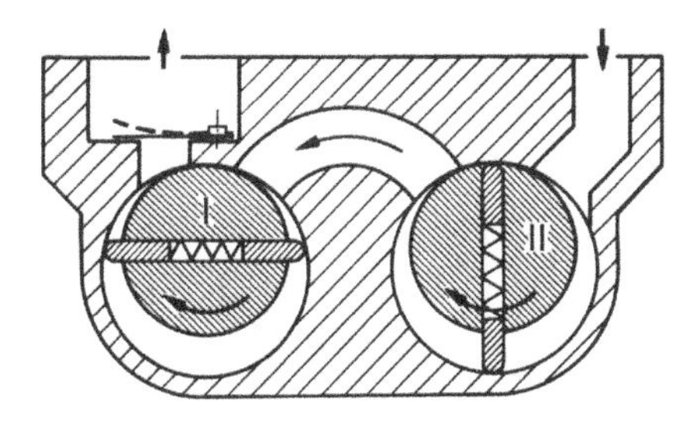

Fig. 10.2 Schematic section through a two-stage rotary pump. Reproduced with permission from Leybold.

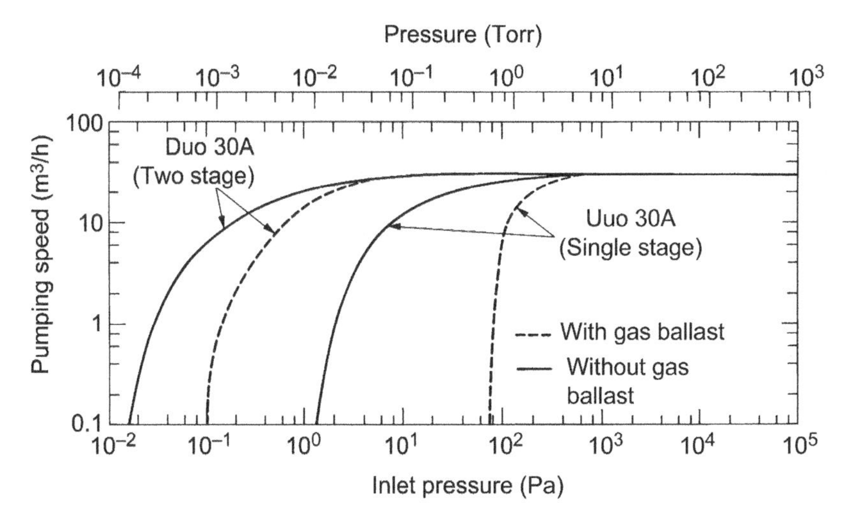

Fig. 10.3 Pumping speed curves for the Pfeiffer UNO 30A and DUO 30A rotary vane pumps. Reproduced with permission from Pfeiffer Vacuum GmbH.

reach an ultimate pressure of 4×10^{-5} Pa when the fluid reservoir was exhausted by another pump. Figure 10.4 shows the effect of prolonged outgassing and the effect of admitting CO_2 to the fluid reservoir after degassing the fluid.

The free-air displacement and the base pressure are two measures of the performance of roughing pumps. This and other relevant parameters of rotary mechanical pumps are described in an ISO Standard [2]. The free air displacement is the volume of air displaced per unit time by the pump at atmospheric pressure with no pressure differential. For the two pumps whose pumping speed curves are shown in Fig. 10.3, this has the value of $30\,m^3/h$ (17.7 cfm) at a pressure of 10^5 Pa (1 atm). At the base pressure of the blanked-off pump, the net speed (forward flow minus backflow) drops to zero because of dissolved gas in the fluid, leakage around the seals, and trapped gas in the volume below the valve. Rotary vane pumps have ultimate pressures in the 3×10^{-3}–1 Pa range; the lowest ultimate pressures are achieved with two-stage pumps. The single- and two-stage pumps

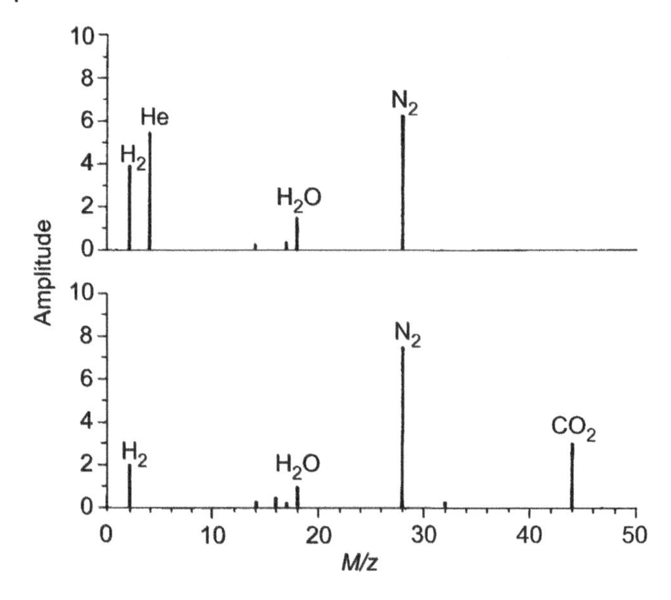

Fig. 10.4 Relative abundance of gases at pump ultimate: Top: After prolonged outgassing of oil and with zero pressure at the exhaust port. Helium pressurization applied intermittently to provide lubricating oil pressure 9.3×10^{-5} Pa ($P = 7 \times 10^{-7}$ Torr). Bottom: After oil had been exposed to carbon dioxide and then exhaust pressure reduced to zero 4×10^{-5} Pa ($P = 3 \times 10^{-7}$ Torr). B.R.F. Kendall [1]/Reproduced with permission from AIP Publishing.

characterized in Fig. 10.3 have ultimate pressures of 1.4 and 1.5×10^{-2} Pa, respectively. These ultimate pressures are obtained with a new pump using clean, low vapor pressure fluid. As the fluid becomes contaminated and the parts wear, the ultimate pressure will increase. Note that there is a distinction between the ultimate pressure of a pump and its lower usable pressure. At its ultimate pressure, its net pumping speed is zero, and it cannot pump a gas load.

When processes evolve large amounts of water, acetone, or other condensable vapors, condensation can occur during the compression stage after the vapor has been isolated from the intake valve. During the compression stroke, vapors may reach their condensation pressure, condense, and contaminate the fluid before the exhaust valve opens. Condensation delays or may prevent the opening of the exhaust valve. If condensation is not prevented, the pump will become contaminated, the ultimate pressure will increase, and gum deposits will form on the moving parts. Some compounds will eventually cause the pump to seize. To avoid condensation and its resulting problems, gas is admitted through the ballast valve. The open valve allows ballast, usually room air, to enter the chamber during the compression stage; the trapped volume is isolated from the intake and exhaust valves. This inflow of gas, which can be as much as 10% of the pump displacement, is controlled by valve 6 in Fig. 10.1. The added gas causes the discharge valve to open before it reaches the condensation partial pressure of the vapor. In this manner the vapor is swept out of the pump and no condensation occurs. The ultimate pressure of a gas ballast pump is not very low with the ballast valve open, as the pump is essentially being short-circuited. Figure 10.3 shows the effect of full gas ballast on the performance of a single- and double-stage pump.

Gas ballast can be used to differentiate contaminated fluid from a leak. If the inlet pressure drops when the ballast valve is opened but drifts upward slowly after the valve is closed, the fluid is contaminated with a high vapor pressure impurity. Additional gas ballast details are covered in reference [3], while Van Atta [4] describes alternative methods for pumping large amounts of water.

10.2 Lobe

The positive displacement blower, or lobe blower is used in series with a rotary fluid-sealed pump to achieve higher speed and lower ultimate pressure in the medium vacuum region than can be obtained with a rotary mechanical pump alone. The dual counter-rotating figure-eight-shaped lobes were developed over 150 years ago by the Roots brothers, and they are widely known today as Roots blowers. The rotors have substantial clearances between themselves and the housing; typically, about 0.2 mm. They rotate in synchronism in opposite directions at speeds of 3000–3500 rpm. These speeds are possible because a fluid is not used to seal the gaps between the rotors and the pump housing. A sectional view of a single-stage lobe blower is given in Fig. 10.5. Methods for characterizing mechanical booster pumps are given in a related ISO standard [5].

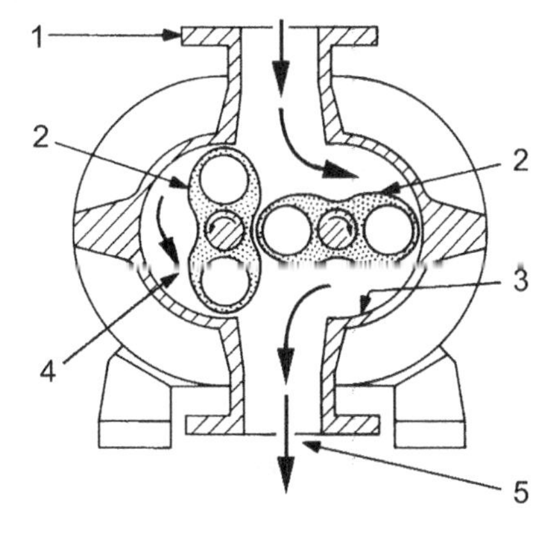

Fig. 10.5 Section through a single-stage lobe blower: (1) inlet, (2) rotors, (3) housing, (4) pump chamber (swept volume), and (5) outlet. Reproduced with permission from Leybold.

The compression ratio, or ratio of outlet pressure to inlet pressure, is pressure dependent and usually has a maximum near 100 Pa. At higher pressures the compression ratio should, theoretically, remain constant. In practice it decreases. Outgassing and the roughness of the rotor surfaces contribute to compression loss at low pressures. Each time the rotor surface faces the high-pressure side, it adsorbs gas. Some of this gas is released when the rotor faces the low-pressure side. The compression ratio K_{omax}, for air for a single-stage Lobe blower of 500-m^3/h displacement is shown in Fig. 10.6. It has a maximum compression ratio of 44. Large pumps tend to have a larger compression ratio than small pumps, because they have a smaller ratio of gap spacing to pump volume. The compression ratio for a light gas such as helium is about 15–20% smaller than the ratio for air. The compression ratio K_{omax} is a static quantity and is measured under conditions of zero flow. The inlet side of the pump is sealed, and a pressure gauge is attached. The outlet side is connected to a roughing pump and the system is evacuated. Gas is admitted to the backing line that connects the blower to the roughing pump. The backing pressure P_b is measured at the blower outlet, and the pressure P_i is measured at the inlet. The compression ratio is given by P_b/P_i.

Lobe blowers generate considerable heat, when pumping gas at high pressures. Heat causes the rotors to expand; if unchecked, rotor expansion could destroy the pump. To avoid overheating, a maximum pressure difference between the inlet and outlet of a lobe blower is specified. This maximum pressure difference is typically 1000 Pa, but that value may be exceeded for a short time without harm to the pump. Lobe blowers are connected as compression or transport pumps.

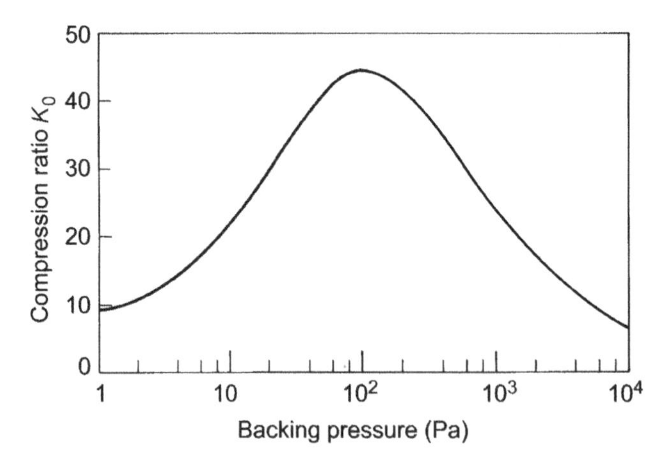

Fig. 10.6 Dependence of the air compression ratio K_{omax} of the Leybold WS500 lobe pump on the backing pressure. Values for helium are about 20% smaller. Reproduced with permission from Leybold.

In compression pumping, the common method, a lobe pump is placed in series with a rotary pump whose rated speed is 2–10 times smaller than its own speed. When pumping is initiated at atmospheric pressure, a bypass line around the lobe pump is opened, or the pump is allowed to free-wheel. All the pumping is done by the rotary pump until the backing pressure is below the manufacturer's recommended pressure difference, at which time the lobe blower is activated, and the bypass valve is closed. Some lobe blowers have this bypass feature built into the pump housing; others are allowed to free-wheel until a pressure sensor activates a clutch between motor and blower. The net speed of a lobe blower of 500-m^3/h capacity backed by a 100-m^3/h rotary piston pump is shown in Fig. 10.7. The speed curve for the mechanical pump alone is shown for comparison. Such lobe blower-rotary pump combinations are often used when speeds of 170 m^3/h or greater are required, because the two pumps cost less than a rotary pump of similar capacity.

The second method, transport pumping, uses a lobe blower in series with a rotary pump of the same displacement. Figure 10.7 also shows the pumping speed of a 60-m^3/h lobe blower backed by a 60-m^3/h rotary vane pump. The pumping speed of the rotary vane pump is shown for comparison. Both pumps are started simultaneously at atmospheric pressure because the critical pressure drop will never be exceeded.

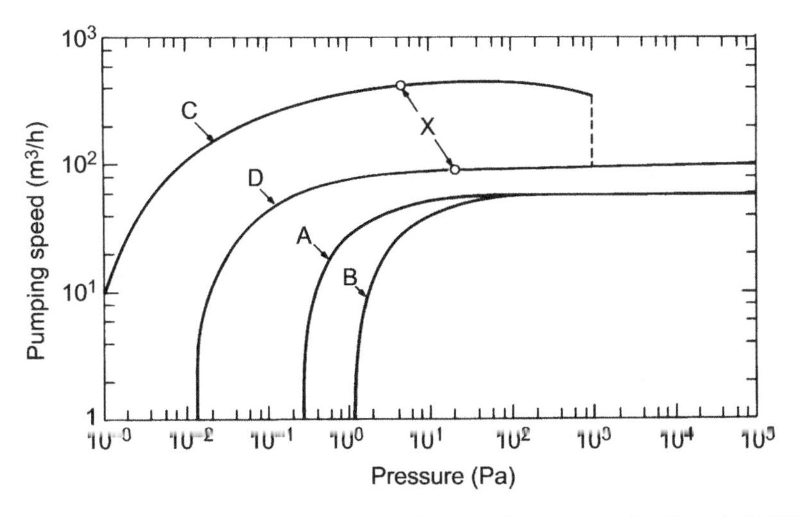

Fig. 10.7 Lobe blower-rotary pump combinations. Transport mode: (A) Leybold RUTA 60 lobe blower and S60 rotary vane; (B) S60 only. Compression mode; (C) Leybold WS500 lobe blower and DK100 rotary piston pump; and (D) DK100 only. Reproduced with permission from Leybold.

Detailed calculations of the effective pumping speed of the lobe blower have been carried out by Van Atta [4]. Here we give approximate formulas for the inlet pressure P_i, and the inlet speed S_i [3]:

$$P_i = P_b\left(\frac{1}{K_{omax}} + \frac{S_b}{S_D}\right) \tag{10.1}$$

$$S_i = \frac{S_b S_D K_{omax}}{S_D + S_b K_{omax}} \tag{10.2}$$

S_D is the pump displacement or speed of the lobe blower at atmospheric pressure. S_b and P_b are, respectively, the speed and pressure of the backing pump. With these approximate equations, the pumping speed curve for the backing pump, the compression ratio K_{omax}, and the lobe blower displacement, a curve of the speed of the lobe pump versus inlet pressure can be calculated. The point on the upper end of the line marked × in Fig. 10.7 is the result of applying Eq.'s (10.1) and (10.2). In this example the inlet pressure P_i, and the inlet speed S_i were calculated for a backing pressure of 20 Pa. At 20 Pa, $K_{omax} = 30$ (Fig. 10.6), and $S_b = 90\,m^3/h$ (Fig. 10.7*d*). Using Eq.'s (10.1), (10.2), and $S_D = 500\,m^3/h$, we obtain $P_i = 4.3$ Pa and $S_i = 422\,m^3/h$.

Lobe pumps are also used to back large diffusion or turbomolecular pumps. For example, a 35-in.-diameter diffusion pump used to evacuate a 2-m^3 chamber is backed by a series combination of a 1300-m^3/h lobe blower and a 170-m^3/h rotary piston pump.

10.3 Claw

Pumps with counter-rotating lobe pairs have been designed using many cross-sectional shapes. The classic Roots pump is the most common. Another design, the Northey hook and claw pump [6], designed in the early 1930's as a compressor, has proven useful for applications that generate debris. It has two irregularly shaped lobes or "claws" that counter-rotate in the same manner as a lobe blower. However, a conventional lobe blower does not compress gas within the pump—it transports gas from inlet to exhaust. Lobe blower compression takes place outside the pump body. The claw pump does compress gas within the pump, as described in the rotational sequence in Fig. 10.8. Each complete revolution of the claw pair compresses a volume of gas before ejection. Claw pumps are more efficient at compressing gas at high pressures than lobe pumps [7] and have at least equal compression ratios at low pressures [8]. The maximum compression is limited by the volume of gas depicted in illustration (*f'*) of Fig. 10.8; this volume returns to the inlet at the end of the cycle. Claw and lobe stages are usually combined to form an efficient

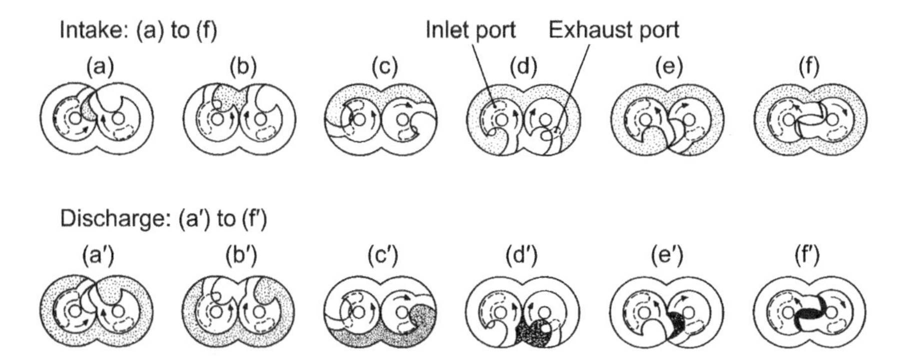

Fig. 10.8 Compression sequence for 1 revolution of a claw stage: (a)–(f) show the intake sequence with the inlet port open from (b)–(d). The compression and exhaust sequence is shown in (a′)–(f′) with the exhaust port open from (c′)–(d′). The carryover volume, which limits the maximum obtainable compression, is shown in (f′). A.P. Troup and N.T.M. Dennis [7]/Reproduced with permission from AIP Publishing.

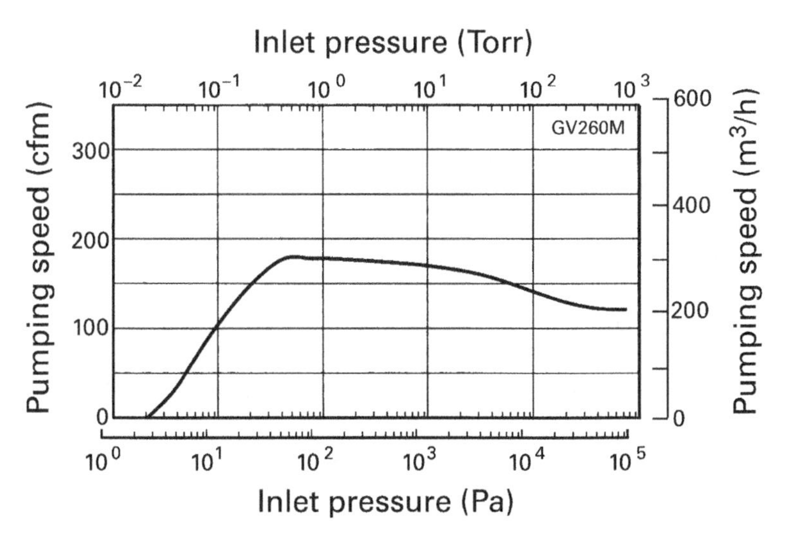

Fig. 10.9 Pumping speed curve for a four-stage booster pump exhausting directly to atmosphere. This pump contains a lobe inlet stage followed by three claw stages. Reproduced with permission from BOC Edwards.

multistage pump that effectively pumps corrosive and abrasive gases [9]. One design uses a lobe inlet stage followed by three claw stages before exhausting to atmosphere; its pumping speed curve is depicted in Fig. 10.9 exhausting to atmosphere. Both lobe and claw pumps require cooling and gas purging [10]. Pumps with combined lobe–claw stages have ultimate pressures in the range of 0.5–1 Pa

(4–7 mTorr) when exhausting directly to atmosphere. They are used to provide oil-free backing for turbomolecular pumps in corrosive or abrasive etching and deposition systems. Because of their shape, claw stages efficiently remove deposited particles. As a result, maintenance intervals can be increased when using these pumps in chemically aggressive process environments.

10.4 Multistage Lobe

Recently, small, multistage versions of the lobe pump have been developed as a low-particle, dry pump alternative to the scroll pump, discussed in the following section. These multistage, miniature lobe pumps, which are available in sizes similar to and larger than scroll pumps, produce significantly less wear-generated particles than do scroll pumps. In addition, there is no pumping or lubricating fluid in the pumping path.

The design and operation of each stage of this pump is identical to those in the lobe pump described in Section 10.3. However, instead of only a single, and often large lobe stage exhausting into another pump, such as the fluid-sealed pump illustrated in Fig. 10.7, the multistage lobe pump incorporates several stages on one shaft. The design of a four-stage multi-lobe pump is illustrated in Fig. 10.10. Because the output pressure of each stage will be higher than its inlet pressure, the volume of each stage becomes increasingly smaller nearer the exhaust, such that the product of the pressure times speed is constant—a demonstration of Boyle's Law and constant gas flow in series components. By connecting several lobe stages in series and operating the pump at higher speeds than an historic lobe pump, the multistage lobe pump can achieve input base pressures similar to the diaphragm, piston, or scroll pump, and as well, exhaust to atmospheric pressure. Pumping speed curves for a representative commercial pump are shown in Fig. 10.11.

The multistage lobe pump has no elastomeric seals between any moving and stationary parts of the pump, e.g., the lobes and castings are not in contact but closely spaced. Therefore, there are no seals to abrade and form particles. Commercial pumps are available with 7 or more stages, and designs with 3 or more lobes on each rotor. The increased rotation speeds of these pumps of up to 15,000 rpm are enabled by precision bearings and shaft seals. To limit potential contamination of the evacuated regions from the bearing/seals, and also to limit toxic or corrosive effluents from entering the bearing regions, some designs allow separate purging of these regions. These pumps will include gas purging and ballast capability to limit condensation within the evacuated regions and flush particles that may form in these regions due to effluent reactions.

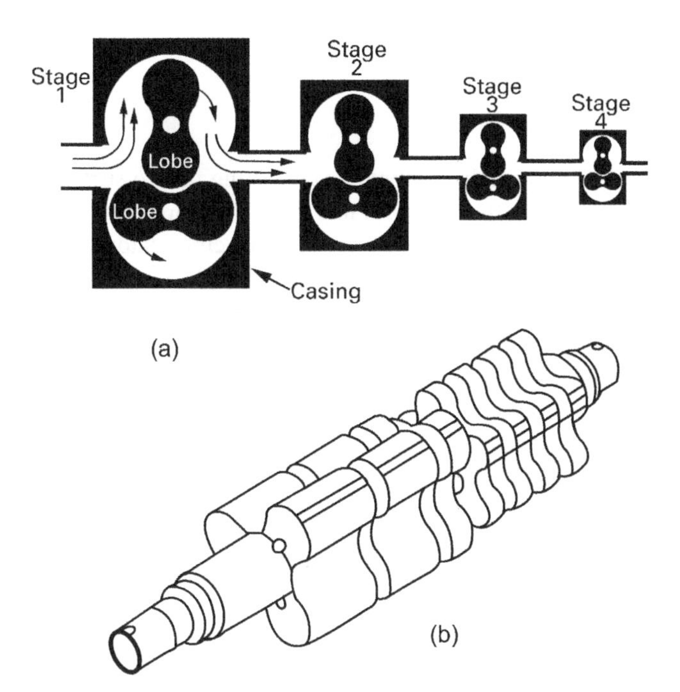

Fig. 10.10 (a) Schematic illustration of a multistate lobe pump with four stages. Reproduced with permission of Gessert Consulting, LLC. (b) Seven-stage, trilobe rotor. Copyright © Reproduced with permission from Edwards Vacuum.

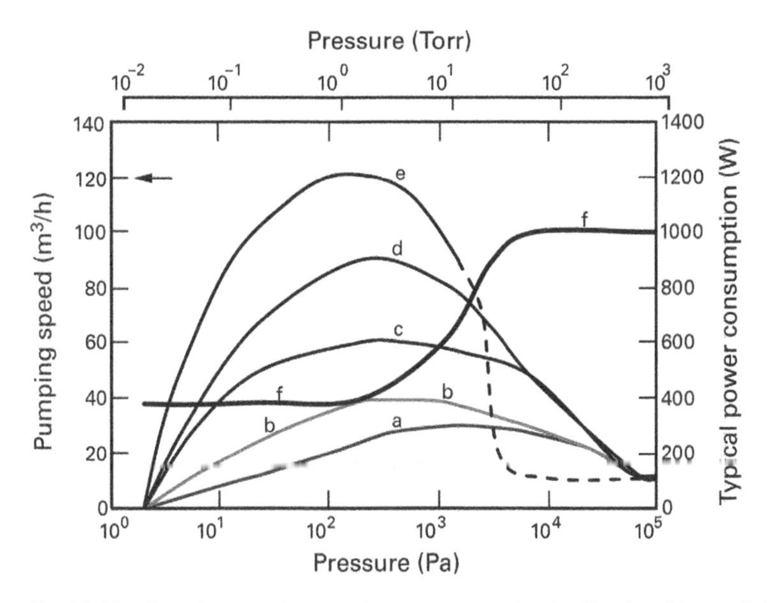

Fig. 10.11 Pumping speed curves for a representative family of multistage lobe pumps. (a) Model nXR30i, (b) nXR40i, (c) nXR60i, (d) nXR90i, (e) nXR120i (dashed; noncontinuous), and (f) Typical Power Consumption. Copyright © Reproduced with permission from Edwards Vacuum.

10.5 Scroll

The scroll pump is a relatively simple compressor consisting of two surfaces, one fixed and one orbiting. Note that the movable plate does not rotate, but rather it orbits. Mirror-image spiral grooves are cut in two facing, stator plates: in turn these plates mesh within a plate containing sets of complementary spiral ridges. Figure 10.12 illustrates cross-sectional and plan views of an orbiting scroll pump. Gas enters the chambers at the periphery and is forced around in a spiral helical path until it reaches the exit port located at the pump center. The pump inlet is located at the periphery, which must be sealed with bellows. The ultimate pressure of a scroll pump is ~1 Pa (10^{-2} Torr), see Fig. 10.13. These pumps are manufactured in small sizes of order 15–40 m^3/h. They find applications for backing turbomolecular drag pumps that have inbuilt molecular drag stages, and they find other applications where small oil-free pumps are needed. The polymer seal that prevents gas flow between the closely spaced edges of the orbiting seal plate and the stator will generate wear particles. One can trap these in a point-of-use 0.02-μm-diameter particle filter mounted on the pump inlet and prevent them from migrating to the process chamber, where they could contaminate the product.

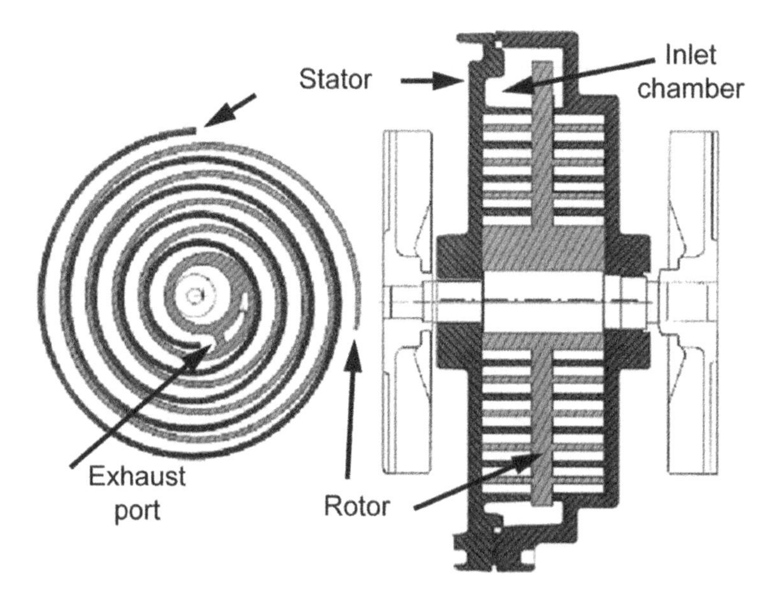

Fig. 10.12 Anest-Iwata orbiting scroll pump mechanism. Right: Plan view of orbiting scroll. Left: Cross-sectional view of stator and orbiting rotor. Reproduced with permission from Synergy Vacuum.

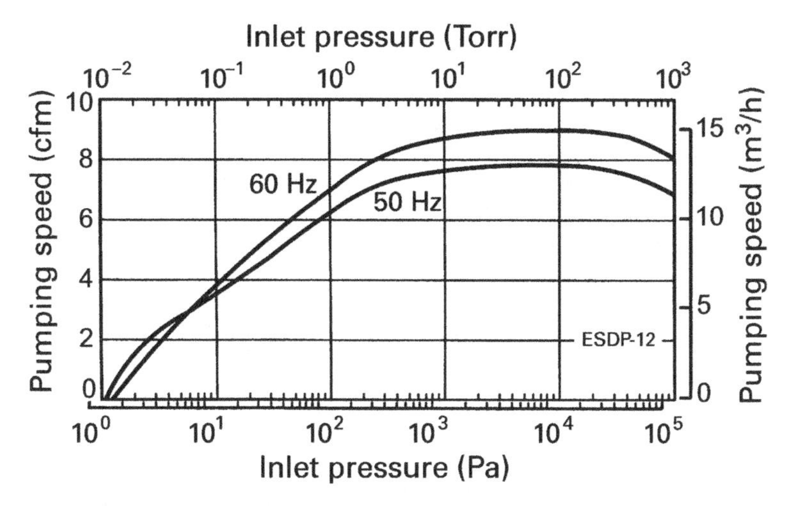

Fig. 10.13 Pumping speed of a model ESDP12 orbiting scroll pump. Reproduced with permission from BOC Edwards.

10.6 Screw

The screw pump is another compressor with a long history of use in fields other than vacuum technology. Screw pumps have been manufactured for oil-free applications, and for pumping abrasive and corrosive gases. They are used for backing magnetic levitated turbomolecular pumps on reactive ion etching and low-pressure chemical vapor deposition systems. The basic design of a screw pump is illustrated in Fig. 10.14. The two uniquely shaped screws, located inside a closely fitting stator, are counter-rotated at equal rotational velocities. Gas enters at the end of the rotating pair and is transported along screws in a trapped region defined by the contact of the screws and the wall. It is expelled into the exhaust plenum at the right-hand end of the screw pair. The rotors may be coated for use in pumping corrosive gases. In some designs, asymmetrical screws rotate at different velocities, provided that the total number of turns is limited. Pumping speed curves for one family of pumps are shown in Fig. 10.15. The ultimate pressure of a screw pump is of order 0.7 Pa (5×10^{-3} Torr) with speeds as high as 2500 m^3/h. Screw pumps are rugged and reliable.

10.7 Diaphragm

Diaphragm pumps, in the form of bellows, are one of the oldest known pumps. They consist of a small chamber designed to have minimal dead space containing a flexible diaphragm connected to a piston, as illustrated in Fig. 10.16.

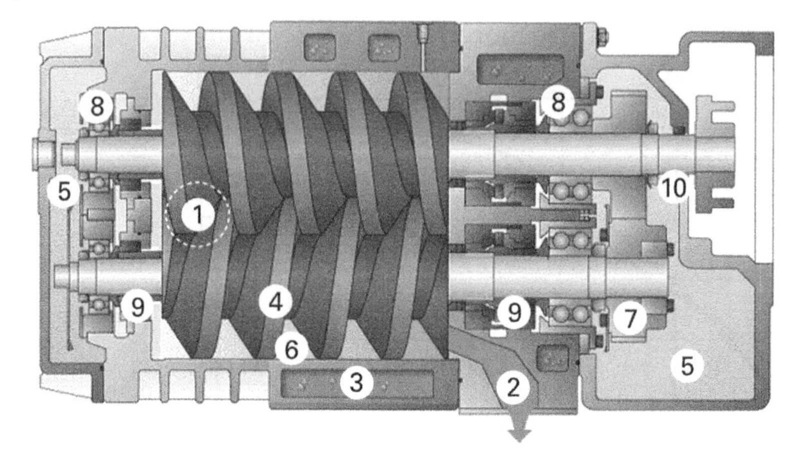

Fig. 10.14 Cross-sectional view of a screw vacuum pump. (1) Inlet, (2) discharge, (3) cooling jacket, (4) screws, (5) gear case, (6) stator housing, (7) drive gears, (8) shaft bearings, (9) shaft seals, and (10) motor shaft seal. Reproduced with permission from Busch Semiconductor Vacuum Group.

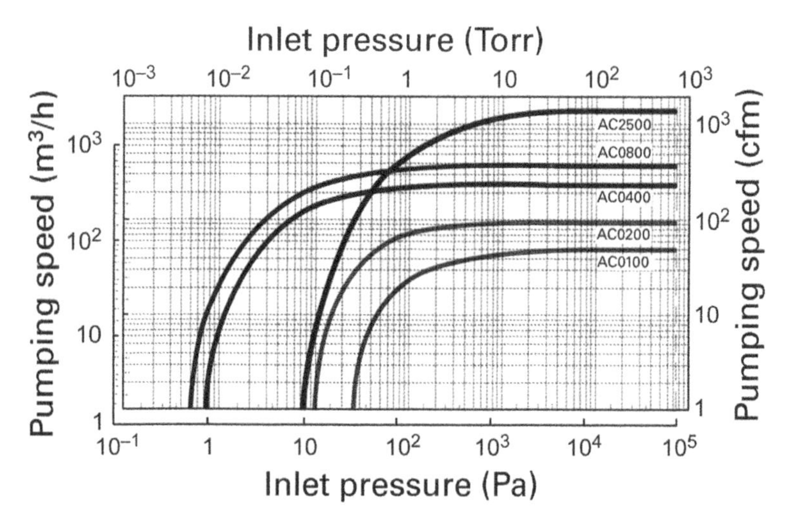

Fig. 10.15 Pumping speed curves for a family of screw pumps. Reproduced with permission from Busch Semiconductor Vacuum Group.

The eccentric shaft moves the piston–diaphragm assembly in rocking, swinging manner so as to draw gas in through the left-hand inlet valve, and a half-stroke later expel the gas through the right-hand exit valve. The pumps are oil-free and may be constructed from chemically inert materials. Up to four stages can be combined to provide ultimate pressures of, respectively, 7000–20,000 Pa (50–150 Torr),

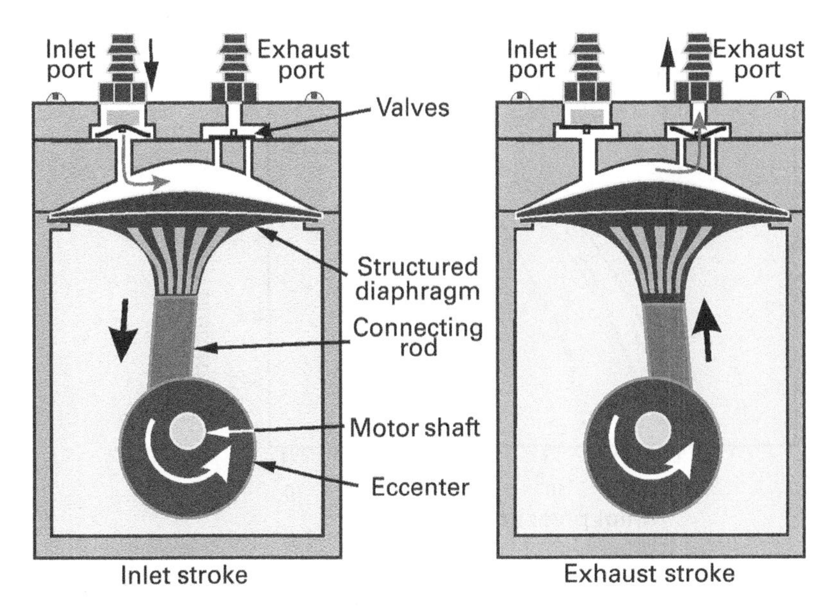

Fig. 10.16 Diaphragm pumping mechanism. Reproduced with permission from KNF Group.

700–1400 Pa (5–10 Torr), 200–400 Pa (1.5–3 Torr), and 50–75 Pa (0.4–0.6 Torr). Sizes do not exceed several m^3/h. Figure 10.17 illustrates the nitrogen and helium pumping speeds for small two-, three-, and four-stage diaphragm pumps. Diaphragm pumps are used for medium vacuum chemically aggressive applications, as oil-free backing pumps for turbomolecular-drag pumps—and to make espresso!

10.8 Reciprocating Piston

Many of the earliest mechanical vacuum pumps were piston pumps. Although these historic pumps used fluid as a sealant and lubricant (e.g., water, greases, and/or oils), the advent of low vapor-pressure, low-friction materials such as PTFE, has allowed this design to be reapplied to dry vacuum pump applications. Compared to a diaphragm pump, a significant advantage of a reciprocating piston pump is that the stroke of the piston can be longer, providing a much higher pumping speed while retaining the cleanliness of a high-performance diaphragm pump. One design for a piston pump with two pistons is shown in Fig. 10.18*a*,*b*. A representative multistage pumping speed curve is sketched in Fig. 10.18*c*, suggesting an ultimate pressure in the range of ~3 Pa (20 mTorr). Other designs are available that allow more than two pistons, and/or for pistons to pump on both the forward and backward stroke and producing an ultimate base pressure of the

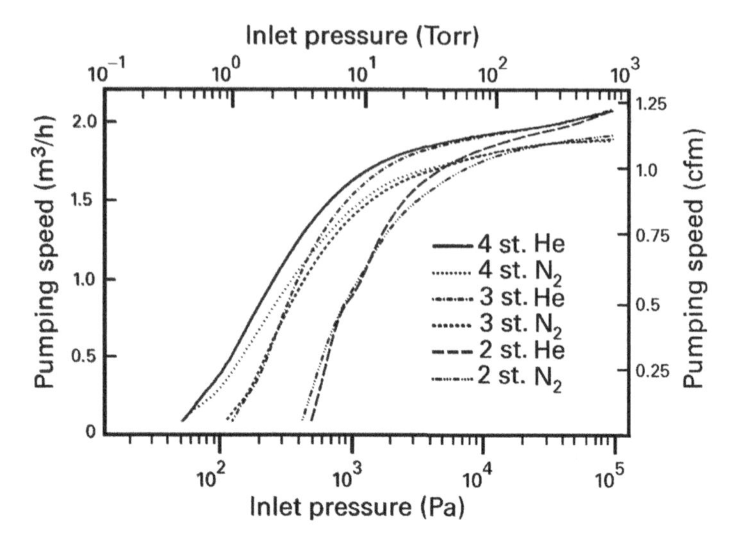

Fig. 10.17 Pumping speed curves for four-stage, three-stage, and two-stage diaphragm pumps for nitrogen and helium. F.J. Eckle et al. [11]/Reproduced with permission from Elsevier.

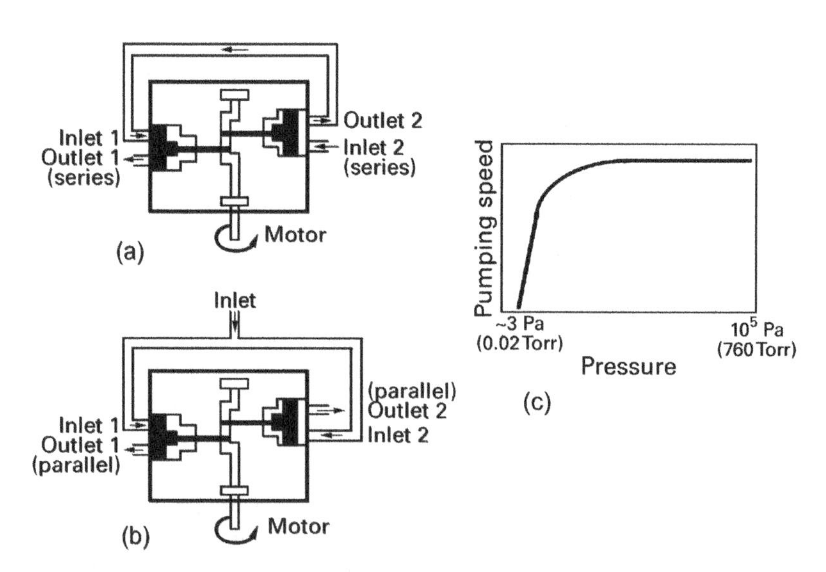

Fig. 10.18 Schematic illustration of a reciprocating piston pump configured for: (a) multistage operation, (b) single-stage operation, and (c) the associated pumping-speed curve. Reprinted with permission of Gessert Consulting, LLC.

order of 1.3 Pa (10 mTorr) [12]. A dry piston pump configured for single-stage operation, Fig. 10.18*b* would have a higher pumping speed, but demonstrate a higher base pressure on the order of 65 Pa (500 mTorr). For both diaphragm and piston pumps, it is often left to the end user to configure the pump for single or multistage operation. Similar to diaphragm pumps, scheduled maintenance involves replacement of worn components at vendor-specified time intervals of cumulative operation.

10.9 Mechanical Pump Operation

There are several common rules for operating mechanical pumps. Their exhaust should be vented outside the building. Most lubricated pumps are supplied with an oil mist separator, but it does not adequately remove the oil vapors. Laboratories and plant safety rules require the use of an outside vent. The vent hose should not run vertically from the exhaust connection, because water or other vapors, which have condensed on cold pipe walls, will flow downward into the pump exit, and contaminate the fluid. A satisfactory solution to this problem is the addition of a sump at the exhaust connection to collect the vapors before they can flow into the pump. A vane pump must also be vented at the time it is stopped to prevent fluid from being forced back into the vacuum system by external air pressure. Venting is done automatically in most pumps, and it can be achieved in others by the addition of a vent valve above the inlet port. The fluid level in mechanical pumps should be checked frequently, especially those that are used on systems regularly cycled to atmosphere or use gas ballast. An example of the oil lost in a 16 m^3/h rotary pump is: 0.03 cc/h (no gas ballast) and 3.018 cc/h (with gas ballast) [13]. As described in this reference, large pumps consume increased amounts of fluid. Fluid should be changed when the pump performance deteriorates or when it becomes discolored or contaminated with particles. Poor fluid maintenance is the major cause of mechanical pump failure. Flushing the pump and changing the fluid can solve 95% of all pump problems. Mechanical pump fluids are discussed in Chapter 17. Vapor pressure and kinematic viscosity of fluids used in mechanical pumps are plotted in Appendixes F.1 and F.3.

References

1 Kendall, B.R.F., *J. Vac. Sci. Technol.* **21**, 886 (1982).

2 ISO 21360-2:2020, *Vacuum Technology—Standard Methods for Measuring Vacuum-Pump Performance—Part 2*: Positive Displacement Vacuum Pumps, International Standards Organization, Geneva, Switzerland, 2020.

3 Pfeiffer Vacuum, Vacuum Technology and Know How, Chapter 4, Pfeiffer Vacuum, GmBH, Asslar, Germany, 2013, p. 59.

4 Van Atta, C.M., *Vacuum Science and Engineering*, McGraw-Hill, New York, 1965, Chapter 5.

5 ISO-21360-3:2020, *Vacuum Technology—Standard Methods for Measuring Vacuum-Pump Performance*, Mechanical Booster Pumps, International Standards Organization, Geneva, Switzerland, 2020.

6 A. J. Northey, *Rotary Compressor or Vacuum Pump*, U.S. Pat. # 2,097,037 (issued October 26, 1937).

7 Troup, A.P. and Dennis, N.T.M., *J. Vac. Sci. Technol., A* **9**, 2048 (1991).

8 Wycliffe, H., *J. Vac. Sci. Technol., A* **5**, 2608 (1987).

9 Bachmann, P. and Kuhn, M., *Vacuum* **41**, 1827 (1990).

10 Hablanian, M., *Vacuum* **41**, 1814 (1990).

11 Eckle, F.J., Bickert, P., Lachenmann, R., and Wortmann, B., *Vacuum* **47**, 799–801 (1996).

12 Hablanian, M., *High Vacuum Technology–A Practical Guide*, 2nd ed., Marcel Dekker, New York, 1997, pp. 185–190.

13 W. Umrath, *Fundamentals of Vacuum Technology*, Oerlikon-Leybold Vacuum, Publication 00.200.02, Kat.-Nr. 199 90, 2007 p. 141.

11

Turbomolecular Pumps

The axial-flow turbine, or turbopump, as it is known today, was introduced in 1958 by Becker [1]. His design originated from a "baffling" idea with which he had experimented a few years earlier—a disk with rotating blades mounted above a diffusion pump [2]. When it was introduced commercially, the pump had low speed and high cost, as compared to a diffusion pump. It did not backstream hydrocarbons and did not require a trap of any kind. Since its introduction, the turbomolecular pump has undergone rapid development both theoretically and experimentally. The important early theoretical development was the work on blade geometry at MIT in the group headed by Shapiro [3,4]. Practical advances in lubrication drive motors, and fabrication techniques have improved the quality and reliability of turbomolecular pumps. Modern turbomolecular pumps have high pumping speeds, large hydrogen compression, and low ultimate pressures. They do not backstream hydrocarbons, have high exhaust pressures, and can be backed by oil-free pumps. They are well-suited to pump gas cleanly at high flow rates or low pressures.

This chapter reviews the pumping mechanism in the free molecular pressure range, and it discusses the relations between pumping speed, compression, backing pump size, and gas flow. Vacuum lubrication techniques are discussed in Chapter 17. The operation and performance of pumps in a variety of applications are discussed in the Systems chapters.

11.1 Pumping Mechanism

The turbomolecular pump is a molecular turbine that compresses gas by momentum transfer from high-speed rotating blades to gas molecules. It is a high-speed molecular bat. The pump operates at rotor speeds in the range of 24,000–80,000 rpm

A Users Guide to Vacuum Technology, Fourth Edition. John F. O'Hanlon and Timothy A. Gessert.
© 2024 John Wiley & Sons, Inc. Published 2024 by John Wiley & Sons, Inc.

and is driven by solid-state power supplies or motor-generator sets. The relative velocity between the alternate slotted rotating blades and slotted stator blades makes it probable that a gas molecule will be transported from the pump inlet to the pump outlet. Each blade is able to support a pressure difference. Because this compression (pressure ratio) is small for a single stage, many stages are cascaded. For a series of stages, the compression for zero flow is approximately the product of the compressions for each stage. Blades impart momentum to the gas molecules most efficiently in the molecular flow region. The pump exhaust pressure must remain in the molecular or transition regime; therefore, it cannot exhaust to atmosphere. For this reason, a backing pump—typically a rotary vane, screw, scroll, or diaphragm pump—is required.

If the foreline pressure is allowed to increase to a point at which the rear blades are in viscous flow, the rotor will be subjected to an additional torque. The power required to rotate a shaft in steady state is proportional to the product of the rotor speed and the torque. The available power in many pumps is limited by the supply, so that too large an increase in foreline pressure (increase in torque) will cause a sudden reduction in the rotor speed and a loss in gas pumping speed. One design used a constant-speed motor whose power consumption increased in proportion to the gas load or increase in torque. Effects of backing pump size and rules for selecting backing pumps are discussed later in this section.

11.2 Speed–Compression Relations

Continuum methods that give a reasonable account of pump performance have been developed [2]; later, Monte Carlo analysis of single-blade rows provided Kruger and Shapiro [3,4,5] with design insights. Their probabilistic methods were valid for all no-flow situations. The model used to analyze a single rotor disk is shown in Fig. 11.1. This disk, which rotated with a tip velocity near the thermal velocity of air, imparted a directed momentum to a gas molecule on collision. The blades were slotted at an angle to make the probability of a gas molecule being transmitted from the inlet to the outlet much greater than in the reverse direction. The stator disks are slotted in the opposite direction. Γ_1 and Γ_1 are, respectively, the number of molecules incident on the disk per unit time at the inlet and at the outlet. a_{12} is the fraction of Γ_1 transmitted from the inlet (1) to the outlet (2) and a_{21} is the fraction of Γ_2 transmitted from the outlet to the inlet. Now define the net flux of molecules through the blades to be a function of the Ho coefficient W, the ratio of net flux to incident flux. Assuming mass conservation and zero-flow conditions, one obtains:

$$\Gamma_1 W = \Gamma_1 a_{12} - \Gamma_2 a_{21} \frac{\partial^2 \Omega}{\partial u^2} \tag{11.1}$$

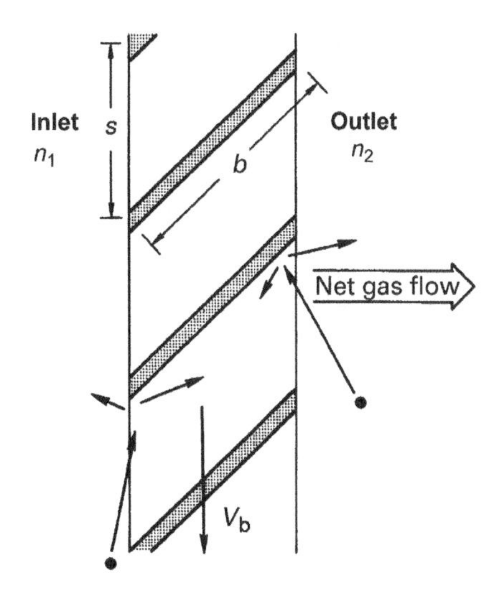

Fig. 11.1 Sectional view of a rotor disk.

or by rearrangement:

$$\frac{\Gamma_2}{\Gamma_1} = \frac{a_{12}}{a_{21}} - \frac{W}{a_{21}} \tag{11.2}$$

If the gas temperature and the velocity distributions are the same everywhere, the ratio Γ_2/Γ_1 will be equal to the pressure ratio P_2/P_1. The ratio of outlet to inlet pressure is called the compression K. Again, for zero-flow conditions, this becomes:

$$\frac{P_2}{P_1} = K = \frac{a_{12}}{a_{21}} - \frac{W}{a_{21}} \tag{11.3}$$

Maximum compression occurs at zero flow, while unity compression occurs at maximum speed or mass flow. This is a general property of a fan. Envision the compression and flow in a household vacuum cleaner as you hold your hand over the inlet (zero flow) and release it (maximum flow). Let us now examine the maximum compression, maximum flow, and the intermediate case for the region between these extremes.

11.2.1 Maximum Compression

Assuming zero flow, $W = 0$, (11.3) reduces to:

$$K = K_{max} = \frac{a_{12}}{a_{21}} \tag{11.4}$$

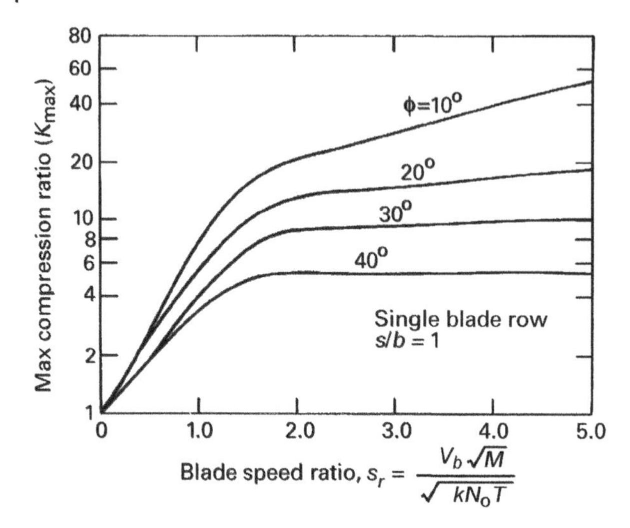

Fig. 11.2 Calculated curve of the compression ratio at zero flow for a single blade row with $s/b = 1$. C.H. Kruger and A.H. Shapiro [4]/Reproduced with permission from Elsevier.

Equation 11.4 states the maximum compression is the ratio of forward to reverse transmission probabilities. To maximize the compression, the ratio a_{12} to a_{21} is maximized. Kruger and Shapiro calculated these forward and reverse transmission probabilities by Monte Carlo techniques. They solved for the transmission probabilities as a function of the blade angle φ, the blade spacing-to-chord ratio s/b, and the blade speed ratio $s_r = V_b(M/2kN_oT)^{1/2}$. Figure 11.2 sketches the results of a calculation for the single-stage compression at zero flow. From this curve, we observe the logarithm of the compression is approximately linear with blade speed ratio for $s_r \leq 1.5$ or:

$$K_{\max} \propto \exp\left[\frac{V_b\sqrt{M}}{\sqrt{2kN_oT}}\right] \tag{11.5}$$

The compression is exponentially dependent on rotor speed and $M^{1/2}$. The constant of proportionality is dependent on the blade angle and s/b. In particular, hydrogen will have a compression much smaller than that for any other gas. For a blade tip velocity of 400 m/s the speed ratio for argon is about unity, and for hydrogen it is about 0.3. From Fig. 11.2 for $\varphi = 30°$ we find this blade velocity corresponds to compressions $K(H_2) = 1.6$ and $K(Ar) = 4$. If two disks (five rotors and five stators) are cascaded, the net compression is approximately 100 and 10^6, respectively. A total of 15 disks would raise $K(H_2)$ to 1000. A stator blade has the same compression and transmission as a rotor. An observer sitting on a stator sees blades moving with the same relative velocity as an observer sitting on a rotor. The linear blade velocity is proportional to the radius and the rotor angular frequency ($V_b = r\omega$). An area closer to the center of the rotor will have a smaller

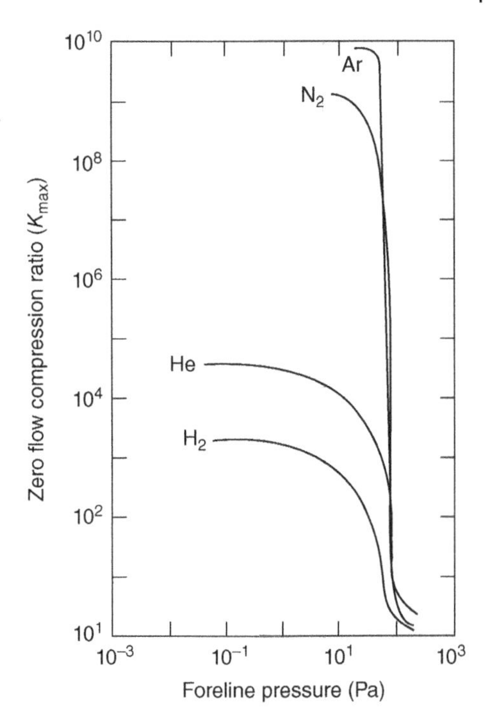

Fig. 11.3 Measured compression ratio for zero flow in a Pfeiffer TPU-400 turbomolecular pump. Reproduced with permission from Pfeiffer Vacuum GmbH.

speed ratio and blade spacing-to-chord ratio. The data of Kruger and Shapiro show that the net effect of these changes is a low compression for the region closest to the rotor axis. Blades designed to have a high $K(H_2)$ should be slotted to a depth of only about 30% of the radius.

Experimental compressions are given for a horizontal-axis, dual-rotor pump in Fig. 11.3. These data were taken in a manner identical to lobe blower compression curves, or diffusion pump forepressure tolerance curves. Gas is admitted to the foreline of a blanked-off pump and the measured compression is the ratio of fore-pressure to inlet pressure. As the foreline or backing line pressure is increased, the rear blades first go into transition flow and then into viscous flow, the rotor speed decreases, and the compression decreases.

11.2.2 Maximum Speed

Maximum speed is achieved when the compression of a blade is unity. However, the model assumed by Kruger and Shapiro holds only for the zero-flow case. Chang and Jou [6] derived an analytic expression for the maximum speed with gas flow given in terms of transmission probabilities. They predicted the maximum speed factor to be:

$$W_{max} = \frac{a_{12} - a_{21}}{1 - a_{21}} \tag{11.6}$$

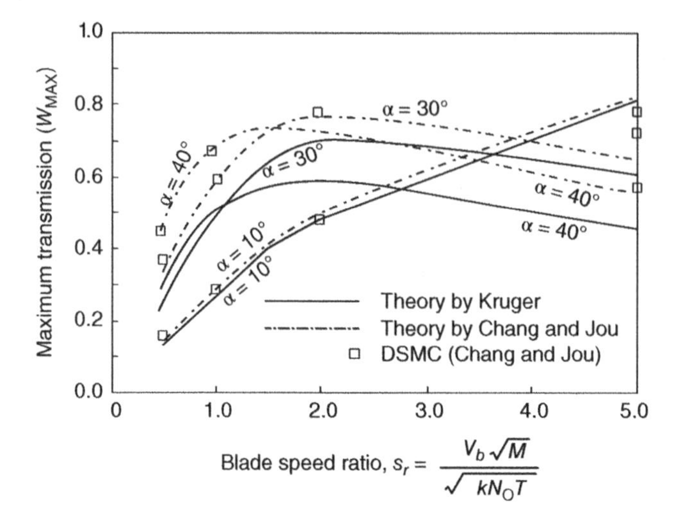

Fig. 11.4 Maximum speed factor (Ho coefficient), single blade row with $s/b = 1$, for three different blade angles. Y-W. Chang and R-Y. Jou [6]/Reproduced with permission from AIP Publishing.

This value is greater than that predicted ($W_{\max} = a_{12} - a_{21}$) by Kruger and Shapiro. To maximize W, a_{12} (forward transmission probability) must be maximized and a_{21} (reverse transmission probability) minimized. The Ho coefficient W for a single blade is given in Fig. 11.4 as a function of blade-speed ratio for a spacing-to-chord ratio $s/b = 1$ and three blade angles. For s_r small, W is reasonably linear with s_r, so that one could write:

$$W \propto \left[\frac{V_b\sqrt{M}}{\sqrt{2kN_oT}} \right] \tag{11.7}$$

Because the molecular arrival rate is proportional to thermal velocity $(kT/m)^{1/2}$, the net pumping speed of the blade is approximately independent of the mass of the impinging molecules, therefore:

$$S \propto V_b \tag{11.8}$$

Monte Carlo methods have been used to calculate the net Ho coefficient for blades in series. These calculations show the Ho coefficient to increase with the number of stages [7], with a saturation in speed after several blades are operated in series.

11.2.3 General Relation

The maximum compression (11.4) occurs only when the pump is pumping no gas (the pump is at base pressure), whereas maximum speed occurs when the pressure ratio is near 1. An operating pump works throughout the region between these extremes. Chang and Jou [6] derived an expression for the transmission factor W within the region between these limits. It is complex, so their approximation, valid in the range $0 < s_1 \ll 1$ is given here:

$$W \cong \frac{a_{12} - \left(\dfrac{n_1}{n_2}\right) a_{21}}{1 - \dfrac{1}{2}\left(1 + \dfrac{n_1}{n_2}\right) a_{21} + \dfrac{1}{2\sqrt{\pi}}\left(\dfrac{n_2}{n_1} - \dfrac{n_1}{n_2}\right) a_{21} s_1^{-1}} \tag{11.9}$$

Equation 11.9 is plotted in Fig. 11.5 for a single blade row for compressions (pressure ratio n_2/n_1) in the range 1–7. The Chang–Jou theory predicted a higher transmission coefficient than predicted by Kruger; however, it agrees with direct simulation Monte Carlo (DSMC) calculations. The differences between the two models become large when blades of open structure are analyzed. Using blades of closed structure (i.e., 10°) the two models give similar results. Figure 11.6 illustrates the speed factors for various gases in a commercial turbomolecular pump.

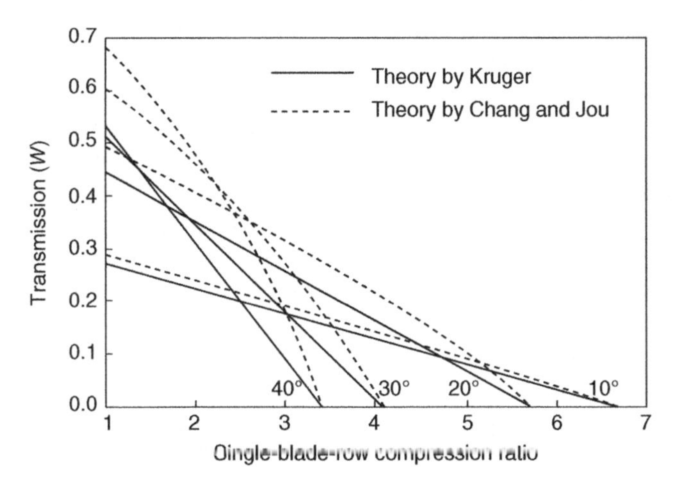

Fig. 11.5 Transmission coefficient versus compression ratio (pressure ratio): Comparison of Kruger's results and Chang-Jou approximate solution for a single blade row with blade speed ratio = 1 and $s/b = 1$. Y-W. Chang and R-Y. Jou [6]/Reproduced with permission from AIP Publishing.

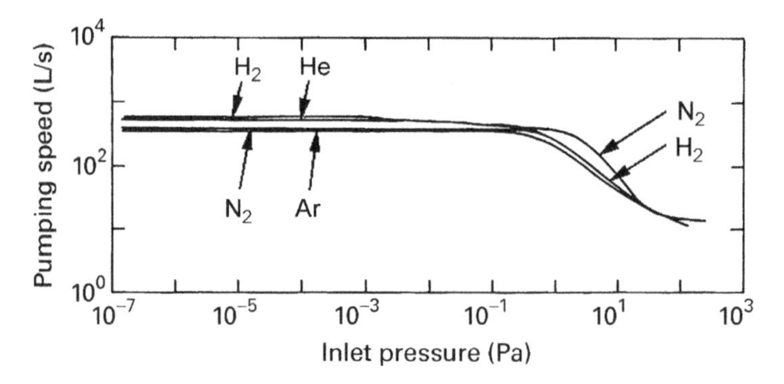

Fig. 11.6 Measured pumping speeds for the Pfeiffer TPU-400 turbomolecular pump. Reproduced with permission from Pfeiffer Vacuum GmbH.

At a nitrogen inlet pressure of 0.9 Pa, the rotor is still running at full speed and the gas throughput is 400 Pa-L/s. This is twice the throughput of a nominal 6-in. diameter diffusion pump operating at an inlet pressure of 0.1 Pa.

11.3 Ultimate Pressure

The ultimate pressure of a turbomolecular pump is determined by the compression for light gases and by the amount of outgassing. This is qualitatively similar to a diffusion pump. (See discussion in Section 12.2.) The main difference between the turbomolecular pump and the diffusion pump is the low hydrogen compression in the turbopump. It is low enough so that the ultimate hydrogen pressure will be determined by $K_{max}(H_2)$ and its partial pressure in the foreline. In some very old pumps, which had a water-vapor compression of less than 10^4, the water-vapor partial pressure may also be compression limited [8]. The partial pressure of all other gases and vapors will be limited by their respective outgassing rates and pumping speeds. During pumping of an unbaked turbomolecular pump, the slow release of water vapor from the blades closest to the inlet may slightly decrease the rate of water removal. The effective compression for water release from the first few blades is much less than K_{max}.

Henning [9] has shown the forepump oil dominates the partial pressure of hydrogen found in a turbomolecular pump system. The ultimate pressure varied from 2×10^{-7} to 5×10^{-7} Pa as a function of the type of oil in the forepump. Because the turbopump oil has a lower vapor pressure than the mechanical pump oil, it will not contribute so much hydrogen to the background as the forepump oil. Henning and Lotz [10] used perfluoropolyether pump fluid for lubricating both the turbomolecular pump and the forepump in the presence of corrosive gases. Using a mass spectrometer, they observed distinct fluorine and hydrogen peaks. This decomposition occurred because the local heating of the bearings caused the oil temperature to exceed the range of thermal stability. They concluded from the presence of hydrogen that the ultimate pressure of the pump was not improved with hydrogen-free fluids. They postulated the limiting pressure was caused by the diffusion of hydrogen through the foreline seals.

Ultimate pressures for baked systems between 2×10^{-8} and 5×10^{-9} Pa are possible with high-compression turbomolecular pumps without the assistance in pumping from cryo baffles or titanium sublimation pumps. Hydrogen will constitute more than 99% if the residual gas at the ultimate pressure [8]. Ultimate pressures of order 10^{-7} Pa (10^{-9} Torr) have been achieved with the use of tandem turbomolecular pumps [11]. Many of the characteristics of turbomolecular pumps are described in a related ISO Standard [12].

11.4 Turbomolecular Pump Designs

A single blade row is inadequate to serve as a high vacuum pump. Multiple-bladed structures with 8–20 disks will provide adequate compression and speed to make a functional pump. As in the diffusion pump, the stages nearest the high vacuum inlet serve a purpose different from those nearest the outlet. The flow through each stage is constant or, stated another way, the product of speed and pressure is a constant. The blades nearest the inlet are designed to have a high pumping speed and a low compression. The blades nearest the foreline are designed to have a high compression and a low pumping speed. For economic reasons it would be impractical to make each blade differently than its neighbor. A compromise results in groups of two or three blade types. The blades in each group are designed for a specific speed and compression. Each group of blades may be considered analogous to a diffusion pump jet. The pump designer may trade-off pumping speed and light gas compression by the choice of s/b ratio and φ. Pumps exhibiting a large compression for hydrogen use blades that are optically opaque (s/b, φ small) compared to those designed to maximize the pumping speed (s/b, φ large).

The earliest design was a horizontal-axis, dual pump structure, with the gas inlet located centrally between the two sets of rotors. This early pump used a three-stage rotor that was abrasive-machined and balanced. The rotor disks were positioned on a cooled hub that was allowed to thermally equilibrate with the disks and hold them rigidly in position. Stator disks were formed in a similar

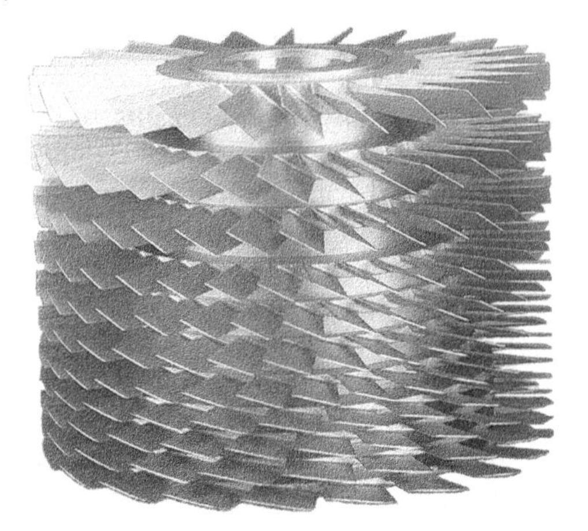

Fig. 11.7 Rotor from a classical vertical turbomolecular pump. Reproduced with permission from Leybold.

manner, cut into half sections, and mounted stage-by-stage as the rotor is moved into the housing. Modern pumps rotate on a vertical-axis as shown in Fig. 11.7. This rotor is machined from a single piece of aluminum, slotted, with three sets of blade angles. In this manner the compression increases and the speed decreases as gas flows toward the foreline. Turbomolecular pump stator blades are constructed from aluminum alloy stampings [7].

The single-rotor, vertical-axis pump has little conductance loss between the inlet flange and the rotor. Pump flanges should not be subjected to a steady or transient twisting moment by using the inlet flange to bear the load of a heavy work chamber, especially a cantilever load, or the impulse of a heavy flange closure [13]. Improper loading can cause premature bearing failure. Turbopumps should be suspended from the system by their inlet flanges. The inlet flange should not be used as a mounting platform for a heavy system.

The maximum rotational speed is ~80,000 rpm, or a blade tip velocity of about 500 m/s. These limits are due to bearing tolerances, thermal coefficients of expansion, and material stress limits. Ball bearings are the component subjected to the greatest wear. Oil, either flowing or in a mist, may be used to lubricate and cool bearings. The oil is, in turn, either water- or refrigeration-cooled. Small-diameter bearings are desired to increase the bearing lifetime. Some pumps use grease-packed bearings [14], although they have been shown to be less reliable than oil-lubricated bearings [15]. Magnetically levitated bearings with extremely low wear rates are common; such a pump is illustrated in Fig. 11.8.

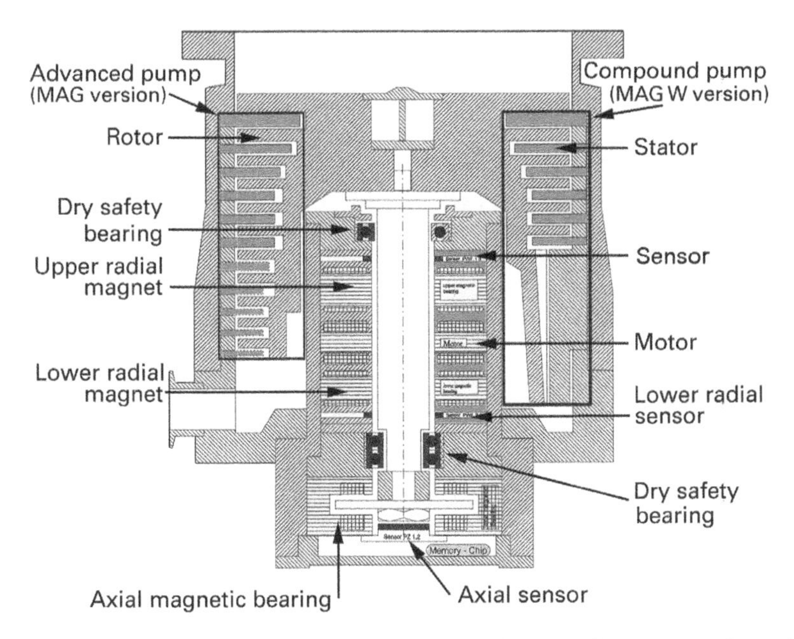

Fig. 11.8 Cross section of a magnetically levitated turbomolecular pump. Reproduced with permission from Leybold Vakuum GmbH.

11.5 Turbo-Drag Pumps

The conventional turbomolecular pump requires a backing pump that can produce exhaust pressures of order 0.1–1 Pa (10^{-3}–10^{-2} Torr). Oil-sealed two-stage rotary mechanical pumps formerly were the pump of choice for this application. In the last decade, cleanliness considerations have dictated the use of completely oil-free pumping systems. Magnetically levitated turbos backed with oil-free mechanical pumps significantly reduce the level of hydrocarbon contamination in turbo systems [16]. Many of the dry pumps discussed in Chapter 10 cannot meet the low-pressure requirements for backing a turbo. As a result, molecular drag pump stages have been incorporated on the axis of the turbo after its exhaust, to compress the exiting gas to a pressure compatible with dry pumps such as the screw, scroll, and diaphragm. One common molecular drag pump, the Holweck pump, consists of either a grooved (threaded) rotating cylinder inside a smooth stator cylinder or vice-versa. The design of one integrated turbomolecular drag rotor is illustrated in Fig. 11.9. Such turbo-drag combinations have foreline tolerances in the range of 4–20 Pa (30–150 mTorr). This allows the use of small low-cost dry backing pumps. The high vacuum pumping speed of a compound turbo-drag pump is similar to that of a pure turbo. See Fig. 11.10. At high flow rates, one observes some flow limitation in the drag stage, and this slightly decreases pumping speed.

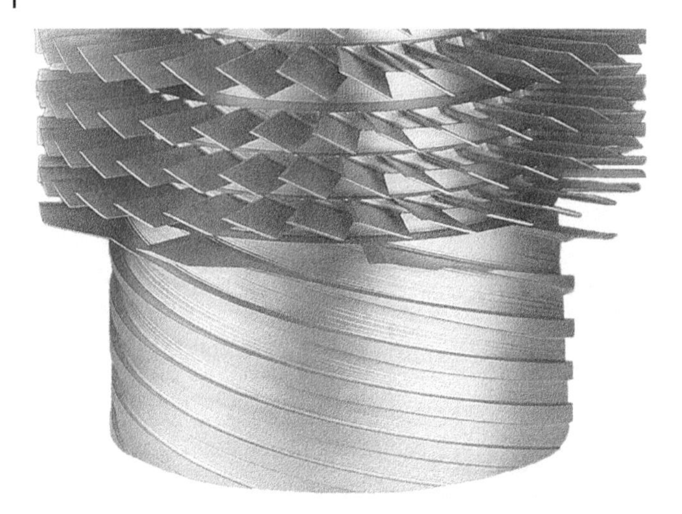

Fig. 11.9 Detail of the Holweck revolving-screw molecular drag stage used in a compound turbo-drag pump. Reproduced with permission from Leybold.

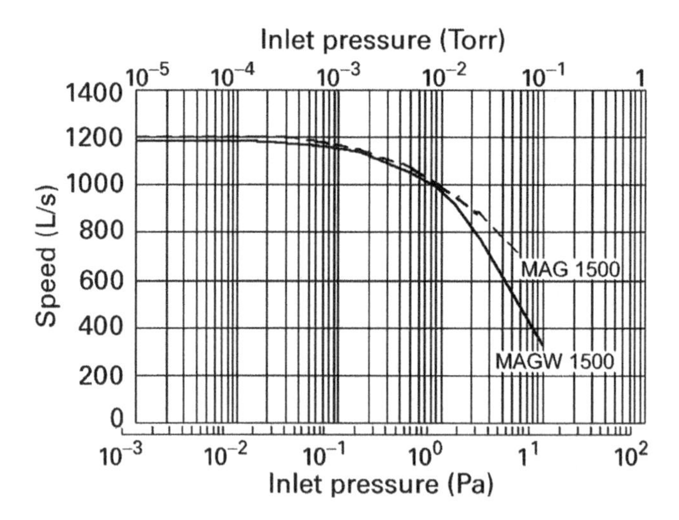

Fig. 11.10 Pumping speeds of a magnetically levitated turbopump (MAG 1500) and a turbo-drag magnetically levitated turbopump (MAG W 1500). Reproduced with permission from Leybold Vakuum GmbH.

References

1 Becker, W., *Vakuum Technik* **7**, 149 (1958).
2 Becker, W., *Vakuum Technik* **15**, 211 (1966).
3 Kruger, C.H. and Shapiro, A.H., in *Rarefied Gas Dynamics. Proceedings of the Second International Symposium on Rarefied Gas Dynamics, Berkeley, CA*, L. Talbot, Ed., Academic, New York, 1961, pp. 117–140.
4 Kruger, C.H. and Shapiro, A.H., *Transactions of the 7th National Symposium of the American Vacuum Society, 1960*, Pergamon Press, New York, 1961, pp. 6–12.
5 C.H. Kruger, *The Axial Flow Compressor in the Free-Molecular Range*, Doctoral Dissertation, Department of Mechanical Engineering, M.I.T., Cambridge, MA, 1960.
6 Chang, Y.-W. and Jou, R.-Y., *J. Vac. Sci. Technol., A* **19**, 2900 (2001).
7 Mirgel, K.H., *J. Vac. Sci. Technol.* **9**, 408 (1972).
8 J. Henning, *Proceedings of the 6th International Vacuum Congress, Kyoto, Japan, Jap. J. Appl. Phys.*, Vol. 13, Suppl. 2–1, H. Kumagai, Ed., p. 5, 1974.
9 Henning, J., *Vacuum* **21** (1971).
10 Henning, J. and Lotz, H., *Vacuum* **27**, 171 (1977).
11 Enosawa, H., Urano, C., Kawashima, T., and Yamanoto, M., *J. Vac. Sci. Technol., A* **8**, 2768 (1990).
12 ISO-21360-4:2020, *Vacuum Technology—Standard Methods for Measuring Vacuum-Pump Performance*. Turbomolecular Pumps, International Standards, Organization, Geneva, Switzerland, 2020.
13 Bernhardt, K.-H., Mädler, M., and Ganschow, O., *Vacuum* **44**, 721 (1993).
14 Osterstrom, G. and Knecht, T., *J. Vac. Sci. Technol.* **16**, 746 (1979).
15 Heldner, M. and Kabelitz, H.-P., *J. Vac. Sci. Technol., A* **8**, 2772 (1990).
16 Conrad, A. and Ganschow, O., *Vacuum* **44**, 681 (1993).

12

Diffusion Pumps

The diffusion pump has been a part of vacuum technology for a century. The discovery of low vapor pressure pumping fluids and the ability to control backstreaming soon made it a practical pump [1]. Today, it is still in wide use, even though ion, cryogenic, and turbomolecular pumps have captured its market for high-technology applications. Because of its long history, it has been the subject of extensive study. Its problems are thoroughly understood, and its performance is, in some cases, understated. Many excellent articles summarize the pump's properties for practical applications [2,3,4] and review its operation [1,5,6,7]. Several texts cover its theory of operation and design [8,9].

Here we review the basic mechanisms of pump speed, throughput, heat effects, backstreaming, baffles, and traps. System problems are treated in later chapters.

12.1 Pumping Mechanism

The name "diffusion pump," first coined by Gaede [10], does not describe the operation of the pump accurately. The diffusion pump is a vapor jet pump, which transports gas by momentum transfer when a fast moving vapor molecule collides with a gas molecule. A motive fluid such as hydrocarbon oil, an organic liquid, or mercury is heated in the boiler until it vaporizes. The vapors flow up the chimney and out through a series of nozzles.

The history of the diffusion pump is filled with innovations. Over a dozen of these are included in the sketch of a metal-bodied diffusion pump illustrated in Fig. 12.1 including: a cooled hood to reduce vapor backstreaming [11,12]; a heater to compensate for heat loss [13]; a streamlined surface to avoid turbulence [14]; multiple jet stages to obtain low pressures [15]; an enlarged upper casing to give a larger pumping aperture and increased speed [16]; a baffle to impede the access to

A Users Guide to Vacuum Technology, Fourth Edition. John F. O'Hanlon
and Timothy A. Gessert.
© 2024 John Wiley & Sons, Inc. Published 2024 by John Wiley & Sons, Inc.

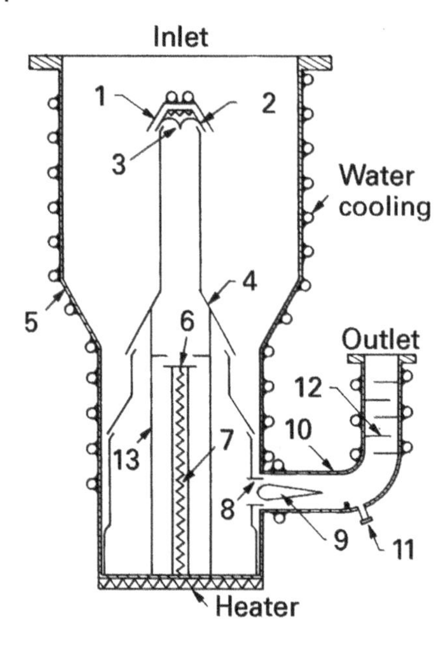

Fig. 12.1 A sectional view of a metal diffusion pump and some of its innovations: (1) cooled hood, (2) nozzle cap heater, (3) streamlined surface, (4) multiple stages, (5) enlarged casing, (6) fluid baffle, (7) heater, (8) lateral ejector, (9) conical body, (10) hot diffuser, (11) catchment and drain, (12) baffle, and (13) concentric chimneys. N.A. Florescu [6]/Reproduced with permission from Elsevier.

the jet of liquid splashed from the boiler below [17]; a heater for superheating the vapor [18]; a lateral ejector stage [19]; a conical body allowing operation against increased forepressure [20]; a hot maintained diffuser for oil purification [21]; a catchment and drain off for highly volatile fractions [22]; outlet baffles to reduce oil loss [15]; and concentric chimneys that allow oil fractionation.

The nozzles—three in Fig. 12.1—direct their vapor streams downward and toward the cooled outer wall where vapor condenses and returns to the boiler. The vapor flow in the jet is supersonic. Gases that diffuse into this supersonic vapor stream are, on average, given a downward momentum and ejected into a region of higher pressure. Modern pumps have several stages of compression—usually three to five for small pumps, and up to seven for large pumps. Each stage compresses the gas to a successively higher pressure than the preceding stage as it transports it toward the outlet.

The boiler pressure in a modern diffusion pump is about 200 Pa (2 Torr). Ideally, the pump cannot sustain a pressure drop any larger than this between its inlet and outlet. The practical maximum value of forepressure tolerated by the pump is less than the boiler pressure. This maximum value called the *critical forepressure*, ranges from 25 to 75 Pa (0.2–0.6 Torr) and is dependent on pump design and boiler pressure. The latter number is typical of modern pumps. The diffusion pump cannot eject gas to atmospheric pressure. It must be "backed" by another pump in order to keep the forepressure (exhaust pressure or foreline pressure) below the critical forepressure. Rotary vane or piston pumps or combinations of rotary and lobe blowers are used as "backing" or "fore" pumps. If the forepressure exceeds the critical value, all pumping action will cease. The pumping action ceases at high pressures because the directed supersonic vapor stream no longer extends from the jet to the wall but is ended in a shock front close to the jet [5]. Those vapor molecules beyond the shock front are randomly directed and cannot stop gas molecules from returning to the inlet. As the critical forepressure is exceeded, the inlet pressure will rise sharply and uncontrollably in response to the cessation of pumping. The critical forepressure should never be exceeded. In newer pumps the inlet pressure and the pumping speed will be unaffected by the value of forepressure as long as it is below the critical value and the gas throughput is low. At maximum throughput the critical forepressure will be reduced to about 3/4 of its normal value [2]. The amount of reduction is a function of the pump design, heater power, and pump fluid.

Each stage of the vapor pump has a characteristic speed and pressure drop. Since the jets are in series, the gas flow Q is the same through each stage. The flow, $Q = S\Delta P$, is the product of the speed of the jet times the pressure drop across the jet. The top jet has the largest speed (and the largest aperture) and the lowest pressure drop. The vapor density in the top jet is less than that in the lower jets. Because the gas flow through a series of jets is the same, each successive jet can have a larger pressure drop and a smaller pumping speed. The last jet has the greatest pressure drop. Many pumps use a vapor ejector as the last stage because it is an efficient gas compressor in this pressure region. The combination of jets and ejector produces a pump with a higher forepressure tolerance than is possible with vapor jets alone. Fractionating pumps [23] have concentric chimneys and boilers with long fluid flow paths that allow light fractions to be preferentially directed to the lower jets after condensation. Degassing of the fluid is accomplished by maintaining a section of the ejector walls at an elevated temperature [21]. Pumps with these and other advances, which use heavy fluids produced by molecular distillation, can pump to 5×10^{-5} Pa (3×10^{-7} Torr) without trapping and $<5\times10^{-9}$ Pa ($<3\times10^{-11}$ Torr) when trapped with liquid nitrogen.

12.2 Speed–Throughput Characteristics

The four operating regions of the diffusion pump are the constant speed, constant throughput, mechanical pump, and compression ratio regions. They are graphically illustrated in Fig. 12.2. In its normal operating range, the diffusion pump is a constant speed device. Its efficiency for pumping gas molecules, the Ho coefficient, also called the capture probability, is about 0.5 for the pump alone, but is approximately 0.3 when a trap and valve is used. The usual operating range for constant speed is about 10^{-1} Pa (10^{-3} Torr) to $<10^{-9}$ Pa (10^{-11} Torr) for most gases.

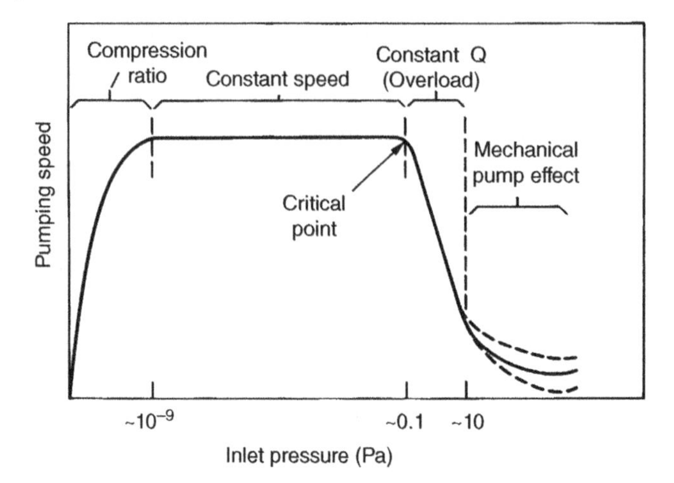

Fig. 12.2 Typical diffusion pump speed curve for a given gas. Four regions are visible: (1) compression ratio limit; (2) normal operation—constant speed, (3) first jet overloaded—nearly constant throughput, and (4) mechanical pump limit. M.H. Hablanian [3]/Reproduced with permission from IOP Publishing.

The maximum or limiting inlet pressure is called the "critical inlet pressure" and it corresponds to the point at which the top jet fails. In a 6-in. diffusion pump the top jet becomes unstable at pressures of ~0.1 Pa (10^{-3} Torr), the middle jet at pressures of about 3 Pa (0.03 Torr), and the bottom jet at pressures of about 40 Pa (0.4 Torr) [24].

The gas throughput in the constant speed range is the product of the inlet pressure and the speed of the pump at the inlet flange. It rises linearly with pressure until the critical inlet pressure is reached. Above that pressure the pump throughput is constant until the jets all cease to function. At higher pressures the throughput again increases in accordance with the speed of the backing pump. The maximum usable throughput of the diffusion pump corresponds to the product of the inlet speed and the critical inlet pressure. If that pressure is exceeded, the backstreaming may increase and jet instabilities will appear. These instabilities make pressure control difficult. The maximum throughput should not be exceeded in the steady state, although it often happens for short periods of time during crossover from rough to high vacuum pumping.

Exceeding the critical forepressure in a well-designed pump is usually the result of equipment malfunction, while the critical inlet pressure is easily exceeded by incorrect operation. If the pump is equipped with a sufficiently large forepump, the critical forepressure can still be exceeded if a leak occurs in the foreline, the mechanical pump oil level is too low, the mechanical pump belt is loose, or a section of the diffusion pump heater is open. The critical inlet pressure can be

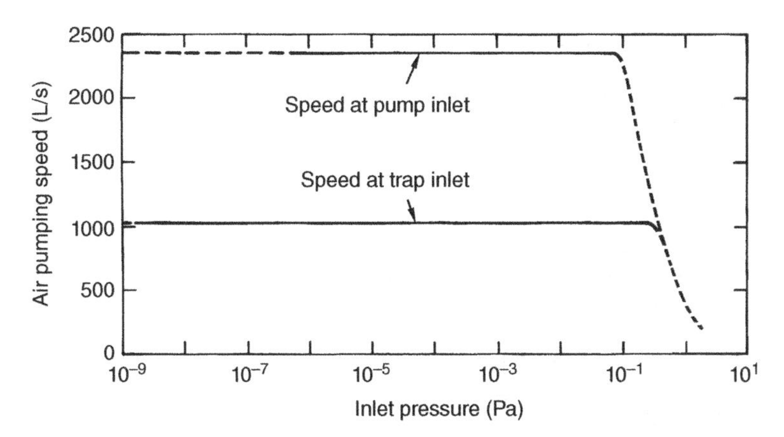

Fig. 12.3 Air pumping speed of the Varian VHS-6, 6-in. diffusion pump with and without a liquid nitrogen trap. Reproduced with permission from Varian Associates.

exceeded by operational error, but otherwise the top jet will continue to pump unless there is a partial heater failure or a large leak.

The speed does not remain constant to extremely low pressures, but it decreases toward zero as shown in the compression ratio region of Fig. 12.2. This curve decreases at low pressures because of the large but finite compression ratio of the diffusion pump jets. The pump whose hypothetical speed curve for one gas is shown in Fig. 12.2 has an ultimate pressure of 10^{-10} Pa (10^{-12} Torr). If, at that point, its forepressure were 1 Pa, its compression ratio would be 10^{10}, a value similar to those quoted in the literature. Figure 12.3 shows the air pumping speed for a 6-in. diffusion pump with and without an LN_2 baffle. All diffusion pumps have some small reverse flow of the gas being pumped; and although this reverse flow is exceedingly small for heavy gases, it may exist for light gases under certain conditions. Because of their high thermal velocity and small collision cross-section, the compression ratio of light gases, such as hydrogen and helium, is lower than that of heavy gases. Figure 12.4 sketches the relative pumping speeds of several gases and vapors as a function of their inlet pressure, and it illustrates the effect of a low compression ratio for hydrogen. The compression ratio for heavy gases will be about 10^8–10^{10}. For light gases it can be small enough (10^3–10^6) in some pumps so that a small foreline concentration can be detected at the inlet [1,25]. It is this phenomenon that explains why hydrogen emanating from an ion gauge in the foreline can be detected at the inlet. The operation of the counterflow leak detector [26] is based on this principle. That is, the detector is located at the inlet, and the test piece is appended to the foreline. The compression ratio for heavy gases is adequate to produce a low pressure in the detector, while at the same time allowing helium to back diffuse and be counted. The counterflow leak detector is described in Chapter 24.

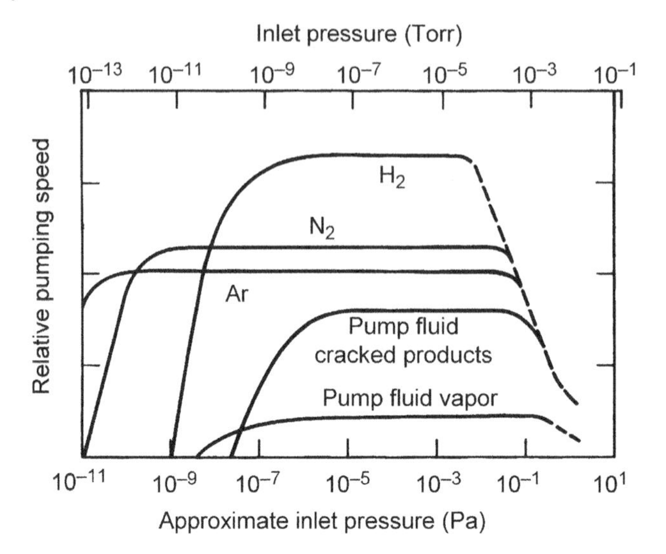

Fig. 12.4 Diffusion pump performance for individual gases. S.G. Burnnet and M.H. Hablanian [25], /Reproduced with permission from The Institute of Environmental Sciences.

The ultimate or limiting pressure in a diffusion-pumped system can be a result of reaching the compression ratio limit or wall outgassing. For the ideal pump with zero outgassing above the top jet and in the work chamber, and using a perfect baffle to collect all oil vapor fragments, the ultimate pressure would be the sum of each of the partial pressures in the foreline divided by their respective compression ratios:

$$P_u = \frac{P_{f1}}{k_1} + \frac{P_{f2}}{k_2} + \frac{P_{f3}}{k_3} \cdots \tag{12.1}$$

For the case in which the base pressure of the system is achieved in the pump's constant speed region, the ultimate pressure is the sum of each independent gas flow Q_i, divided by the pumping speed for each gas S_i:

$$P_u = \frac{Q_1}{S_1} + \frac{Q_2}{S_2} + \frac{Q_3}{S_3} \cdots \tag{12.2}$$

In practice the ultimate pressure is usually determined by Eq. (12.2). In some situations, it can be due to gases dissolved in the fluid, the lightest fractions of pump fluid that are released by the trap, or the compression limit. For example, the partial pressure of H_2 and perhaps He may be determined by the compression ratio Eq. (12.1). A single pumping speed curve (Fig. 12.2), representative of all gases, cannot be drawn because the pumping speed is not the same for all gases.

See Fig. 12.4. The pumping speed is greater for light gases, but not in proportion to $m^{1/2}$ as predicted by the ideal gas law. Usually, He will be pumped about 20% faster than N_2.

12.3 Boiler Heating Effects

The effect of boiler heat input variation is summarized concisely in Fig. 12.5. The general trends are that the oil temperature, forepressure tolerance, and throughput increase with boiler power, while pumping speeds decrease at high heat inputs because of the increased density of oil molecules in the vapor stream [7]. It is not possible to optimize the pumping speed for all gases at the same heater power because of the differences in mass and thermal velocity. Each gas reaches maximum speed at a different input power. The pumping speed is a function of the momentum transfer between fluid and gas molecules. Heavy fluid molecules often have lower pumping speeds than light molecules unless the boiler temperature is adjusted. Excessively increasing the boiler temperature also hastens fluid degradation [3]. Pumps can be filled with any fluid; however, some older pumps, especially those designed to work with older fluids, do not provide enough heat input for heavy fluids. The maximum throughput is directly proportional to boiler power. They are dimensionally equivalent; 1000 Pa-L/s = 1 W. For pumps of efficient design, this can be as high as 150 Pa-L/s per kW of boiler

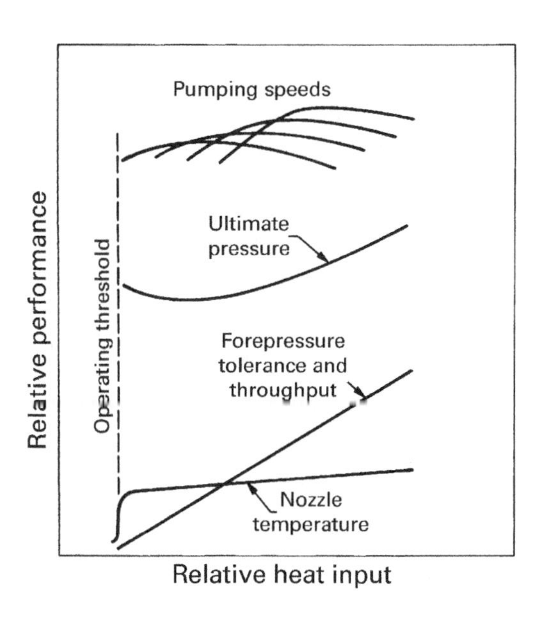

Fig. 12.5 Effect of heat input variations on various diffusion pump parameters. M.H. Hablanian [3]/Reproduced with permission from IOP Publishing.

power [2]. A straight-sided pump with a 200-mm-diameter boiler and throat has the same maximum throughput as a pump with a 200-mm-diameter boiler and an expanded top like the pump shown in Fig. 12.1. The maximum speed of a pump in the high vacuum region is proportional to its inlet area.

12.4 Backstreaming, Baffles, and Traps

For the purpose of this discussion, backstreaming is defined as the transport of pumping fluid and its fractions from the pump to the chamber. Hablanian [27] observed that discussion of backstreaming must not be limited to the pump, but must include the trap, baffle, and ductwork as well because all affect the transfer of pumping fluid vapors from the pump body to the chamber. First, consider the contributions from the pump. Five sources of steady-state backstreaming have been identified: (1) evaporation of fluid condensed on the upper walls of the pump, (2) premature boiling of the condensate before it enters the boiler, (3) the over divergence of the oil vapor in the top jet, (4) leaks in the jet cap, and (5) evaporation of fluid from the heated lip of the top jet [12]. The backstreaming from (1) can be reduced by the use of low vapor pressure fluids and added trapping over the pump. Most pump designs eliminate sources (2) and (4). The use of a water-cooled cap [1,12] directly over the top jet assembly substantially reduces (3) and (5), which were found to be the major causes of fluid backstreaming. With these precautions the backstreaming can be reduced to ~10^{-3} (mg/cm^2)/min a short distance above the pump inlet.

Further reduction of the backstreaming is possible by use of a baffle or a trap. The words "trap" and "baffle" are often misused. Operationally, a trap is a pump for condensable vapors, and a baffle is a device that condenses pump fluid vapors and returns the liquid to the pump boiler. Today the two words are often used imprecisely, and when the baffle is cryogenically cooled the distinction disappears. Pump-fluid molecules or fragments may find their way through the trap by creeping along the walls, by colliding with gas molecules, and by reevaporation from surfaces. Creep can be prevented by the use of traps with a creep barrier—a thin membrane extending from the warm, outer wall to the cooled surface [28] or by the use of autophobic fluids such as pentaphenylsilicone or pentaphenylether. Backstreaming due to oil–gas collisions is a linear function of pressure up to the transition region and a function of the trap and pump design. For one 200-mm-diameter diffusion pump and trap combination, the peak value was found to be 3×10^{-6} (mg/cm^2)/min at a pressure of 5×10^{-2} Pa [29]. At higher pressures the backstreaming rate was decreased by the flushing action of the gas. In normal operation it traverses this region quickly. The maximum integrated backstreaming rate from oil–gas collisions is small enough so that contamination from this source is of no concern in an unbaked system.

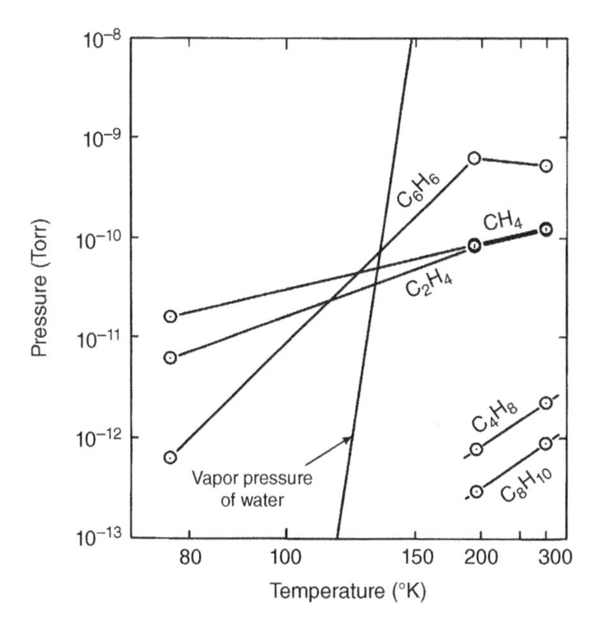

Fig. 12.6 Residual gas analysis of selected mass fragments backstreaming from a diffusion pump filled with DC-705 fluid. Plotted from data reported in C.M. Gosselin and P.J. Bryant [30].

The problem of reevaporation is subtle. The vapor pressures of diffusion pumps vary widely. See Appendix F.2. Two high-quality fluids (pentaphenyl silicone and pentaphenylether) have very low room temperature vapor pressures. Some decomposition of the fluid does occur in the boiler, and some light fractions are generated. Residual gases in a DC-705 charged diffusion pump showed backstreaming of light, intermediate, and heavy fragments [30]. See Fig. 12.6. Light fractions (methane, ethane, and ethylene) were not effectively trapped even on a liquid-nitrogen-cooled surface because of their high vapor pressures. The very heavy fragments (e.g., C_8H_{10}) were quite effectively trapped with only a water-cooled baffle. The partial pressure of an intermediate weight fragment C_6H_6 was reduced by a factor of 1000, when the trap was cooled from 25 to −196°C. When using low vapor pressure fluids, such as pentaphenylsilicone or pentaphenylether, the basic operational difference between a liquid nitrogen trap and a cold-water baffle is the ability of the LN_2 trap to pump C_6H_6 and to trap, at least partially, some of the lightweight fractions.

The quantitative effects of various trap, baffle, and creep barrier combinations are summarized in Table 12.1 [27]. It was noted that the addition of the chevron water baffle between the liquid nitrogen trap and the pump is not much better than the addition of a piece of straight pipe or elbow of the same length.

Table 12.1 Diffusion Pump Backstreaming

Conditions	Duration of Test (h)	Backstreaming Rate (mg/cm^2)/min
(1) Without baffle	165	1.6×10^{-3}
(2) With liquid nitrogen trap	170	5.3×10^{-6}
(3) Same as (2)	380	6.5×10^{-6}
(4) Item (3) plus water baffle	240	2.8×10^{-7}
(5) Item (4) plus creep barrier	240	8.7×10^{-8}
(6) Same as (5)	337	1.2×10^{-7}

Measurements made with a 6-in.-diameter diffusion pump (NRC HS6–1500), DC-705 pump fluid and liquid-nitrogen-cooled collectors.
Source: M.H. Hablanian [27]/Reproduced with permission from AIP Publishing.

The Herrick effect [31], and the fluid burst resulting from the formation and collapse of the top jet are two transient phenomena that produce backstreaming. The Herrick effect is the ejection of frozen fluid droplets from the surface of a fluid-covered trap during the initial stages of cooling with liquid nitrogen. These fluid droplets ricochet off the walls and land on samples or fixtures. A well-designed cold cap and water-cooled baffle followed by a continuously operating liquid nitrogen trap will operate for more than a year without collecting excessive amounts of fluid on the trap. The transient backstreaming from the top jet during warm-up and cooldown of the pump is well-documented [12,19,32]. Figure 12.7

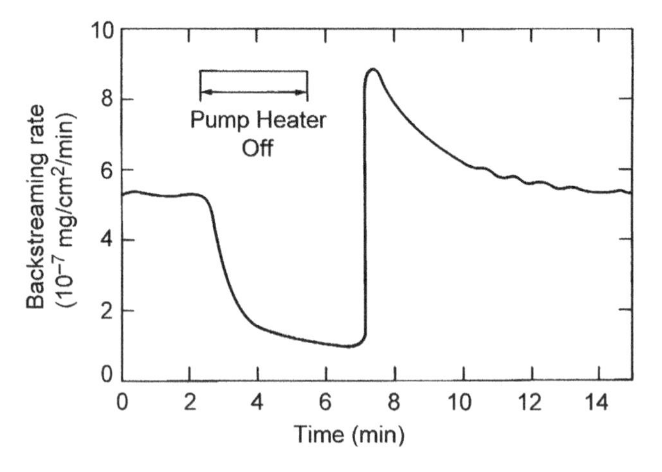

Fig. 12.7 Backstreaming of the parent peak ($M/z = 446$; Convalex-10) over a liquid nitrogen trap during cooldown and start-up of a diffusion pump. G. Rettinghaus and W.K. Huber [32]/Reproduced with permission from AIP Publishing.

shows an RGA trace of the parent molecule, $M/z = 446$ of Convalex-10 [32]. The backstreaming decreases as the fluid is cooled and reaches a peak of about twice the steady-state rate during heating. A total backstreaming of 6×10^{-4} mg/cm^2 was measured for a complete start–stop cycle [29]. This backstreaming can be avoided by continuous operation of the diffusion pump or by using the gas-flushing techniques.

By use of high quality, low vapor pressure fluids such as DC-705, Santovac-5, and a continuously operating liquid nitrogen trap, the contamination due to pump fluid backstreaming can be reduced to extremely small levels. The lowest value of backstreaming shown in Table 12.1 corresponds to a contamination rate of one monolayer per year in a bell jar 500 mm high and 350 mm diameter. This level of organic contamination is below that produced by O-rings and other sources [2]. Fluid backstreaming in a diffusion pump operating at high vacuum is only one source of organic backstreaming. Additional concerns that relate to specific systems are discussed in the Systems chapters.

References

1. Hablanian, M.H. and Maliakal, J.C., *J. Vac. Sci. Technol.* **10**, 58 (1973).
2. Hablanian, M.H., *Solid State Technol.* **17**, 37 (1974).
3. M.H. Hablanian, *Proceedings of the 6th International Vacuum Congress, Kyoto, Japan, Jpn. J. Appl. Phys.*, Vol. 13, Suppl. 2–1, H. Kumagai, Ed., p., 25 (1974).
4. Singleton, J.H., *J. Phys. E.* **6**, 685 (1973).
5. Florescu, N.A., *Vacuum* **10**, 250 (1960).
6. Florescu, N.A., *Vacuum* **13**, 569 (1963).
7. G. Töth, *Proc. 4th Int. Vac. Congr., (1968), Institute of Physics and the Physical Society, London*, 300 (1969).
8. Dushman, S. "Vacuum Pumps", Chapter 3", in *The Scientific Foundations of Vacuum Technology*, 2nd ed., J.M. Lafferty, Ed., Wiley, New York, 1962, Chapter 3.
9. Power, B.D., *High Vacuum Pumping Equipment*, Reinhold, New York, 1966.
10. W. Gaede, German Pat. 286,404 (filed September 25, 1913).
11. M. Morand, U.S. Pat. 2,508,765 (filed July 27, 1947; priority France, September 25, 1941).
12. Power, B.D. and Crawley, D.J., *Vacuum* **4**, 415 (1954).
13. C.G. Smith, U.S. Pat. 1,674,377 (filed September 4, 1924).
14. W.A. Giepen, U.S. Pat. 2,903,181 (filed June 5, 1956).
15. G. Barrows, Brit. Pat. 475,062 (filed May 12, 1936).
16. J.R. O. Downing, U.S. Pat. 2,386,299 (filed July 3, 1944).
17. B.D. Power, Brit. Pat. 700,978 (filed January 25, 1950).
18. J.R.O. Downing and W.B. Humes, U.S. Pat. 2,386,298 (filed January 30, 1943).
19. R.B. Nelson, U.S. Pat. 2,291,054 (filed August 31, 1939).

20 J.J. Madine, U.S. Pat 2,366,277 (filed March 18, 1943).
21 Nöller, N.G., Reich, G., and Bächler, W., *1957 Fourth National Symposium on Vacuum Technology Transactions*, W.G. Matheson, Pergamon, NY, 1958, p. 6.
22 B.B. Dayton, U.S. Pat. 2,639,086 (filed November 30, 1951).
23 C.R. Burch and F. E. Bancroft, Brit. Pat. 407,503 (filed January 19, 1933).
24 Lamont, L.T., Jr., *J. Vac. Sci. Technol.* **10**, 251 (1973).
25 Burnett, S.G. and Hablanian, M.H., *J. Environ. Sci.* **5**, 7 (1964).
26 For example, The Porta Test® leak detector model manufactured, Varian Associates, Palo Alto, CA.
27 Hablanian, M.H., *J. Vac. Sci. Technol.* **6**, 265 (1969).
28 Milleron, N., *Trans. 5th Nat. Vac. Symp. (1958)*, Pergamon, New York, 1959, p. 140.
29 Rettinghaus, G. and Huber, W.K., *Vacuum* **24**, 249 (1974).
30 Gosselin, C.M. and Bryant, P.J., *J. Vac. Sci. Technol.* **2**, 293 (1963).
31 Hablanian, M.H. and Herrick, R.F., *J. Vac. Technol.* **8**, 317 (1971).
32 Rettinghaus, G. and Huber, W.K., *J. Vac. Sci. Technol.* **9**, 416 (1972).

13

Getter and Ion Pumps

Getter and ion pumps are capture pumps. They operate by capturing gas molecules and binding them to a surface. The physical and chemical forces that bind molecules to surfaces are sensitive to gas species, and all gases are not pumped equally well. As a result, two or more capture processes are usually combined to pump a wide range of active and noble gases effectively.

Getter and ion pumps are often referred to as clean pumps. They are clean in the sense that they do not backstream heavy organic molecules as do diffusion or oil-sealed mechanical pumps; however, they can generate hydrogen, methane, carbon oxides, and inert gases, and produce particles. Entrainment pumps do contaminate, but not in the traditional sense. Certain gases will displace previously adsorbed gases. When stimulated, carbon in metals can react with surface water vapor to produce methane or carbon oxides. Hydrogen and some other gases may be poorly pumped, displaced, or thermally released from surfaces on which they were adsorbed. Capture pumps may produce, or not pump, one or more gas species. The labeling of these gases as contaminants depends on the application.

In this chapter we review the titanium sublimation pump (TSP), the non-evaporable getter pump (NEG), and the ion pump.

13.1 Getter Pumps

Many reactive metals pump large quantities of active gases rapidly, because they getter (react with) these gases. Gases either react to form a surface compound, e.g., TiO, or like hydrogen diffuse into the bulk of the getter. The pumping speed of a surface getter is determined by the sticking coefficient of the gas. In a surface getter pump, there is little diffusion into the bulk. It is often cooled to enhance

A Users Guide to Vacuum Technology, Fourth Edition. John F. O'Hanlon and Timothy A. Gessert.
© 2024 John Wiley & Sons, Inc. Published 2024 by John Wiley & Sons, Inc.

the sticking coefficient, and at reduced temperatures, in-diffusion is slow. The pumping speed of bulk getter material is limited by diffusion of gas through surface compounds, so bulk getters are usually operated at elevated temperatures to enhance the dissolution and diffusion of surface compounds. The TSP is a surface pump; the NEG pump is a bulk material pump.

13.1.1 Titanium Sublimation

Many metals, including molybdenum, niobium, tantalum, zirconium, aluminum, and titanium, are surface getters for active gases. They become the active surface of a vacuum pump when they are deposited on a surface in a thin film layer. Titanium is the choice for commercial pumps because it can be sublimed (changed from the solid to vapor state without being a liquid) at much lower temperatures than most other metals, is inexpensive, and pumps a variety of gases. Figure 13.1 depicts one form of TSP. An alternating current heats the filament, which sublimes the titanium and deposits it on adjacent walls. Pumping modules are fabricated with three or four separately heated filaments to extend the time between filament replacements. Active gases are captured on the fresh titanium surface, which is cooled with water or liquid nitrogen. Because Ti reacts with pumped gases, a fresh titanium layer must be deposited periodically to ensure continuous pumping. The pumping characteristics of titanium differ for the active gases, the intermediate gases, and the chemically inactive gases. The active gases (carbon oxides, oxygen, water vapor, and acetylene) are pumped with high sticking coefficients. Water dissociates into oxygen and hydrogen, which are then pumped separately. The temperature of the film has no major effect on the pumping speed

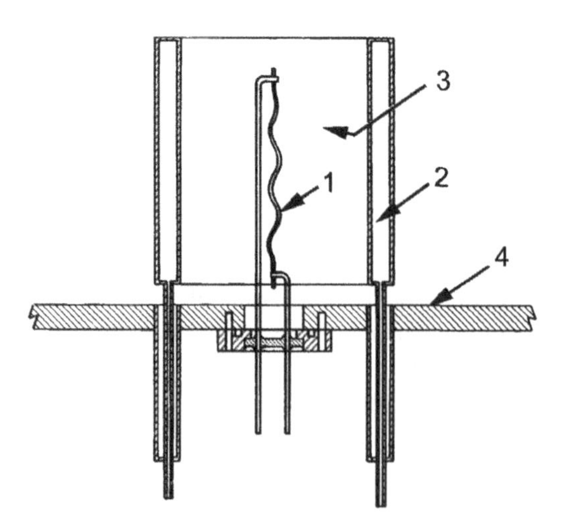

Fig. 13.1 Schematic of a basic titanium sublimation pump. (1) Titanium alloy filament, (2) coolant reservoir, (3) titanium deposit, and (4) vacuum wall.

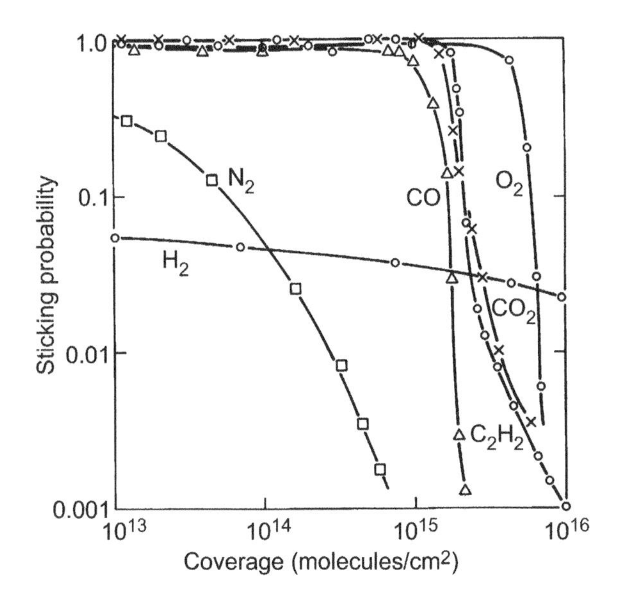

Fig. 13.2 Room-temperature sorption characteristics for pure gases on batch evaporated clean titanium films. Reproduced with permission from A.K. Gupta and J.H. Leck [1]. Copyright 1975, Pergamon Press, Ltd.

of these gases because the sticking coefficients are generally near unity in the range 77–300 K. Sticking coefficients of intermediate gases (H_2 and N_2) are low at room temperature but increase at 77 K. After adsorption, H_2 will diffuse into the underlying film. Chemically inactive gases such as He and Ar, are not pumped. Methane behaves as an inactive gas; it is only slightly sorbed on Ti at 77 K. Figure 13.2 gives the room temperature sorption characteristics of several active and intermediately active gases [1]. Sticking coefficients are the highest for all gases on clean films, and for the very active gases, they remain so until near saturation.

The replacement of one previously sorbed gas by another gas creates a memory effect, and results in actual sticking coefficients that depend on the nature of the underlying adsorbed gas. Gupta and Leck [1] observed a definite order of preference in gas replacement. Table 13.1 illustrates the order in which active gases replace less active gases. Oxygen, the most active gas, can replace all other gases, whereas all other active gases displace methane, which is bound only by van der Waals forces.

Gas replacement is a major cause of the large differences in measured sticking coefficients, especially when the films were not deposited under clean conditions. Harra [2] has reviewed the sticking coefficients and sorption of gases on titanium films measured in several independent studies and tabulated their average values in Table 13.2. These coefficients represent the average of the values obtained in

Table 13.1 Order of Preference of Gas Displacement on Titanium Film

	Displaced Gas[a]				
Pumped Gas	**CH_4**	**N_2**	**H_2**	**CO**	**CO_2**
CH_4	—	N	N	N	N
N_2	Y	—	N	N	N
H_2	Y	Y	—	N	N
CO	Y	Y	Y	—	N
CO_2	Y	Y	Y	Y	—

[a] Y, Yes; N, No.
Source: Adapted from A.K. Gupta and J.H. Leck [1].

Table 13.2 Initial Sticking Coefficient and Quantity Sorbed for Various Gases on Titanium

	Initial Sticking Coefficient		Quantity Sorbed[a] ($\times 10^{15}$ molecules/cm^2)	
Gas	(300 K)	(78 K)	(300 K)	(78 K)
H_2	0.06	0.4	8–230[b]	7–70
D_2	0.1	0.2	6–11[b]	—
H_2O	0.5	—	30	—
CO	0.7	0.95	5–23	50–160
N_2	0.3	0.7	0.3–12	3–60
O_2	0.8	1.0	24	—
CO_2	0.5	—	4–24	—
He	0	0		
Ar	0	0		
CH_4	0	0.05		

[a] Fresh film thickness of 10^{15} Ti atoms/cm^2.
[b] The quantity of hydrogen or deuterium sorbed at saturation may exceed the number of Ti atoms/cm^2 in the fresh film through diffusion into the underlying films at 300 K.
Source: Adapted from Harra [2].

different laboratories and under different conditions. They should be more representative of those in a typical operating sublimation pump of unknown history, than are those measured under clean conditions.

TSPs operate at pressures below 10^{-1} Pa (10^{-3} Torr). Above that pressure, surface compounds inhibit sublimation. A typical pumping speed curve is sketched

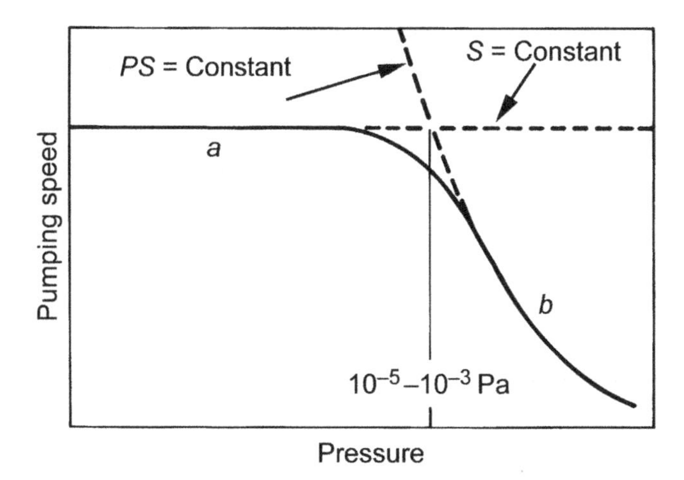

Fig. 13.3 Characteristic pumping speed versus pressure for a TSP: (a) Speed determined by the getter area, sticking coefficient, gas species, and inlet conductance; (b) speed determined by pressure and rate of sublimation.

in Fig. 13.3. At low pressures there are few collisions between titanium atoms and gas molecules until the titanium atoms reach the surface. Below 10^{-4} Pa (10^{-6} Torr), more titanium is sublimed than is needed when the filament is operated continuously. This results in a constant pumping speed that is determined by the surface area of the film and the conductance of any interconnecting tubing. At pressures greater than 10^{-4} Pa (10^{-6} Torr), titanium–gas collisions occur before the titanium strikes the surface, and the pumping speed is determined by the rate of titanium sublimation. Gas throughput is proportional to the titanium sublimation rate, therefore, the pumping speed will decrease as $1/P$, as shown in Fig. 13.3.

Precise calculation of the pumping speed in the low-pressure region is not easy, because of sticking coefficient uncertainty and geometry. In molecular flow the pumping speed S of the geometry shown in Fig. 13.1 is given approximately by:

$$\frac{1}{S} = \frac{1}{S_i} + \frac{1}{C_a} \tag{13.1}$$

where S_i is the intrinsic speed of the surface and C_a is the conductance of the aperture at the end of the cylindrical surface on which the titanium is deposited. This conductance can be ignored if the film is deposited on the walls of the chamber. If a valve or connecting pipe is used, the appropriate series conductance should be added. The intrinsic speed is approximately:

$$S_i(\text{L/s}) = 1000A\frac{v}{4}s' \tag{13.2}$$

Here, A is the area of the film (m^2), v is the gas velocity (m/s), and s' is the sticking coefficient. Cooling to 77 K provides little additional pumping speed in devices whose speed is conductance limited.

At high pressures the pumping speed is determined by the rate of titanium sublimation. This theoretical maximum throughput is related to the titanium sublimation rate (TSR) by the relation [3]:

$$Q(\text{Pa-L/s}) = \frac{10^8 V_o \text{TSR}(\text{atoms/s})}{nN_o} = \frac{10^{-18}}{n}\text{TSR}(\text{atoms/s}) \tag{13.3}$$

V_o is the normal specific volume of the gas, N_o is Avogadro's number, and n is the number of titanium atoms that react with each molecule of gas; $n = 1$ for CO and $n = 2$ for N_2, H_2, O_2, and CO_2 [2]. For Ti, Eq. (13.3) becomes:

$$Q(\text{Pa-L/s}) = \frac{1.25\times10^{-2}}{n}\text{TSR}(\mu\text{g/s}) \tag{13.4}$$

This theoretical throughput can be reached only when the titanium is fully reacted with the gas. The corresponding pumping speed is obtained by dividing the throughput by the pressure in the pump.

TSP's is used in combination with other pumps that pump inert gases and methane. Small TSP's are used continuously for short periods to aid in crossover between a sorption pump and an ion pump. Large TSP's have been developed for use in conjunction with smaller ion pumps for long-term, high-throughput pumping. TSP's are used intermittently for long periods at low pressures for high-speed pumping of reactive gases. At low pressures the film needs to be replaced only periodically to retain the pumping speed. Titanium is sublimed only until a fresh layer is deposited. The pump is then turned off until the film saturates. Figure 13.4 sketches the pressure rise with decrease in pumping speed as the titanium film saturates in a typical ion-pumped system [3]. P_i is the initial pressure with the ion pump and sublimator operating, and P_f is pressure with only the ion pump operating. Not shown on these curves is the pressure burst of H_2, CH_4, and C_2H_6 from Ti during sublimation. The methane and ethane are formed in a reaction between hydrogen and carbon impurities in the hot filament [1,4]. After sublimation begins, hydrogen is pumped. Methane and ethane are marginally pumped on surfaces held at 77 K.

Commercially available TSPs use directly heated filaments, radiantly heated sources, or electron-beam-heated sources. The most commonly used source is a directly heated filament with a low-voltage ac power supply. Filaments were first made from titanium twisted with tantalum or tungsten and later from titanium wound over niobium and tantalum wire [5]. Because of thermal contact problems, the sublimation rate proved to be unpredictable. Modern pumps use filaments fabricated from an alloy of 85% Ti and 15% Mo [6,7,8]. This filament has an even sublimation rate and a long life. A typical filament 15-cm long can be operated

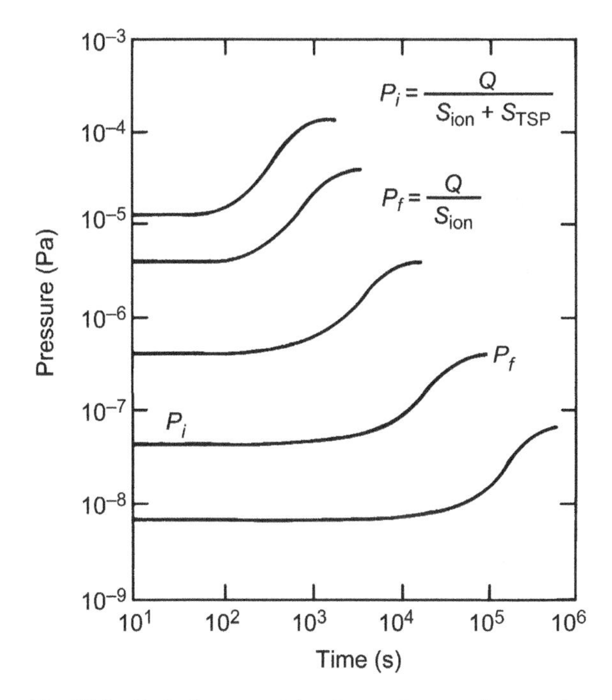

Fig. 13.4 Typical pressure rise due to decrease in pumping speed as a titanium film saturates. Reproduced with permission from General Characteristics of Titanium Sublimation Pumps, B.E. Keitzmann, 1965, Varian Associates, 611 Hansen Way, Palo Alto, CA 9430.

at sublimation rates of 30–90 μg/s. Large TSPs have been constructed with radiantly heated titanium at sublimation rates as high as 150 μg/s [9,10] and with electron-beam-heated, rod-fed sources at sublimation rates ranging from 300 μg/s to 0.15 g/s [11]. Electron-beam-heated sources do not operate well at pressures higher than 10^{-3} Pa and for most applications are too expensive to operate at pressures below 10^{-5} Pa. They are an excellent high-speed pump in the intermediate region. Radiantly heated sources are best for high-speed pumping in the very high vacuum region.

13.1.2 Non-evaporable Getters

A NEG pumps by surface adsorption followed by bulk diffusion. Its speed for pumping active gases is determined by the gas diffusion rate into the bulk. For this reason, NEG pumps are operated at high temperatures. They do not pump inert gases or methane because these gases do not adsorb on the surface. One effective getter for vacuum use is an alloy of 84% Zr and 16% Al [12,13,14,15,16]. This alloy, when heated to 400°C, has a pumping speed of ~0.3 L-s^{-1}-cm^{-2} (N_2), ~1 L-s^{-1}-cm^{-2} (CO_2, CO, O_2), and 1.5 L-s^{-1}-cm^{-2} (H_2) [12]. At room temperature H_2 is pumped at about half the speed that it is pumped at 400°C, provided that no oxide or nitride diffusion barriers exist. Other gases are not pumped at room temperature because the surface compounds quickly form diffusion barriers. All gases except hydrogen are pumped as stable compounds and are entrapped permanently [13]. Hydrogen is pumped as a solid solution and may be released by heating above 400°C. The diffusion of carbon monoxide, carbon dioxide, nitrogen, and oxygen has been shown to obey a simple parabolic rate law [17]. NEG pump speeds have been shown to be constant below 10^{-3} Pa and to decrease as $P^{-1/2}$ above 5×10^{-1} Pa (4×10^{-3} Torr) [18].

To operate a Zr–Al getter pump, the chamber is evacuated to a pressure below 1 Pa, after which the pump is activated by heating to 800°C to allow surface atoms to diffuse within the bulk. The temperature is then reduced to 400°C. The activation step is repeated each time the pump is cooled and released to atmosphere. Another alloy (Zr–70%, V–24.6%, Fe–5.4%) has demonstrated high getter efficiency at room temperature after activation at 500°C [19,20].

NEG pumps have been used as appendage pumps on small systems, as well as primary pumps in large UHV systems, fusion machines, and particle accelerators. Appendage pumps equipped with getter cartridges fabricated from a plated steel coated with Zr–Al alloy have pumping speeds as high as 10–50 L/s [21]. One getter ion pump package has a combined pumping speed of 1000 L/s [22]. NEG pumps have a large capacity for pumping hydrogen [16]. The NEG pump has been used in fusion machines [19,23,24], because it can operate without a magnetic field, and it has a high hydrogen pumping speed at room temperature. With the assistance of an ion pump to handle the methane and argon, the NEG can reach base pressures of 10^{-9} Pa. The CERN large positron collider uses 27 km of linear NEG pump with a speed of 500 L/s per meter of chamber [24]. Getter pumping was in Tokamaks to control the density of hydrogen plasmas and remove chemically active impurities. The use of glow discharge cleaning has been shown to have no deleterious effects on the operation of a Zr–Al NEG [25]. Hseuh and Lanni [26] have established a worst-case pressure of less than 3×10^{-9} Pa in an accelerator storage ring using a linear Zr–V–Fe alloy and lumped ion pumps 10 m apart. Getter pumps are now finding other applications, such as for purifying gases used in semiconductor device processing equipment. NEG pumps are described in a related ISO Standard [27].

13.2 Ion Pumps

The development of the ion pump has made it possible to pump to the ultrahigh vacuum region without concern for heavy organic contamination. This pump exploits a phenomenon formerly considered detrimental to vacuum gauge operation: pumping gases by ions in Bayard-Alpert and Penning gauges. Ions are pumped easily because they are more reactive with surfaces than neutral molecules and if sufficiently energetic can physically embed themselves in the pump walls. If the ions were generated in a simple parallel-plate glow discharge, the pumping mechanism would be restricted to a rather narrow pressure range. Above about 1 Pa the electrons cannot gain enough energy to make an ionizing collision and below about 10^{-1}–10^{-2} Pa (10^{-3}–10^{-5} Torr) the electron mean free path becomes so long that the electrons collide with a wall before they encounter a gas molecule. Ions can be generated at lower pressures, if the energetic electrons can be constrained from hitting a wall before they collide with a gas molecule. This confinement can be realized with certain combinations of electric and magnetic fields.

The pumping action of a magnetically confined dc discharge was first observed by Penning [28] in 1937, but it was not until two decades later that Hall [29] combined several Penning cells and transformed the phenomenon into a functional pump. Some elemental forms of the (diode) sputter-ion pump are shown in Fig. 13.5 and include the ring anode cell [28], the long anode cell [31], Ti cathode [32], and multicell anodes [29]. Each Penning cell is approximately 12 mm in diameter × 20 mm long with a 4-mm gap between the anode and the cathode. Modern pumps are constructed of modules of cells arranged around the periphery of the vacuum wall with external permanent magnets of 0.1–0.2-T strength and cathode voltages of ~5 kV.

The electric fields present in each Penning cell trap the electrons in a potential well between that the two cathodes and the axial magnetic field forces the electrons into circular orbits that prevent their reaching the anode. This combination of electric and magnetic fields causes the electrons to travel long distances in

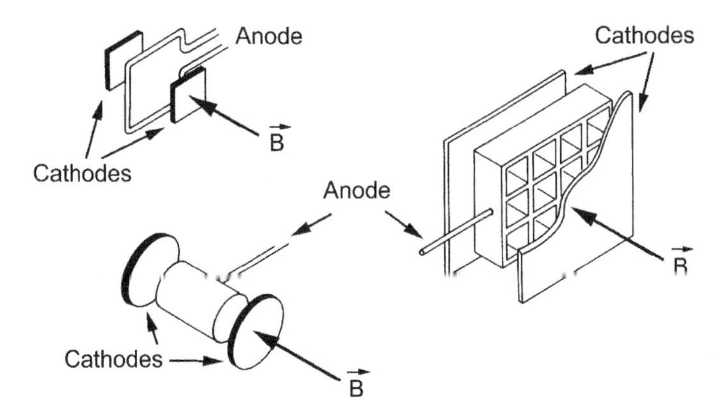

Fig. 13.5 Early forms of the diode sputter-ion pump. (Upper Left) Ring anode cell. (Lower Left) long anode cell; same cell with Ti cathodes. (Right) multicell anode. Reproduced with permission from D. Andrew [30]. Copyright 1969, The Institute of Physics.

oscillating spiral paths before colliding with the anode and results in a high probability of ionizing collisions with gas molecules. The time from the random entrance of the first electron into the cell until the electron density reaches its steady state value of ~10^{10} e/cm^3 is inversely proportional to pressure. The starting time of a cell at 10^{-1} Pa (10^{-3} Torr) is nanoseconds, while at 10^{-9} Pa (10^{-11} Torr) it is 500 s [33]. The ions produced in these collisions are accelerated toward the cathode, where they collide, sputter away the cathode, and release secondary electrons that in turn are accelerated by the field. Many other processes occur in addition to the processes necessary to sustain the discharge; for example, a large number of low-energy neutral atoms are created by molecular dissociation, and some high-energy neutrals are created from energetic ions by charge neutralization as they approach the cathode, collide, and recoil elastically.

The mechanism of pumping in an ion pump is dependent on the nature of the gas being pumped and is based on one or more of the following mechanisms: (1) precipitation or adsorption following molecular dissociation; (2) gettering by freshly sputtered cathode material; (3) surface burial under sputtered cathode material; (4) ion burial following ionization in the discharge; and (5) fast neutral atom burial. (Ions are neutralized by surface charge transfer and reflected to another surface where they are pumped by burial.) Rutherford, Mercer, and Jepsen [34] explained the first four mechanisms, and Jepsen explained the role of elastically scattered neutrals [35]. These mechanisms are illustrated in Fig. 13.6.

Organic vapors, active gases, hydrogen, and inert gases are pumped in distinctly different ways. There are a few generalities. Initially, gases tend to be pumped

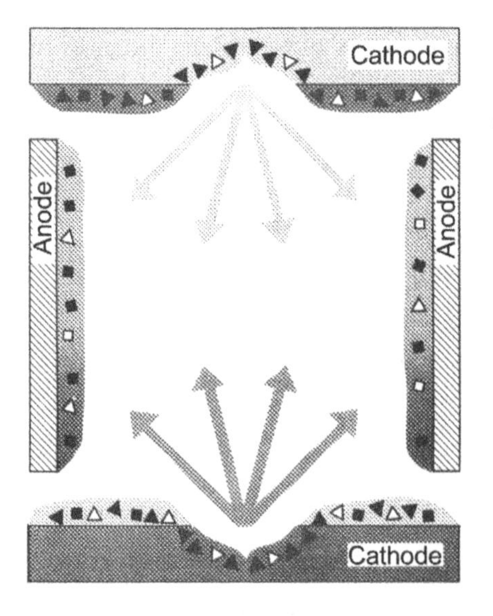

Fig. 13.6 Schematic diagram showing sputter deposition and pumping mechanisms in a Penning cell: ▪ Chemically active gases buried as neutral particles; ► chemically active gases ionized before burial; □ inert gases buried as neutral particles; Δ inert gases ionized before burial. Reproduced with permission from D. Andrew [30]. Copyright 1969, The Institute of Physics.

rapidly and their partial pressure decays to a steady state [35,36,37]. In steady state, reemission rates equal pumping rates. This is more pronounced with noble than with active gases. Pumping speeds cannot be uniquely defined for a gas independent of the composition of other gases being pumped simultaneously. The sputter ion pump is capable of reemitting any pumped gas. This reemission or memory effect complicates the interpretation of some experiments. Organic gases are easily pumped by adsorption and precipitation after being dissociated by electron bombardment [34].

Active gases such as oxygen, carbon monoxide, and nitrogen are pumped by reaction with titanium, which is sputtered on the anode surfaces, and by ion burial in the cathode. These gases are easily pumped because they form stable titanium compounds [34].

Hydrogen behaves differently. Its low mass prevents it from sputtering the cathode significantly. It behaves much like it does in a TSP. It is initially pumped by ion burial and neutral adsorption [35,38] and diffuses into the bulk of the titanium and forms a hydride. Sustained pumping of hydrogen at high pressures will cause cathodes to warp [34] and release gas as they heat. The hydrogen pumping speed does not rate limit unless cathode surfaces are covered with compounds that prevent indiffusion. The pumping of a small amount of an inert gas, say argon, cleans the surfaces and allows continued hydrogen pumping [39], whereas a trace amount of nitrogen will reduce the speed by contaminating the surface [38].

Noble gases are not pumped efficiently in a diode pump. They are pumped by ion burial in the cathodes and by reflected neutral burial in the anodes and cathodes. The noble gas pumping on the cathodes is mostly in the area near the anodes where the sputter build-up occurs. Because most of the neutrals are reflected with low energies in the diode pump, their pumping speed in the anode or other cathode is low; for example, argon is pumped only at 1–2% of the active gas speed.

Argon, in particular, suffers from pumping instability. Periodically the argon pressure will rise as pumped gas is released from the cathodes. Figure 13.7 illustrates some of the geometrical constructions that were devised as a solution to the problem of low argon pumping speed and its periodic reemission. Brubaker [40] devised a triode pump with a collector surface that operated at a potential between the anode and cathode (Fig. 13.7*a*). Its function was to collect low-energy ions that could not sputter, and Hamilton [41] showed it worked equally well when the collector surface was held at anode potential (Fig. 13.7*b*,*c*). In the triode pump the argon pumping speeds are as high as 20% of the nitrogen speed. This high-speed (high implantation rate) results from the high energy of the neutrals, which are scattered at small angles from the cathode walls with little energy loss. Sputtering is much more efficient at these small angles than at normal incidence and sputtering of titanium on the collector is more efficient than in the diode pump. The slotted cathode [42] attempts to accomplish this sputtering with one less electrode than

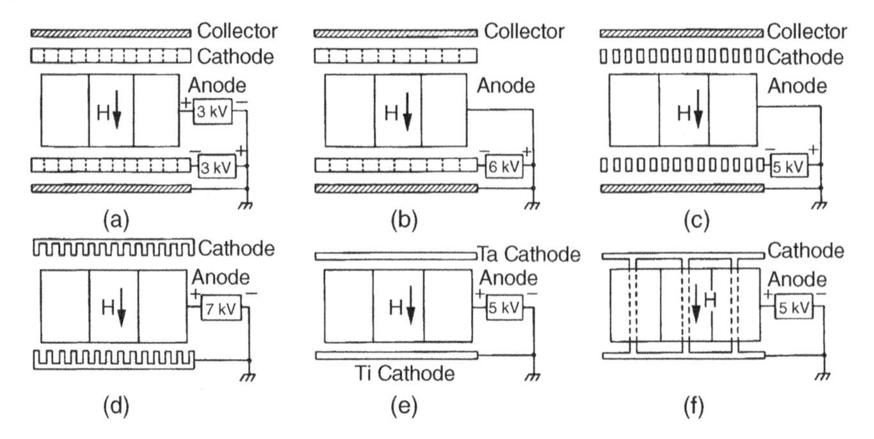

Fig. 13.7 Pump designs for inert gas pumping: (a) Brubaker's triode pump; (b) Hamilton's triode pump; (c) The Varian Noble Ion Pump triode pump; (d) the slotted cathode diode; (e) the differential ion pump; and (f) the magnetron pump. Reproduced with permission from D. Andrew [30]. Copyright 1969, The Institute of Physics.

the triode (Fig. 13.7*d*). This pump has an argon pumping speed of 10% of the speed for air. The differential cathode design [43] is sketched in Fig. 13.7*e*. A tantalum cathode replaced one titanium cathode. In this manner the recoil energy of the scattered noble gas neutrals, which depends on the relative atomic weight of the cathode material and gas atom, is increased [44,45]. This gives more effective noble gas pumping than in the diode pump. An argon-stable magnetron structure [30] is depicted in (Fig. 13.7*f*). The central cathode rod is bombarded by a high flux of ions at oblique angles of incidence. The sputtering of the rod creates a flux that continually coats the cathode plates and the impinging ions and results in a net argon speed of 12% of the airspeed [30]. Among the designs discussed here for increasing the argon pumping speed and reducing or eliminating its instability, the triode and differential diode are in most widespread use.

The operating pressure range of the sputter-ion pump extends from 10^{-2} to below 10^{-8} Pa (10^{-4}–$<10^{-10}$ Torr). Characteristic pumping-speed curves of a diode and a triode pump are shown in Fig. 13.8. If starting is attempted at high pressures, say 1 Pa (0.01 Torr), a glow discharge will appear, and the elements will heat and release hydrogen. As the pressure is reduced, the glow discharge extinguishes, and the speed rapidly increases. At low pressures the speed decreases because the sputtering and ionization processes decrease. The exact shape of a pumping speed curve is a function of the magnetic field intensity, cathode voltage, and cell diameter-to-length ratio. As the pressure decreases, the ionization current decreases. At high pressures, the ion current is approximately proportional to the pressure, and may be used as a gauge. However, at low pressures, there will be no

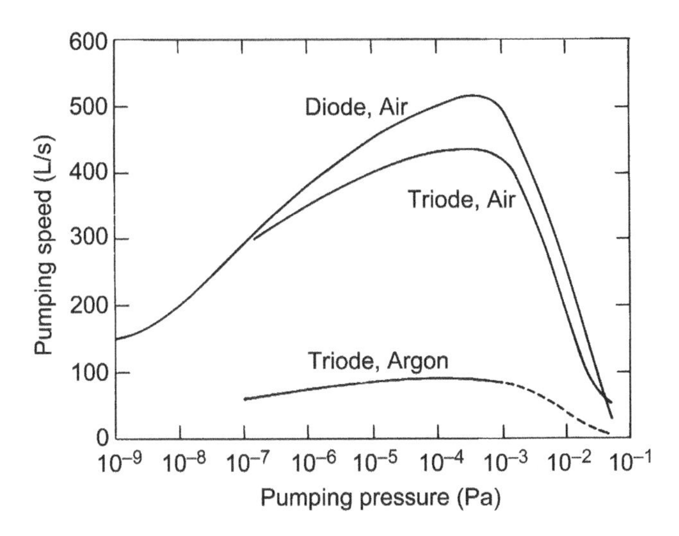

Figure 13.8 Pumping speeds for air and argon in a 500-L/s Varian diode Vac Ion pump and a 400-L/s Varian triode Vac Ion pump. Speeds measured at the inlet of the pump. Reproduced with permission from General Characteristics of Titanium Sublimation Pumps, B.E. Keitzmann, 1965, Varian Associates, 611 Hansen Way, Palo Alto, CA 9430.

relation between the total ion pump current and the actual pressure. Spurious pump currents prevent the ion current from being used as a gauge at low pressures.

The lifetime of a diode pump is a function of the time necessary to sputter through the cathodes. A typical value is 5000 h at 10^{-3} Pa (10^{-5} Torr) or 50,000 h at 10^{-4} Pa (10^{-6} Torr). The triode, which pumps slightly better than the diode at high pressures, is also easier to start at high pressures and has a lifetime of less than half the diode. In both pumps the life may be shorter due to electrode shorting caused by loose flakes of titanium.

Sputter ion pumps have the advantage of freedom from hydrocarbon contamination and ease of fault protection but do suffer from the reemission of previously pumped gases, particularly hydrogen, methane, and noble gases.

References

1 Gupta, A.K. and Leck, J.H., *Vacuum* **25**, 362 (1975).

2 Harra, D.J., *J. Vac. Sci. Technol.* **13**, 471 (1976).

3 B.E. Keitzmann, *General Characteristics of Titanium Sublimation Pumps*, Varian Associates, Palo Alto, CA, 1961.

4 Holland, L., Laurenson, L., and Allen, P., *Transactions of the 8th National Vacuum Symposium of the American Vacuum Society, 1961*, L. Preuss, Ed., Pergamon, New York, 1962, p. 426.

5 Clausing, R.E., *Transactions of the 8th National Vacuum Symposium of the American Vacuum Society, 1961*, Pergamon, New York, 1962, p. 345.
6 Kuzmin, A.A., *Prib. Tekh. Eksp.* **3**, 497 (1963).
7 McCracken, G.M. and Pashley, N.A., *J. Vac. Sci. Technol.* **3**, 96 (1966).
8 Lawson, R.W. and Woodward, J.W., *Vacuum* **17**, 205 (1967).
9 Harra, D.J. and Snouse, T.W., *J. Vac. Sci. Technol.* **9**, 552 (1972).
10 Harra, D.J., *J. Vac. Sci. Technol.* **12**, 539 (1975).
11 Smith, H.R., Jr., *J. Vac. Sci. Technol.* **8**, 286 (1971).
12 della Porta, P., Giorgi, T., Origlio, S., and Ricca, F., *Transactions of the 8th National Vacuum Symposium of the American Vacuum Society, 1961*, Pergamon, New York, 1962, p. 229.
13 Giorgi, T.A. and Ricca, F., *Nuovo Cimento Suppl.* **1**, 612 (1963).
14 Kindl, B., *Nuovo Cimento Suppl.* **1**, 646 (1963).
15 Kindl, B. and Rabusin, E., *Nuovo Cimento Suppl.* **5**, 36 (1967).
16 Lange, W.J., *J. Vac. Sci. Technol.* **14**, 582 (1977).
17 Parkash, S. and Vijendran, P., *Vacuum* **33**, 295 (1983).
18 Cecchi, J.L. and Knize, R.J., *J. Vac. Sci. Technol., A* **1**, 1276 (1983).
19 Boffito, C., Ferrario, B., della Porta, P., and Rosai, L., *J. Vac. Sci. Technol.* **18**, 1117 (1981).
20 Boffito, C., Ferrario, B., and Martelli, D., *J. Vac. Sci. Technol., A* **1**, 1279 (1983).
21 della Porta, P. and Ferrario, B., *Proc. 4th Int. Vac. Congr. (1968)*, Vol. 1, Institute of Physics and the Physical Society, London, 1968, p. 369.
22 S.A.E.S. Getters USA, Buffalo, NY.
23 Moenich, J.S., *J. Vac. Sci. Technol.* **18**, 1114 (1981).
24 Benvenuti, C., *Nuc. Instrum. Methods* **205**, 391 (1983).
25 Dylla, H.F., Cecchi, J.L., and Ulrickson, M., *J. Vac. Sci. Technol.* **18**, 1111 (1981).
26 Hseuh, H.C. and Lanni, C., *J. Vac. Sci. Technol., A* **1**, 1283 (1983).
27 ISO Standard 21360-5:2020, *Part 5: Measuring the Performance of NEG Vacuum Pumps*, International Standards Organization, Geneva, Switzerland, 2020.
28 Penning, F.M., *Physica* **4**, 71 (1937).
29 Hall, L.D., *Rev. Sci. Instrum.* **29**, 367 (1958).
30 Andrew, D., Sethna, D.R., and Weston, G.F., *4th Int. Vac. Cong. (1968)*, Institute of Physics and the Physical Society, 1968, p. 337.
31 Penning, F.M. and Nienhuis, K., *Philips Tech. Rev.* **11**, 116 (1949).
32 Guerswitch, A.M. and Westendrop, W.F., *Rev. Sci. Instrum.* **25**, 389 (1954).
33 Craig, R.D., *Vacuum* **19**, 70 (1969).
34 Rutherford, S.L., Mercer, S.L., and Jepsen, R.L., *1960 Trans. 7th Natl. Vac. Symp.*, Pergamon, New York, 1961, p. 380.
35 Jepsen, R.L., 1968 *Proc. 4th Intl. Vac. Congr.*, Institute of Physics and the Physical Society, London, 1969, p. 317.
36 Dallos, A. and Steinrisser, F., *J. Vac. Sci. Technol.* **4**, 6 (1967).

37 Dallos, A., *Vacuum* **19**, 79 (1969).
38 Singleton, J.H., *J. Vac. Sci. Technol.* **8**, 275 (1971).
39 Singleton, J.H., *J. Vac. Sci. Technol.* **6**, 316 (1969).
40 Brubaker, W.M., *6th Natl. Vac. Symp. (1959)*, Pergamon, New York, 1960, p. 302.
41 Hamilton, A.R., *Transactions of the 8th National Vacuum Symposium of the American Vacuum Society, 1961*, Vol. 1, Pergamon, New York, 1962, p. 338.
42 Jepsen, R.L., Francis, A.B., Rutherford, S.L., and Keitzmann, B.E., *1960 7th Natl. Vac. Symp.*, Pergamon, New York, 1961, p. 45.
43 Tom, T. and Jones, B.D., *J. Vac. Sci. Technol.* **6**, 304 (1969).
44 Baker, P.N. and Laurenson, L., *J. Vac. Sci. Technol.* **9**, 375 (1972).
45 Denison, D.R., *J. Vac. Sci. Technol.* **14**, 633 (1977) see its reference #1.

14

Cryogenic Pumps

Cryogenic pumping involves capturing molecules on a cooled surface by weak van der Waals or dispersion forces. In principle, any gas can be pumped, provided that the surface temperature is low enough for incident molecules to remain on the surface after losing kinetic energy. In practice, it is difficult to capture helium, neon, and hydrogen. Cryogenic pumping is clean. The only gas or vapor contaminants are those not pumped or released from pumped deposits. Unlike the ion pump, the cryopump does not retain condensed and physically adsorbed gases after the pumping surfaces have been warmed. Proper precautions must be taken to vent the pumped gas load.

Cryopumps are used in a wide range of applications and in many forms. Liquid nitrogen-cooled molecular sieve pumps are used as roughing pumps. Liquid nitrogen traps are used between diffusion pumps and chambers to pump oil and water vapor. Liquid nitrogen or liquid helium "cold fingers" are used in high vacuum chambers to augment other pumps. Liquid cryogens or closed-cycle helium gas refrigerators are used to cool high- and ultrahigh vacuum cryopumps. Turbomolecular, TSP, or NEG pumps are sometimes appended to cryopumps to improve their pumping speed for hydrogen and deuterium. Cryopumping is the only form of pumping by which extremely large pump speeds (10^7 L/s) can be realized.

Cryogenic pumping is not a new technique. The theory and techniques of pumping on cooled surfaces have been a part of vacuum technology much longer than the helium gas refrigerator; however, knowledge of both is necessary to understand the operation of a He-gas cryopump. Several reviews of cryopumping have been published. Hands [1] reviewed small refrigerator-cooled cryopumps and very large pumps used for fusion experiments. Bentley [2] explained the operation of the Gifford–McMahon refrigerator, and Haefer [3] discussed the mechanisms of cryogenic pumping, given system calculations, and examples of pumps

A Users Guide to Vacuum Technology, Fourth Edition. John F. O'Hanlon and Timothy A. Gessert.
© 2024 John Wiley & Sons, Inc. Published 2024 by John Wiley & Sons, Inc.

and applications. Welch [4] has written a comprehensive treatise on capture pumping technology. In this chapter we review the mechanisms of cryocondensation and cryosorption on which all cryogenic pumping is based; we discuss pumping speed, ultimate pressure and saturation effects, refrigeration techniques, and pump characteristics. We discuss system operation and regeneration techniques in Chapters 18 and 19. Issues relating to cryopump operation at high flow rates are discussed in Chapter 18 and 22.

14.1 Pumping Mechanisms

Low-temperature pumping is based on cryocondensation, cryosorption, and cryotrapping. In Chapter 4, we defined the equilibrium or saturated vapor as the pressure at which the flux of vapor particles to the surface equals the flux of particles leaving the surface and entering the vapor phase, provided that all the molecules, solid, liquid, and vapor are at the same temperature. The arriving molecules are attracted to condensation sites on the liquid or solid, where they are held for some residence time after which they vibrate free and desorb into the vapor phase. The vapor pressure and residence time are temperature dependent. As the temperature is reduced, the vapor pressure is reduced, and the residence time is increased. See, for example, Table 4.1. Tables of vapor pressures of the common gases are given in Appendix B.5. Cryocondensation becomes a useful pumping technique when a surface can be cooled to a temperature at which the vapor pressure is so low, and the residence time is so long, that the vapor is effectively removed from the system. Liquid nitrogen is an excellent pump for water vapor, because the vapor pressure of water at 77 K is 10^{-19} Pa. The probability that an atom will condense on collision with a cold surface is called the condensation coefficient. Condensation coefficients of many gases at reduced temperatures lie between 0.5 and 1.0 [5–7].

Any solid surface has a weak attractive force for at least the first monolayers of gas or vapor. Figure 14.1 describes a typical relationship between the number of molecules adsorbed and the pressure above the adsorbed gas for Xe, Kr, and Ar on porous silver at 77.4 K [8]. These adsorption isotherms tend toward a slope of one at very low pressures. This shows the number of adsorbed atoms goes to zero linearly with the pressure. The sorption sites become increasingly populated as the pressure increases. The limiting sorption capacity is reached after a few monolayers have been deposited. A typical monolayer can hold about 10^{15} atoms/cm^2; however, the actual number is material dependent. The data shown in Fig. 14.1 saturate at 2×10^{19} atoms, because the surface area is larger than 1 cm^2. At the vapor pressure, condensation begins, and the surface layer increases in thickness. The thickness of the solid deposit is limited only

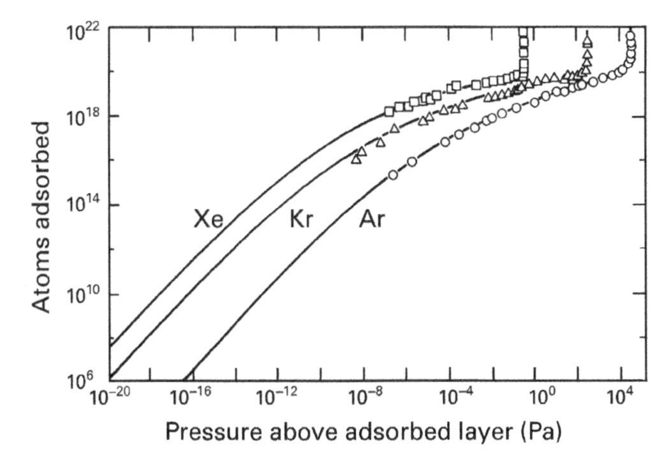

Fig. 14.1 Adsorption isotherms of xenon, krypton, and argon on porous silver adsorbent at 77 K. The lines represent plots of an analytic solution, and the points are experimental. Reproduced with permission from J.P. Hobson [8]. Copyright 1969, The American Chemical Society.

by thermal gradients in the solid and by thermal contact with nearby surfaces of different temperatures. The density and thermal conductivity of the solid frost are a function of its formation temperature. It will decrease as the condensation temperature decreases.

A curve similar to Fig. 14.1 may be measured for each temperature of the sorbate. The effect of temperature on the adsorption isotherm is illustrated in Fig. 14.2

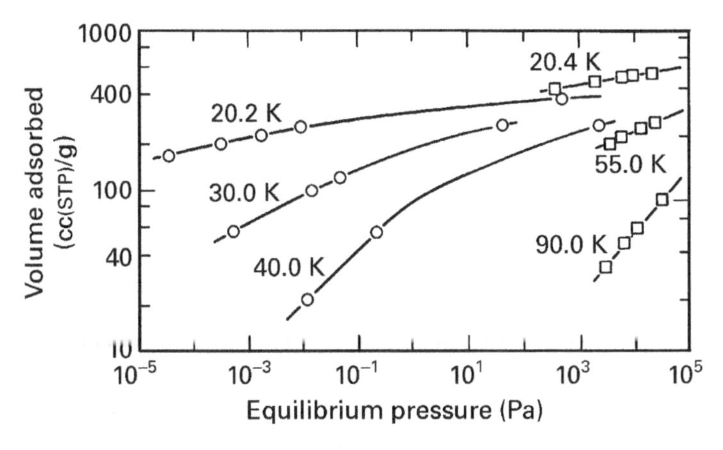

Fig. 14.2 Adsorption of hydrogen on coconut charcoal at low pressures. ○ Gareis and Stern [8], □ Van Dingenan and Van Itterbeek [9]. Reprinted with permission from S.A. Stern and J.T. Mullhaupt [10]. Copyright 1965, The American Vacuum Society.

for hydrogen on a bed of activated charcoal. Gas adsorption at a given pressure is increased if the temperature is reduced, because the probability of desorption is less than at higher temperatures.

Adsorption is an important phenomenon because it allows a substance to be pumped to a pressure far below its saturated vapor pressure. For gases such as helium, hydrogen, and neon, this is the only mechanism by which pumping takes place. The data in Fig. 14.1 show ultimate pressures ranging from 10^{-1} to 10^{-12} of the saturated pressure. The ultimate pressure is a function of the surface coverage. The surface coverage can be minimized by pumping a small quantity of gas or by the generation of a large surface area with porous sorbents such as charcoal or zeolite. Adsorption isotherms have been measured for many materials and several data sources are given in reference [8]. The adsorbent properties of charcoal and molecular sieve, which are most interesting for cryopumping, have been the subject of considerable investigation [9,11–18]. In some cases, the isotherms do not approach zero with a slope of one, which suggests that the sorbent was not completely equilibrated. Some analytical expressions for adsorption isotherms have been published [8,18,19].

The last mechanism of low-temperature pumping has been given the name cryotrapping [20] and has been studied in some detail [3,20–23]. Cryotrapping is simply the dynamic sorption of one gas within the porous frozen condensate of another. A gas, which would not normally condense, will sorb if it arrives simultaneously with another condensable gas. Some of the noncondensable gas molecules are adsorbed on the surface of the condensed gas microcrystallites, while others are incorporated within the crystallites [3]. Cryocondensation takes place only between certain pairs of gases. Examples are hydrogen in argon, helium in argon, and hydrogen in carbon monoxide. The cryotrapping of hydrogen in argon is important in sputtering, whereas the pumping of helium in argon is important in fusion work.

Figure 14.3 illustrates the cryotrapping of hydrogen by argon. In this experiment a diffusion pump and a cryosurface pumped in parallel on a known hydrogen gas flow. This resulted in a steady-state hydrogen pressure for zero argon flow whose magnitude was determined by the hydrogen flow rate and diffusion pump speed. As the argon flow was increased, the cryosurface began to pump the hydrogen and reduce its partial pressure. These data show that the efficiency of pumping hydrogen (the hydrogen/argon trapping ratio) is much higher at 5 than at 15 K. At a temperature of more than 23 K, cryotrapping of hydrogen in argon did not occur. The density of the solid argon deposit decreased with condensation temperature; porous argon contained more hydrogen sorption sites than dense argon [22]. Thermal cycling of the argon to a higher temperature irreversibly increased its density and evolved the previously cryotrapped hydrogen.

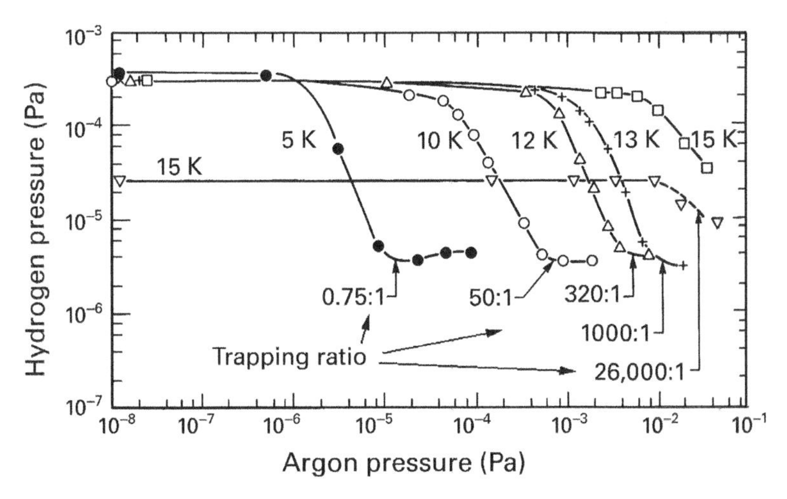

Fig. 14.3 Cryotrapping of hydrogen on solid argon at various temperatures. The drop in hydrogen pressure corresponds to the onset of cryotrapping at a particular argon pressure. Adapted from J. Hengevoss and E.A. Trendelenburg [21].

14.2 Speed, Pressure, and Saturation

In Chapters 2, 3, and 7 we outlined kinetic theory and introduced the concepts of gas flow, conductance, and speed. If the system is isothermal, these ideas can be used to predict the performance of a pump or system. If the temperature varies throughout the system, as it will near a cryopump, these notions must be applied with care; some are subject to misinterpretation, while others are simply not true. We stated, for example, that the mean free path was pressure dependent. Strictly speaking, it is particle-density-dependent. The pressure in a closed container will increase if the temperature is increased, but the mean free path will not change because the particle density remains constant. Such a misunderstanding can easily develop because we normally work with constant temperature systems and associate pressure change with particle density change.

The definitions of conductance and speed require the throughput, Q to be constant in a series circuit [24]. The throughput is constant only in an isothermal system. See Section 3.2. Throughput has dimensions of energy. If the system is not isothermal, energy is being added to the gas stream as it flows through a warm pipe and removed when it flows through a cool region. However, particle flow remains constant, unless frozen to a cold surface or emitted from a hot surface. It is this concern that directs us to formulate the behavior of a cryogenic pump in terms of particle flow rather than throughput.

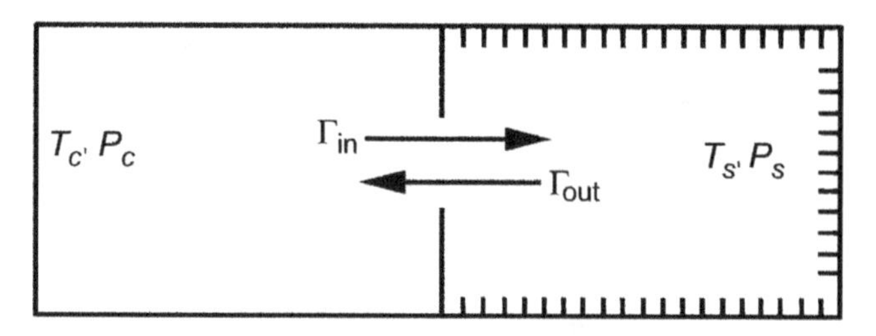

Fig. 14.4 Cryogenic pumping model. The gas in the pump has a temperature, T_s equal to the pumping surface, and a pressure, P_s, which is in equilibrium with the gas condensed or adsorbed on the pumping surface.

The diagram in Fig. 14.4 describes a chamber with gas at pressure, P_c, and temperature, T_c, connected by area, A to a cryogenic pump whose surfaces are cooled to temperature, T_s, and in which the gas is in thermal equilibrium with the surface. The temperature of the gas in the chamber is assumed to be greater than the gas in the pump. The net flux of particles into the pump is $\Gamma_{net} = \Gamma_{in} - \Gamma_{out}$. This may be written:

$$\Gamma_{net} = \frac{An_c v_c}{4} - \frac{An_s v_s}{4} = \frac{AP_c v_c}{4kT_c} - \frac{AP_s v_s}{4kT_s}$$
$$\Gamma_{net} = \frac{AP_c v_c}{4kT_c}\left[1 - \frac{P_s}{P_c}\left(\frac{T_c}{T_s}\right)^{1/2}\right] \tag{14.1}$$

In this derivation we have assumed the condensation coefficient is unity. Equation 14.1 may be simplified by observing that the term outside the brackets is Γ_{in}. The maximum particle flow into the pump corresponds to $\Gamma_{out} = 0$, or $\Gamma_{in} = \Gamma_{max}$. Equation 14.1 may be written as:

$$\frac{\Gamma_{net}}{\Gamma_{max}} = \left[1 - \frac{P_s}{P_c}\left(\frac{T_c}{T_s}\right)^{1/2}\right] \tag{14.2}$$

Now define:

$$P_{ult} = P_s\left(\frac{T_c}{T_s}\right)^{1/2} \tag{14.3}$$

and express Eq. (14.2) as:

$$\frac{\Gamma_{net}}{\Gamma_{max}} = c\left[1 - \frac{P_{ult}}{P_c}\right] \tag{14.4}$$

where the condensation coefficient, c, is now included. Eq. (14.3) is the thermal transpiration equation discussed in Chapter 2 (2.37). It relates the ultimate pressure, P_{ult}, in the chamber of our model, to the pressure over the surface. If the pump is a condensation pump, P_s is the saturated vapor pressure. If the pump is a sorption pump, P_s is the pressure obtained from the adsorption isotherm, knowing the fractional surface coverage and temperature of the sorbent.

The ultimate pressure for the cryosorption or cryocondensation pump modeled in Fig. 14.4 can be determined from Eq. (14.3) by use of the proper value of P_s. The ultimate pressure for cryocondensation pumping is equal to the saturated vapor pressure multiplied by the thermal transpiration ratio $(T_c/T_s)^{1/2}$. It is a constant during operation of the pump, provided that the temperature of the cryosurface does not change. The ultimate pressure for cryosorption pumping will increase with time because the saturation pressure over the sorbent is a function of the quantity of previously pumped gas. In either case the ultimate pressure in the chamber will be greater than the saturated vapor pressure or adsorption pressure by the transpiration ratio. For $T_c = 300\,K$ and $T_s = 15\,K$ the ratio is $P_{ult} = 4.47 P_s$.

Equation 14.4 may also be used to characterize the speed of the pump because $\Gamma_{net}/\Gamma_{max}$ is proportional to S_{net}/S_{max}. The pumping speed of a cryocondensation pump is constant and maximum when $P_{ult} \ll P_c$, regardless of the quantity of gas pumped. All gases except H_2, He, and Ne have a saturated vapor pressure of less than 10–20 Pa at 10 K.

The pumping speed of a cryosorption pumping surface is affected by its prior use, because the saturation pressure of the surface increases as the sites become filled. For high-vapor-pressure gases such as H_2 and He the pumping speed on a molecular sieve at 10–20 K can actually diminish from S_{max} to zero as the clean sorbent gradually becomes saturated with gas during pumping. Figure 14.5 illustrates the expected behavior of speed (linear scale) and ultimate pressure (log scale) as a function of the quantity of gas being pumped (log scale) for both cryosorption and cryocondensation pumping. For both cases the net speed goes to zero as the chamber pressure reaches the ultimate pressure.

The simple model presented here is valid for predicting the performance of a cryogenic pump connected to the chamber by an aperture or pumping port when all the gas in the pump is cooled to the temperature of the pumping surface. Unfortunately, practical pumps do not meet these criteria. A chamber completely immersed in a liquid cryogen has an ultimate pressure given by P_s; that is, the gas temperature in Equation (14.3) is $T_c = T_s$. Moore [25] has shown the ultimate pressure in a system consisting of a parallel cryopanel and warm wall to be:

$$P_{ult} = \frac{P_s}{2}\left[1 + \left(\frac{T_1}{T_2}\right)^{1/2}\right] \tag{14.5}$$

where T_1 is the temperature of the warm wall and T_2 is the panel temperature. Space chambers are constructed with large inner cryopanels. The gas density and

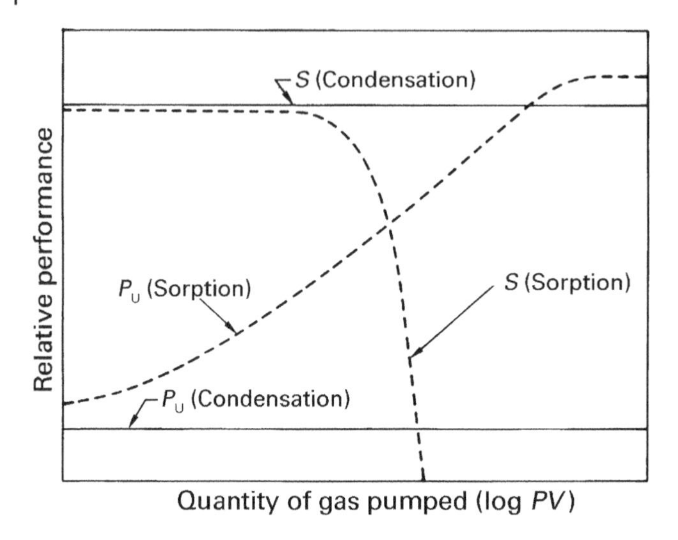

Fig. 14.5 Relative variation of pumping speed and ultimate pressure versus quantity of gas pumped for cryosorption pumping and for cryocondensation pumping.

temperature in those pumping systems are neither uniform nor in equilibrium, and the pressure measured depends on the orientation of the gauge [25]. The same is true for a cryogenic pump that contains surfaces cooled to several different temperatures. The model does not account for the heat carried to the cooled surfaces by the gas or by radiation other than to imply an equal amount of energy must be removed from a cooled surface by the cryogen or refrigerant so that its temperature remains constant. The effects of thermal loading on the pumping surfaces are discussed in Section 14.4. This simple model is sufficient to understand conceptually the speed–pressure relationships for the cryocondensation and cryosorption pumping of individual gases.

In most pumping requirements the pump must adsorb or condense a mixed gas load. Cryotrapping is one instance where the pumping of one gas aids the pumping of another. For example, the pumping speed of hydrogen in the presence of an argon flux may be higher than predicted by cryosorption. Its speed may not decrease to zero when the sorbent is completely covered. Cryosorption pumping of mixed gases may also cause desorption of a previously pumped gas, or reduced adsorption of one of the components of the mixed gas. Water vapor will inhibit the pumping of nitrogen [26], and CO has been shown to replace N_2 and Kr on Pyrex glass [27]. This is similar to the gas replacement phenomenon for chemisorption that occurs in TSPs; a gas with a small adsorption energy tends to be replaced or pumped less efficiently than a gas with a large adsorption energy. Hobson [28] and Haefer [3] have reviewed single gas adsorption processes in cryopumps, while Kidnay and Hiza [29] have summarized the literature on mixture isotherms.

14.3 Cooling Methods

Cryogenic pumping surfaces are cooled either by direct contact with liquid cryogens, or by gases in an expansion cooler. Liquid helium and liquid nitrogen are used to cool surfaces to 4.2 and 77 K, respectively. Other cryogens such as liquid hydrogen, oxygen, and argon are used to obtain different temperatures for specific laboratory experiments. In a liquid-cooled pump heat is removed from the cooled surfaces to an intermittently filled liquid storage reservoir or to a coil through which the liquid cryogen is continuously circulated. In a two-stage closed-cycle gas refrigerator pump, gaseous helium is cooled to two distinct temperature ranges, 10–20, and 40–80 K. Cryopumping surfaces are attached to these locally cooled heat sinks. Both methods of removing heat, liquid cooling, and gas cooling require mechanical refrigeration but in different ways. Liquid cryogens are most economically produced in large refrigerators at a central location and distributed in vacuum-insulated Dewar flasks, while helium gas refrigerators are economical for locally removing the heat load of a small cryogenic pump. The liquid cryogen requirements of a large cryogenic pumped space chamber, or large manufacturing facility, warrant installation of a liquefier at the point of use.

Systems using liquid cryogens are often called open-loop systems, because the boiling liquid is usually allowed to escape into the atmosphere. This is not necessary and not economical for helium. Helium gas recovery systems have been in use for decades, because of the expense of the world's dwindling helium supply. Helium gas refrigerators are examples of closed-loop systems. Warm gas is returned to the compressor after absorbing heat at low temperatures. That said, closed-cycle systems have losses and need to be repressurized periodically.

Many thermodynamic cycles have been developed for the achievement of low temperatures [30–33]. Some produce liquid helium or other liquid cryogens, some cool semiconducting and superconducting devices, and others produce refrigeration of useful capacity at temperatures ranging from 100 K to a few degrees above liquid helium temperature. It is the latter class of refrigerators that is used to cool cryogenic pumps. The refrigerator must be reliable, simple, and easy to manufacture and operate. Cycles embodying these attributes have been developed by Gifford and McMahon [32–37] and by Longsworth [38–40]. These two cycles are variants of a cycle developed by Solvay [30,33] in 1887. Except for some Stirling cycle machines [41], almost every cryogenic pump in use today is operated on one of these two cycles.

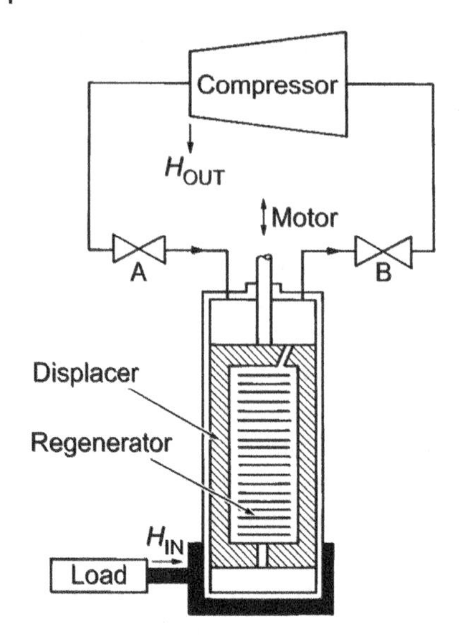

Fig. 14.6 Schematic representation of a single-stage Gifford–McMahon helium gas refrigerator. Adapted with permission from CTI-Cryogenics, Kelvin Park, Waltham, MA 02154.

Figure 14.6 illustrates a basic one-stage Gifford–McMahon refrigerator. The helium compressor is located remotely from the expander, and is connected to the expander by two flexible, high-pressure hoses. Within the expander is a cylindrical piston or displacer made from an insulating material. The piston is called a displacer [35] because the regions at each end are connected to give them little pressure difference.

Inside the displacer is a regenerator: a single-channel heat exchanger through which the gas flows at different times in alternate directions. It is tightly packed with a metal of high heat capacity and large surface-area-to-volume ratio. Alloys of lead or copper (or alloys whose heat capacity is high at cryogenic temperatures) in the shape of shot or screen are used to pack the regenerator. In the steady state the regenerator will have a temperature gradient. Ambient-temperature helium entering from the warm end will give heat to the metal, and cold gas entering from the cooler end will absorb heat from the metal. Even though the regenerator is tightly packed, there is not much flow resistance. The regenerator can transfer thermal energy from the incoming to the outgoing helium quickly and with great efficiency.

The operation of the Gifford–McMahon refrigerator can be understood by following the helium through a complete steady-state cycle. High-pressure gas from the outlet of the compressor is admitted to the regenerator through valve A while the displacer is at the extreme lower end of the cylinder. See Fig. 14.6. During the time that valve A is open, the displacer is raised. The incoming gas passes through the cold regenerator and is cooled as it gives heat to the regenerator. At this point in the cycle the gas temperature is about the same as the load. Valve A is then closed before the displacer reaches the top of its stroke. Further movement of the displacer forces the remainder of the gas through the displacer. The exhaust valve B is now opened to allow the helium to expand and cool. The expanding helium has performed work. It is this work of expansion which causes the refrigeration effect. No mechanical work is done since expansion did not occur against a piston. Heat flowing from the load, which is intimately coupled to the lower region of the cylinder walls, warms the helium to a temperature somewhat below that at which it entered the lower cylinder area. As the gas flows upward through the regenerator, it removes heat from the metal and cools it to the temperature at which it was found at the beginning of the cycle. The displacer is now pushed downward to force the remaining gas from the end of the cylinder out through the regenerator where it is exhausted back to the compressor at ambient temperature. A single-stage machine of this design can achieve temperatures in the 30–60 K range.

Lower temperatures can be achieved with two-stage machines. The first, or warm, stage operates in the range 30–100 K, while the second, or cold, stage operates in the range 10–20 K. The exact temperatures depend on the heat load and capacity of each stage. A heat-balance analysis of the refrigeration loss has been performed by Ackermann and Gifford [42]. In the Gifford–McMahon refrigerator the gas is cycled with poppet valves; the valves and the displacer are moved by a motor, and all are located on the expander. A Scotch yoke displacer drive is used because it applies no horizontal force to the shaft.

Figure 14.7 illustrates the expander developed by Longsworth [39,40]. As on the Gifford–McMahon refrigerator, the remotely located compressor is connected to

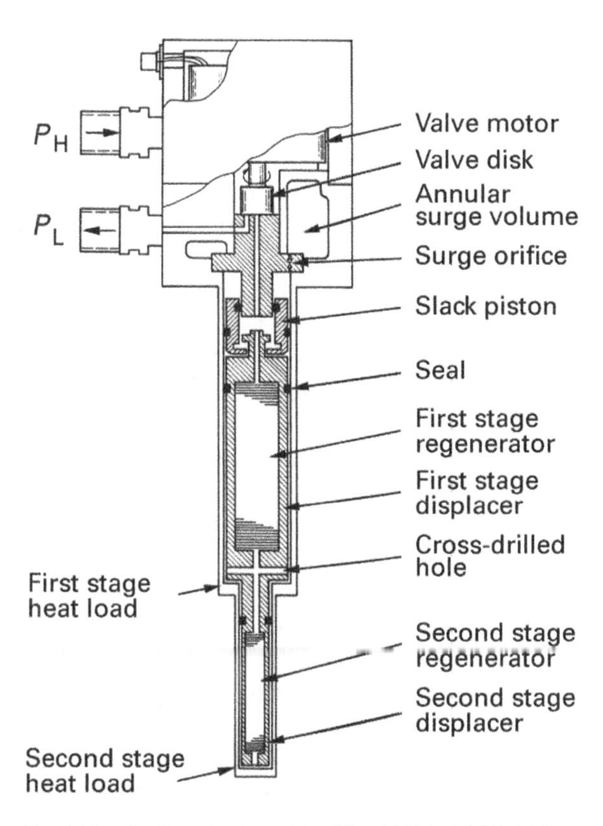

Fig. 14.7 Sectional schematic of the API Model DE-202 expansion head. Reproduced with permission from Air Products and Chemicals, Inc., PA 18105.

the expander by hoses of high-pressure capacity. The helium is cycled in and out of the expansion head through a motor-driven rotary valve. The expander shown here contains a two-stage displacer and two regenerators. The displacer is gas driven instead of motor driven, as in the original Gifford–McMahon cycle. A slack piston is incorporated to improve timing. Surrounding the valve stem is an annular surge volume. This volume is maintained at a pressure intermediate to the supply and exhaust pressures by a capillary tube connected to the regenerator inlet line; it provides the reference pressure for pneumatic operation of the displacer.

In the steady state the operating cycle proceeds as follows [43]: The valve is timed to admit high-pressure helium gas through the stem into the volume below the slack piston and in the regenerators while the displacer is in its lowermost position. Because the pressure over the slack pistons is less than the inlet pressure, the piston compresses this gas as it moves upward. The gas then bleeds into the surge volume through the surge orifice at a constant rate. The surge orifice is like a dashpot; it controls the speed of the displacer. As the displacer moves upward, high-pressure gas flows through the regenerators and is cooled in the process. The inlet valve stops the flow of high-pressure gas just before the displacer reaches the top. This slows the displacer and expands the gas in the displacer. The exhaust valve opens and the gas in the displacer expands as it is exhausted to the low-pressure side of the compressor. The slack piston moves downward suddenly until it contacts the displacer, after which it moves at constant velocity as gas flows from the surge volume into the space over the slack piston. The expansion of gas in the displacer causes it to cool below the temperature of the regenerator. This is the refrigeration effect. Like the Gifford–McMahon cycle, this cycle does no mechanical work on the displacer because both ends are at the same pressure. The exiting gas removes heat from the regenerator. Before the end of its stroke the displacer is decelerated by closure of the exhaust valve. This completes one cycle of expander operation. Heat is absorbed at two low temperatures and released at a higher temperature.

The compressor used in either of the two Solvay cycle variants is depicted in Fig. 14.8. It uses a reliable oil-lubricated, air-conditioning-type compressor with an inlet pressure of approximately 7×10^5 Pa (100 psig) and an outlet pressure of about 2×10^6 Pa (300 psig). After the gas is compressed, the heat of compression is removed by an air or water after cooler. Oil lubrication can be used without contaminating the cold stages, because it is removed by a two-stage separator and adsorber. Traces of oil entering the regenerator can cause problems. The oil must be cooled before entering the adsorber, as hot oil vapors are not adsorbed on charcoal. It is imperative that the adsorber cartridge be packed tightly and remain cool or it will not adsorb oil vapor.

The most important attribute of small Solvay-type refrigerators is reliability. The low-pressure differential across the seals in the displacers means light pressure loading and long life. Also contributing to long removal life is the use of

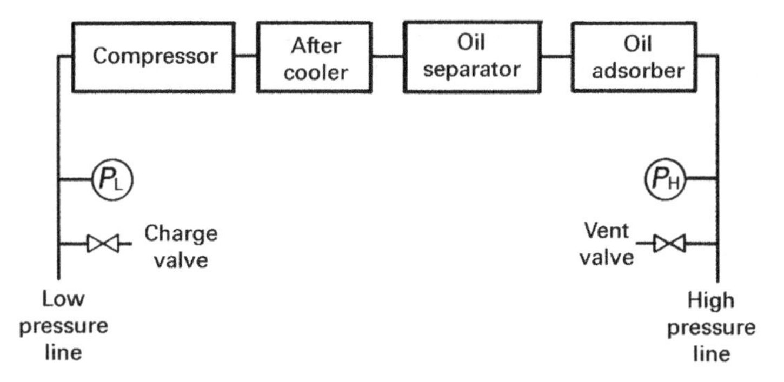

Fig. 14.8 Block diagram of a remotely located helium gas compressor.

room-temperature valves, a reliable compressor, and oil techniques. The result is a compact refrigerator that can be isolated from compressor vibration and attached to a vacuum chamber in any attitude.

The coefficient of performance of a refrigerator is defined as the ratio of heat removed to work expended in removing the heat. For an ideal Carnot cycle, the most efficient of all possible cycles, this is [44]:

$$\frac{H_{\text{out}}}{W_{\text{in}}} = \frac{T_1}{T_2 - T_1} \tag{14.6}$$

where heat is being absorbed at T_1 and released at T_2. For $T_2 = 300\,\text{K}$ a Carnot cycle would require a heat input of 2.9 W to remove 1 W at 77 K and would need a heat input of 14 W to remove 1 W at 20 K. In practice the efficiency of a refrigerator is defined as the ratio of ideal work to actual work. The refrigerators described here have efficiencies of ~3–5% [35,45].

There is interdependence between the refrigeration capacity of the two stages of a cryogenic refrigerator. There is a balance between the heat flow to each stage and the heat removed by each stage. Increasing the load on the warm stage will cause the cold stage to warm slightly also. Each manufacturer will have data on individual compressor performance. The emphasis has been on producing machines with increased capacity in the first stage to isolate radiant heat effectively from the second stage.

14.4 Cryopump Characteristics

In the previous sections we discussed the speed–pressure characteristics of some ideal pumping surfaces. The characteristics of a real cryogenic pump may differ significantly from those ideal cases. A detailed prediction of real pump performance requires a more complete model. The effects of thermal gradients between pumping surfaces and refrigerator, gas and radiant-heat loading, and the geometrical isolation of condensation and sorption stages are three important effects that have not been considered in the ideal model. In the remainder of this section the gas-handling characteristics of rough sorption pumps and refrigerator and liquid-cooled high vacuum pumps are related to the materials, geometry, and heat loading of the pumping surfaces.

14.4.1 Sorption Pumps

In the early 1900's, Dewar used refrigerated sorption pumping to evacuate an enclosed space. Sorption pumping, as we know it today uses high-capacity artificial zeolite molecular sieves and liquid nitrogen. A unique feature of cryosorption rough pumping is its ability to pump to 10^{-1} Pa without introducing hydrocarbons into the chamber.

A sorption pump designed for rough pumping is illustrated in Fig. 14.9. It consists of an aluminum body that contains many conducting fins and is filled with an adsorbent. A polystyrene foam or metal vacuum Dewar filled with liquid nitrogen surrounds the entire canister. Adsorbent pellets are loosely packed in the canister and do not make good thermal contact with its liquid-nitrogen-cooled walls. To improve the thermal contact pumps with internal arrays of metal fins are

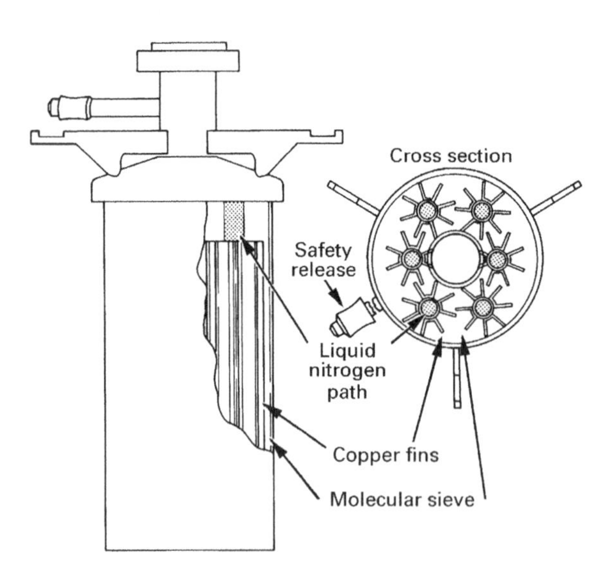

Fig. 14.9 Typical liquid nitrogen-cooled sorption pump. Reproduced with permission from Ultek Division, Perkin-Elmer Corp., Palo Alto, CA 94303.

constructed. Even so, the interior of the pump is not in equilibrium with the liquid nitrogen bath, especially during pumping.

A common adsorbent is Linde 5A molecular sieve. This sieve, with an average pore diameter of 0.5 nm, exhibits a high capacity for the constituents of air at low pressure. Figure 14.10 illustrates the adsorption isotherms in a pump containing a charge of 1.35 kg of molecular sieve. The adsorptive capacity for nitrogen is quite high in the range 10^{-3}–10^{5} Pa, while the capacity for helium and neon is quite low in this range. Mixed gas isotherms (e.g., neon in air) will show even less pumping capacity for the least active gas (neon), because it will be displaced by any active gas. The preadsorption of water vapor will greatly reduce the capacity for all gases. As little as 2 wt% water vapor is detrimental to pump operation [26]. These isotherms are not valid during dynamic pumping because the incoming gas will warm the sieve nonuniformly. They do represent the equilibrium condition of a real pump immersed in a liquid nitrogen bath.

The neon pressure in air is 1.2 Pa (9 mTorr), and this limits the ultimate pressure. Figure 14.11 illustrates the time dependence of the air pressure in a 100-L chamber for single-stage and two-sequential-stage pumping with pumps, each containing 1.35 kg of molecular sieve and prechilled at least 15 min. More than

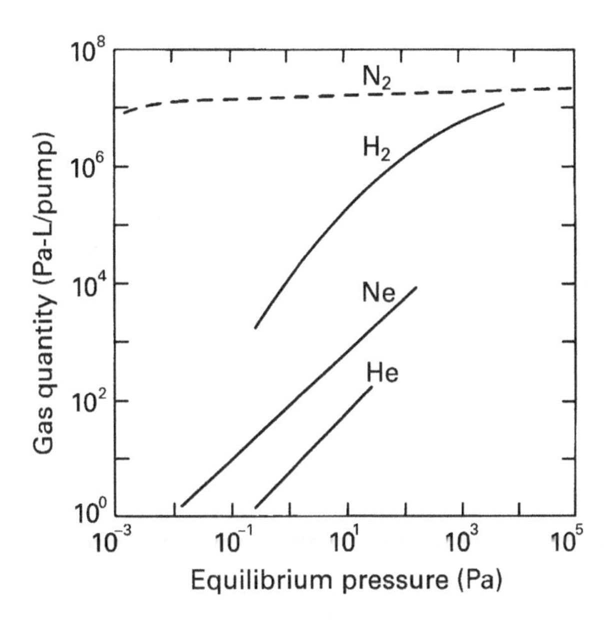

Fig. 14.10 Adsorption isotherms of nitrogen, hydrogen, neon, and helium at 77.3 K in a sorption pump charged with 1.35 kg of molecular sieve. Reproduced with permission from F. Turner [46]. Copyright 1973, Varian Associates, 611 Hansen Way, Palo Alto, CA 94303.

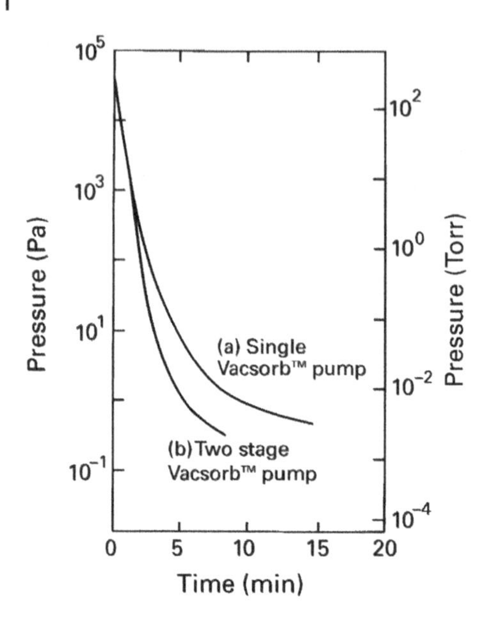

Fig. 14.11 Pumping characteristics of a 100-L, air-filled chamber with (a) one sorption pump and (b) two sequentially staged sorption pumps. Reproduced with permission from F. Turner [46]. Copyright 1973, Varian Associates, 611 Hansen Way, Palo Alto, CA 94303.

50% of the residual gas present after sorption pumping of air was found to be neon [47]. Staged roughing can reduce the ultimate pressure attainable by sorption pumping with a single pump. In staged roughing, one pump is used, until the chamber pressure reaches approximately 1000 Pa (~10 Torr). At that time, the valve connecting that pump to the system is quickly closed and a second pump is connected to the chamber. Figure 14.11 shows the ultimate pressure to be lower and the pumping speed higher than when only one pump was used. The improved pumping characteristic is a result of adsorbate saturation and neon removal. The first pump removed 10^7 Pa-L (99%) of air, including 99% of the neon, while the second removed only 10^5 Pa-L. The ultimate pressure of the second pump is less than the first stage because it pumped a smaller quantity of gas and because most of the neon was swept into the first pump by the nitrogen stream and trapped there when the valve closed. The valve needs to be closed quickly at the crossover pressure to prevent back-diffusion of the neon. Alternatively, a carbon vane or water aspirator pump may be used in place of the first sorption roughing stage.

The ultimate pressure attainable with a sorption pump is a function of its history, in particular the bake treatment and the kinds of gases and vapors. Pressures of an order of 1 Pa (7.5 mTorr) are typical. Multistage pumping performance has been characterized by several researchers [46–49], while Dobrozemsky and Moraw [50] have measured sorption pumping speeds for several gases in the pressure range of 10^{-4}–10^{-1} Pa.

Pumps are normally baked to a temperature of 250°C for about 5 h with a heating mantle or reentrant heating element. Miller [51] measured a water vapor desorption maximum in Linde 5A in the range 137–157°C. All other gases desorb well at room temperature. Baking is required to release water vapor and obtain the full capacity of the sieve.

Each sorption pump has a safety pressure release valve. At no time should the operation of this valve be hindered. Gases will be released when the pump is warmed to atmosphere and when it is baked. A single sorption pump of the size described requires about 5–8 L of liquid nitrogen for initial cooling.

14.4.2 Gas Refrigerator Pumps

A typical high vacuum cryogenic pumping array for a two-stage helium gas refrigerator is depicted in Fig. 14.12. The outer surface is attached to the first (warm, or 80 K) stage and the inner pumping surface is attached to the second (cold, or 20 K) stage. Indium gaskets are used to make joints of high thermal conductance. In practice the temperature of the stages is a function of the actual heat load and

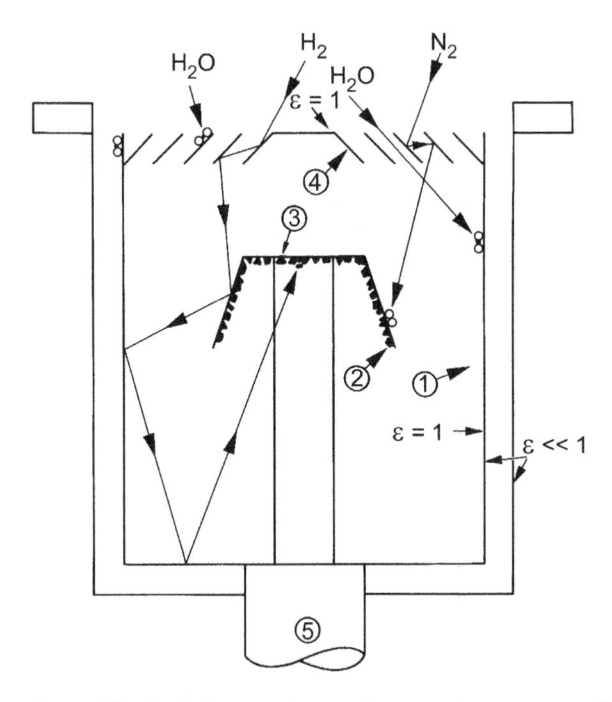

Fig. 14.12 Typical cryogenic pumping array for a two-stage helium gas refrigerator: (1) First-stage array, (2) second-stage cryosorption pump, (3) second-stage cryocondensation pump, (4) chevron baffle, and (5) refrigerator head.

thermal path. The warm stage pumps water vapor and sometimes CO_2. First stage pumping of CO_2 is a function of the temperature and the partial pressure. The vapor pressure of CO_2 on a 77 K surface is 10^{-6} Pa; CO_2 will be pumped only if its concentration is large enough or the surface temperature is adequately low. The second stage contains a surface that pumps differently on its two surfaces: the outer surface is a cryocondensation pump, while the inner surface is a cryosorption pump. The cryosorption surface is necessary to pump helium, hydrogen, and neon. It is shielded from the inlet aperture as much as possible to increase the probability that other gases will be condensed on the cryocondensation surface and to prevent direct radiation from an external heat source. Charcoal is the most commonly used sorbent because it can be degassed at room temperature. It has a greater capacity and is less affected by impurities than molecular sieve. Molecular sieve must be degassed at 250°C, and this is incompatible with the use of indium gaskets.

The temperature of the two pumping surfaces is determined by the total heat flux to the surfaces. Heat is leaked to these surfaces through the expander housing, radiated from high-temperature sources, removed from the incoming gas, and conducted from the warm walls by bouncing gas molecules. Thermal radiation emanates from nearby 300 K surfaces and internal sources such as plasmas, electron beam guns, and baking mantles. Each condensing molecule releases a quantity of heat equal to its heat of condensation plus its heat capacity. Heat can be conducted from a warm to a cold surface by gas–gas collisions if the pressure is high enough so that Kn < 1. In the high vacuum region, the radiant flux is much larger than the gas enthalpy or gas conductance. At a pressure of 10^{-4} Pa (7.5×10^{-7} Torr), a 1000-L/s pump condenses 0.1 Pa-L/s of nitrogen or a heat load of 0.6 mW. At this pressure, gas conductance can also be ignored.

The radiant heat flow between two concentric spheres or cylinders [51] is given by:

$$\frac{H}{A} = \frac{\sigma \varepsilon_1 \varepsilon_2}{\varepsilon_2 + \frac{A_1}{A_2}(1 - \varepsilon_2)\varepsilon_1} \left(T_2^4 - T_1^4\right) \tag{14.7}$$

where the subscripts 1,2 refer to, respectively, the inner and outer surfaces. If $\varepsilon_1 = \varepsilon_2 = 1$, the heat flow will be maximum and:

$$\frac{H}{A} = \sigma\left(T_2^4 - T_1^4\right) \tag{14.8}$$

This yields a heat flow of 457 W/m^2 from the 300 K surface to the 20 K surface. A typical first-stage area of 0.1 m^2 for a small cryogenic pump would then absorb a heat load of 45.7 W. Clearly, this would overload the refrigerator. The radiant heat load is reduced by reducing the emissivity of the inside surface of the vacuum wall and the outside surface of the first stage. If the walls are plated with nickel ($\varepsilon = 0.03$) [52], this heat load could be reduced to 0.7 W. Even if these surfaces became contaminated so that $\varepsilon_1 = \varepsilon_2 = 0.1$, the heat flow would be only 2.4 W.

The first stage isolates the second stage from the radiant heat load and pumps water vapor. Some condensable gas deposits, however, have the property of drastically altering the emissivity of the surface on which they condense. As little as 20 μm of water ice on a polished aluminum substrate at 77 K causes the emissivity to increase from 0.03 to 0.8 [53]. Other data [54] and calculations [55] agree. CO_2 also absorbs thermal energy [54,55].

The two purposes of the first stage—namely pumping water vapor and thermally shielding the second stage—seem contradictory. If the first stage pumps water vapor, its emissivity will rise and it will absorb excess heat and overload the refrigerator. This problem is overcome by keeping the outer wall close to the vacuum wall; the water vapor is now pumped along the upper perimeter of the chevron, the chevron itself, and the interior wall of the first stage. It cannot reach the lower portion of the outside wall. Alternatively, for special cases the exterior may be wrapped with multilayer reflective insulation.

The entrance baffle is designed to prevent radiation from illuminating the second stage. It does this by absorbing radiation in the chevron and allowing transmitted radiation to see the blackened inside wall, where it is absorbed. The entrance baffle also impedes the flow of gases to the second stage. An opaque baffle reduces the radiant loading on the second stage but also reduces the pumping speed for all gases [56]. Good baffle designs are a compromise between radiant heat absorption and pumping speed reduction. One array with an optical transmission of 7×10^{-4} and a molecular transmission of 0.24 was calculated to have the optimum balance between heat adsorption and speed loss [57]. In some pumps the chevrons are painted black; in others they are highly reflective. It matters little because both will soon be "blackened" with water vapor. Figure 14.13 illustrates a cutaway view of one commercial cryogenic pump. Cryopump design problems, including the effect of temperature gradients through the deposit and in the arrays on long-term pump performance, have been reviewed by Hands [58]. Lee and Lee have analyzed the cryopump first-stage array and concluded that a 1-mm-thick oxygen-free electronic (OFE®) copper shield resulted in optimum cooling time; if aluminum is used, it should be 1.75× as thick as copper [59].

Approximate pumping-speed calculations can be made for each species if the temperatures and geometry are known; for example, gases pumped on the second stage must first pass through the chevron baffle, where they are cooled. If the effective inlet area, inlet gas temperature, second-stage area, and species are known, the approximate speed and ultimate pressure can be estimated. The pumping speed of a gas in a cryogenic pump is not only related to

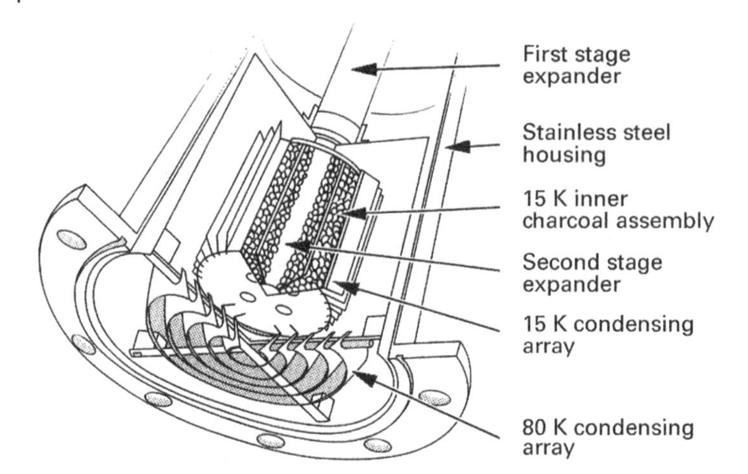

Fig. 14.13 Cutaway view of the Cryo-Torr 8 cryogenic pump. Reproduced with permission from CTI Cryogenics, Kelvin Park, Waltham, MA 021543.

the size of the inlet flange and refrigeration capacity. It is also dependent on the pumping array (relative sizes of the warm and cold stages), gas species, and history (ice, hydrogen, and helium load). All pumping speeds will fall off near 10^{-1} Pa (10^{3} Torr) as the refrigerator becomes overloaded. Because of these effects, it is not possible to draw an illustration analogous to Fig. 12.4, that would describe gas refrigerator-cooled pumps. A representative clean pump of 250-mm-diameter inlet flange diameter might have $S(H_2) = 4000$ L/s; $S(N_2) = 1200$ L/s; $S(Ar) = 1000$ L/s; $S(H_2) = 1200$ L/s; $S(CO_2) = 900$ L/s; and $S(He) = 700$ L/s. The only valid generalization is that helium pumping speed is usually small and has a low saturation value. Hydrogen pumping speeds are larger than those of helium but will be nil if the sorbent has saturated or its temperature is greater than 20 K.

When the pump becomes saturated, it should be shut down, while the cryo surfaces warm and the pump is regenerated. Regeneration techniques are discussed in Chapter 19. The vaporizing gases exit through the safety valve or manually operated valve, which is connected to a mechanical pump. All cryogenic pumps must have a properly functioning safety release to allow the escape of condensed and adsorbed gases and vapors if the pump fails, is shut down, or loses power. If there were no safety valve, the pump would become a bomb when warmed.

Recommended practices for measuring the performance and characteristics of closed-loop helium gas refrigerator cryopumps have been published by the American Vacuum Society [60]. The ISO Standard for cryopumps describes many performance details [61].

14.4.3 Liquid Cryogen Pumps

Three-stage cryogenic pumps that use liquid helium (4.2 K), gaseous helium boil-off (20 K), and liquid nitrogen are effective in pumping all gases, especially when the molecular sieve is bonded to the third stage [62]. Liquid nitrogen and liquid helium pumps are used to pump both large [63–65] and small vacuum chambers [16,66]. Liquid cryogen pumps have a higher pumping speed for hydrogen and helium than do gaseous helium refrigerator pumps because the stages are colder, and their temperatures are more stable. Additionally, their capacity for pumping He and H_2 is greater due to the increased sticking coefficient at 4.2 K [67]. Ultimate pressures of 10^{-11} Pa (10^{-13} Torr) have been reported [63].

Liquid cryogen pumps must be individually designed to meet the diverse requirements of applications ranging from small analytical and molecular beam epitaxy systems to very large particle beamlines and chambers for testing space modules. NASA's Thermal Vacuum Chamber A [68] illustrated on the cover, contains within is vacuum wall, a 56-ft-diameter × 90-ft-high LN_2 shroud, and a 45-ft-diameter × 80-ft-high inner liquid helium shroud with a combined pumping speed of 2×10^7 L/s for condensable gases. Both the Apollo space module and the fully extended James Webb telescope were tested in this chamber.

References

1 Hands, B.A., *Vacuum* **32**, 603 (1892).
2 Bentley, P.D., *Vacuum* **30**, 145 (1980).
3 Haefer, R., *J. Phys. E: Sci. Instrum.* **14**, 159 (1981).
4 Welch, K., *Capture Pumping Technology*, Pergamon, Oxford, 1991.
5 Dawson, J.P. and Haygood, J.D., *Cryogenics* **5**, 57 (1965).
6 Eisenstadt, M.M., *J. Vac. Sci. Technol.* **7**, 479 (1970).
7 Brown, R.F. and Wang, E.S., in *Proceedings of the 1964 Cryogenic Engineering Conference (Sections A–L)*, Adv. Cryog. Eng., **10**, K.D. Timmerhaus, Ed., Plenum, New York, 1965, p. 283.
8 Hobson, J.P., *J. Phys., Chem.* **73**, 2720 (1969).
9 Gareis, P.J. and Stern, S.A., *Bulletin de l'Institut International du Froid, Annexe 1966-5*, l'Institut International du Froid, Paris, 1966, p. 429.
10 Stern, S.A. and Mullhaupt, J.T., *J. Vac. Sci. Technol.* **2**, 165 (1965).
11 Van Dingenan, W. and Van Itterbeek, A., *Physica* **6**, 49 (1939).
12 Garies, P.J. and Stern, S.A., *Cryog. Eng. News* **26**, 85 (1967).
13 Stern, S.A. and DiPaolo, F.S., *J. Vac. Sci. Technol.* **4**, 347 (1967).
14 Grenier, G.E. and Stern, S.A., *J. Vac. Sci. Technol.* **3**, 334 (1966).

15 Kidnay, A.J., Hiza, M.J., and Dickenson, P.F., in *Proceedings of the 1967 Cryogenic Engineering Conference Stanford University Stanford, California August 21–23, 1967, Adv. Cryog. Eng.*, **13**, K.D. Timmerhaus, Ed., Plenum, New York, 1968, p. 39.

16 Powers, R.J. and Chambers, R.M., *J. Vac. Sci. Technol.* **8**, 319 (1971).

17 Johannes, C., in *A Collection of Invited Papers and Contributed Papers Presented at National Technical Meetings During 1970 and 1971, Adv. Cryog. Eng.*, **17**, K.D. Timmerhaus, Ed., Plenum, New York, 1972, p. 307.

18 Hobson, J.P., *J. Vac. Sci. Technol.* **3**, 281 (1966).

19 Redhead, P.A., Hobson, J.P., and Kornelsen, E.V., *The Physical Basis of Ultrahigh Vacuum*, Chapman and Hall, London, 1968, p. 37.

20 R.L. Chuan, Univ. South Calif., Eng. Center Rep. 56–101, 1960

21 Hengevoss, J. and Trendelenburg, E.A., *Vacuum* **17**, 495 (1967).

22 Hengevoss, J., *J. Vac. Sci. Technol.* **6**, 58 (1969).

23 Boissin, J.C., Thibault, J.J., and Richardt, A., *Le Vide, Suppl.* **157**, 103 (1972).

24 Lewin, G., *J. Vac. Sci. Technol.* **5**, 75 (1968).

25 Moore, R.W., Jr., *Transactions of the 8th National Vacuum Symposium of the American Vacuum Society, 1961*, L. Preuss, Ed., Pergamon Press, New York, 1962, p. 426.

26 Stern, S.A. and DiPaolo, F.S., *J. Vac. Sci. Technol.* **6**, 941 (1969).

27 Tuzi, Y., Kobayashi, M., and Asao, K., *J. Vac. Sci. Technol.* **9**, 248 (1972).

28 Hobson, J.P., *J. Vac. Sci. Technol.* **10**, 73 (1973).

29 Kidnay, A.J. and Hiza, M.J., *Cryogenics* **10**, 271 (1970).

30 Collins, S.C. and Canaday, R.L., *Expansion Machines for low Temperature Processes*, Oxford University Press, Oxford, 1958.

31 Barron, R., *Cryogenic Systems*, McGraw-Hill, New York, 1966.

32 Bridwell, M.C. and Rodes, J.G., *J. Vac. Sci. Technol., A* **3**, 472 (1985).

33 R. Radebaugh, *Applications of Closed-Cycle Cryocoolers to Small Superconducting Devices*, NBS Special Publication 508, U.S. Department of Commerce, National Bureau of Standards, Washington, DC, 1978, p. 7.

34 W.E. Gifford, Refrigeration Method and Apparatus, U. S. Pat. 2,966,035 (1960).

35 Gifford, W.E. and McMahon, H.O., in *Proceedings of the Eleventh International Congress of Refrigeration, Munich, August-September 1963*, Prog. Refrig. Sci. Technol, **1**, M. Jul and A. Jul, Eds., Pergamon, Oxford, 1965, p. 105.

36 Gifford, W.E., in *Prog. Cryog.*, K. Mendelssohn, Ed., Vol. **3**, Academic, New York, 1961, p. 49.

37 Gifford, W.E., in *Proceedings of the 1965 Cryogenic Engineering Conference Rice University Houston, Texas August 23–25, 1965*, Adv. Cryog. Eng., **11**, K.D. Timmerhaus, Ed., Plenum, New York, 1966, p. 152.

38 R.C. Longsworth, Refrigeration method and apparatus, U.S. Pat. 3,620,029, (1971).

39 Longsworth, R.C., in *Proceeding of the 1970 Cryogenic Engineering Conference, The University of Colorado Boulder, Colorado June 17–19, 1970, Adv. Cryog. Eng.*, **16**, K.D. Timmerhaus, Ed., Plenum, New York, 1971, p. 195.
40 Longsworth, R.C., in *Adv. Cryog. Eng.*, K.D. Timmerhaus, Ed., Vol. **23**, Plenum, New York, 1978, p. 658.
41 For example, Type K-20 Series Cryogenerator, manufactured by, N. V. Philips Gloeilampenfabrieken, Eindhoven, Netherlands.
42 Ackermann, R.A. and Gifford, W.E., in *Proceeding of the 1970 Cryogenic Engineering Conference The University of Colorado Boulder, Colorado June 17–19, 1970, Adv. Cryog. Eng.*, **16**, K.D. Timmerhaus, Ed., Plenum, New York, 1971, p. 221.
43 Longsworth, R.C. "Capture Pumping Technology", Chapter 5", in *Cryopumping*, 2nd ed., North Holland, 2001, p. 325.
44 Sears, F.W., *An Introduction to Thermodynamics, The Kinetic Theory of Gases and Statistical Mechanics*, Addison-Wesley, Reading, Ma, 1953, p. 84.
45 T.R. Strobridge, NBS Technical Note 655, U. S. Department of Commerce, National Bureau of Standards, Washington, D. C., 1974.
46 F. Turner, Cryosorption Pumping, Varian Report VR-76, Varian Associates, Palo Alto, CA, 1973.
47 Turner, F.T. and Feinleib, M., *Transactions of the 8th National Vacuum Symposium of the American Vacuum Society, 1961*, L. Preuss, Ed., Pergamon Press, New York, 1962, p. 300.
48 Cheng, D. and Simpson, J.P., in *Proceedings of the 1964 Cryogenic Engineering Conference (Sections A–L), Adv. Cryog. Eng.*, **10**, K.D. Timmerhaus, Ed., Plenum, New York, 1965, p. 292.
49 Vijendran, P. and Nair, C.V., *Vacuum* **21**, 159 (1971).
50 Dobrozemsky, R. and Moraw, G., *Vacuum* **21**, 587 (1971).
51 Miller, H.C., *J. Vac. Sci. Technol.* **10**, 859 (1973).
52 *Ibid*, p. 348.
53 Caren, R.P., Gilcrest, A.S., and Zierman, C.A., in *Proceedings of the 1963 Cryogenic Engineering Conference University of Colorado College of Engineering and National Bureau of Standards Boulder Laboratories Boulder, Colorado August 19–21, 1963, Adv. Cryog. Eng.*, **9**, K.D. Timmerhaus, Ed., Plenum, New York, 1964, p. 457.
54 Moore, B.C., *Transactions of the 9th National Vacuum Symposium Transactions of the ninth National Vacuum Symposium of the American Vacuum Society, October 31 – November 2, 1962*, Macmillan, New York, 1962, p. 212.
55 Tsujimoto, S., Konishi, A., and Kunitomo, T., *Cryogenics* **22**, 603 (1982).
56 Lee, J.W. and Lee, Y.K., *Vacuum* **44**, 697 (1993).
57 Benvenuti, C., Blechschmidt, D., and Passarde, G., *J. Vac. Sci. Technol.* **19**, 100 (1981).

58 Hands, B.A., *Vacuum* **26**, 11 (1976).

59 Lee, J.W. and Lee, J.Y., *Vacuum* **42**, 457 (1991).

60 Welch, K.M., Andeen, B., de Rijke, J.E., Foster, C.A., Hablanian, M.H., Longsworth, R.C., Millikin, W.E., Jr., Sasaki, Y.T., and Tzemos, C., *J. Vac. Sci. Technol., A* **17**, 3081 (1999).

61 ISO-21360-6:2020, *Vacuum Technology — Standard Methods for Measuring Vacuum-Pump Performance*: Part 6, Cryopumps. International Standards, Organization, Geneva, Switzerland, 2020.

62 Halama, H.J. and Aggus, J.R., *J. Vac. Sci. Technol.* **12**, 532 (1975).

63 Benvenuti, C., *J. Vac. Sci. Technol.* **11**, 591 (1974).

64 Benvenuti, C. and Blechschmidt, D., *Jpn. J. Appl. Phys. Suppl.* **2** (Pt. 1), 77 (1974).

65 Halama, H.J., Lam, C.K., and Bamberger, J.A., *J. Vac. Sci. Technol.* **14**, 1201 (1977).

66 Schafer, G., *Vacuum* **28**, 399 (1978).

67 Day, C. and Schwenk-Ferrero, A., *Vacuum* **53**, 253 (1999).

68 NASA, Image of NASA Johnson Space Center's Thermal Vacuum Chamber A. https://www.nasa.gov/offices/setmo/facilities/thermal_vacuum_chamber_a

Part IV

Materials

Knowledge of the basic properties of materials from which vacuum systems are fabricated is essential to operate and maintain vacuum systems properly. One of the most troublesome properties of materials used in vacuum applications is the release of gas from solids at low pressures. Chapter 15 discusses the origins of gas released from metals, glasses, ceramics, and polymers commonly used in systems. It also discusses how surface and dissolved gas can be removed and how surfaces can be kept clean. However, some cleaning details specific to ultraclean systems are discussed in Chapter 20. Chapter 16 discusses techniques for joining and sealing materials, both permanently and temporarily; valves are a complex but absolutely necessary component. Chapter 17 describes both fluids used in pumps and lubricants used with vacuum components. There is nothing to vacuum; it is all in the packaging.

A Users Guide to Vacuum Technology, Fourth Edition. John F. O'Hanlon and Timothy A. Gessert.
© 2024 John Wiley & Sons, Inc. Published 2024 by John Wiley & Sons, Inc.

15

Materials in Vacuum

A superficial view of a high or ultrahigh vacuum pumping system gives an impression of simplicity: clean, polished metal or glass surfaces, viewports, electrical and motion feedthroughs, piping, and pumps. A close examination reveals that many requirements are placed on materials in vacuum environments and these requirements sometimes conflict. The chamber walls must support a load of 10,335 kg/m^2, a load that is present on the surfaces of all vacuum systems, even those merely roughed to 1000 Pa. Metals must be easy to machine and connect with demountable seals, or join permanently by welding, brazing, or soldering. Methods are needed for sealing glasses, ceramics, and other insulators to metals to form optical and electrical feedthroughs. The outgassing load from fixture and chamber surfaces must be reduced to obtain low pressures. High-vacuum systems contain a large internal surface that cannot be baked; thus, they should be fabricated from materials with low outgassing rates. Baking is necessary to reduce the outgassing rate and reach the lowest possible pressure in ultrahigh vacuum systems. High-temperature baking reduces the maximum stress limit and increases the strain or deformation of stressed parts, and thermal decomposition limits the temperature at which some materials can be baked. The interdependence of contamination control, cleaning, joining, construction, and application need to be clearly understood when choosing materials for use in a vacuum environment.

In this chapter, we review the outgassing and structural properties of metals, glasses, ceramics, and polymeric materials used in the construction of vacuum systems. Stainless steel is the dominant material of construction for sophisticated processes, and its structural properties are discussed in some detail. Several excellent papers review materials that are used in high and ultrahigh vacuum applications [1–4].

Vaporization, permeation, and outgassing are important properties to vacuum performance. Vapor pressure determines the useful upper-pressure limit for any

A Users Guide to Vacuum Technology, Fourth Edition. John F. O'Hanlon and Timothy A. Gessert.
© 2024 John Wiley & Sons, Inc. Published 2024 by John Wiley & Sons, Inc.

material. Desorption from vacuum chamber surfaces provides an additional gas load to the system. Certain materials are rather impermeable to gases, whereas others allow certain gases to permeate from the atmospheric side. These effects, collectively referred to as outgassing, limit the ultimate system pressure. In this chapter we review important vacuum-related properties of metals, glasses, ceramics, and polymers. The use of these materials in the fabrication of seals, valves, and other components is described in Chapter 16.

15.1 Metals

Metals are used in the chamber and to form its walls. Metals used for vacuum chamber walls should be joinable and sealable and should have high strength, low permeability to atmospheric gases, low outgassing rate, and low vapor pressure. Metals used within the chamber should have a low outgassing rate and low vapor pressure. Specific vacuum uses, such as filaments, radiation shields, and thermal sinks, will add other constraints. In this section, we discuss the vaporization, permeation, and outgassing properties of several metals and some structural properties of aluminum and austenitic stainless steel.

15.1.1 Vaporization

Most metals have a sufficiently low vapor pressure for vacuum use. The vapor pressures of the elemental metals are found in Appendix C.6. Certain metals should not be used in vacuum construction because their vapor pressures are high enough to interfere with normal vacuum baking procedures. Alloys containing zinc, lead, cadmium, selenium, and sulfur, for example, have unsuitably high vapor pressures for vacuum applications. Zinc is a component of brass, cadmium is commonly used to plate screws, and sulfur and selenium are used to make the free machining grades 303S and 303Se stainless steel. These materials should not be used in vacuum system construction. Vapor pressures of the elements are given in Appendix C.6.

15.1.2 Permeability

For gases that do not dissociate, gas permeability is proportional to the product of gas solubility and diffusion constant, as shown in Eq. (4.19). Hydrogen is one of the few gases that permeate metals to a measurable extent. Because hydrogen dissociates on adsorption, its permeation rate is proportional to the square root of the pressure difference. See Section 4.5.2. Figure 15.1 plots the temperature dependence of hydrogen permeation through several metals. Hydrogen permeation is the least in Al. Other metals through which hydrogen permeates are, in order of increasing permeability, Mo, Ag, Cu, Pt, Fe, Ni, and Pd. The addition of chrome to iron allows the formation of a chrome oxide barrier that reduces hydrogen permeation. The influx will be greater than this value if rusting occurs on the external wall because the reaction of water vapor with iron creates a high local surface H_2 partial pressure.

Techniques for measuring outgassing are described in Chapter 4. Experimental techniques used to measure the permeation and diffusion constants in solids and have been described elsewhere along with the permeability values of several metals [1].

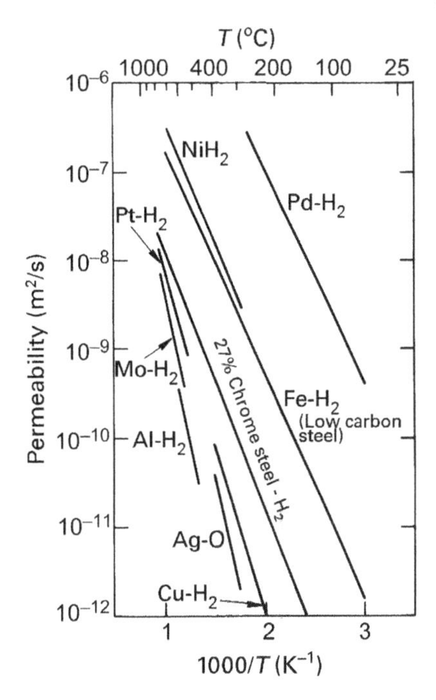

Fig. 15.1 Permeation constant of H_2 through various metals as a function of temperature. Reproduced with permission from F.J. Norton [5]. Copyright 1962, Pergamon Press, Ltd.

15.1.3 Outgassing

The gas load in vacuum fixtures and chamber walls is adsorbed on metal surfaces and affects system performance if it is not removed. Gas is dissolved in a metal during its initial melting and casting. It consists mainly of hydrogen, oxygen, nitrogen, and carbon oxides. Gas is also physisorbed and chemisorbed on the interior surfaces from exposure to ambient atmosphere. It consists of a large quantity of water vapor, with carbon oxides, oxygen, and some nitrogen. The nature and quantity of the adsorbed layer is also a function of the metal, the gas used to vent the system to atmosphere, and the time and extent to which it was exposed to the surrounding air.

In SI the outgassing rate (quantity of gas evolved per unit time per unit surface area) has units of Pa-m/s. These units can seem incorrect, but they are a result of dividing the quantity of gas evolved per unit time (Pa-m^3/s) by the unit surface area (1 m^2). Conversion factors are given in Appendix A.3. The pressure in a chamber with net outgassing rate, q, and area, A, when pumped at a speed, S, is given by:

$$P(\text{Pa}) = 1000\frac{q(\text{Pa-m/s})A(\text{m}^2)}{S(\text{L/s})} = \frac{q(\text{Pa-m/s})A(\text{m}^2)}{S(\text{m}^3/\text{s})} \quad (15.1)$$

The factor of 1000 is included when the pumping speed is expressed in L/s rather than m^3/s.

It is important to note that published outgassing, such as given in this chapter and in Appendix C, are net outgassing rates and not true outgassing rates. Hobson described a model of a hollow sphere made from the material being measured [6]. q_{net}, the net outgassing rate per unit area of the material, was obtained by properly measuring the flux evolving from the inside of the sphere, Q_T, and dividing it by the area of the sphere, A. However, the nonzero pressure inside the sphere indicated there was a flux incident on its surface. For the case where the sticking coefficient $s > 0$, the total or true outgassing rate per unit area of the material was shown to be $q_{true} = q_{net} + \Gamma_s$. When $s = 0$, the net outgassing is the same as the true outgassing. In Chapter 4, we noted that measured sticking coefficients of water on stainless steel range from 0.001 to 0.2. Hobson cited examples of a proton storage ring and a space shuttle wake shield where a nonzero sticking coefficient caused a very significant difference in the calculated pressures near the outgassing material.

15.1.3.1 Dissolved Gas

Release of dissolved gas from inside metals can be eliminated by rendering it immobile, reducing its initial concentration, or erecting a barrier to its passage. Dissolved gas can be rendered immobile by completely immersing a system in liquid helium. At 4.2 K the outdiffusion flux for any gas is so small, Eq.'s (4.3) and (4.5), that no special precautions need to be taken. The initial concentration can be substantially reduced by vacuum melting, by first degassing parts in a vacuum furnace, or by *in situ* bake of the completed system. A barrier to this outgassing flux can be created by incorporating a layer of metal such as copper, which has low permeability, or by forming an oxide barrier such as chrome oxide on stainless steel. An oxide barrier to hydrogen diffusion can be formed by an air or oxygen bake or by a multistep chemical treatment such as Diversey [7] cleaning. The latter method leaves some water vapor on the surface. Less thorough cleaning methods are needed after the system has been initially treated. Either glow discharge cleaning or a vacuum bake can be used to clean a chamber after each exposure to ambient. The nature and duration of the cleaning depend on the materials, construction, and desired base pressure.

Vacuum melting is an excellent technique for removing dissolved gas under certain conditions. It is expensive and used for specialized applications that require hydrogen and oxygen-free material in small quantities such as certain internal parts and charges for vacuum evaporation hearths.

Vacuum firing of components and subassemblies will effectively remove the dissolved gas load in cleaned and degreased parts. Hydrogen firing is traditionally used for this purpose because it reduces surface oxides. It has the disadvantage of incorporating considerable gas in the metal at high temperatures, which can slowly outdiffuse at low temperatures.

Vacuum or inert gas firing is preferred for vacuum components, especially those in ultrahigh vacuum systems. The maximum firing temperatures for several metals are given in Table 15.1. Iron and steel are today not fired at temperatures over 800°C. See Section 20.2.2. Copper and its zinc-free alloys can be fired at 500°C.

It is important to use oxygen-free electronic (OFE®) copper, formerly known as OFHC® copper, because the copper oxide found in electrolytic tough pitch (ETP) copper will react with hydrogen to form water vapor during baking. The steam will generate voids, which will create porous leaky regions. OFE copper is formed by melting and casting ETP copper under a protective coating of carbon monoxide that reduces copper oxide and prevents contact with atmospheric oxygen.

Based on a single diffusion constant, the time to depletion is $t = d^2/(6D)$ [9]. A 1-h bake of stainless steel at 1000°C would be equivalent to a 2500-h bake at 300°C. Calder and Lewin [10] calculated the time required to reach an outgassing rate of 10^{-13} Pa-m/s for a stainless steel sample 2 mm thick with an initial exposure to 4×10^4 Pa. Their results are shown in Table 15.2 for hydrogen diffusion in stainless steel at temperatures of 300–635°C. This model is valid for high hydrogen outgassing rates in stainless steel. Hydrogen outgassing in UHV chambers is discussed in Section 20.3. Jousten [11] drew different conclusions from vacuum-fired 316LN stainless steel. He concluded that the hydrogen outgassing rate decreased with the addition of several 100°C baking cycles, and that the outgassing rate

Table 15.1 Firing Temperatures for Some Common Metals

Material	Firing Temp. (°C)
Tungsten	1800
Molybdenum[a]	950
Tantalum	1400
Platinum	1000
Copper and alloys[b]	500
Nickel and alloys (Monel, etc.)	750–950
Iron, steel, stainless steel	1000

[a] Embrittlement takes place at higher temperatures. The maximum firing temperature is 1760 °C.
[b] Except zinc-bearing alloys, which cannot be vacuum fired at high temperatures because of excessive zinc evaporation.
Source: Reproduced with permission from A. Guthrie [8]. Copyright 1963, John Wiley & Sons, New York.

Table 15.2 The Theoretical Time to Reach an Outgassing Rate of 10^{-13} Pa-m/s in Stainless Steel; Single Diffusion Constant Model

t (s)	*D* (m^2/s)	*T* (°C)
10^6 (11 DAYS)	3.5×10^{-12}	300
8.6×10^4 (24 h)	3.8×10^{-11}	420
1.1×10^4 (3 h)	3.0×10^{-10}	570
3.6×10^3 (1 h)	9.0×10^{-10}	635

Source: Adapted from A. Guthrie [8].

decreased with reduced baking temperature. The lowest observed rate was 5×10^{-12} Pa-m/s (4×10^{-15} Torr-L-s^{-1}-cm^{-2}). This result was attributed to oxide formation following exposure to humid air. Most likely, this result was due to the large concentration of hydrogen in deep traps that was released after the high-temperature bake, but not released after a low-temperature bake.

When baking, is not good practice to fire both a screw and the part containing the tapped hole because the screw will bind when tightened.

15.1.3.2 Surface and Near-Surface Gas

Adsorbed gas may be removed from metal surfaces by thermal desorption, chemical cleaning, or energetic particle bombardment. The procedures used for cleaning metal vacuum system parts depend on the system application. The outgassing rate of unbaked and uncleaned stainless steel is of order 10^{-5} Pa-m/s after 10 h of pumping, depending on its effective surface area. An unbaked system with a 0.5-m^3 work chamber may have as much as 6 m^2 of internal tooling and wall area. If the high-vacuum pump has a base plate pumping speed of 2000 L/s, the pressure after 10 h of pumping will be 3×10^{-5} Pa, not a good base pressure for such a system. The outgassing rate of stainless steel must be reduced by a factor of 10–100 over its untreated value to be suitable for high-vacuum applications and by a factor of 10^4–10^5 for ultrahigh vacuum applications. Unbaked systems with base pressures of ~10^{-6} Pa are generally cleaned by chemical or glow discharge cleaning techniques and perhaps a mild bake at temperatures of 40–80°C, whereas ultrahigh vacuum chambers are baked to ~150°C or use a form of glow discharge cleaning such as Ar–O_2 or H_2–He glow discharge cleaning. Some large collider systems cannot be baked because of their physical size.

Dylla et al. has measured the unbaked outgassing rates of 304L stainless steel when prepared by several cleaning methods [12]. Their data, shown in Fig. 15.2, follow the well-reported $1/t$ dependence. All cleaning methods produced nearly identical outgassing rates; differences could be attributable to differences in atmospheric air

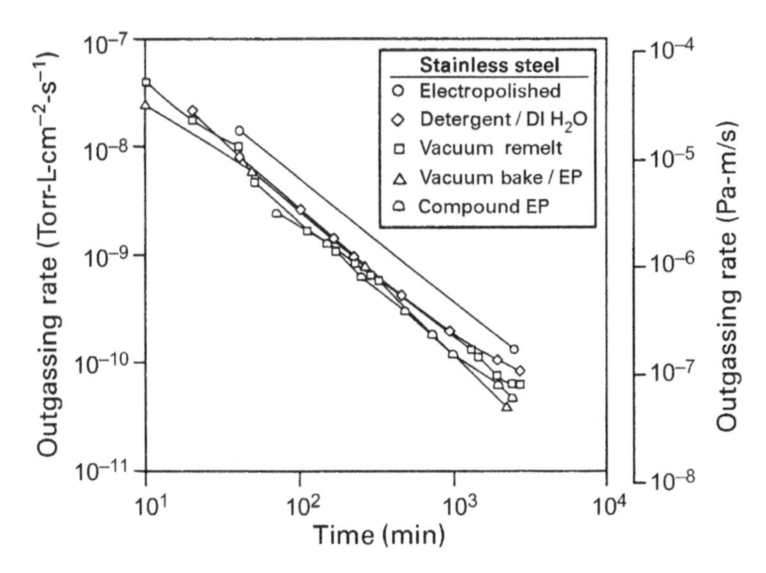

Fig. 15.2 Measured outgassing rates from 304L stainless steel for five different surface cleaning treatments. Reproduced with permission from H.F. Dylla et al. [12]/AIP Publishing.

pre-exposure times [13] and effective surface areas. These experiments found the dominant adsorbed material (>85%) to be water vapor. The 304L pretreatment methods shown in Fig. 15.2 included detergent (Alconox®) and DI water, electropolishing, vacuum remelting, vacuum baking/electropolishing, and compound electropolishing. Commercial electropolishing uses phosphoric acid-based solutions. Compound electropolishing refers to a simultaneous mechanical and electropolishing that results in a mirror-like surface with an extremely low surface roughness [14].

Initial surface preparation and cleaning steps such as Blanchard grinding, gross removal by wiping and brushing, high-pressure water jets, and immersion in hot (>100°C) perchloroethylene vapor, trichloroethylene, or 1,1,1-trichloroethane often precede detergent cleaning. A number of detergents such as Almeco 18® [15] and Almeco P20® [16] have been used. Gross removal of material by bead blasting with either alumina or glass beads results in rough and contaminated surfaces that impede cleaning. This procedure was suggested as a way to avoid use of phosphorous-based cleaning solutions. Anecdotally, it was suggested that beads did not break if the air pressure was 40 psig or less. This was proven incorrect; both alumina and glass beads were observed to fracture at pressures of 40 psig and less. Energy dispersive X-ray (EDX) analysis of a low-pressure bead blasted surface, illustrated in Fig. 15.3, shows that silicon bead fragments were buried in the treated surface. For these reasons, use of this method is often limited to cleaning removable deposition shielding.

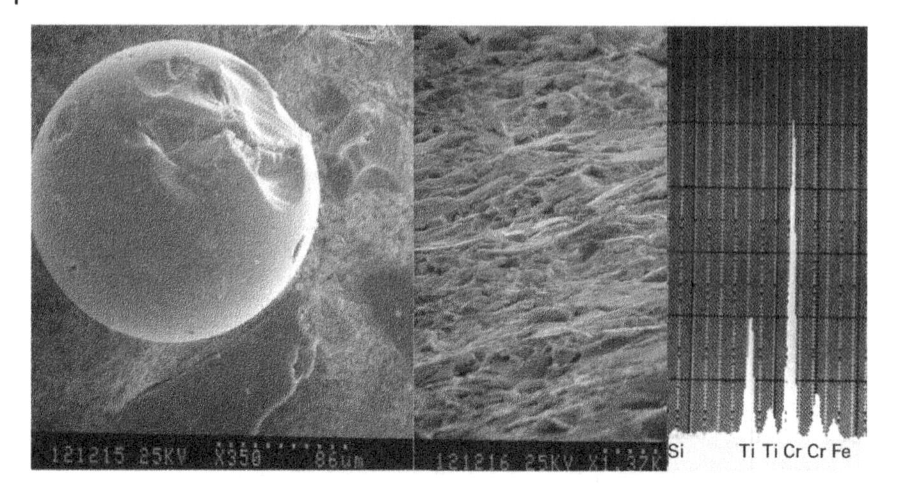

Fig. 15.3 Analysis of stainless steel surface cleaned with 40-psig silica beads. Left-SEM bead image; Center, SEM image of treated surface; Right, EDX surface analysis. J.F. O'Hanlon, unpublished.

Other chemical cleaning procedures, such as Diversey cleaning, reduce hydrogen outgassing from the near-surface region by formation of a chrome-oxide-rich passivation layer. Water vapor was found to be the major outgassing product released from a Diversey cleaned surface [17].

Vacuum baking of stainless steel has been extensively studied because of the wide use it enjoys in vacuum system construction. Nuvolone [18] has systematically compared several of these treatments on 316L stainless steel under identical pre-cleaning and measurement conditions. This work, which is consistent with earlier work, is described in Table 15.3. The lowest outgassing rate was obtained with oxidation in pure oxygen at 2700 Pa. The oxide barrier effectively reduced the hydrogen-outgassing rate. Consistent with recent work, Nuvolone's results show that surfaces cleaned by an 800°C vacuum bake or a 400°C air bake can be stored for long periods before use, provided that they are given a low-temperature bake after assembly.

Reduction of thermal desorption and ultimate pressures have been the main motivator for improved surface cleaning studies; however, the accelerator community requires stringent cleaning to reduce particle-induced outgassing [19]. Glow discharge cleaning has been used for conditioning the beam tubes in the Brookhaven colliding beam accelerator [20] and at the CERN intersecting storage ring [21]. Ar–O_2 plasmas can also remove material by sputtering. In large systems, sputtering is undesirable, because it deposits materials on insulators and windows. Hydrogen glow discharge cleaning has been used at 400 eV. The atomic hydrogen ions therefore have energies not greater than 200 eV, which is below

Table 15.3 Outgassing Rates of 316L Stainless Steel After Different Processing Conditions

		Outgassing Rates (10^{-10} Pa-m/s)				
Sample	**Surface Treatment**	H_2	H_2O	CO	Ar	CO^2
A	Pumped under vacuum for 75 h	893	573	87	—	13.
	50 h vacuum bakeout at 150°C	387	17	6	—	0.4
B	40 h vacuum bakeout at 300°C	83	0.7	2.2	—	0.01
C	Degassed at 400°C for 20 h in a vacuum furnace (6.5×10^{-7} Pa)	19	0.3	0.44	0.16	0.11
D	Degassed at 800°C for 2 h in a vacuum furnace (6.5×10^{-7} Pa)	3.6	—	0.07	—	0.05
	Exposed to atmosphere for 5 mo, pumped under vacuum for 24 h	—	73	67	—	13
	20-h vacuum bakeout at 150°C	3.3	—	0.08	—	0.04
E	2 h in air at atmospheric pressure at 400°C	17	—	1.12	—	0.4
	Exposed to atmosphere for 5 mo, pumped under vacuum for 24 h	—	80	69	—	33
	20-h vacuum bakeout at 150°C	17	0.75	0.37	—	0.17
F	2 h in oxygen at 27,000 Pa at 400°C	600	253	—	123	—
	20-h vacuum bakeout at 150°	5.2	0.09	0.4	0.51	—
G	2 h in oxygen at 2700 Pa at 400°C	—	20	13	8.7	—
	20-h vacuum bakeout at 150°C	—	0.9	0.64	0.45	—
H	2 h in oxygen at 270 Pa at 400°C	—	16	52	19	—
	20-h vacuum bakeout at 150°C	5.7	3.2	0.36	2	—

All samples were first degreased in perchloroethylene vapor at 125 °C, ultrasonically washed for 1 h in Diversey 708 cleaner at 55°C, rinsed with clean water, and dried.
Source: Adapted from Nuvolone [18].

sputtering threshold [22]. Desorption of a dilute hydrogen isotope was enhanced by a high constant pressure of hydrogen, and this considerably speeded surface cleaning [23]. See Section 4.3.1. Operators of large systems favor low-temperature cleaning procedures, as they reduce the cost and time to reach base pressure. Numerous procedures have been surveyed by the AVS Recommended Practices Committee [24].

Aluminum is also used for construction of large vacuum systems. Traditionally, aluminum has been considered to be a material with a large outgassing rate, due to its thick, porous oxide. However, for unbaked chamber use, Fig. 15.4 indicates

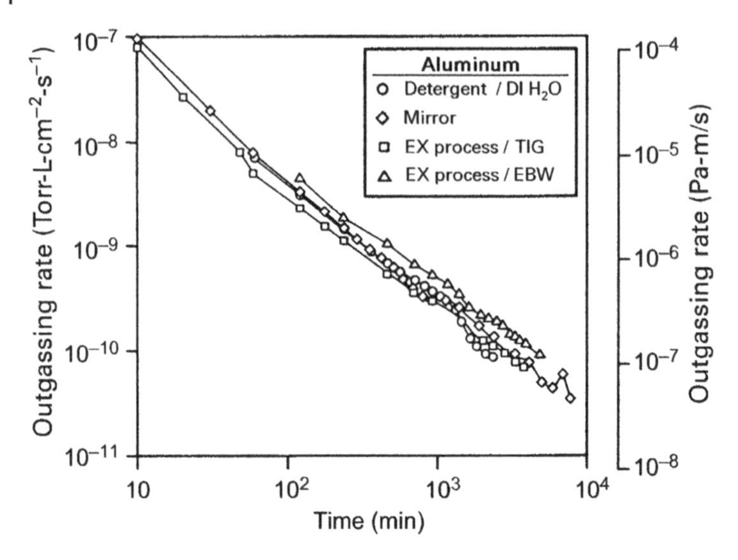

Fig. 15.4 Measured outgassing rates from 6063 aluminum for four different surface cleaning treatments. Reproduced with permission from H.F. Dylla et al. [12]/AIP Publishing.

little difference between samples cleaned with detergent and DI water, machined in ethanol, or extruded in a dry $Ar–O_2$ ambient TIG or EB welding [13]. This is consistent with the observation that water vapor was the most dominant constituent released from the surface of unbaked aluminum. The outgassing rates from vacuum-baked aluminum are as low as those measured from stainless steel.

Although exceedingly low outgassing rates ($<10^{-11}$ Pa-m/s) can be reached after long pumping times in large accelerator systems, a more difficult problem is faced by builders of systems for semiconductor and storage media manufacturing. In these systems, one must reach ultrahigh vacuum conditions within 24–36 h after maintenance, without baking over a temperature of ~80°C. Thus, the central issue is management of water vapor: its adsorption, residence, and desorption and removal by the pumps. The outgassing data given in Fig. 15.2 and Table 15.3 for stainless steel, and in Fig. 15.4 for aluminum, are complemented with information in Appendices C.1 and C.2. The ranges of values given in these tables and graphs represent the different conditions under which the samples were prepared and show the differences in measurement techniques.

Aluminum foil is frequently used to cover open flange faces or to wrap cleaned parts until final assembly. Aluminum foil that has been prepared for vacuum use will not contaminate the cleaned parts, whereas ordinary household "food grade" aluminum foil will have a range of contaminants including residues of its vegetable-based rolling oil [25]. See Fig. 15.5 and Table 15.4 for aluminum.

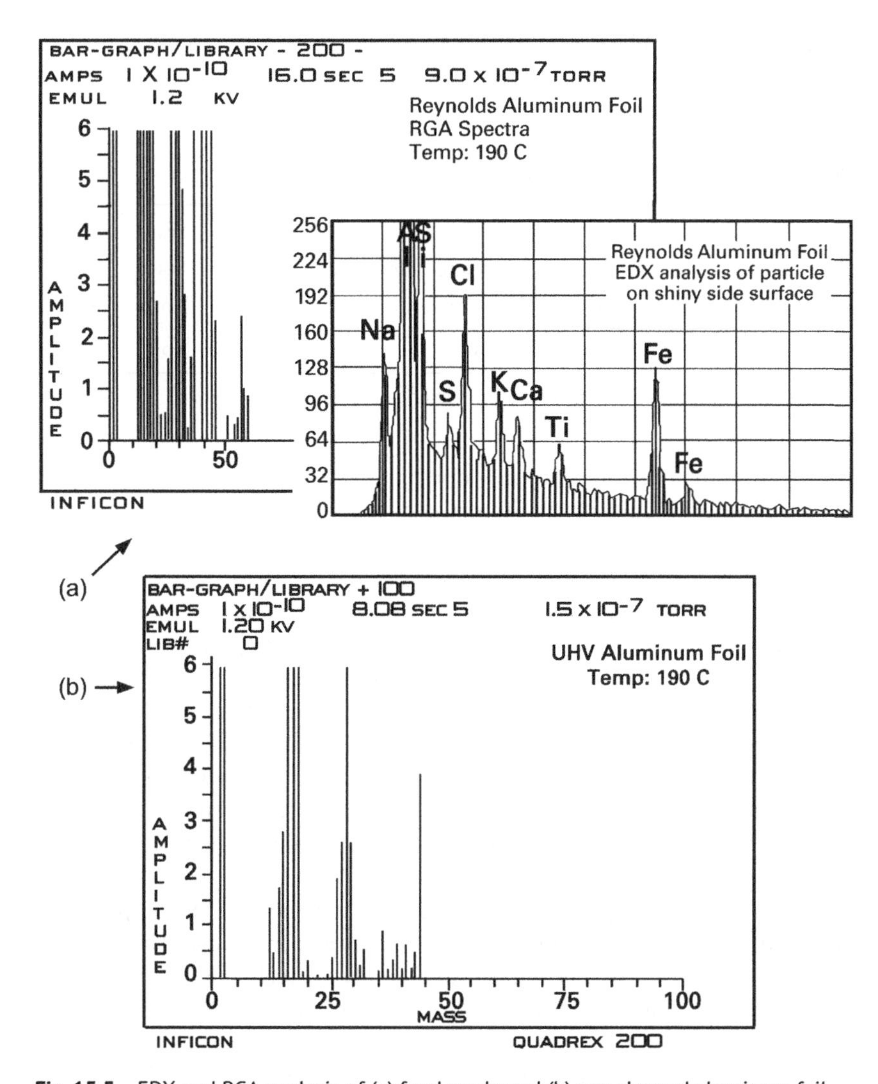

Fig. 15.5 EDX and RGA analysis of (a) food grade and (b) pre-cleaned aluminum foil. Reproduced with permission of the AVS, from V. Rao [25]. Copyright 1993.

15.1.4 Structural Metals

Aluminum and stainless steel are the two metals most commonly used in the fabrication of sophisticated vacuum chambers. Aluminum is inexpensive and easy to fabricate, but hard to join to other metals. It is often used in the fabrication of vacuum collars for glass bell jar systems which are sealed with elastomer O-rings, as well as in some internal fixtures. It is difficult to seal with a metal

Table 15.4 Selected Properties of Common Aluminum Alloys

	% Alloying Element[a]							
Alloy	**Cu**	**Si**	**Mg**	**Cr**	**Common Forms**	**TIG Weld**	**Bend**	**Vacuum**
4043	—	5.0	—	—	Weld filler	Yes	Yes	Yes
5052	—	—	2.5	0.25	Sheet, angle, and tube	Yes	Yes	Yes
6061	0.25	0.6	1.0	0.2	Sheet, angle, and tube	Yes	No	Yes
Cast	Proprietary				Jig Plate	No	No	No

[a] Welding Alcoa Aluminum, Aluminum Co. of America, Pittsburgh, PA, 1966.

gasket. Aluminum was largely bypassed in historic vacuum system construction because of these difficulties. The properties of a few common alloys are given in Table 15.5. Cast tool and jig plate are readily available. It cannot be welded and should never be used in a vacuum system, because it is porous.

Aluminum has been reexamined for use in the construction of chambers for very large high-energy particle accelerators and storage rings [26]. Its high electrical and thermal conductivity and low cost are an asset in the construction of beam tubes. Its residual radioactivity is less than that of stainless steel, because its atomic number is less than that of iron, chrome, or nickel. In this application, explosively bonded aluminum-to-stainless steel sections have been used to make the transition to stainless steel flanges [26]. Ishimaru [27] eliminated stainless steel altogether by designing a system of flanges and bolts of high-strength aluminum alloy for use with aluminum O-rings.

For ordinary laboratory high-vacuum systems, stainless steel is the preferred material. It has high yield strength, is easy to fabricate, and stable. The stainless steels used in vacuum systems are part of a family of steels characterized by an iron–carbon alloy that contains greater than 13% chrome. The 300 series austenitic steel is used in vacuum and cryogenic work because it is corrosion resistant, easy to weld, and nonmagnetic. One alloy in this series is an "18-8" steel that contains 18% chrome and 8% nickel. To this basic composition additions and changes are made to improve its properties. The low outgassing rate and oxidation resistance of stainless steel are due to the formation of a Cr_2O_3 layer on the surface. Figure 15.6 outlines some of the 300 series alloys and their characteristics. Appendix C.8 contains additional properties and applications of the series.

Types 304 and 316 are the most commonly used grades. Type 303 is easy to machine but is not used in vacuum systems because it contains either sulfur, phosphorous, or selenium. For some applications such as cryogenic vacuum vessels, it is desirable to reduce the thickness of the structural steel walls to reduce

Table 15.5 Properties of Some Glasses Used in Vacuum Applications

			Soda			Lead
Property	**Fused Silica**	**Pyrex 7740**	**7720[a]**	**7052[a]**	**0080**	**0120**
Composition						
SiO_2	100	81	73	65	73	56
B_2O_3	—	13	15	18	—	—
Na_2O	—	4	4	2	17	4
Al_2O_3	—	2	2	7	1	2
K_2O	—	—	—	3	—	9
PbO	—	—	6	—	—	29
LiO	—	—	—	1	—	—
Other	—	—	—	3	9	—
Viscosity Characteristics						
Strain point °C	956	510	484	436	473	395
Annealing point °C	1084	560	523	480	514	435
Softening point °C	1580	821	755	712	696	630
Working point °C	—	1252	1146	1128	1005	985
Expansion coefficient $\times 10^{-7}$/°C	3.5	35	43	53	105	97
Shock temperature, 1/4-in. plate °C	1000	130	130	100	50	50
Specific gravity	2.20	2.23	2.35	2.27	2.47	3.05

[a] 7720 glass is used for sealing to tungsten and 7052 glass is used for sealing to Kovar.
Source: Reproduced with permission from Corning Glass Works, Corning, NY.

heat losses. This can be accomplished without loss of other properties by the use of a nitrogen-bearing alloy such as 304LN or 316LN, or by cold stretching and annealing. The maintenance of corrosion resistance, methods of increasing strength, and reduction of porosity on AISI 300 series stainless steels have been discussed by Geyari [28] and should be thoroughly understood by anyone needing stainless steel for a unique application where it is to be stressed to its limits.

Although stainless steel has many desirable characteristics for vacuum use, it is expensive, and not needed for some low-technology uses. Decorative coating is one such application where chambers constructed from epoxy paint coated cold-rolled steel will suffice. Systems designed to produce shiny coatings on items like toys and inexpensive consumer products can be produced in chambers with base

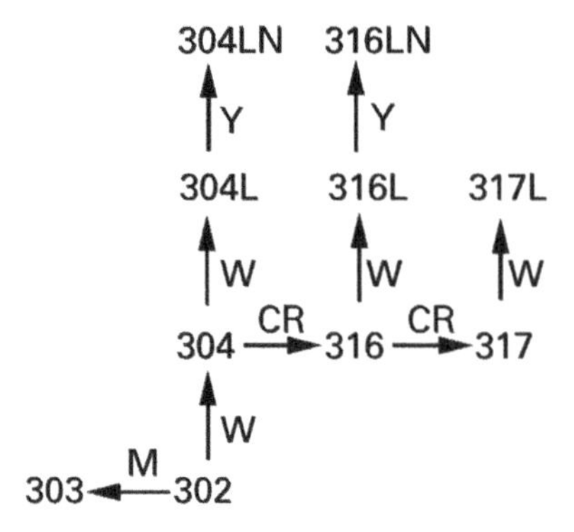

Fig. 15.6 Stainless steels used in vacuum equipment (AISI designation). CR = corrosion resistance, W = ease of welding, Y = yield strength, and M = ease of machining. Reproduced with permission from C. Geyari [28]. Copyright 1976, Pergamon Press, Ltd.

pressures of order 10^{-3} Pa (10^{-5} Torr). Stainless steel isn't needed to attain this vacuum level, and deposited film purity isn't a concern.

Another important use of very large vacuum systems constructed from non-stainless steel is for deposition of active and passive semiconductor layers used in terrestrial photovoltaic module production.

15.2 Glasses and Ceramics

Glass is an inorganic material that solidifies without crystallizing. The common glasses used in vacuum technology are formulated from a silicon oxide base to which other oxides have been added to produce a product with specific characteristics. Soft glasses are formed by the addition of sodium and calcium oxides (soda-lime glass) or lead oxide (lead glass). Hard glasses are formed by the addition of boric oxide (borosilicate glass). Table 15.5 lists the chemical composition and physical properties of glasses often encountered in vacuum applications.

The physical properties of a glass are best described by the temperature dependence of the viscosity and expansion coefficient. Specific viscosity values describe its important properties because glass has no defined melting point. At the strain point ($10^{15.5}$ Pa-s), stresses are relieved in hours. The annealing temperature is defined at a viscosity of 10^{14} Pa-s at which stresses are relieved in minutes. The softening point viscosity is about $10^{8.6}$ Pa-s, and the working point corresponds to a viscosity of 10^5 Pa-s. Glass is brittle, and because of its high thermal expansion and low tensile strength, it can shatter if unequally heated. Its expansion coefficient and thickness are important when selecting material for glass-to-glass or glass-to-metal seals.

How viscosity and thermal expansion change with temperature determines the suitability of a glass for a specific application. Borosilicate glasses are used whenever the baking temperature exceeds 350°C, while fused silica is required for temperatures higher than 500°C. Thermal expansion coefficient and shear strength determine the maximum temperature gradient that a glass can withstand and to what it can be sealed. For example, a borosilicate glass dish can be heated to 400°C in a few minutes, while a large, 1-in.-thick vessel fabricated from lead glass will require 24 h to reach the same temperature without cracking. Glasses are used for bell jars, Pirani gauges, U-tube manometers, McLeod gauges, ion gauge tubes, cathode ray tubes, controlled leaks, diffusion furnace liners, viewports, seals, feedthroughs, and electrical and thermal insulation.

Ceramics as a group are polycrystalline, nonmetallic inorganic materials formed under heat treatment with or without pressure. Ceramics are mechanically strong, have high dielectric breakdown strength, and low vapor pressure. They include glass-bonded crystalline aggregates and single-phase compounds, such as oxides, sulfides, nitrides, borides, and carbides. Ceramics contain entrapped gas pores and are not so dense as crystalline materials. Their physical properties improve as their density approaches that of the bulk. Alumina is made with densities that range from about 85% to almost 100% of its bulk density. Most ceramics have a density of ~90% of the bulk. Important physical properties of ceramics include their compression and tensile strength and thermal expansion coefficient. High-density alumina, for example, has a tensile strength 4–5× greater than that of glass, and has a compression strength 10× greater than its tensile strength. The properties of some ceramics are listed in Table 15.6. Alumina is the most commonly used ceramic in applications such as high-vacuum feedthroughs and internal electrical standoffs. Machinable glass ceramic also finds wide application in the vacuum industry for fabricating precise and complicated shapes. It is a recrystallized mica ceramic whose machinability is derived from the easy cleavage of the mica crystallites.

Table 15.6 Physical Properties of Some Ceramics

Ceramic	Main Body Composition	Expansion Coefficient ($\times 10^{-7}$)	Softening Temperature (°C)	Tensile Strength (10^6 kg/m^2)	Specific Gravity
Steatite	$MgOSiO_2$	70–90	1400	6	2.6
Forsterite	$2MgOSiO_2$	90–120	1400	7	2.9
Zircon porcelain	ZnO_2SiO_2	30–50	1500	8	3.7
85% alumina	Al_2O_3	50–70	1400	14	3.4
95% alumina	Al_2O_3	50–70	1650	18	3.6
98% alumina	Al_2O_3	50–70	1700	20	3.8
Pyroceram 9696[a]	Cordierite ceramic	57	1250	14[b]	— 2.6
Macor 9658[a]	Fluro-phlogopite	94	800	10[b]	2.52

[a] Reprinted with permission from Corning Glass Works, Corning, NY.
[b] Modulus of rupture.
Source: Adapted from Weston [4].

Borides and nitrides have found applications in vacuum technology. Evaporation hearths are made from titanium diboride and titanium nitride, alone or in combination. They are available in machinable or pyrolytically deposited form. Forsterite ceramics ($2MgO$: SiO_2) are used in applications where low dielectric loss is needed, and beryllia (BeO) is used when high thermal conductivity is necessary. Beryllia must be machined while carefully exhausting the dust because it is extremely hazardous to breathe. Machine operators must adhere to strict safety procedures including the wearing of properly designed and fitted face masks. Reviews of ceramics and glasses have been published by Espe [29] and Kohl [30].

Permeation of gas through glasses and ceramics occurs without molecular dissociation. The permeation constant, which is given by Eq. (4.19), depends on the molecular diameter of the gas and the microstructure or porosity of the glass or ceramic [31]. Figure 15.7 contains permeation rate data for several gases through silicon oxide glasses [32–36]. With the exception of deuterium and hydrogen, it shows that the permeation rate decreases as molecular diameter increases. The measured permeation rate of hydrogen was much larger than predicted because of surface reactions and solubility effects [31]. The permeation of a gas through a glass depends on the size of the pores in relation to the diameter of the diffusing species. Permeation is minor through a crystalline material such as quartz but increases with lattice spacing. Nonnetwork forming Na_2O, which is added to SiO_2 to form soda glass, fills these voids, and causes the permeation rate to decrease. See Fig. 15.8. Figure 15.9 gives permeation rates for several glasses [32,37,38]. Shelby [39] has reviewed the diffusion and solubility of gases in glass. The thermal properties and helium permeation of Corning Macor® machinable glass ceramic have been reviewed by Altemose and Kacyon [40].

Gases are physically and chemically soluble in molten glass. The gas on the surface of glass is primarily water with some carbon dioxide. Water vapor may exist on glass in layers as thick as 10–50 monolayers [41]. The first bake of glass releases considerable surface water, while the second and succeeding bakes release structural water [42]. See Fig. 15.10. This release of structural water is proportional to $t^{1/2}$ and indicates a diffusion-controlled process [42]. A high-temperature bake should completely eliminate outgassing of water from glass, because all the surface water is released in a high-temperature bake and the diffusion constant of water vapor is negligible at room temperature. At 25°C, gas evolution from glass is dominated by helium permeation. Outgassing rates of some unbaked ceramics and glasses are given in Appendix C.3. In addition, Colwell [43] has tabulated the outgassing rates of more than 80 untreated refractory and electrical insulating materials used in the construction of vacuum furnaces.

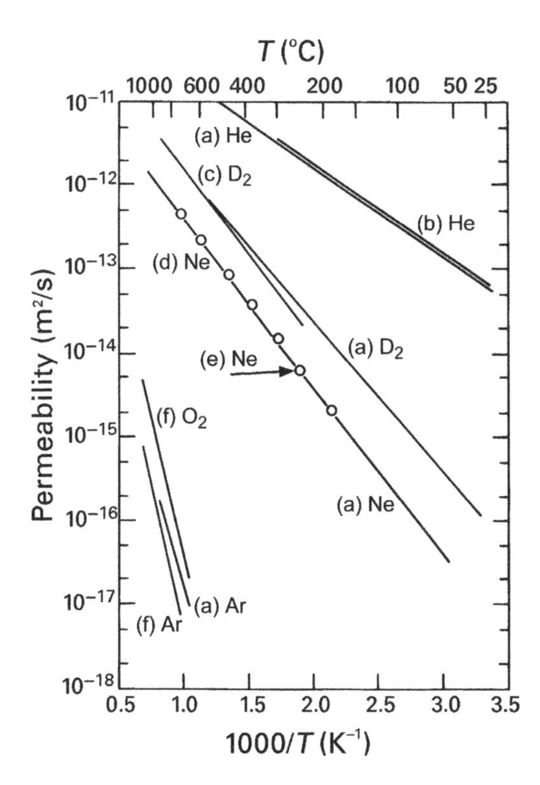

Fig. 15.7 Permeability of He, D_2, Ne, Ar, and O_2 through silicon oxide glasses. Data from (a) reference [32]; (b) reference [33]; (c) reference [34]; (d) reference [35]; (e) reference [36]; and (f) reference [31]. Reproduced with permission from W.G. Perkins [1]. Copyright 1973, AIP Publishing.

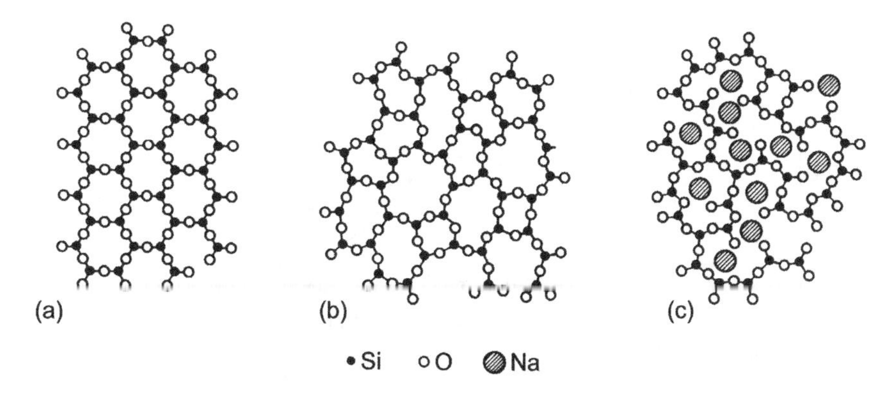

Fig. 15.8 (a) Atomic arrangement in a crystalline material possessing symmetry and periodicity; (b) the atomic arrangement in a glass; (c) the atomic arrangement in a soda glass. Reproduced with permission from the F.J. Norton [31]. Copyright 1953, John Wiley & Sons.

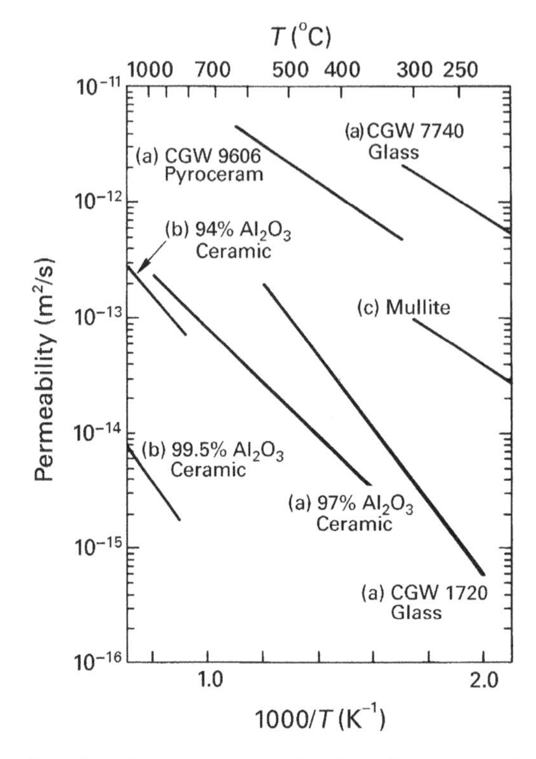

Fig. 15.9 Helium permeability through a number of glasses and ceramics. After Perkins. Data from: (a) reference [33]; (b) reference [34]; (c) reference [35]. Reproduced with permission from W.G. Perkins [1]. Copyright 1973, AIP Publishing.

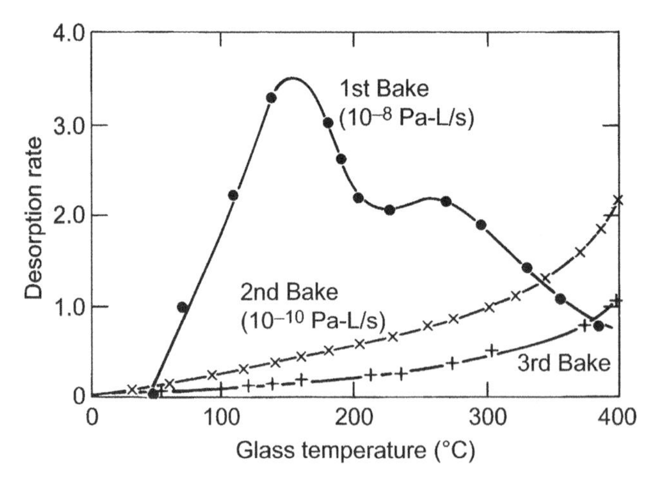

Fig. 15.10 Desorption of water from a Pyrex glass surface of 180 cm^2 at increasing temperature. Adapted with permission from K. Erents and G. Carter [42]. Copyright 1965, Pergamon Press, Ltd.

15.3 Polymers

Polymeric materials find applications in several areas. Generic, trade, and chemical names of common materials are given in Table 15.7. Elastomers such as Buna-N, viton, silicone, and Kalrez are used to form O-ring gaskets for static, sliding, and rotary seals. Other polymers such as Vespel and Kapton, are used for high-voltage vacuum feedthroughs. Other properties of elastomers, which are important when these materials are used as seals, are discussed in Chapter 16.

Permeation and outgassing are important physical properties of elastomers. Gases diffuse through voids in the intertwined polymer chains by a thermally activated process. The dimensions of the voids are larger than those in a glass or metal, and as a result, the permeation constants are significantly larger. Comparison of the permeation rates of gases through polymers with those for

Table 15.7 Generic, Trade, and Chemical Names of Polymer Materials Frequently Used in Vacuum

Generic	Trade	Chemical
Fluoroelastomer	Viton[a], Fluorel[b]	Vinylidene fluoride–hexafluoropropylene copolymer
Buna-N (nitrile)		Butadiene–acrylonitrile
Buna-S		Butadiene–styrene copolymer
Neoprene		Chloroprene polymer
Butyl		Isobutylene–isoprene copolymer
Polyurethane	Adiprene[a]	Polyester or polyether di-isocyanate copolymer
Propyl	Nordel[a]	Ethylene–propylene copolymer
Silicone	Silastic[d]	Dimethylpolysiloxane polymer
Perfluoro-elastomer	Kalrez[a]	Tetrafluoroethylene–perfluoromethylvinyl ether copolymer
PTFE	Teflon[a], Halon[e]	Tetrafluoroethylene polymer
PCTFE	Kel-F[b]	Chlorotrifluoroethylene copolymer
Polyimide	Vespel[a], Envex[c]	Pyromellitimide polymer

[a] E. I. du Pont de Nemours and Company.
[b] 3-M Company.
[c] Rogers Corporation.
[d] Dow Corning Corporation.
[e] Allied Chemical Company.
Source: Reproduced with permission from R.N. Peacock [44]. Copyright 1980, The American Vacuum Society.

glasses shows that the diffusion process in a polymer is not as sensitive to molecular diameter as it is in a glass. This implies that the diffusion of air and other heavy gases through polymers is a serious problem, while helium is the only gas of any consequence to diffuse through glass. Helium certainly diffuses rapidly through elastomeric materials, but its concentration in air is about 5 ppm, so that its net flux is much less than other atmospheric gases. Elastomers will swell when in contact with certain solvents used for cleaning and leak detection. This swelling or increased spacing between molecules results in an increased permeability [2].

Laurenson and Dennis [45] have studied the temperature dependence of gas permeation in three elastomers. Their results, presented in part in Fig's 15.11 and 15.12, have two important conclusions. One is that the change of permeation with temperature is not the same for all gases so the ratio of permeation constant

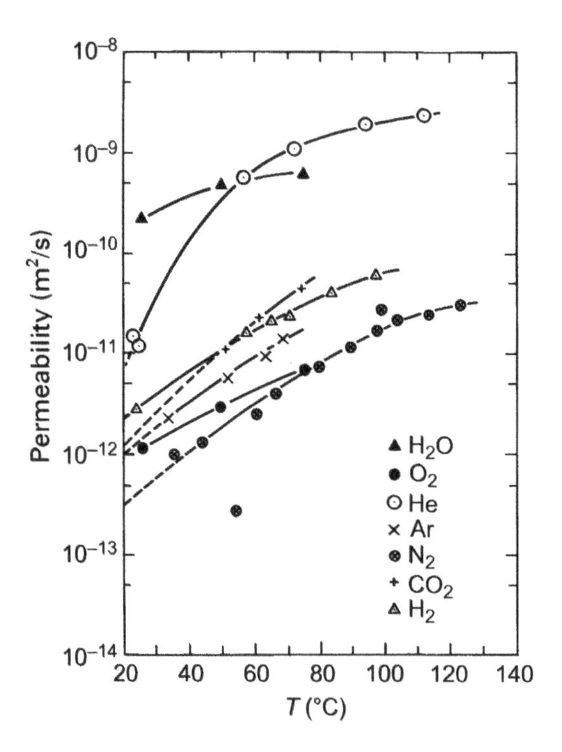

Fig. 15.11 Relation between permeation and temperature for seven common gases through Viton. Reprinted with permission from L. Laurenson and N.T.M. Dennis [45]. Copyright 1985, The American Vacuum Society. The added data for H_2O and O_2 were reported by C. Ma et al., *Journal of the IES*, March/April 1995, p. 43, using tracer gas measurements in ultrapure N_2.

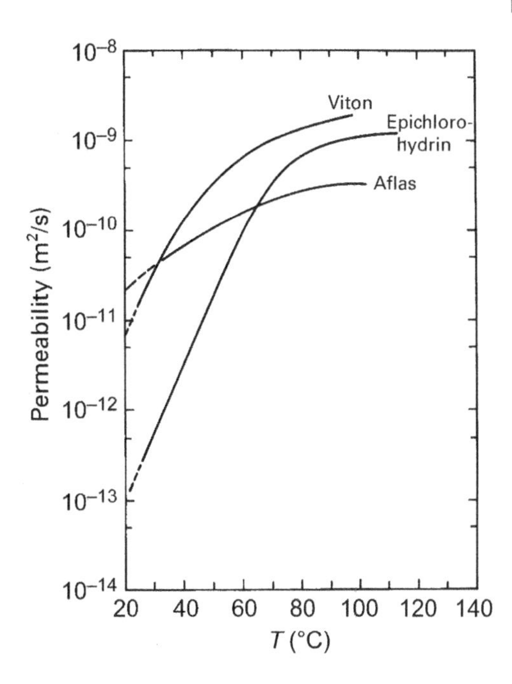

Fig. 15.12 A comparison of the helium permeability through three elastomers: Viton, epichlorohydrin, and Aflas, over the range 20–100°C. Reproduced with permission from L. Laurenson and N.T.M. Dennis [45]. Copyright 1985, AIP Publishing.

between two gases at room temperature is not the same at elevated temperature. Second, the change of permeation with temperature for a given gas is not the same for all materials. The material with the lowest permeability to a given gas at room temperature is not necessarily the best material to use at 100°C. Permeation rates for some materials commonly used in vacuum systems are tabulated in Appendix C.6.

The outgassing of unbaked elastomers has been studied extensively [1,46–62] and shown to be dominated by the evolution of water vapor. The data in Fig. 15.13 illustrate the time dependence of the outgassing rates of several elastomers. Baking will reduce this gas load, as revealed in the mass scans of Figs. 9.7 and 9.8 were recorded during the heating of Buna-N and Viton. These mass scans show that plasticizers, which were added to the polymer before vulcanization, were released as the temperature increased. Unreacted polymer also evolved. At a higher temperature, the elastomer began to decompose. de Csernatony [59,60] reviewed the elastomers Kalrez, and the current Viton E60C. Viton E60C appears to be similar in its outgassing to Viton A, while Kalrez has a low outgassing rate and can withstand temperatures up to 275°C.

Outgassing data for elastomers commonly used in vacuum are given in Appendixes C.4 and C.7. Schalla [63] gives data on the outgassing of cellular foams and stranded elastic cords used in the medium vacuum range, whereas

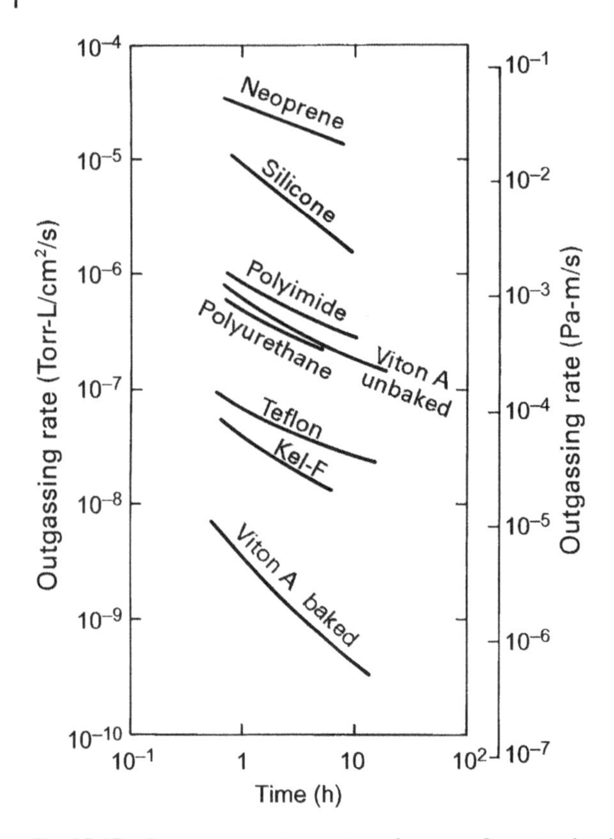

Fig. 15.13 Room temperature outgassing rates for several polymer materials. The values shown are selected from the literature as typical. The rates decline between one and two decades from one to ten hours of pumping. Moderate baking is even more helpful as a comparison of the two Viton curves shows. See reference numbers [49,50,58,59]. Reproduced with permission from R.N. Peacock [44]. Copyright 1980, The American Vacuum Society.

Glassford et al. [64,65] have reported the outgassing rates of aluminized mylar and other multilayer insulation materials. Outgassing characteristics of several optically black solar absorbing coatings [66,67], as well as elastomers and carbon foam [63], have been measured.

Sigmond [68] noted that the pertinent outgassing properties were not dominated by the polymer, but by water and by plasticizers, fillers, and stabilizers of proprietary composition and quantity that were altered from lot-to-lot without the consumer's knowledge. A volatile organic of mass number 149 was found to be present in ethyl alcohol stored in polyethylene bottles and in samples stored in test tubes closed with plastic caps [68]. This organic vapor turned out to be a fragment of dibutyl phthalate, a commonly used plasticizer. Because this fragment was easily removed by a 100°C bake, it will contaminate materials stored in plastic bags or boxes or touched by plastic gloves. Parks and coworkers [69] measured the outdiffusion of one plasticizer, 3,5,-di-tert-butyl-4-hydroxytoluene (BHT) from polypropylene wafer carrier boxes and found its diffusivity to be D (m^2/s) = $1855 \exp(-11{,}986/T)$.

References

1 Perkins, W.G., *J. Vac. Sci. Technol.* **10**, 543 (1973).
2 Elsey, R.J., *Vacuum* **25**, 299 (1975).
3 Elsey, R.J., *Vacuum* **25**, 347 (1975).
4 Weston, G.F., *Vacuum* **25**, 469 (1975).
5 Norton, F.J., *Trans. 8th Natl. Vac. Symp. (1961)*, 1962, p. 8.
6 Hobson, J.P., *J. Vac. Sci. Technol.* **16**, 84 (1979).
7 N. Milleron and R.C. Wolgast, "Guidlines for the Fabrication of Vacuum Systems and Components," Chapter 9 in: *Methods of Experimental Phys.* 14, G.L. Weissler and R.W. Carson, Eds., Academic, New York, 1979, p. 444.
8 Guthrie, A., *Vac. Technol.*, 277 (1963).
9 Rogers, W.A., Buritz, R.S., and Alpert, D.L., *J. Appl. Phys.* **25**, 868 (1954).
10 Calder, R. and Lewin, G., *Brit. J. Appl. Phys.* **18**, 1459 (1967).
11 Jousten, K., *Vacuum* **49**, 359 (1998).
12 Dylla, H.F., Manos, D.M., and LaMarche, P.H., *J. Vac. Sci. Technol., A* **11**, 2623 (1993).
13 Li, M. and Dylla, H.F., *J. Vac. Sci. Technol., A* **11**, 1702 (1993).
14 Ohmi, T., *Microcontamination* **6**, 49 (1988).
15 Benvenuti, C., Canil, G., Chiggiato, P., Collin, P., Cosso, R., Guérin, J., Ilie, S., Latorre, D., and Neil, K.S., *Vacuum* **53**, 317 (1999).
16 Herbert, J.D. and Reid, R.J., *Vacuum* **47**, 693 (1996).
17 Barton, R.S. and Govier, R.P., *Vacuum* **20**, 1 (1970).
18 Nuvolone, R., *J. Vac. Sci. Technol.* **14**, 1210 (1977).
19 Dylla, H.F., *Vacuum* **47**, 647 (1995).
20 Hseuh, H.C., Chou, T.S., and Christianson, C.A., *J. Vac. Sci. Technol., A* **3**, 518 (1985).
21 Calder, R., Grillot, A., LeNormamd, F., and Mathewson, A., in *Proceedings of the Seventh International Vacuum Congress and the Third International Conference on Solid Surfaces: Of the International Union for Vacuum Science, Technique and Applications, September 12–16, 1977*, R. Dobrozemsky, Ed., Vienna, Austria, IUVSTA, 1977, p. 231.
22 Staib, P., Dylla, H.F., and Rossnagel, S.M., *J. Vac. Sci. Technol.* **17**, 291 (1980).

23 Knize, R.J. and Cecchi, J.L., *J. Vac. Sci. Technol., A* **1**, 1273 (1983).

24 Sasaki, Y.T., *J. Vac. Sci. Technol., A* **9**, 2025 (1990).

25 Rao, V., *J. Vac. Sci. Technol.* **11** (4), 1714 (1993).

26 Govier, R.P. and McCracken, G.M., *J. Vac. Sci. Technol.* **7**, 552 (1970).

27 Ishimaru, H., *J. Vac. Sci. Technol.* **15**, 1853 (1978).

28 Geyari, C., *Vacuum* **26**, 287 (1976).

29 Espe, W., *Materials of High Vacuum Technology*, Vol. **2**, Pergamon Press, New York, 1966.

30 Kohl, W.H., *Handbook of Materials and Techniques for Vacuum Devices*, Reinhold, New York, 1967.

31 Norton, F.J., *J. Am. Ceram. Soc.* **36**, 90 (1953).

32 Perkins, W.G. and Begeal, D.R., *J. Chem. Phys.* **54**, 1683 (1971).

33 Swets, D.E., Lee, R.W., and Frank, R.C., *J. Chem. Phys.* **34**, 17 (1961).

34 Lee, R.W., *J. Chem. Phys.* **38**, 448 (1963).

35 Frank, R.C., Swets, D.E., and Lee, R.W., *J. Chem. Phys.* **35**, 1451 (1961).

36 Shelby, J.E., *Phys. Chem. Glasses* **13**, 167 (1972).

37 Miller, C.F. and Shepard, R.W., *Vacuum* **11**, 58 (1961).

38 R.H. Edwards, *Permeation of Monatomic Gasses Through Aluminum Oxide*, M.S. Thesis, University of California, Berkeley, 1966.

39 Shelby, J.E. "*Molecular Solubility and Diffusion*", in *Treatise on Materials Science and Technology: Glass II*, M. Tomozawa and R.H. Doremus, Eds., Vol. 17, Academic, New York, 1979.

40 Altemose, V.O. and Kacyon, A.R., *J. Vac. Sci. Technol.* **16**, 951 (1979).

41 Todd, B.J., *J. Appl. Phys.* **26**, 1238 (1970).

42 Erents, K. and Carter, G., *Vacuum* **15**, 573 (1965).

43 Colwell, B.H., *Vacuum* **20**, 481 (1970).

44 Peacock, R.N., *J. Vac. Sci. Technol.* **17**, 330 (1980).

45 Laurenson, L. and Dennis, N.T.M., *J. Vac. Sci. Technol., A* **3**, 1707 (1985).

46 Barton, R.S. and Govier, R.P., *J. Vac. Sci. Technol.* **2**, 113 (1965).

47 Addis, R.R., Jr., Pensak, L., and Scott, N.J., *Transactions of the 7th National Vacuum Symposium of the American Vacuum Society, 1960*, Pergamon, New York, 1961, p. 39.

48 R. Geller, *Le Vide*, No. 13, 71, (1958).

49 Fluk, M.M. and Horr, K.S., *Transactions of the National Vacuum Symposium of the American Vacuum Society. 9th, Oct. 31–Nov. 2, 1962, Los Angeles, California*, G. Bancroft, Ed., Macmillan, New York, 1963, p. 224.

50 Munchhausen, M. and Schittko, F.J., *Vacuum* **13**, 548 (1963).

51 Hait, P.W., *Vacuum* **17**, 547 (1967).

52 Blears, J., Greer, E.J., and Nightengale, J., in *Adv. Vac. Sci. Technol.*, E. Thomas, Ed., Vol. **2**, Pergamon, 1960, p. 473.

53 de Csernatony, L., *Vacuum* **16**, 13 (1966).

54 de Csernatony, L., *Vacuum* **16**, 129 (1966).
55 de Csernatony, L., *Vacuum* **16**, 247 (1966).
56 de Csernatony, L., *Vacuum* **16**, 427 (1966).
57 de Csernatony, L. and Crawley, D.J., *Vacuum* **17**, 55 (1967).
58 Edwards, T.L., Budge, J.R., and Hauptli, W., *J. Vac. Sci. Technol.* **14**, 740 (1977).
59 de Csernatony, L., in *Proceedings of the Seventh International Vacuum Congress and the Third International Conference on Solid Surfaces: Of the International Union for Vacuum Science, Technique and Applications, September 12–16, 1977*, R. Dobrozemsky, Ed., Vienna, Austria, IUVSTA, 1977, p. 259.
60 de Csernatony, L., *Vacuum* **27**, 605 (1977).
61 Dayton, B.B., *Sixth National Symposium on Vacuum Technology Transactions, Oct. 7–9, Philadelphia, PA*, C.R. Meissner, Ed., Pergamon, New York, 1960, p. 101.
62 Thieme, G., *Vacuum* **17**, 547 (1967).
63 Schalla, C.A., *J. Vac. Sci. Technol.* **17**, 705 (1980).
64 Glassford, P.M. and Liu, C.-K., *J. Vac. Sci. Technol.* **17**, 696 (1980).
65 Glassford, P.M., Osiecki, R.A., and Liu, C.-K., *J. Vac. Sci. Technol., A* **2** (3), 1370 (1984).
66 Erickson, E.D., Berger, D.D., and Frazier, B.A., *J. Vac. Sci. Technol., A* **3**, 1711 (1985).
67 Erickson, E.D., Beat, T.G., Berger, D.D., and Frazier, B.A., *J. Vac. Sci. Technol., A* **b**, 206 (1984).
68 Sigmond, T., *Vacuum* **25**, 239 (1975).
69 Ho, Y.-M., in *13th Annual IEEE/SEMI Advanced Semiconductor Manufacturing Conference, Advancing the Science and Technology of Semiconductor Manufacturing, ASMC-2002*, H.G. Parks and B. Vermere, Eds., IEEE, Boston, 30 April 2002, p. 314.

16

Joints Seals and Valves

In the 1960's, a discussion of joining and sealing techniques would have been incomplete without a description DeKhotinsky cement, Glyptal, and black wax. The era of sealing valve stems by wrapping with string soaked in grease has also long passed. Materials technology has advanced to the level where glass-to-metal seals, demountable elastomer gaskets, metal-to-metal gaskets, brazed joints, welded joints, highly reliable valves, and motion feedthroughs have been standard practice or catalog items for decades. Reliable joints, seals, and components are so common that we take them for granted and this laissez-faire attitude sometimes contributes to their misuse. Occasionally, we inappropriately use a rubber hose as a flex connector, bake a brass valve, grease a static elastomer seal, bake Viton to an excessively high temperature, try to save money by reusing lightly tightened copper gaskets, or apply "stop leak" to a leaky weld. Singleton [1] once remarked that in the early days of vacuum technique, some workers preferred the clear variety of Glyptal leak sealant so as to hide their mistakes. Today's mistakes can be hidden with a higher degree of sophistication with vacuum epoxy and a coat of paint.

In this chapter, we review welded and brazed metal joints, metal, glass, and ceramic joints, elastomer and metal-sealed flanges, valves, and motion feedthroughs. We emphasize proper selection and use of joining and sealing techniques and not design. Historical reviews of sealing and joining techniques have been published by Roth [2–4].

16.1 Permanent Joints

Welding, brazing, and soldering are used to make permanent joints in vacuum chambers, pumping lines, and components. The technique one chooses depends on materials joined and the thermal and vacuum environment for which they are

A Users Guide to Vacuum Technology, Fourth Edition. John F. O'Hanlon and Timothy A. Gessert.
© 2024 John Wiley & Sons, Inc. Published 2024 by John Wiley & Sons, Inc.

designed. Stainless steel, titanium, and aluminum weld easily. Flame-sealed glass joints, glass-to-metal, and ceramic-to-metal joints are widely used in high and ultrahigh vacuum (UHV) system construction.

16.1.1 Welding

Metals can be joined permanently by welding—local melting—of closely mating pieces. Welding of vacuum components is most commonly done by the tungsten inert gas (TIG) process to avoid oxidation. A detailed review of TIG and other welding techniques is given in the ASTM *Metals Handbook* [5]. TIG welding is not the only technique; plasma-arc and electron-beam welding are also used. However, these techniques are not as readily available or applicable for all types of joints. TIG welding, also known as heliarc, or argon arc welding, is a technique for forming clean, oxide-free, leak-tight joints by flooding the area immediately around the arc (including inside a tube or pipe) with an inert gas, usually argon. TIG welding is used in vacuum fabrication to join materials such as stainless steel, aluminum, nickel, copper, and titanium. It is not suitable for joining alloys with high-melting-point components, e.g., brass, and certain aluminum or stainless-steel alloys. Stainless steel is the most common material used in the construction of high- and ultrahigh vacuum vessels. Any 300 series alloy, with the exception of 303S and 303Se, is easily welded. It is more difficult to weld aluminum than stainless steel, but skilled welders can make leak-tight joints. Dissimilar metals can be welded; combinations of stainless steel, copper, nickel, and Monel can be joined by TIG welding. Aluminum and stainless can be joined by explosion bonding. Rosendahl [6] and Geyari [7] have written excellent reviews of welding austenitic stainless steels. Carbon dioxide shield gas produces reasonably clean welds that are clean enough to fabricate external fixtures—a cost-conscious alternative to argon that is quite popular with home welders.

Figure 16.1 depicts several acceptable vacuum welding techniques. Welding from the vacuum side or through-welding from the atmospheric side, along with welding walls of equal thickness, are elements of good welding practice. It is important for the weld to be clean on its vacuum side. If the weld is made on the atmospheric surface, but does not penetrate completely, a gap may be created, which is impossible to clean. For strength, some joints need to be welded on the atmospheric side as well. If this is necessary, it should be a discontinuous, or "skip" weld. See for example, the welds in Fig. 16.1 indicated by dashed lines. If the outer weld was continuous and the inner joint had a leak, it could not be detected; trapped gas would constitute a slow leak, known as a "virtual" leak. In small-diameter tubing, the weld cannot be made on the vacuum side. In this case, a through weld will prevent the formation of regions where cleaning solvents can collect. Continuous butt orbital welding is a superior process for joining long

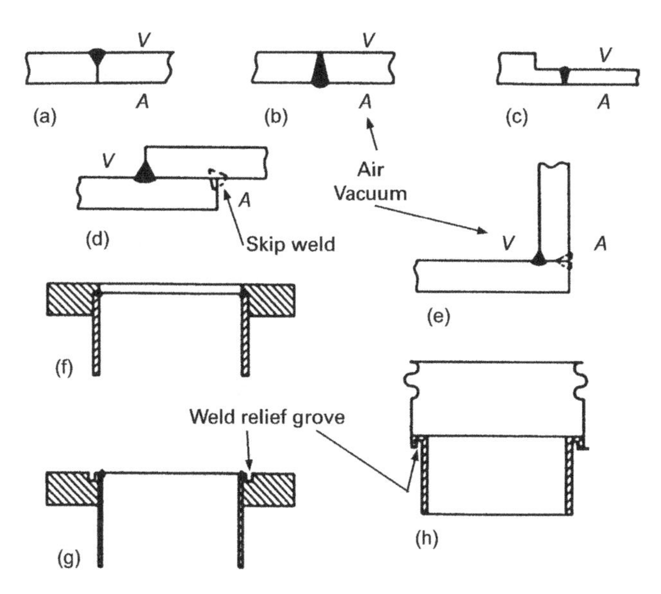

Fig. 16.1 Examples of welded joints: (a) Butt-welded from the vacuum side; (b) through-welded from the atmospheric side; (c) joining parts of unequal thickness; (d) lap weld illustrating proper tack welds on the atmospheric side; (e) corner weld; (f) thick-wall tubing-to-flange weld. The flange is machined so that the bead depth is approximately equal to the wall thickness. (g) Thin-wall tubing-to-flange weld joint showing relief groove; (h) bellows-to-pipe adapter. The right side is shown before welding. The raised lip (0.025 cm high × 0.035 cm thick) provides filler metal for the weld; a copper heat sink is attached to the bellows end section during welding.

lengths of tubing. This technique forces the tubing ends together, while they are continuously purged with Ar or an Ar–H_2 mixture during welding. Pulsed dc power is applied to a rotating, gas-shielded tungsten tip, such that the melt zone completely penetrates the tube weld. The small hydrogen content in the purge gas is chosen to match the hydrogen content of the steel. Large pipes can be chemically cleaned following welding. Ultraclean gas delivery piping tubing is not bent to form turns, as bending releases particles. Instead turns are fabricated from machined corner blocks that are butt welded to straight lengths.

It is also important for the two surfaces to be as equal in thickness as possible. A weld relief groove should be used to match thicknesses when welding a flange to a thinner tube. This is particularly important in the welding of thin-wall tubing and bellows. See Fig. 16.1. These two concepts—welding walls of equal thickness and welding from the vacuum side or completely through to the vacuum side—are incorporated in all properly designed and executed weld joints.

Carbide precipitation and inclusions in ASIA 300 series stainless steels potentially present problems during fabrication. Carbon that has precipitated at grain

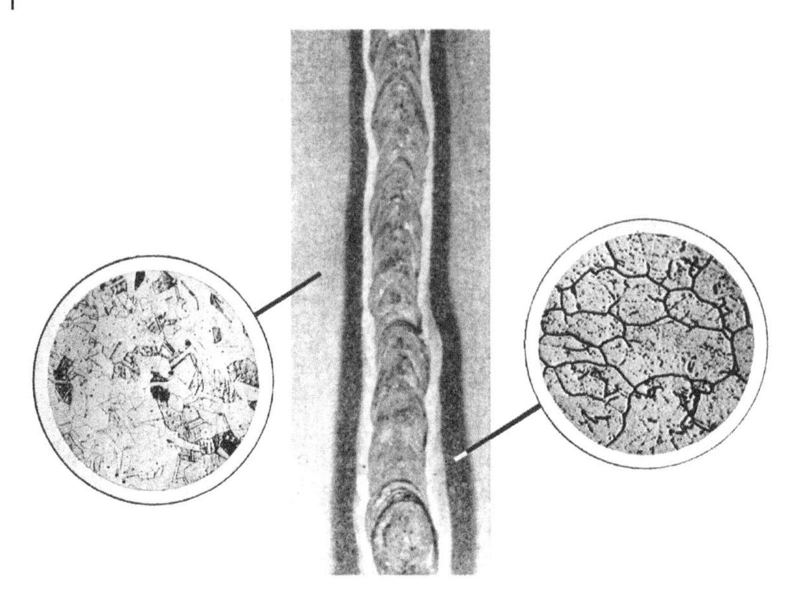

Fig. 16.2 The appearance and location of carbide-rich zones bordering a weld bead in 304 stainless steels. The sample has been etched to make the carbide-rich regions visible. The dark zones on either side of the weld are regions in which the metal cooled from 900 to 500°C in more than 5 min. The right inset shows a magnified view of the carbide-rich zone. The left inset shows a magnified view of the grain structure of normal annealed 304 stainless steel. Reproduced with permission from Allegheny Ludlum Steel Corp.

boundaries, from welding or improper cooling after annealing, removes with it a substantial fraction of the chrome from nearby regions. See Fig. 16.2. The nearby regions then contain less than 13% chrome and are no longer stainless steel. They are subject to corrosion if exposed to a corrosive atmosphere. The formation of microscopic cracks is also a concern for stainless steel subjected to low temperatures. A crack in a carbide-rich zone can cause a leak in a cold trap, cold finger, or cryopumping array. These are difficult to leak check, as the leak may disappear when the component is warmed to room temperature.

Carbide precipitation may be prevented by use of an alloy containing a low carbon content, a stabilized alloy, or a minimum-heat welding technique. A good solution is the use of low-carbon steel alloys such as 304L and 316L; however, they require more nickel and are more expensive to manufacture than their higher-carbon counterparts. Titanium, niobium, and tantalum form carbides more easily than chrome, so an alternative solution is the use of an alloy such as 321, 347, or 348 that is stabilized with one of these elements. A minimum heat weld will also reduce the time that the metal weld region spends in the dangerous 500–900°C region. Minimum heat welds are simplified by use of weld relief grooves like those

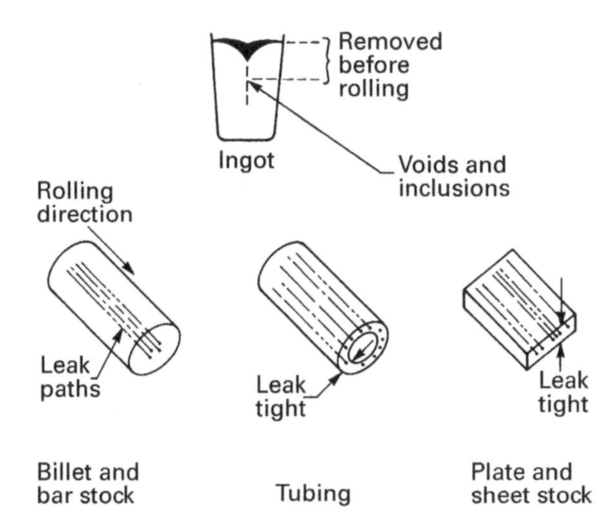

Fig. 16.3 Schematic-inclusions in steel during casting and rolling. Reproduced with permission from Varian Report VR-39, Stainless Steel for Ultra-high Vacuum Applications, V. A. Wright, Copyright 1966, Varian Associates, 611 Hansen, Way, Palo Alto, CA.

shown in Fig. 16.1. An 18-8 alloy of 0.06% carbon will not precipitate carbides at the grain boundaries if it is cooled from 900 to 500°C in less than 5 min. As the carbon content is reduced, the steel can remain in the critical temperature region for additional time without carbide precipitation.

A most important concern of the UHV user, and occasionally the high vacuum user, is the proper fabrication and joining of stainless-steel components to eliminate leaks through inclusions in the metal. Grease and other impurities mask these minute inclusions, which occur in the process of making steel, until the steel wall is thoroughly baked. At that time tiny leaks will appear. The impurities in a cooling ingot distribute themselves at the top and center. See Fig. 16.3. The portion with most of the oxide and sulfide impurities is removed before rolling, and any remaining impurities are stretched into long narrow leak paths. These inclusions are parallel to the direction of rolling. It is important to know the inclusion direction when selecting the raw stock from which components are to be made. In particular, inclusions can cause virtual leaks when sheet or tubing is welded on the atmospheric side without full penetration. Figure 16.4 illustrates the origin of leak paths in high-vacuum flanges made of plate stock. To avoid such potential leaks, modern flanges are made from bar stock. Also, the material is usually forged to break up the long filamentary inclusions and reduce the possibility of such leak paths. For critical applications a section from each end of the billet will be individually examined and certified prior to fabrication.

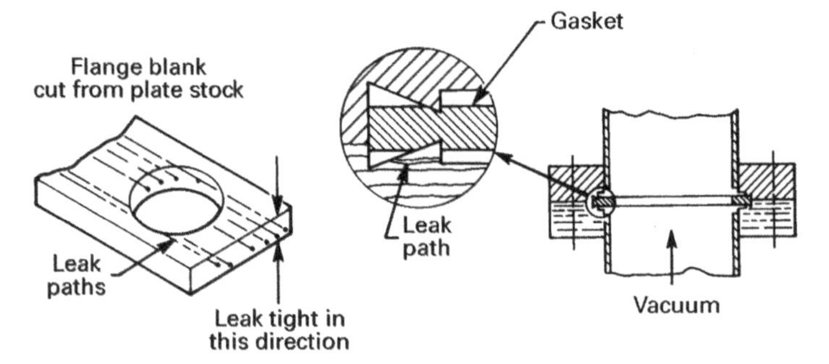

Fig. 16.4 Porosity in high-vacuum flanges. Reproduced with permission from Varian Report VR-39, Stainless Steel for Ultra-high Vacuum Applications, V. A. Wright, Copyright 1966, Varian Associates, 611 Hansen, Way, Palo Alto, CA.

Nonmagnetic steels are used to fabricate some electron-beam and low-cost chambers. One must take care not to use filler welding rods, whose constituents can produce a porous weld.

16.1.2 Soldering and Brazing

Soldering and brazing are techniques for joining metal parts with a filler metal whose melting point is lower than the melting point of the parts to be joined. The definitions of soldering and brazing are not universal. One definition of brazing states that it is done with fillers whose melting point is greater than 450°C, while soldering uses fillers that melt at lower temperatures. The term "hard soldering" is obsolete. Brazing and soldering are used where alloy or metal combinations cannot be welded, or where warping of parts during welding would produce unacceptable distortion. Soldering is little used in vacuum technique, because low-melting-point solders contain high-vapor-pressure materials, which are unsuitable for use in baked systems. One exception is the use of indium alloy solders for sealing thermocouple feedthroughs.

Furnace brazing in forming gas, hydrogen, or vacuum is a common method for joining large pieces. Torch brazing introduces unacceptably large local thermal stresses in welded joints that would warp large pieces, whereas furnace brazing heats parts isothermally. Bellows were traditionally brazed to heavier tubing before joints of the type illustrated in Fig. 16.1*h* were developed. Brazing can be done by torch, by dipping, by induction heating, or by heating in a furnace. An excellent review of practical brazing techniques is contained in the pamphlet by Peacock [8].

The parts to be brazed or soldered should be closely machined and have a large area of overlap. The filler has a lower shear stress than the metals being joined, and a large surface is necessary to place the stress on the metal rather than on the

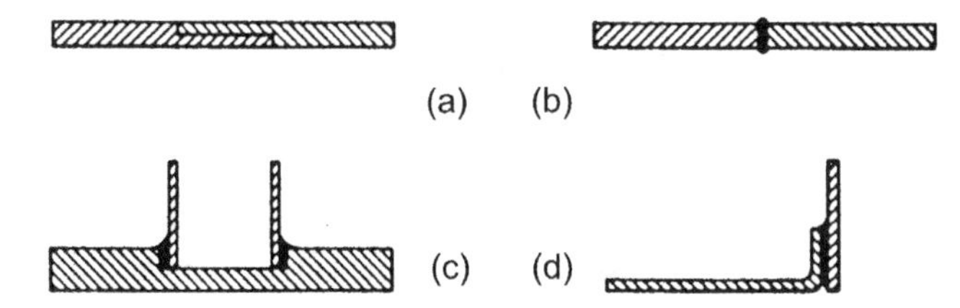

Fig. 16.5 Examples of brazed joints: (a) One form of a strong butt-lap joint; (b) a weak butt joint in which the stress is placed on a small filler area; (c) a poor tube-to-flange joint with excessively large clearances; (d) a strong corner joint.

filler. The metal parts need to fit uniformly so the filler can flow into the joint by capillary action. Too large a gap will result in voids, but if the gap is too narrow, the filler may not flow in a gap. Figure 16.5 illustrates proper and improper construction of brazed and soldered joints. Of course, the filler should melt at a temperature below the melting point of the work.

Pure metal and alloy braze fillers are available in a wide range of melting points ranging from −40 (mercury for cryogenic joints) to 3180°C (rhenium). Several copper–silver and copper–gold alloys with melting points in the range 800–1000°C can be used for brazing stainless steel. Copper, and other alloys are also used in the construction of valves, diffusion pump casings, feedthroughs, and internal fixtures. Cu–Ag is a well-known braze filler for stainless steel; however, Cu–Au alloys have a very low surface tension and readily flow between parts with 0.001-in. gaps. Tables of braze alloys and applications are found in Kohl [9], the American Welding Society Brazing Manual [10], and the Handy and Harmon Brazing Book [11].

The vapor pressure of some metals is high enough to preclude their use in baked vacuum systems. Older solders were lead-based. Brazing alloys with melting points under 700°C contained Pb, Zn, Cd, or P. Such alloys are limited to torch brazing of assemblies that will never be baked. If components containing these elements are baked, the high-vapor-pressure components will vaporize and diffuse throughout the vacuum system. In the same manner, they will contaminate a brazing furnace. In addition to the vapor pressure limitation, there are well-known metal incompatibilities. Au and Al bond to form an intermetallic compound known as the "purple plague." Au and Ag cannot be used in contact with Hg, and brazing alloys containing Ag cannot be used in contact with Fe–Ni–Co alloys such as Kovar, without causing intergranular corrosion.

16.1.3 Joining Glasses and Ceramics

The techniques for joining glasses and ceramics recognize the different expansion coefficient, tensile strength, and shear strength of each material. Glasses have expansion coefficients ranging from 105×10^{-7}/°C for soda-lime glass to

3.5×10^{-7}/°C for fused quartz. Glasses are weak in tension and strong in compression; both their tension and compression strengths are weaker than those of ceramics or metals. Here we review techniques for glass-to-glass seals and for glass- and ceramic-to-metal seals.

Glasses can be fused in a flame or joined in a furnace by frit or cane solder glass if their expansion coefficients differ by less than 10%. Frit seals are made with ground solder-glass slurry whose solvent evaporates as the frit melts. Pre-shaped cane glass seals are made from frit or cane solder glass. The parts to be sealed with solder glass are aligned, loaded with a suitable weight, and thermally cycled to the melting temperature of the solder glass. The solder glass and the parts need to have the same expansion coefficient, but the solder-glass melting point must be less than the sealed parts. A review of solder glasses has been published by Takamori [12].

Glass pairs with widely dissimilar expansion coefficients such as quartz and Pyrex, or lead glass and Pyrex, cannot be directly sealed. Such combinations are joined with a graded seal. A graded seal, also known as a step seal, is made by successively joining glasses whose expansion coefficient differs by about 10–15%. Adjacent glasses must also have nearly equal solidification temperatures; otherwise, large thermal sterss can be frozen in during cooling.

Glasses can be sealed to metals if the metal has the same expansion coefficient as the glass, if the metal holds the glass in compression, or if the metal is very thin so that it can plastically deform. Glass does not adhere directly to metal (except for platinum), but rather to the metal oxide. The metal oxide must be stable and adhere well to the parent metal. Common glass-to-metal seals use matching expansion coefficients. For example, platinum, or an iron–nickel–chrome alloy such as Sealmet 4, seals to lead glass. Tungsten seals to 7720 and 3320 uranium glass, and Kovar seals to 7052 glass. Ring seals are sometimes used where a metal band or ring of appropriate expansion coefficient contracts on cooling to hold the glass in compression. A seal for joining glass to a feathered, deformable copper edge—the Housekeeper seal—has been out of vogue for many years. However, a stainless steel and 7052 glass version designed by Benbenek and Hoenig [13] is commercially available. A cross-sectional view is shown in Fig. 16.6*a*. Construction details, given in the original paper, have been amplified by Rosebury [14]. The direct sealing of glass to stainless steel avoids the corrosion problems associated with the use of iron–nickel–cobalt alloys. Espe [15] describes the alloys used in glass-to-metal seals, while Kohl [9] and Rosebury [14] discuss glass-to-metal seals.

Ceramic-to-metal seals are more easily made and stronger than glass-to-metal seals. Ceramics have high compression strength, so it is not necessary to have as close an expansion coefficient match between ceramic and metal as it is between glass and metal. The high compression strength of a ceramic results in a very rugged seal. Ceramic seals are made by firing a thin layer of refractory metal on the

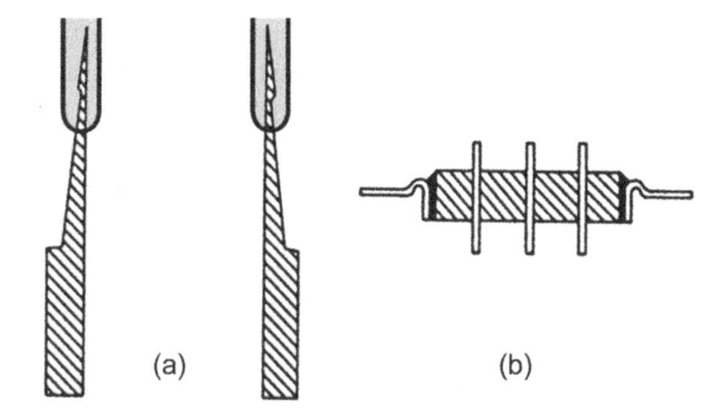

Fig. 16.6 (a) Tapered stainless tubing for fabricating a 7052 or Pyrex glass-to-stainless Housekeeper seal. Reproduced with permission from J.E. Benbenek and R.E. Honig [13]. Copyright 1960, AIP Publishin.g (b) Ceramic-to-metal seal.

ceramic at high temperatures. The metal is then furnace-brazed to the refractory using filler with a lower melting point than the refractory. For example, alumina is brazed to Kovar via a fired molybdenum paste. See Fig. 16.6*b*.

16.2 Demountable Joints

The most reliable, long-lived connection which can maintain its integrity in vacuum and related environments is a permanent, welded joint. Special-purpose welded metal joints can be cut open, and glass can always be cut and reworked. However, most vacuum systems have a requirement for easy access, which cannot be met in this way. Demountable joints are therefore, a practical necessity.

A demountable seal must be designed and constructed to form a leak-tight joint and maintain it until the time it is to be opened. Forming a leak-tight joint requires an initial contact force large enough to make the joined material merge and fill the irregularities in the harder of the two materials. Maintaining the joint requires a force sufficient to overcome the effects of differential expansion and long-term plastic flow. Depending on the relative expansion coefficients of the two materials, a joint may develop an unwanted leak if it is heated or cooled [4]. The restoring force must be maintained by the stored energy in the joint. Stored energy is a fundamental requirement in a demountable seal. (There are one or two cases where two metals form a diffusion joint or bond, and the force can be removed, e.g., copper–steel shear seals; however, they have to be pulled apart.) A joint consists of three components: the seal or gasket material, the flange pair, and the clamping means. The required energy may be stored in any one of the components or shared between them [8]. In addition to maintaining the seal force, the joint must be able to withstand the thermal, chemical, and radiation environment of the system. It must be constructed from materials whose vacuum properties (vapor pressure, permeation, and outgassing rate) are compatible with the vacuum operating range.

Elastomeric compounds are excellent choices for the seal or gasket material. They store elastic energy and conform to fit surface irregularities of the flange. In addition, they are resistant to many chemicals and can be reused. Unfortunately, elastomers have high permeation rates for atmospheric gases and cannot withstand high baking temperatures. For these systems, metal is the only acceptable gasket material. In this section, we review the techniques for demountable joints using both elastomers and metal gaskets. Elastomer gaskets have been reviewed by Peacock [16], while metal seals have been reviewed by Roth [4].

16.2.1 Elastomer Seals

Elastomer seals between metal or glass flanges are made by deforming the elastomer freely between two flat surfaces or confining it in a groove of rectangular, triangular, or dovetail cross section. Elasticity, plasticity, hardness, compression set, seal loading, outgassing, and gas permeation are important attributes of an elastomer seal. A summary of various elastomer seal properties is given in Table 16.1. Elastomers are formed by first grinding the starting polymer, mixing with plasticizers and stabilizers, and vulcanizing it to a state that is largely elastic. These materials are incompressible, that is, any deformation or compression in one direction must be accompanied by motion in another, so that the total volume remains constant. If the compound is completely elastic, it will return to its exact shape after the force has been removed. If it has some degree of plastic behavior, it will flow and not return to its original shape. The measure of its shape change is called compression set.

Compression set occurs in an elastomer that has been deformed for a long time. It is more pronounced in elastomers that have been heated, because compression set is strongly temperature dependent. Figure 16.7 gives comparative compression set data for five elastomers. The time effects of compression set in Kalrez 1050 and Viton A and E-60C are described in Table 16.2. From these data, it might appear that a Kalrez or Viton seal would leak excessively after baking. Peacock [16] comments that many of the problems with compression set in perfluoropolyether may be due to improper groove design. As this material is heated, it expands; expansion can induce excessive compression set if inadequate room for expansion is left in the groove.

Adequate seal loading is rarely a problem with an elastomer. However, excessive compression can limit the upper baking temperature, because of thermal expansion. O-rings are typically compressed 15–20% of their diameter. For a nominal 0.318-cm-diameter O-ring of 75 Shore hardness, this translates into a seal

Table 16.1 A Summary of Various Mechanical and General Considerations Regarding the Selection of Polymer Seal Materials

Seal Material	Lin. Coeff of Thermal Exp ($\times 10^{-5}$/°C)	Max Oper. T (°C)	Cold Flow at T_{op}	Gas Permeation	Wear/Abrasion Resistance	Prime Seal Appl'n
Fluoroelastomer						
Viton E-60C	16	150	Good	Mod	Good	*a*
Viton A	16	150	Fair	Mod	Good	*a*
Buna-N	23	85	Good	Mod	Very good	*b*
Buna-S	22	75	Good	High	Good	*c*
Neoprene	24	90	Good	Mod	Very good	*d,e*
Butyl	19	—	Good	Mod	Good	*f*
Polyurethane	3–15	90	Poor	Mod	Excellent	*g,h*
Propyl	19	175	Good	High	Very good	*h*
Silicone	27	230	Poor	Very high	Poor	*i*
Perfluoroelastomer	23	275	Poor	—	Excellent	*j*
Teflon	5–8	280	Very poor	Mod	Excellent	*j*
Kel-F	4–7	200	Good	Low	Very good	*j*
Polyimide	5	275	Good	Mod	Very good	*j,k*

(*a*) Generally used vacuum seal; (*b*) best all-around low cost; (*c*) little vacuum application; (*d*) oil resistance; (*e*) low cost; (*f*) specific chemical application; (*g*) radiation resistant; (*h*) mechanical properties; (*i*) electrical applications; (*j*) chemical resistance; (*k*) high temperatures.
Source: Reproduced with permission from R.N. Peacock [16]. Copyright 1980, AIP Publishing.

force of about 2.7 kg/cm of O-ring length [16]. The initial pressure will be slightly reduced with time as the elastomer undergoes some plastic deformation, but it will not affect the integrity of the vacuum seal. Because sealing is determined by contact pressure, the compressive force, deformation, and percentage groove filling must all decrease as the hardness of the O-ring, its expansion coefficient, or both are increased. A chord compression of 20% is typical for Viton O-rings, whereas Kalrez should not be compressed more than 12%.

O-ring manufacturers can supply tables of groove depths and widths with information for a range of O-ring sizes. Sessink and Verster [17] observed the empirical tabulations from nine sources to vary widely. They found chord compressions up to 38% of the chord diameter and found groove filling ratios (chord cross-section area/groove cross-section area) of 74–102%. In their study, they found the general

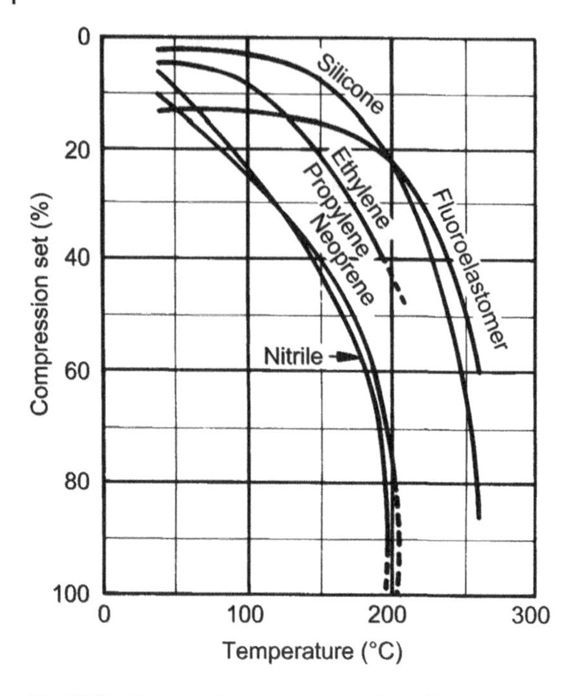

Fig. 16.7 Compression set as a function of temperature measured according to ASTM D-395 (22 h). Reproduced with permission from R.N. Peacock [16]. Copyright 1980, The American Vacuum Society.

Table 16.2 Compression Set for Viton and Kalrez for Various Times and Temperatures Measured per ASTM D-395B

		Compression Set (percent)		
Temperature (°C)	**Time (h)**	**Kalrez 1050**	**Viton A**	**Viton E-60C**
24	70	20	21	8
100	70	32	—	—
204	70	71	63	13
204	360	—	90	30
204	960	—	—	55
204	7200	—	—	100
232	70	71	—	30
260	70	6	—	—
288	70	74	—	100

Source: Reproduced with permission from R.N. Peacock [16]. Copyright 1980, AIP Publishing.

criterion for high-vacuum sealing to be a minimum initial contact pressure of 13 kg/cm^2 for gaskets in the hardness range 60–75 Shore [17,18]. The tables listed in the various references are not all for the same elastomer. Expansion coefficient, temperature range, and hardness each affect the groove depth and width. It is best to obtain a table from an O-ring supplier to verify the groove dimensions for a particular elastomer. Outgassing and permeation data for elastomers are given in Chapter 15 and in Appendix C. Outgassing data are shown in Fig. 15.13 and Appendices C.4 and C.7. Water vapor is the major constituent to degas from unbaked Viton [19].

Permeation data are given in Appendix C.5. Outgassing rates for single, unbaked Viton gaskets of $q = 3.67 \times 10^{-6}/t^{0.41}$ Torr-L/s per linear cm of 5.3-mm-diameter have been measured [20]. This is close to the inverse square root dependence associated with normal permeation. After a very long time, this value will stabilize at the steady-state permeation rate. Using both measured permeation constants and measured and empirically determined solubility values, one can construct a plot of the approximate permeation through a single Viton O-ring exposed to atmosphere at room temperature. Figure 16.8 shows calculated air flux through a Viton O-ring assuming that its interior was devoid of gas and that the presence of water vapor, the dominant component, does not affect the solubility of other gases. Note that the *time* to reach steady state is determined by the diffusion constant, whereas the steady-state *flux* is determined by the *product* of the diffusion constant and solubility. Water vapor has a small diffusion constant in Viton, and for this reason

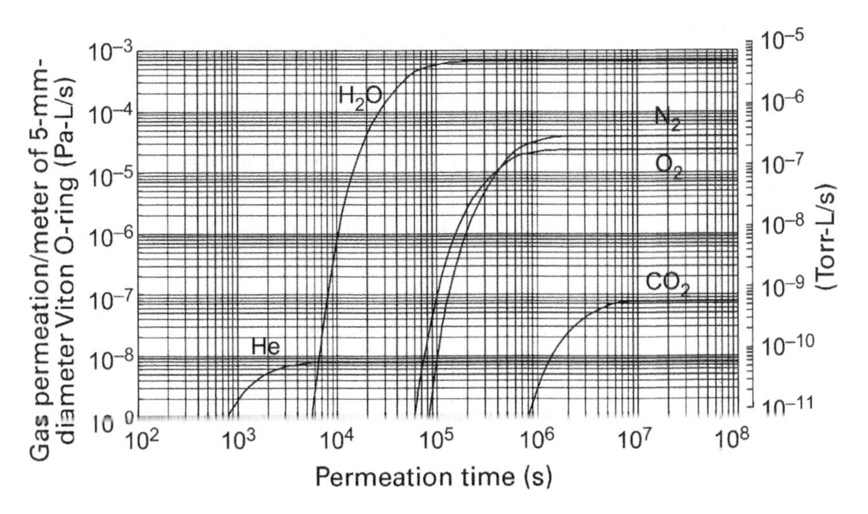

Fig. 16.8 Calculated permeation rates for the components of atmospheric air through Viton, using measured permeation constants, and calculated values for solubility or diffusion constants. These calculations assume no interaction between individual constituents.

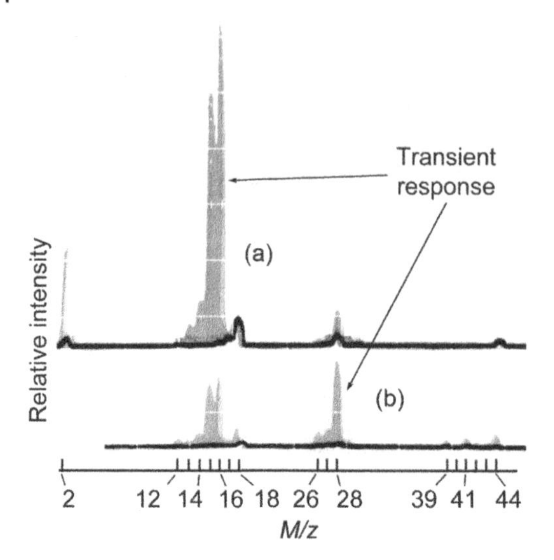

Fig. 16.9 Measured mass-resolved gas release from two valves with Viton seats during closure of each valve. Black scans are background; gray scans are transient gas release. (a) prior exposure to methane, (b) prior exposure to carbon monoxide.

it can take a day to reach steady state. The high solubility of water vapor in Viton doesn't affect its diffusion time only its total flux, as depicted in Fig. 16.8.

High gas solubility in elastomers has another unpleasant side effect; gases are released when the elastomer is squeezed. This effect is shown in Fig. 16.9 for a Viton-gasketed butterfly valve. Here we see methane gas released from the valve's Viton gasket during closure. The lower trace demonstrates carbon monoxide release during a different valve closure. In each case the gasket had been exposed to the respective source gas for some time prior to its closure. Elastomer gaskets are sponges.

A 480-fold reduction in the outgassing of an immersed Viton O-ring was observed after a 16-h, 100°C bake, but when the O-ring was used as a seal, the reduction was only 6-fold [21]. These measurements illustrate how permeation, and not outgassing, dominate the long-term (base) pressure in a vacuum system containing a large number of O-rings. Gas permeation can limit the time available for leak checking and limit the ultimate pressure in a system containing a large number of O-rings. Detailed studies of the permeation of several gases through uncompressed Viton, Aflas, and epichlorohydrin at several temperatures are presented in Fig's. 15.11 and 15.12 [22].

Seal materials degrade by radiation exposure. Peacock [16] gave a brief review of gamma radiation damage. He noted the trend in elastomers was to become brittle, take a large compression set and increase in hardness after being subjected to

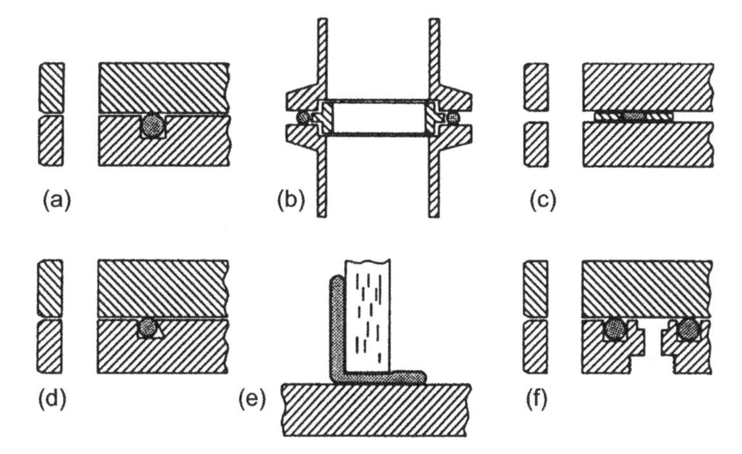

Fig. 16.10 Elastomer seal forms: (a) Rectangular groove, (b) ISO-KF flange with centering ring, (c) confined gasket, (d) dovetail groove, (e) L-gasket, (f) double gasket with differential pumping port.

high doses of gamma rays. Teflon appeared affected at lower doses than other elastomers, whereas polyimide, e.g., Vespel, withstood considerably more flux than other elastomers. Wheeler and Pepper [23] showed that a high X-ray flux (8×10^6 rad/s) decomposed Teflon into saturated fluorocarbon gases and polymer fragments short enough to desorb from the surface.

Gaskets can be shaped in many ways. Figure 16.10 illustrates several ways that gaskets are used between metal and metal-to-glass joints. The rectangular O-ring groove, Fig. 16.10*a*, is a common joint in the United States. The ISO-KF (klein-flansch) with a centering ring, Fig. 16.10*b*, first became the European standard; it is now universal. The confined gasket, Fig. 16.10*c*, is commercially available and useful, especially on noncircular joints. A half- or full-dovetail groove, Fig. 16.10*d*, is especially useful for vertical doors. An L-gasket, Fig. 16.10*e*, is freely squeezed between a glass pipe or bell jar and a metal surface. A differentially pumped gasket, Fig. 16.10*f*, is used to reduce permeation between atmosphere and vacuum; it is a practical alternative to a metal gasket for some applications [20].

Cleaning an elastomer by a solvent wash is an ineffective way to reduce outgassing. A simple vacuum bake is most effective. An unbaked Viton O-ring will have an initial outgassing rate of 10^{-3} Pa-m/s. See Appendix C.4. After a 4-h bake at 150°C and 12 h of pumping, this value is reduced to 4×10^{-7} Pa-m/s. Re-exposure to atmosphere will result in increased water outgassing. O-ring grooves need to be carefully cleaned and both groove and ring wiped clean with a lint-free cloth. Grease is not needed to make a static seal between an elastomer and a metal surface. It will cause pressure bursts as trapped gas pockets are released. Occasionally we use grease on a main door seal flange that has become scratched with misuse.

One need only apply a *very thin* coating of grease with lint-free cloth. Finger oils have a very high vapor pressure, and they are an unacceptable substitute for grease; cleanroom gloves and lint-free cloth are best practice.

The most commonly used O-ring elastomers are Buna-N and Viton E-60C. However, much-published data are for Viton A, but this compound has not been used for many years. Viton E-60C has a lower room-temperature outgassing rate and less compression set than Viton A. Buna-N is used for low-cost applications, while Viton is used where a moderate bake and low outgassing are needed. A 200°C bake will release adsorbed gases, unreacted polymer, and plasticizers from Viton [16]. It cannot be used at this temperature because of compression set; at higher temperatures it will decompose. Practically, there is no reason to bake Viton gaskets higher than, say, 50–80°C. Silicone has an unusually high permeation rate; it is infrequently used as a gasket material in high-vacuum systems. Silicone compounds are formulated for range of high-temperature applications. Polyimide has a low outgassing rate [24,25], but adsorbs large amounts of water when reexposed. Hait [26] and Edwards et al. [27] describe flange seals made from thin polyimide films. Elastomer gaskets are widely used in systems that pump to the 10^{-6}-Pa (10^{-8}-Torr) range.

Commercial Viton O-rings are not compounded from 100% Viton; they contain dyes, mold release agents, and a variable degree of filler material. See comments at end of Section 15.3. It is understood that the use of the trademark requires a minimum content of the copolymer, but beware, the percentage and nature of the remaining materials vary with the source, resulting in variable quality. Several manufacturers now produce high-quality, reliable Viton O-ring gaskets fabricated from extremely pure material.

Elastomers shed large quantities of fine particles during normal use. After 20 min agitation in a 1:6 solution of isopropyl alcohol and ultrapure water, particle counts of $1–2\times10^4/cm^2$ surface area $\geq$0.3-μm diameter were measured for Viton, silicone, and nitrile O-rings (A.M. Arif, personal communication, Dec. 1994).

On occasion, it may be necessary to fabricate an O-ring using bulk O-ring stock, appropriate glue, and procedures from the 1950's. Because of their wide-spread use in many other applications, "O-ring Splicing Kits" are widely available. Further, if the proper diameter of O-ring stock is unavailable, the splicing kit can sometimes be used to produce a smaller-circumference O-ring from an O-ring with larger circumference.

16.2.2 Metal Gaskets

The thermal, radiation, outgassing, and permeation properties of elastomers make them unsatisfactory for many seal applications. Metal gaskets must be used in high-quality UHV systems and are often used in high-vacuum systems on parts that do not have to be frequently opened to reduce the total outgassing load.

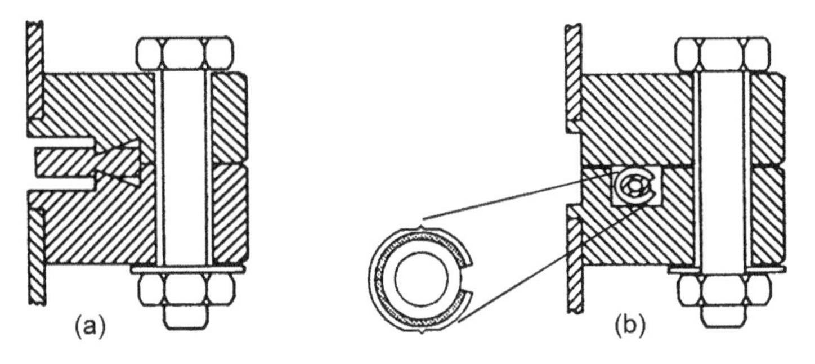

Fig. 16.11 Metal gasket seals: (a) ConFlat type knife-edge seal; (b) Helicoflex Delta seal.

Many metal-sealed gaskets have been designed. Two are illustrated in Figure 16.11. In these flange designs, the gasket is deformed to fill irregularities in the surface of the mating flange. The copper gasket seal has proven to be a popular bakeable seal. The most widely used commercial design is the ConFlat® [28] flange, Fig. 16.11*a*. It consists of two symmetrical flanges each containing a work-hardened knife-edge. The flanges are tightened until they touch and "capture" a section of the copper ring between the knife-edge and the outer shoulder. The copper gasket was designed to contact the outer flange shoulder; however, gaskets may be improperly fabricated with a smaller OD, thus eliminating the force needed to make a proper seal. The knife-edge flange is universal and available in a circular geometry up to 30-cm diameter. High-quality versions fabricated from 321 or 347 stainless steel, with an additive to prevent grain growth, can be repeatedly baked up to 450°C with excellent reliability. The Helicoflex® "Delta" gasket, Fig. 16.11*b*, consists of an internal spring surrounded by inner and outer linings. The inner lining distributes the spring force uniformly along the outer sealing lining. The materials of the spring and linings can be chosen to address a range of vacuum-process considerations.

Large flanges are difficult to keep leak-free if the knife-edges lose alignment. Knife-edges can become misaligned by differential thermal expansion during normal baking and warping during welding, as shown in Fig. 16.12*a* [29]. Gaskets are formed from OFE copper, which begins to lose hardness at 200°C; recrystallization begins at 300°C [29]. An alternative design for a copper gasket has been proposed [30,31]. The taper seal gasket illustrated in Fig. 16.12*b* uses half the sealing force, has no copper inside the knife-edge, and can use a chain clamp instead of bolts [30]. Analysis of the internal forces showed that a large amount of sealing energy in the conventional ConFlat is wasted in the process of pushing the knife-edge into the gasket [31]. This energy generates intense stress in the gasket immediately under the tip of the knife-edge, resulting in stress-induced crystallization and loss of elasticity. Elastic energy storage is necessary to maintain any seal.

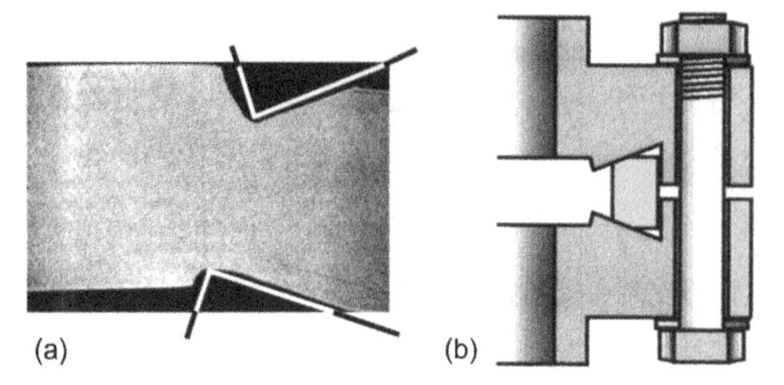

Fig. 16.12 (a) Distortion in a copper gasket caused by uneven thermal expansion of the flange, or weld-induced distortion of the knife-edge flanges. W. Unterlerchner [29]/ Reproduced with permission from American Vacuum Society. (b) A taper gasket designed for use in the ConFlat knife-edge flange. Reproduced with permission from S. Kurokouchi et al. [30]. Copyright 2001, AIP Publishing

In an elastomer seal, energy is stored in the gasket. However, copper cannot store energy if high stresses induce crystallization. The taper seal gasket contacts the outer flange edge; energy can be stored in gasket and in the bolts or chain clamp.

16.3 Valves and Motion Feedthroughs

The components used for control, isolation, or motion all use some form of an elastomer or metal seal in combination with a moveable vacuum wall, usually a metal bellows. Valve seals are adaptations of demountable seals described in the previous section. Conductance, leak rate, vacuum range, baking temperature, radiation, or corrosive gas exposure, and the need for line-of-sight view in the open position are all variables that influence valve design and selection. Transmitting linear or rotary motion through a vacuum wall is complicated. Dynamic sealing or flexible walls, along with lubrication, are now required. The sealing techniques and materials are also dependent on the vacuum range, torque, and the rotational speed at which the motion will be transmitted. Weston [32] has extensively reviewed valves and feedthroughs for UHV applications.

In this section, we discuss small and large valves, special-purpose valves, and motion feedthroughs.

16.3.1 Small Valves

We define a small valve as one in which the throat diameter is less than, say, 6 cm and in which the sealing force is directly applied via a hand lever or screw or a pneumatically operated shaft. Figure 16.13 identifies the components of a simple, hand-operated, elastomer-sealed right-angle valve. Most small valves are right angle, because that is a simple mechanical way of directly applying the sealing force in a direction normal to the valve seat. All valves, small or large, should be constructed from materials whose outgassing load is low enough so that it does not contaminate the process at the operating pressure. We desire the valve to have a maximum conductance for gas flow and to form a leak-tight seal between the plate and the seat and between the stem and the bonnet. We also desire the valve to be reliable and maintenance-free and have a long operating life. The valve design illustrated in Fig. 16.13 is typical of an obsolete-style valve that was used in roughing lines. Its basic problem, even if made from stainless steel, is its unreliable O-ring stem seal. Today this design is used only in valves designed to release a medium- or low-vacuum component to atmosphere. In that application, the stem seal is never exposed to vacuum.

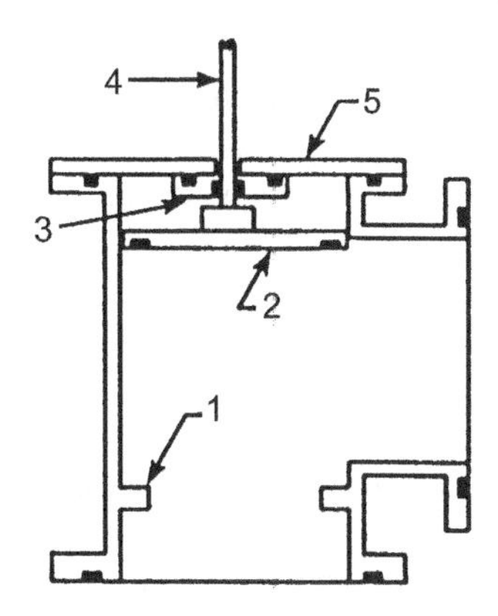

Fig. 16.13 Components of a simple elastomer-sealed valve: (1) Valve seat; (2) valve plate; (3) bonnet; (4) stem; (5) stem seal.

In Fig. 16.14*a* we illustrate a right-angle valve in which the O-ring stem seal has been replaced with a stainless-steel bellows. The linear motion needed to open

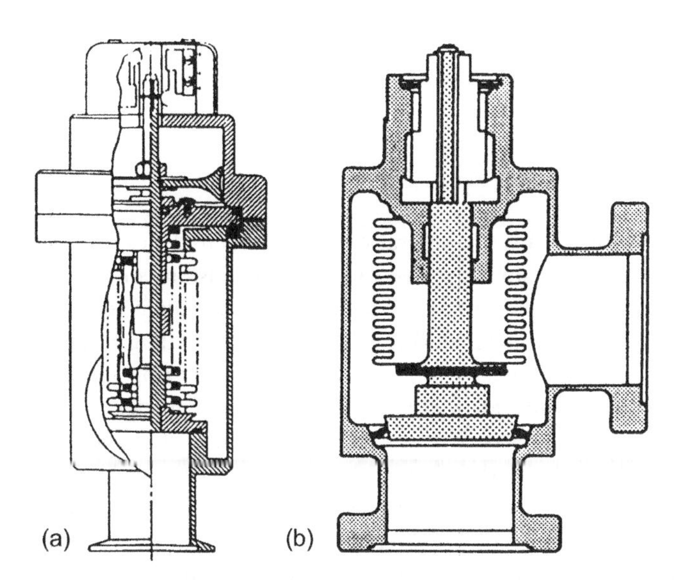

Fig. 16.14 Small right-angle valves: (a) Elastomer-sealed, pneumatically driven valve with a bellows shaft seal. Reprinted with permission of HPS division of MKS Corp., 5330 Sterling Drive, Boulder, CO 80301. (b) All metal valve with metal sealed seat. Reproduced with permission of VAT Inc. 600 W. Cummings Park, Woburn, MA 01801.

and close the valve is transmitted to the valve plate via the bellows. The remaining elastomer seals are static. The valve in this illustration is pneumatically driven and can be baked to 150°C in the closed position and to 200°C in the open position. It is used on systems pumped to the 10^{-7}-Pa range. Bellows-sealed valves of this type have lifetimes ranging from 20,000 to over 200,000 cycles, and valve plate leak rates of 10^{-12}–10^{-10} Pa-L/s (10^{-14}–10^{-12} Torr-L/s) A less expensive valve of this type is constructed from brass with a brass bellows and Buna-N seals for non-corrosive applications in the medium vacuum range. Butterfly valves, which operate by rotating a valve plate through 90°, are made with an elastomer O-ring seal on the outer edge of the valve plate. They are available in hand-operated versions.

Elastomer seals prevent the use of this valve in UHV systems baked to temperatures over 200°C. An all-metal welded body and metal plate seals are necessary for systems baked at high temperatures. The first all-metal valve was developed by Alpert [33] in the early 1950s. Today, several designs are commercially available. All use some form of copper or gold valve plate and a stainless-steel knife-edge seat. Designs with a sapphire valve plate and gold-plated stainless-steel seats are also available. The design illustrated in Fig. 16.14*b* employs a coated steel valve plate and a seat fabricated from a steel with a high elastic limit in the shape of a conical Belleville washer. The body of this valve is welded, and it is joined to the system by copper knife-edge flanges. Line-of-sight versions of valves such as those illustrated in Fig. 16.14 are available. These valves are constructed by mounting the valve seat at a 45° angle to the connecting tubing. Because of the geometry, a line-of-sight version requires a seat diameter over 1.5 times the pipe diameter. The distance between flanges reduces their conductance to a value less than a right-angle version.

16.3.2 Large Valves

Large valves differ from small valves in several ways. The required closing force scales with the seal area. For elastomer-sealed valves this area is proportional to the O-ring chord diameter and the valve plate diameter. A 5-cm-diameter valve sealed with a 3.2-mm-chord-diameter O-ring requires a total sealing force of 400 N. A 15-cm-diameter valve with a 6-mm-chord diameter O-ring requires about 2000 N, whereas a 30-cm valve with an 8-mm-chord diameter O-ring requires a sealing force of about 5000 N. If the valve is to remain closed against atmospheric pressure, the body must withstand an even greater force. A 30-cm valve must withstand a force of 7500 N at atmospheric pressure. These forces place stringent design requirements on the valve body. The valve seat must remain flat under these forces at room temperature and while being baked. Direct line of sight transmission through the valve is required for some applications. Physical vapor deposition sources are often isolated by a valve during substrate changing.

Particle accelerators and storage rings require clear aperture valves in the beamlines. Multichambered, in-line, thin-film deposition systems use large line-of-sight valves, not always circular, to pass wafers between isolated chambers in device processing. Other applications require an isolation valve to have a high open conductance, long life, ease of maintenance, and an ability to be baked. These valves range in sizes to 1.2-m diameter for both high vacuum and UHV.

The most common large valve design is the gate valve. Gate valves are so common that some think they are synonymous with large valves. Gate valves are made in sizes as small as 2 cm; however, they are not the only large valves. Figure 16.15 illustrates an older style design. The gate travels horizontally until the valve plate is centered under the seat. Continued translational motion of the shaft causes hinges to move the valve plate upward to a closed position. The bonnet seal is either a grease-packed double O-ring or, in better-quality valves, a bellows. Inexpensive varieties of this design are made with cast aluminum bodies while higher quality versions are made with stainless steel.

A highly reliable design is illustrated in Fig. 16.16. In this design the shaft moves the valve plate in and aligns it with the seat. At the end of the valve plate travel, the shaft forces ball bearings out of detents and expands the valve plate against the seat and the backing plate against the valve housing. The use of multiple balls is one advantage of this design. By increasing the number of points at which the force is applied to the seal plate, the wear on each point can be reduced to a low value. These two gate valve designs are available in sizes to 30 cm; however, larger elastomer-sealed gate valves are manufactured. Metal-sealed gate valves are currently available in diameters of 30–40 cm. The reliability of such a valve drops drastically as the diameter increases beyond 30 cm. Baking of a large metal-sealed

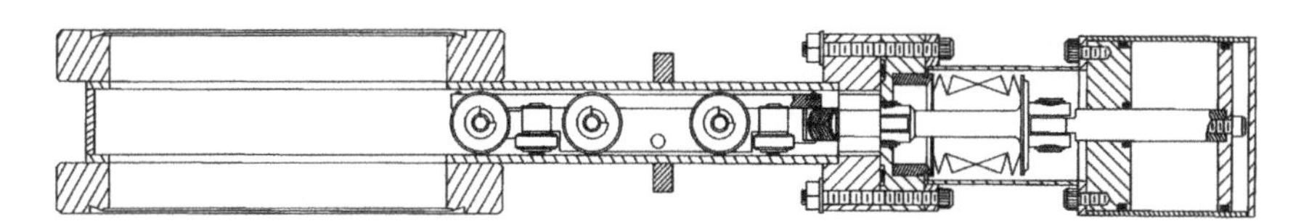

Fig. 16.15 Bellows-sealed gate valve. Reproduced with permission from High-Vacuum Apparatus, Inc., 1763 Sabre St., Hayward, CA 94545.

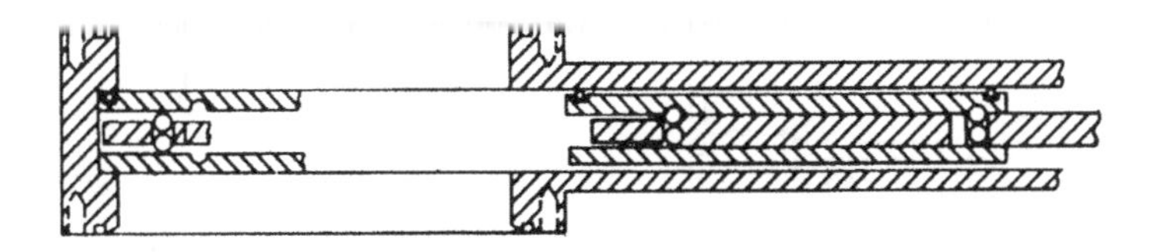

Fig. 16.16 Sealing concept used in the VAT gate valve.

valve is limited to about 200°C. Large metal-sealed valves have the same problems as large knife-edge flanges.

Other designs use a swinging plate shaped like a Belleville washer, which is forced into position by a wedge pushing on the center, or a coaxial bellows to seat the valve plate against a knife-edge located in the opposite face.

Aluminum-bodied, grease-lubricated valves are acceptable on unbaked systems that pump to 10^{-5} Pa (10^{-7} Torr). Stainless-steel valves with bellows shaft seals whose moving parts are lubricated with a low-vapor-pressure solid lubricant can be baked to 150°C with the valve closed or 200°C open and can be used on UHV systems. Viton is the most common elastomer used in large valve seals, although others have experimented with polyimide [34]. In all gate valves, elastomer- and metal-sealed, the valve plate must be seated in a direction orthogonal to the drive motion with a force up to several thousand newtons. Moving parts need to be lubricated, and the valve interior needs to be pumped. For these reasons the gate valve is the most complicated piece of machinery in a vacuum system. Stainless-steel valves can have plate seal leak rates of order 2×10^{-8}–2×10^{-9} Pa-L/s (10^{-10}–10^{-11} Torr-L/s) with lifetimes ranging from 10,000 to 50,000 cycles with the typical value on the low side of this range. They can fail by seal leaks, bellows leaks, and wear of moving parts. Bellow's failure is the most common problem. Old-style gate valves were often unkindly referred to as "a machine shop in a vacuum system."

Elastomer-sealed poppet and slit valves are made in large sizes. These styles are illustrated in Fig. 16.17. Poppet valves up to 1.2-m diameter are commercially

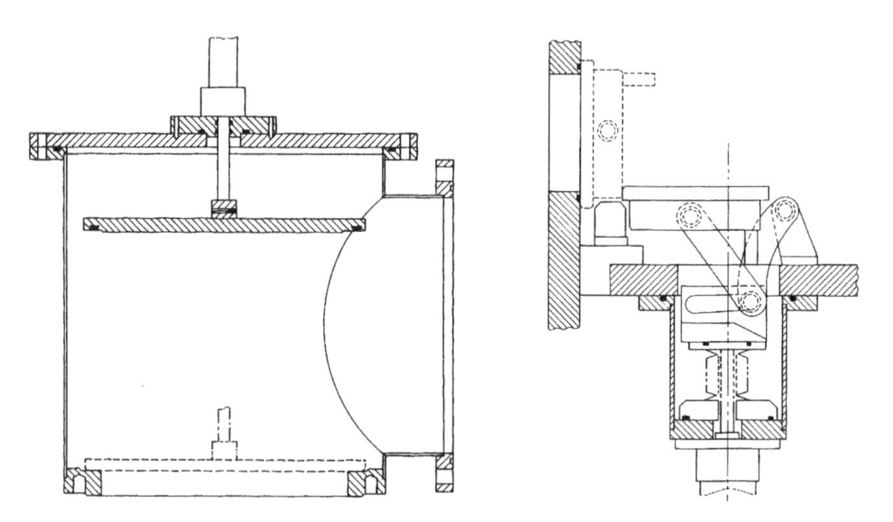

Fig. 16.17 Poppet (left) and (right) slit valves. Reproduced with permission from CVC Products, Inc. 525 Lee Rd., Rochester, NY.

available. Poppet valves use a simpler mechanism than gate valves, because the operating force is applied via a large, compressed air piston in line with the valve plate motion. The slit valve pivots about an axis parallel to the valve plate surface. These valves are much deeper than a gate valve and their conductance is less. Most large diffusion pumps require an elbow to be connected to a chamber, so the poppet does not reduce the overall conductance as much as we first assume. In some small ion-pumped systems, the poppet valve serves both to isolate and baffle the pump entrance. In-line chambers generally cannot be connected closer than needed for a flap valve. Any loss in conductance is more than offset by increased reliability and lower cost. Poppet valves are best used to seal a pump under vacuum while the chamber is vented. In this way the atmospheric pressure forces the valve plate closed. Slit valves are also constructed in rectangular shapes. For example, a 5- × 30-cm valve might be used on an in-line sputtering system for isolating chambers and load locks. The plates are constructed from stainless steel, use elastomer gaskets, and are designed for operation in the high-vacuum region. Butterfly valves are also fabricated in large diameters.

Large elastomer and metal gasket valves are available for a number of applications, baked and unbaked. They are commercially available in small sizes and custom-made in larger diameters. Double-sided all-metal valves have been custom-made [35]. Valves can be purchased with contacts to indicate the full-open or closed position. The technology for fabricating large-diameter, all-metal valves that are bakeable to temperatures >300°C is not simple. Poppet valves are made to fit large diffusion pumps. Special-purpose rectangular slit valves are usually custom-designed for a particular application. As the valve size increases, so does cost, the probability of failure, maintenance, and baking difficulty. Ishimaru et al. [36] devised a bakeable valve with no elastomer or metal knife-edge seal. It makes use of polished metal sealing surfaces that are forced together by compressed air-filled bellows.

16.3.3 Special-Purpose Valves

The valves described in the previous sections all serve to connect and isolate components of various diameters. There are special cases where we wish to control the conductance between the two chambers with a partially open valve or control the rate with which gas enters a chamber.

Controlling gas flow in the pressure range 1–10 Pa (10^{-2}–10^{-1} Torr) is necessary to regulate the pressure in a sputtering or etching chamber. The pressure is regulated with a throttle valve located in the pumping line adjacent to the pump. Typically, the high-vacuum pump will be operating at a maximum pressure of, say, 1×10^{-1} Pa (1×10^{-3} Torr), whereas the deposition or etch process may require 3 Pa (2×10^{-2} Torr). The throttle valve can be closed enough to provide this pressure drop. For a chamber using an argon flow of 100 Pa-L/s, (1 Torr-L/s), this corresponds to a valve conductance of ~35 L/s, in contrast to the fully open conductance of >5000 L/s for a 15-cm-diam valve in the molecular flow region.

The techniques used to throttle gas flow can be simple. For example, a hole in the plate of a gate or butterfly valve. The valve is opened to pump to the base pressure and closed to place the desired throttling conductance in series. This does not allow for adjustment of the closed conductance. Several arrangements have been designed to allow control of the closed conductance. A butterfly valve can close against an external shaft stop. This design and others, which use venetian blinds, rotating pie-shaped segments, or a small iris, are commercially available. Lehmann et al. [37] have designed a large iris valve which has a variable aperture and electronic control. Fifteen-centimeter valves can be throttled to have a conductance that is adjustable in the range 1–200 L/s. Throttle valves are typically made with elastomer-sealed flanges and are not bakeable. The operating characteristics of a throttle valve are usually displayed as a plot of conductance versus pressure drop over the range of adjustment. The techniques used to control the closed position of the throttle range from a manual adjustment of the stop setting, to closed-loop pressure control systems in which the error signal from a pressure gauge operates a motor.

Gas flow control valves, sometimes called leak, or metering valves are used to admit controlled quantities of gas from a source external to the vacuum. The external source may be a high-pressure (100- to 150-kPa-gauge), or reduced-pressure gas distribution line. The simplest form of metering valve is a needle valve. When it is closed, the needle is seated against a hollow tapered cone of soft metal. Another design used a cylinder in which a fine spiral groove is turned. A screw moves the cylinder inside a hollow mating piece and effectively changes the length of the capillary through which the gas flows. These designs are not bakeable; however, they are commercially available in sizes that span 1–4 decades of flow and cover the range 1×10^{-3}–2×10^{5} Pa-L/s (1×10^{-5}–2×10^{3} atm-cc/s). Leak valves in which a sapphire flat is pressed into a metal knife-edge are designed to be baked. A sapphire-metal leak can control flows as low as 10^{-8} Pa-L/s (10^{-6} Torr-L/s). Fixed helium leaks made from glass capillary tubing are used to calibrate RGAs.

16.3.4 Motion Feedthroughs

Vacuum systems would have few applications if there were no way of transmitting motion to the vacuum environment. Rotary and linear motion are necessary to operate pumps and valves, move samples and sources, open and close shutters, and perform many specialized tasks. Rotary and linear motion feedthroughs are characterized by the torque, the speed at which it is transmitted, and the operating

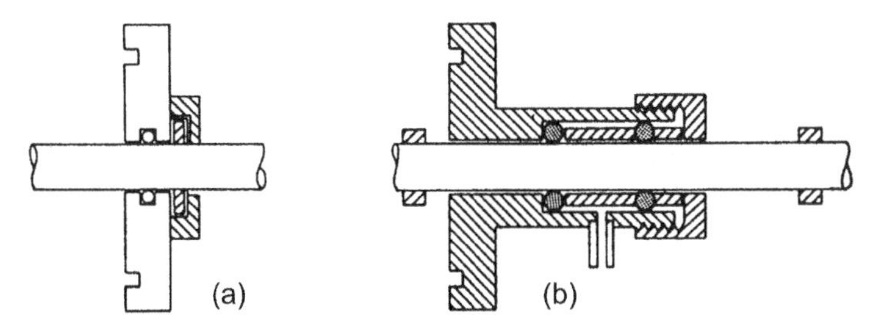

Fig. 16.18 (a) Rotating and (b) translating elastomer sealed feedthroughs.

pressure. Rotary and linear feedthroughs for the medium and high-vacuum region usually use elastomer seals.

Figure 16.18 depicts two basic forms of dynamic elastomer seal. The rotary seal shown in 16.18*a* is one form of a simple hand or low-speed rotary seal. In this sketch the O-ring groove is cut into the housing while the shaft contains a retaining ring to prevent linear shaft motion. Alternatively, the O-ring groove may be cut into the shaft. A better version of this feedthrough uses ball bearings on both sides of the O-ring. O-ring manufacturers can provide tables of groove dimensions for these designs using common elastomers. Groove dimensions are somewhat different than those used static seals. Unlike static seals, dynamic O-ring seals need to be greased. The only exception is Teflon. The properties and choices of grease for use in vacuum are discussed in Chapter 17. An improved rotary seal, which can also be used for translation, is shown in Fig. 16.18*b*. It is a double-pumped seal and can be used in the high-vacuum range. Replacing machined grooves with sleeves reduces the fabrication cost.

These two elastomer feedthroughs are only two of a large number of elastomer seals. Today it is often easier to substitute a metal seal rather than design a high-quality elastomer seal. One exception is for applications which require a high rotational speed and high torque in the very high-vacuum or near UHV region where baking may be necessary. For this application a Teflon seal is superior to a greased elastomer. Teflon cold flows and must be spring-loaded to avoid leaking. Seal gaskets with a C-shaped cross-section have been designed with special springs located inside the C-ring [38–40]. These seals can be used singly, or in pairs which are differentially pumped and baked to 250°C. They can be rotated at high speed and transmit high torque. Differentially pumped rotary flange seals have been designed that use two large Teflon C-rings [41]. This design allowed one 15-cm-diameter flange to rotate with respect to another with a dynamic leak rate of 10^{-6} Pa-L/s (10^{-8} Torr-L/s).

Two techniques for transmitting motion through a wall, which do not make use of an elastomer or metal seal, are the differentially pumped seal and the magnetic liquid seal. The differentially pumped seal shown in Fig. 16.19 uses a closely

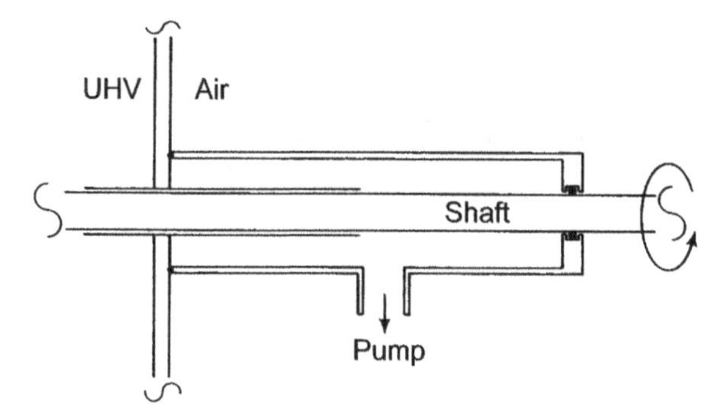

Fig. 16.19 Bakeable differentially pumped motion feedthrough. The inner seal is a low-conductance, differentially pumped gap between a hollow cylinder and a round shaft. The outer seal could be either an O-ring gasket or a magnetic fluid seal.

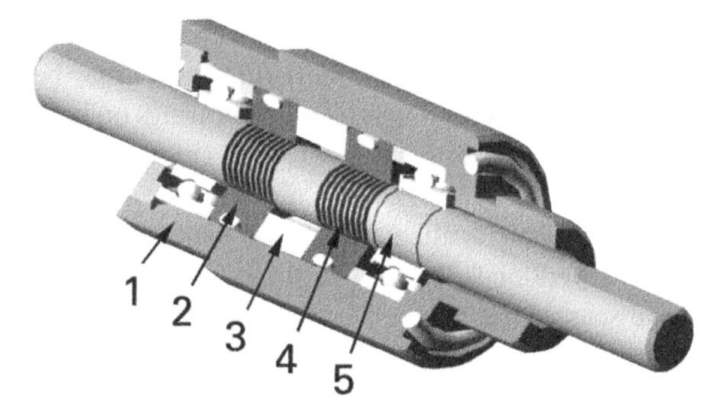

Fig. 16.20 Ferrofluidic® seal: (1) Housing; (2) pole piece; (3) magnet; (4) Ferrofluid; (5) shaft. Reproduced with permission from Ferrotec (USA) Corporation, Nashua, NH 03060.

spaced surface to create a low conductance between the vacuum chamber and the differentially pumped chamber. The inner parts of this seal can be baked to high temperature without any problems.

The magnetic-liquid seal depicted in Fig. 16.20 allows rotary motion to be transmitted on a solid magnetically permeable shaft, by sealing the gap between the housing and the shaft with a magnetic liquid [42,43]. The magnetic liquid consists of a low-vapor-pressure fluid, e.g., perfluoropolyether, and fine magnetic particles held in suspension with a surfactant. The magnetic lines of force concentrate the magnetic liquid at the ridges machined on the shaft. This seal allows high speed (5000 rpm) and high torque. It is useful in unbaked systems, which are pumped to the 10^{-5}-Pa region, but air permeates, and it may outgas fluid fragments.

Except for the differentially pumped feedthroughs, none of the motion feedthroughs we have described are suitable for the UHV where baking over 100°C is required. For these conditions, all-metal linear and rotary motion feedthroughs, the differentially pumped feedthrough, or magnetic couplings are required. Magnetic couplings transmit torque through a thin metal vacuum wall to a magnetic shaft within the vacuum chamber. Magnetic couplings are an excellent way to transmit motion through walls of UHV chambers as long as cooling water or electrical connections are not required. If water or electrical cables must be provided to the rotating assembly, the only purely UHV option is the differentially pumped feedthrough of Fig. 16.19. An example of a magnetic coupling is illustrated in Fig. 16.21. In this assembly, an external motor rotates a magnet that is coupled to a follower located within the UHV environment. The load is attached to the end of the shaft. Some form of lubrication is required for the follower bearings, and this can be chosen to match the required rotational speeds and loads.

Cross-section views of two types of bellows, hydroformed and welded bellows, are shown in Fig. 16.22. Hydroformed bellows are made by hydraulically stretching rolled and welded thin-wall tubing, or deep-drawn thin-wall cups. Welded bellows are made by sequentially welding a series of thin-wall diaphragms. In both designs, 304L, 316L, and 321 are the most commonly used materials for vacuum applications. Welded bellows are more expensive and more flexible than hydroformed bellows; however, fewer convolutions are needed. The limiting extension,

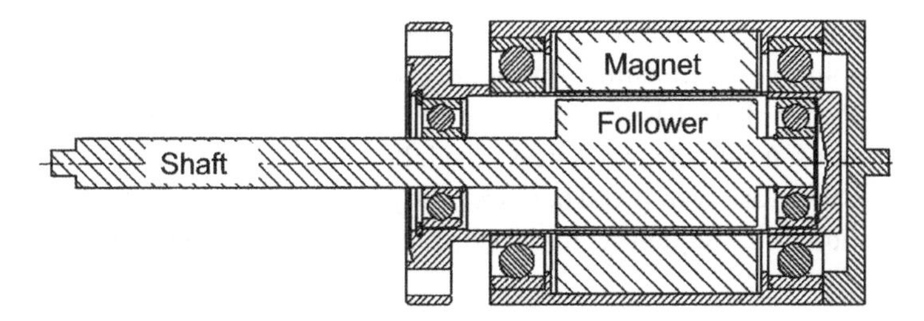

Fig. 16.21 Magnetically coupled UHV feedthrough mounted on a ConFlat flange. An external, rotating magnet couples torque through the stainless-steel vacuum wall to a magnetic follower and shaft within the vacuum chamber. Reproduced with permission from Veeco, St. Paul Division, 4900 Constellation Drive, St. Paul, MN 55127.

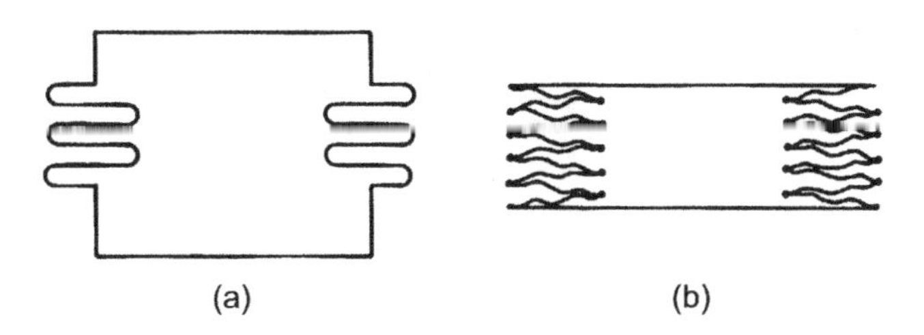

Fig. 16.22 (a) Hydroformed and (b) welded bellows.

compression, and ultimate lifetime of a bellows depend on the stresses encountered at the ends of the stroke. These in turn are dependent on the inner-to-outer radii ratio, material, how much the material has been cold worked, etc. A designer chooses the extension, compression, and the total pitched bellows length to keep the stresses in the bellows below a design value. This design value will be low when long life is a requirement. A few rough generalizations can be made; bellows are often designed to work only in compression but work well in both compression and extension. Stroke ratios ranging from 1/3:2/3 (extension–compression) to 1:1 (extension–compression) have been used in specific designs. The total stroke length for a bellows, used in a vacuum application, ranges from 20–33% of the pitched length for hydroformed bellows, to 80% of the pitched length for edge welded bellows.

Hydroformed bellows are adequate for many applications, especially where the bending or compression is small. Edge-welded bellows find many applications where a long stroke is needed in a very short space. Bellows were traditionally furnace-brazed onto a small pipe section for welding onto larger assemblies. Mating pieces machined with heat relief grooves have resulted in elegant pipe joints and valve bodies.

A long bellows section can be used for translational motion as described in Fig. 16.23*a*. Another solution uses a rotating bellows to drive a lead screw to convert rotary motion outside the chamber into linear motion inside the chamber.

One ingenious solution for transmission motion is shown in Fig. 16.23*b*. This design illustrates the problems basic to all rotary bellows feedthroughs. Since the shaft is not continuous, it cannot carry a heavy load. Ball bearings are required on both sides of the seal. A variety of metal bellows rotary feedthroughs are commercially available and used frequently in UHV shutters and manipulators for

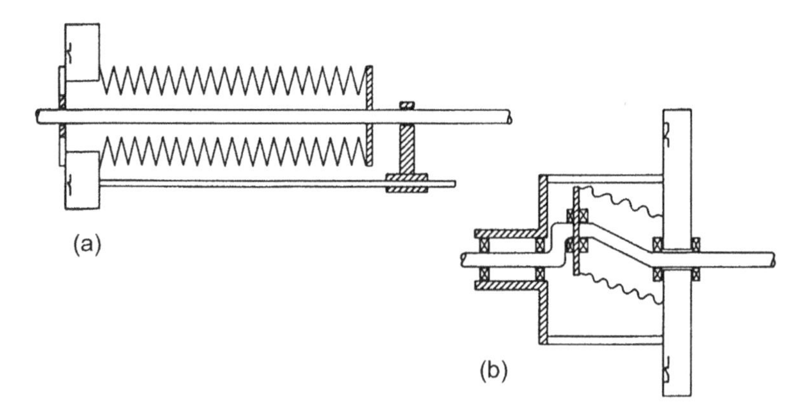

Fig. 16.23 (a) Translating and (b) rotating feedthroughs using metal bellows.

precision movement of small parts, as well as rotation of shutters and substrate holders. Bellows need to be carefully cleaned, but not with solvents, as they are difficult to remove. A vacuum bake is effective. Baking temperatures should not exceed those required for brazing, should it have been used for assembly.

Outgassing has been observed from stainless steel during bending and sliding motion such as that found in sliding vacuum manipulators, motion feedthroughs, and valves. Hydrogen was released from the bulk of stainless steel during flexing and while flexed, whereas CH_4, CO, and CO_2 were released from the surface during flexing [44]. Hydrogen and small quantities of CH_4 and CO were observed to desorb from stainless steel during sliding friction in UHV [45].

References

1 Singleton, J.H., *J. Vac. Sci. Technol., A* **2**, 126 (1984).
2 Roth, A., *Vacuum Sealing Techniques*, Pergamon Press, Oxford, 1966.
3 Roth, A., *J. Vac Sci. Technol.* **9**, 14 (1972).
4 Roth, A., *J. Vac. Sci. Technol., A* **1**, 211 (1983).
5 American Society for Metals, Metals Handbook, *8th ed., Welding and Brazing*, Vol. 6, T. Lyman, Ed., American Society for Metals, 1971, p. 113.
6 C.H. Rosendahl, *Sheet Metal Industries*, February 1970, p. 93.
7 Geyari, C., *Vacuum* **26**, 287 (1976).
8 R.N. Peacock, *Vacuum Joining Techniques*, HPS Corporation, 1898 S. Flatiron Ct. Boulder, CO, 1981.
9 Kohl, W.H., *Handbook of Materials and Techniques for Vacuum Devices*, Reinhold, New York, 1967.
10 American Welding Society Committee on Brazing and Soldering, *Brazing Manual*, Reinhold, New York, 1960.
11 Handy and Harmon Corp, *Brazing Book*, Handy and Harmon Co., White Plains, New York.
12 Takamori, T., in *Treatise on Materials Science and Technology: Glass II, 17*, M. Tomozawa and R.H. Doremus, Eds., Academic, New York, 1979, p. 117.
13 Benbenek, J.E. and Honig, R.E., *Rev. Sci. Instrum.* **31**, 460 (1960).
14 Rosebury, F., *Handbook of Electron Tube and Vacuum Techniques*, Addison-Wesley, Reading, MA, 1965.
15 Espe, W., *Materials of High Vacuum Technology*, Vol. 2, Pergamon Press, New York, 1966.
16 Peacock, R.N., *J. Vac. Sci. Technol.* **17**, 330 (1980).
17 Sessink, B. and Verster, N., *Vacuum* **23**, 319 (1973).
18 The relation between International Rubber Hardness Degrees and Young's Modulus is given in: *ASTM D1415-62T*. This standard is available from the

American Society for Testing Materials, 1916 Race St., Philadelphia, PA 19103. Shore A degrees are approximately IRH degrees. Typical values of Young's modulus are $E = 35$, 54 and 68 kg/cm^2 corresponding to 60, 70, and 75 Shore A degrees.

19 de Csernatony, L., *Vacuum* **16**, 13 (1966).

20 Johnson, M.L., Manos, D.M., and Provost, T., *J. Vac. Sci. Technol., A* **15**, 763 (1997).

21 de Csernatony, L., *Vacuum* **16**, 129 (1966).

22 Laurenson, L. and Dennis, N.T.M., *J. Vac. Sci. Technol., A* **3**, 1707 (1985).

23 Wheeler, R. and Pepper, S.V., *J. Vac. Sci. Technol.* **20**, 226 (1982).

24 de Csernatony, L., in *Proceedings of the Seventh International Vacuum Congress and the Third International Conference on Solid Surfaces of the International Union for Vacuum Science, Technique and Applications, September 12–16, 1977*, R. Dobrozemsky, Ed., IUVSTA, Vienna, Austria, 1977, p. 259.

25 de Csernatony, L., *Vacuum* **27**, 605 (1977).

26 Hait, P.W., *Vacuum* **17**, 547 (1967).

27 Edwards, T.W., Budge, J.R., and Hauptli, W., *J. Vac. Sci. Technol.* **14**, 740 (1977).

28 Wheeler, W.R. and Carlson, M., *Trans. 8th Natl. Vac. Symp., and Proc. 2nd. Int. Congr. Vac. Sci. Technol., 1961*, Pergamon, New York, 1962, p. 1309.

29 Unterlerchner, W., *J. Vac. Sci. A* **5**, 2540 (1987).

30 Kurokouchi, S., Morita, S., and Okabe, M., *J. Vac. Sci. Technol., A* **19**, 2963 (2001).

31 Fend, H., *Vacuum* **47**, 527 (1996).

32 Weston, G.F., *Vacuum* **34**, 619 (1984).

33 Alpert, D., *J. Appl. Phys.* **24**, 860 (1953).

34 Yokokura, K. and Kazawa, M., *J. Vac. Soc. Jpn.* **24**, 399 (1981).

35 Foerster, C.L. and McCafferty, D., *J. Vac. Sci. Technol.* **18**, 997 (1981).

36 Ishimaru, H., Kuroda, T., Kaneko, O., Oka, Y., and Sakurai, K., *J. Vac. Sci. Technol., A* **3**, 1703 (1985).

37 Lehmann, H.W., Curtis, B.J., and Fehlman, R., *Vacuum* **34**, 679 (1984).

38 *Bal-Seal* ring seal, BAL Engineering Corp., Lutz, FL.

39 *Omni-Seal*, Omni-Seal, Garden Grove, California.

40 C-Ring, metal ring seals, Fluorocarbon Co., Herfordshire, UK.

41 Silverman, P.J., *J. Vac. Sci. Technol., A* **2**, 76 (1984).

42 R.E. Rosensweig and R. Kaiser, *Office of Adv. Res. and Tech., NASA CR 1407*, August 1969.

43 Raj, K. and Grayson, M.A., *Vacuum* **31**, 151 (1981).

44 Řepa, P. and Orálek, D., *Vacuum* **53**, 299 (1999).

45 Nevshupa, R.A., de Segovia, J.L., and Deulin, E.A., *Vacuum* **53**, 295 (1999).

17

Pump Fluids and Lubricants

Organic liquids are the operating fluids of diffusion pumps. They provide a vacuum seal between moving surfaces in fluid-sealed mechanical pumps, as well as lubricate and cool bearings and sliding surfaces in screw, scroll, lobe, and turbo pumps. Rotary- and linear-motion vacuum feedthroughs and moving parts within a vacuum chamber require lubrication to reduce friction and prevent wear. In this chapter, we review properties of organic and inorganic fluids, and inorganic solids used as either pumping fluids or lubricants.

17.1 Pump Fluids

Highly refined mineral oils and synthetic esters, silicones, ethers, and fluorocarbons are widely used in vacuum technology. Mineral oils lack many properties of the ideal fluid. When used in a diffusion pump, their ultimate pressure is unacceptably high for some applications. They are not stable in oxygen, have some tendency to sludge and foam, and do not offer adequate protection when used as boundary layer lubricants. Synthetics were developed to overcome these shortcomings. Many fluids with low vapor pressures, high viscosity indexes, a high degree of oiliness, and chemical inertness have been synthesized and widely used.

17.1.1 Fluid Properties

Vapor pressure and lubricating ability are their two most important properties. Low vapor pressure is necessary to avoid oil vapor transport to the vacuum chamber, and mechanical pump fluids need to be good lubricants. In Section 17.2 we discuss three rheological properties: absolute and kinematic viscosity, and

A Users Guide to Vacuum Technology, Fourth Edition. John F. O'Hanlon and Timothy A. Gessert.
© 2024 John Wiley & Sons, Inc. Published 2024 by John Wiley & Sons, Inc.

viscosity index (V.I.). In this Section we review vapor pressure and other physical and chemical properties of pump fluids.

17.1.1.1 Vapor Pressure

Regardless of its other qualities, a pump fluid, lubricant, or additive is of no use if its vapor pressure is so high that it contaminates the vacuum system or process. A minimum vapor pressure is necessary; any further reduction will improve performance, simplify trapping, and reduce contamination. For example, a mechanical pump fluid should have a vapor pressure less than 0.1 Pa (10^{-3} Torr) at its operating temperature, whereas a diffusion pump requires a fluid whose room-temperature vapor pressure is in the range 10^{-3}–10^{-7} Pa (10^{-5}–10^{-9} Torr). Most of us will never measure the vapor pressure of a pump fluid, however, we do need to know how, when—and if—available data were measured.

Many techniques have been devised for the measurement of oil vapor pressure [1–3]. Knudsen effusion [1,2] is reliable and accurate, but there is no standard for this technique. In the Knudsen technique, a small sample of fluid is heated to a constant temperature in a partly-filled cell with a small orifice. At equilibrium, the rate of vaporization from the fluid surface is equal to the rate of arrival. This equilibrium pressure is the fluid's vapor pressure. A tiny hole in the top of the cell allows a small fraction of the vapor to effuse from the cell into a vacuum without changing the internal pressure. The surface area of the liquid needs to be at least 10 times the area of the hole in order to maintain the liquid at its equilibrium vapor pressure, and diameter of the opening must be less than the mean free path of the heated molecules. A typical cell size is 1.2×10^{-2} m in diameter by 4×10^{-3} m high, with a 3×10^{-3}-m-diameter orifice. Outside the cell, the pressure must be $<\sim10^{-3}$ Pa to prevent molecules from returning to the opening. If these criteria are met, the vapor pressure can be calculated from Eq's. (2.9) and (2.13):

$$P=\left(\frac{dm}{dt}\right)\frac{1}{aA}\left(\frac{2\pi kT}{m}\right)^{1/2} \tag{17.1}$$

or

$$P(\text{Pa})=2.278\times10^{4}\left(\frac{dm}{dt}\right)\frac{1}{aA}\left(\frac{T}{M}\right)^{1/2} \tag{17.2}$$

where a is the transmission probability of the orifice. When the orifice thickness is much less than its diameter, $a=1$; otherwise, the appropriate transmission probability must be taken from Table 3.2. The vapor pressure is calculated from the known cell temperature and weight loss dm/dt. The fractional weight loss is small. For example, a 10^{-4}-kg sample of pentaphenyl ether with a molecular weight of 447 and a vapor pressure of 1 Pa at 200°C will lose weight at the rate of 3×10^{-10} kg/s from an effusion orifice of area 7×10^{-6} m^2. Vapor pressure measurements are made over a range of temperatures corresponding to 0.2–10% weight loss in 600–900 s. The temperature range will be different for each fluid, but in all cases, 80–100°C is near the minimum temperature for adequate sensitivity. All data reported below that temperature are extrapolated data. The Clapeyron equation:

$$\left(\frac{dP}{dT}\right) = \frac{\Delta H}{T\Delta V} \tag{17.3}$$

gives the vapor pressure–temperature relationship for many substances. ΔH is the heat of vaporization of one kmol of substance, ΔV is the volume change per kmol during vaporization. When the specific volume of the gas phase is much greater than that of the liquid phase and when ΔH is independent of T, the solution is:

$$\log P_v = A - \frac{B}{kT}\frac{dy}{dx} \tag{17.4}$$

This solution is known as the Clausius–Clapeyron equation. If the heat of vaporization is a constant, B is a constant, and P_v versus $1/kT$ will plot as a straight line for temperatures below about half the critical temperature. Hickman [4] measured vapor pressures of several polyethers using a tensimeter and calculated their Clausius–Clapeyeron constants. Figure 17.1 sketches the behavior of one 5-ring compound, an early version of Santovac-5. It is important to understand that vapor pressures reported at room and cryogenic temperatures are determined

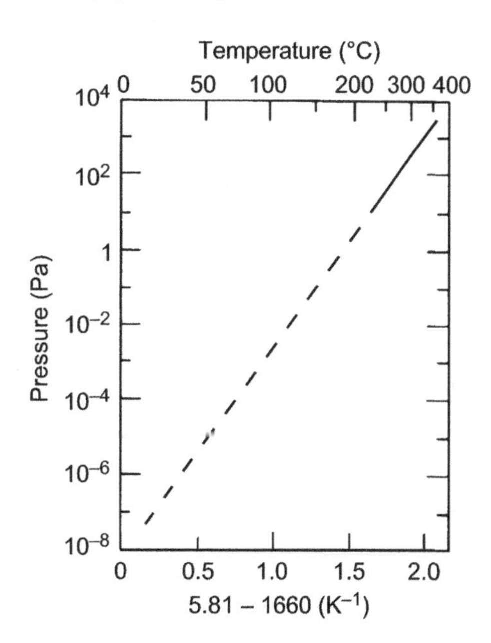

Fig. 17.1 Vapor pressure of a pentaphenyl ether calculated using Clausius-Clapeyron constants given in Ref. [4]. The solid line indicates the temperature range over which data were measured.

by extrapolation from higher temperature measurements. Laurenson and Troup [5] used a quartz crystal microbalance to measure the vapor flux from a Knudsen cell and measured vapor pressures of mechanical pump fluids to 20°C.

Equation 17.4 is obeyed by single chemical compound fluids. All fluids are not single chemical compounds. Polymers and other mixtures contain molecules of various weights. Polysiloxanes and perfluoropolyether (PFPE) fluids are examples of fluids that contain molecules of similar structure and variable weight. A hydrocarbon mineral oil is an example of a fluid, which has an added complication; it contains three distinct structures: paraffins, naphthenes, and aromatics. Each fluid contains a distribution of molecular weights, and the published molecular weight is an average.

The concept of vapor pressure has little meaning, when applied to polymers with a broad molecular weight distribution such as polysiloxane or polyether. A single chemical compound has a distinct boiling point at a particular pressure. Polymers of broadly varying molecular weights with a large deviation about the mean molecular weight have no unique vapor pressure. The measured vapor pressure is dominated by the vapor pressure of the lower molecular weight fractions. When pump fluid is heated, light fractions preferentially evaporate, and the composition of the residual fluid changes slowly and systematically.

Evaporation of light fractions is known to affect vapor pressure measurements [4,5]. If a mixture contains a large fraction of lightweight molecules, as do some polymers, a linear Clausius–Clapeyron equation is not obtained except at high temperatures. Extrapolating these data to low temperatures by means of a straight-line approximation is invalid, because it gives a vapor pressure characteristic of the fluid minus its lightweight impurities. Fractionating diffusion pumps systematically eject the lighter-weight fractions to the fore pump. Extrapolating high-temperature effusion data from diffusion pump fluids to lower temperatures yields a vapor pressure that is characteristic of the fluid after it has been operated for some time in a fractionating pump or has been carefully distilled. Extrapolating vapor pressures of some mechanical pump fluids to low temperatures is not valid, because the light fractions remain, and their vapor pressures are higher than the value predicted by the straight-line approximation.

The Knudsen technique has been used to characterize the vapor pressures of pump fluids; however, measurements have been performed historically by few laboratories, and in some cases, original measurements have been republished for generations.

As an alternative to the measurement of vapor pressure, the ultimate, or blank-off pressure in a specific pump is used to characterize diffusion pump fluids. The ultimate pressure is dependent on several factors including the pump fluid, power, pump design, and cooling rate. The vapor pressure and ultimate pressure are not the same, but they do seem to correlate. The ultimate pressure will usually be within a half an order of magnitude higher than the true vapor pressure [3,4,6–9]. Methods for measuring the volatilization rate of vacuum lubricants are described in ASTM Standard D2715 [10].

17.1.1.2 Other Characteristics

Color, pour point, flash and fire points, and gas solubility are properties that further characterize lubricants and pump fluids. Fluid color is an aid in identification. The color of hydrocarbon pump oils will vary from clear to medium yellow and is characterized by a standard such as the Saybolt color index [11]. Naphthenes are darker than paraffins. Pure straight-chain paraffins and most synthetics are transparent, but additives and dyes will cloud or color fluids.

The pour point is the lowest temperature at which a fluid will flow. Vacuum pump fluids that have been dewaxed behave like Newtonian fluids to very low temperatures. The viscosity at the pour point is of order 10^5–10^6 mm^2/s [12]. Flash and fire points are, respectively, the temperatures at which a fluid will burn momentarily and continuously in the presence of a flame [13]. The autoignition temperature is the temperature at which the fluid will ignite spontaneously. The ignition properties of organic fluids are a function of the vapor quantity, the surface-to-volume ratio of the test cell, and the spark energy [14]. They are reproducible, but test conditions may not be identical to operating conditions.

The solubility of gas in liquid obeys Henry's law and is directly proportional to absolute pressure. Gas solubility increases rapidly with fluid viscosity in the 5–50-mm^2/s range. The increase in solubility with viscosity becomes less rapid for 100 mm^2/s oils, and then ceases to increase at ~2000–3000 mm^2/s (L. Laurenson, private communication, January 1983). The solubility of air at one atmosphere (volume %) has found to be 7–10% for hydrocarbon oil, and 20% for silicone. The solubility of individual gases in mineral oil (20°C) is: CO_2–80%; N_2–6%; and He–2% [15]. The solubility of Fomblin Y-25 at 20°C was found to be Air–28%; H_2–8%; He–4%; CO_2–327%; ethylene–122%. Its SF_6 solubility was 350% at 0°C [16]. Tritium was found to be 8% soluble in polyphenyl ether and PFPE, and 7% soluble in mineral oil [17]. Dissolved gas is undesirable in mechanical pump fluid. It increases both pumping time and base pressure [15]. High gas solubility can also cause fluid foaming and allow a dissolved gas to react with gases that have been previously pumped. Dissolved gas reduces oil viscosity, whereas bubbles in oil increase its viscosity and slow its removal [18].

17.1.2 Fluid Types

Many types of mechanical and diffusion pump fluids are in use, and each has its own attributes. The first diffusion pumps used mercury, and soon after, mineral oils. In 1929 Burch [19] distilled low-vapor-pressure mineral oils and proposed their use as an alternative to mercury. Later, Hickman [4] distilled synthetic esters and ethers for vacuum use. In the ensuing decades many refined hydrocarbon oils and synthetic fluids have been developed. Some were displaced by superior products, eliminated for safety, or were never accepted, because they offered no advantage.

17.1.2.1 Mineral Oils

The mineral oils used in vacuum pumps are mixtures of paraffin, naphthene, and aromatic hydrocarbons. Paraffins, $C_nH_{(2n+2)}$, are straight- or branched-chain hydrocarbon structures containing only single bonds. The high-boiling-point paraffins make excellent lubricants. They are stable at high temperatures, fluid at low temperatures, and have reasonably constant viscosity over a wide temperature range (high V.I.). They are adhesive enough for high shear loads but are not stable in O_2 at high temperatures. Paraffins have many possible isomers, for example, $C_{10}H_{22}$ has 75, with differing properties. Aromatic compounds contain phenyl groups with straight- or branched-chain structures. They form sludge at high temperatures and have an undesirably low V.I. Naphthenes contain rings and chains with no double bonds. Naphthenes have properties between those of paraffins and aromatics. Carbon analysis shows the typical "paraffin mineral oil" to be composed of ~65% paraffin, 30% naphthene, and 5% aromatics. Carbon analysis gives weight percent carbon in singly bonded chains. All this carbon is not from paraffin structures, because other structures contain some single bonds.

Preparation of a vacuum fluid begins with vacuum-distilled base oil that is further purified by solvent extraction and dewaxing [20–22]. The oils supplied to the vacuum fluid distiller are either single cut with one peak in the molecular weight distribution, or blends made from two of the relatively few refined single cuts. These are either a light (30–40 mm^2/s), a medium (60–70 mm^2/s), or a heavy (100–120 mm^2/s) oil.

Oil is purified by additional distillations. The distillation conditions are chosen to produce fluids of the desired viscosity and vapor pressure. Distilled fluids have a distribution of molecular weights. Stripping both tars and low-molecular-weight ends produces a single-cut fluid with a distribution of molecular weights centered about the mean molecular weight. Blending produces a fluid with more than one peak in the molecular weight distribution. Viscosity is inversely proportional to vapor pressure, so the vapor pressure of these broad cuts will differ widely. It is not possible to produce a fluid of extremely low vapor pressure from a blend of two oils with widely differing vapor pressures.

The base oil has characteristics unique to its origin; as a result, all mineral oils are not the same. Not only does the origin determine the paraffin/naphthene/aromatic ratio and the impurities, but it determines the type of paraffin isomers. For example, the amount of sulfur and other impurities in mineral oil varies with the geographical origin of the crude. For many applications residual sulfur is detrimental, but it does inhibit oxidation [23]. Refiners of the highest-quality, single-cut mineral oils select their base stock from a single source, and they test each lot for uniformity.

For this discussion, we arbitrarily divide vacuum pump mineral oils into four "grades": mechanical pump, diffusion pump, fully saturated paraffin, and inhibited fluids. The "mechanical pump" grade is, loosely, composed of blended fluids or single-cut fluids that have not been refined to remove light ends and tars. Vapor pressure requirements in a mechanical pump are not as severe as in a diffusion pump. This grade is typically used in mechanical pumps used for rough pumping chambers and backing turbomolecular and diffusion pumps. "Diffusion-pump" grade fluids may be characterized as having a single peak in the molecular weight with narrow mass dispersion. The vapor pressure will be the lowest for those cuts with the highest average molecular weight. Hydrocarbon diffusion pump fluids are used in a variety of high vacuum pumping applications but are not suitable for many ultrahigh vacuum applications.

Fully saturated paraffin oil, or white oil, is made by catalytically hydrogenating paraffin oil. This fluid will be somewhat more stable in the presence of mild corrosive gases than ordinary mineral oil because its lack of dangling bonds reduces its reactivity. White oil was sometimes used in mechanical pumps that handle halogens, e.g., plasma etching systems. Today, inert fluorocarbon fluids are used.

Dyes have been added to mechanical pump fluids to give them a distinctive color. They serve no functional purpose. Most additives have high vapor pressures and are useful for a limited number of applications. Examples are diffusion pumps used for decorative coating and mechanical pumps used for pumping corrosives.

Mineral oils lack many properties of the ideal fluid. When used in a diffusion pump, their ultimate pressure can be unacceptably high. They are not stable in oxygen, have a tendency to sludge and foam, and do not offer adequate protection in boundary layer lubrication region. Synthetic fluids overcome many of these shortcomings.

17.1.2.2 Esters

Esters are chemicals formed by the reaction of an organic acid and an alcohol. Esters used in vacuum pump fluids all contain the same ester chemical bond but have differing structures and rather widely varying properties. Sebacate esters are organic esters that were originally developed as jet engine and aircraft instrument lubricants and today used to lubricate turbomolecular pumps. Older ester fluids, especially those containing phosphorous, are no longer used because they are not environmentally safe.

17.1.2.3 Silicones

The unique character of the silicon–oxygen bond yields fluids with useful properties. Silicones, or siloxane polymers, are made up of repeated silicon–oxygen groups with silicon bonds to side groups. The type of side groups (methyl, phenyl, alkyl, chloro, etc.) and the number of silicon atoms determine their properties and usefulness. The large central silicon atom allows the phenyl and methyl side groups great mobility. The high flexibility of the siloxane chain accounts for the high V.I. of silicones. As a class, silicones have the highest V.I. of any fluid. Trisiloxanes and polysiloxanes are two commonly used pump fluids.

Trisiloxanes are used as diffusion pump fluids. They do not adhere to steel and cannot be used as lubricants in mechanical pumps [24]. They are manufactured by controlled hydrolysis of silanes and addition of phenyl groups, followed by distillation. The first silicone diffusion pump fluids, DC-702 and DC-703 [25], were mixtures of closely related molecular species with similar boiling points. Further separation leads to the isolation of two specific chemical compounds, tetraphenyl tetramethyl trisiloxane, DC-704 [9], and pentaphenyl trimethyl trisiloxane, DC-705 [6]. DC-705 has one of the lowest vapor pressures of any diffusion pump fluid.

17.1.2.4 Ethers

An ether is a derivative of a water molecule in which the hydrogen has been replaced by alkyl or aryl groups. Polyphenyl ethers were synthesized for use as diffusion pump fluids, and later found use as high-temperature jet engine lubricants. Hickman [4] was the first to use them in a diffusion pump. He found the five-ring phenyl ether to be stable and have extremely low vapor pressure. Commercially available fluids are mixed meta- and para-isomers of the pentaphenyl ether, which contain trace impurities of the four-ring compound. The four-ring compound has a high vapor pressure, while the six-ring compounds are either solids or glasses. Pentaphenyl ether is very viscous at low temperatures but is stable and has excellent high-temperature lubricating properties. Its wear, friction, and load capacity are in some cases equal to mineral oil [26]. Its chemical stability and low vapor pressure make it an outstanding fluid for critical diffusion pump applications.

17.1.2.5 Fluorochemicals

Fluorochemical fluids are characterized by their inertness to a wide range of chemical compounds. Large-scale fluorine fluid chemistry studies were initiated during the Manhattan Project in a search for uranium hexafluoride dilutants, as well as by the Navy in a search for fire-resistant fluids. Partially and fully fluorinated fluids have found use as lubricants for space applications, oxygen compressors, and liquid oxygen systems.

Fluorinated pump fluids, perfluoro alkyl polyethers (PFPEs or PFPE for short), are currently manufactured by two techniques. Fomblin [16,27] fluids are prepared by the UV-stimulated photooxidation of hexafluoropropylene and oxygen. It is a random copolymer of C_3F_6O and COF_2 [28]. Krytox [29,30] fluids are prepared by the polymerization of hexafluoropropylene epoxide. Krytox contains repeating C_3F_6O groups.

Raw PFPEs have a distribution of molecular weights extending as high as 10,000 AMU. They are distilled to yield cuts with average molecular weights in the range 1800–3700 that are suitable for use in mechanical, turbo, and diffusion pumps.

PFPEs are stable Lewis bases that react with few chemicals. They should not be placed in contact with ammonia, amines, liquid fluorine, liquid boron trifluoride, or sodium or potassium metal. PFPE fluids have been shown to decompose when heated (>100°C) in the presence of Lewis acids. The trichlorides and trifluorides of aluminum and boron are examples of Lewis acids that may be generated by or are used in reactive ion etching. If heated to high temperatures, Lewis acids act as depolymerization catalysts and allow release of toxic fragments.

17.1.3 Selecting Fluids

Many fluids are available for use in rotary, lobe, turbo, and diffusion pumps. We describe how pump requirements and possible gas reactions limit the choice of pump fluids. Properties of representative fluids are included.

Fluids used in vane and piston pumps must provide a vacuum seal between the moving surfaces and lubricate the bearings and sliding surfaces. The fluid in a lobe blower is used to lubricate the gear drive. Fluid assists in heat transfer from the bearings to the pump surface or cooling jacket in all three pumps. Fluids should not react with process gases or evolve fragments that could contaminate a process.

17.1.3.1 Rotary, Vane, and Lobe Pump Fluids

These fluids must have a low vapor pressure and be viscous enough to form a film that will fill the gap between the moving surfaces. The viscosity required in a particular pump depends upon the clearances between moving parts, the rotational speed, and the pump operating temperature. Table 17.1 lists typical properties of representative vane, lobe blower, and turbo pump fluids. The viscosity indexes tabulated here were obtained from the manufacturer or calculated from the ASTM standard [32]. Some piston and lobe pumps are machined to fine tolerances, while others are not so closely machined and require a more viscous fluid. Viscosity specifications are available from pump manufacturers. Appendices F.2, F.3, and F.4 give vapor pressures and kinematic viscosities of representative fluids.

Interpreting the vapor pressure data is not so easy. Some were taken on a Knudsen cell, some on an isoteniscope, and others by unstated procedures. Manufacturers' vapor pressure data may be taken from either a "typical," the "worst-case," or a sample product lot. There can be an order-of-magnitude difference between worst-case and typical data.

Table 17.1 Properties of Representative Mechanical and Turbomolecular Pump Fluids

Chemical Type	Representative Trade Name	MW (ave)	Sp.Gr. at 25°C	P_v at 25°C (Pa)	Pour Point (°C)	Viscosity Index	Fire Point (°C)
Mineral Oil							
"Mech Pump"	Balzers P-3[a]	190	0.88	1×10^{-2}	−16	95	295
	Duo-Seal® 1407[b,c]	450	0.88	—	−6.7	>95	240.6
	Inland 19[d]	440	0.88	5×10^{-4}	−15	130	244.5
Cleaning Fluid	Inland FF-10[d]	—	—	—	—	25	221
Ester							
Sebacate	Balzers T-11[a,e]	416	0.92	5×10^{-4}	−60	130	251
	Leybold HE-500[e,f]	430	0.92	2×10^{-3}	−62	130	—
Fluorocarbon							
PFPE	Fomblin® 25/5[g,h]	3250	1.9	3×10^{-4}	−35	120	None
	Fomblin® 06/6[g,h]	1800	1.88	3×10^{-4}	−30	50	None
	Krytox® 1514[g,i]	3000	1.88	3×10^{-4}	−35	100	None
	Krytox® 1506[g,i]	2150	1.86	3×10^{-4}	−45	50	None

[a] Balzers High Vacuum, Fürstentum, Liechtenstein.
[b] Vapor pressure data taken using a trapped McLeod gauge, data are not representative of the fluid.
[c] Sargent-Welch Co., Vacuum Products Div., Skokie IL.
[d] IVACO Inc., Churchville, NY.
[e] Contains additives.
[f] Leybold-Heraeus GmbH, Köln, Germany.
[g] Recommended by manufacturer for oxygen pumping.
[h] Montedison USA Inc. New York, NY.
[i] du Pont and Co., Chemicals and Pigments Department, Wilmington, DE.
Source: Adapted from O'Hanlon [31].

"Mechanical pump" grade mineral oils are satisfactory for routine pumping applications such as backing a diffusion pump or roughing an air-filled chamber. It is common in research and development labs to find "diffusion-pump" grade mineral oils used in mechanical pumps. Their use results in a slightly reduced base pressure and less backstreaming, provided that the pump was flushed several times to eliminate all traces of previous fluid. Ultimately, they degrade thermally [30].

Light hydrocarbon fluids are used for cleaning contaminants from mechanical pumps. These flushing fluids are simply a light-end by-product from the distillation of ordinary mechanical pump oil. They have a high vapor pressure and are thin (about half the viscosity of normal oils) and are good solvents. They clean by causing the pump to run hot. The increased temperature aids in the dissolution of sludge.

PFPE does not oxidize and is completely safe for use when pumping oxygen. It is inert to most corrosive gases. However, corrosive gases can slowly etch the interior of a pump if the acids they generate are not removed continually. Acid neutralizing, and recirculation oil filters are especially important when using PFPE because the fluid is only infrequently changed. Lewis acids are one class of chemicals that can react with PFPE at temperatures >100°C. Oil temperatures in ballasted or viscous flushed pumps can reach 120°C, and vane tip temperatures can be 100°C higher than the bulk oil temperature under any operating condition. Slow decomposition of PFPE results, and a small amount of fluid must be replaced periodically.

17.1.3.2 Turbo Pump Fluids

The requirements placed on turbo pump oil are somewhat different from those of mechanical pump oil. Because the bearing loading is not severe, a high-shear-strength, high-viscosity oil is not required. In full film lubrication the coefficient of friction is proportional to $\eta U/L$. High-speed bearings require low-viscosity oil. Although the average fluid temperature is 70°C, spot heating on the bearings can cause decomposition of fluids with poor vapor pressure. To prevent foaming, it is important that the fluid be vacuum degassed before use. A light-viscosity mineral oil that has been refined to remove both light ends and tars will work in a turbomolecular pump. The small amount of residual tar is of less concern than hydrogen from light ends, which has a low compression ratio and contributes to the background spectrum.

Sebacate esters, modified by the addition of an antioxidant, a rust inhibitor, a V.I. improver, and extreme pressure additive, have been formulated for lubricating high-speed turbomolecular pump bearings. PFPE fluids have been used in many turbo pumps. The manufacturer should be consulted to determine if PFPE is compatible with a specific pump and, if so, what viscosity is required.

17.1.3.3 Diffusion Pump Fluids

The ideal diffusion pump fluid should be stable, should have a low vapor pressure, low specific heat, and a low heat of vaporization, and should be safe to handle, dispose of and use. It should not thermally decompose, entrap gas, or react with its surroundings. Unfortunately, no such fluid exists. The properties of the distilled hydrocarbons and synthetic fluids currently used in diffusion pumps are given in Table 17.2. diffusion pump fluid vapor pressures are given in Appendix F.3.

The ultimate pressure obtainable with a mineral oil is limited by its decomposition on heating. Several "diffusion pump" grades are manufactured for different applications. Lighter "diffusion pump" cuts (average M.W. 300–450) are used in applications where high pumping speed, moderate pressure, and low cost are most important. These fluids are sometimes used in mechanical pumps. Heavier "diffusion pump" grades with an average (M.W. >550) are used to reach low ultimate pressures with reduced backstreaming. Inhibited diffusion pump fluids with improved oxidation resistance are used for vacuum metallizing. All mineral oils oxidize when exposed to air while they are hot.

Table 17.2 Properties of Representative Diffusion Pump Fluids

Chemical Type	Representative Trade Name	MW (ave)	Sp.Gr at 25°C	P_v at 25°C (Pa)	Fire Point (°C)	Latent Heat (J/g)	C_p (J/g/°C)	T_{boiler} 100 Pa (°C)
Mineral Oil								
	Apiezon® C[a]	574	0.87	1×10^{-6}	293	217.6	1.92	269
	Balzers-71[b]	280	0.88	3×10^{-6}	325	—	1.88	180
	Invoil® 20[c]	450	0.88	5×10^{-5}	259	170	1.88	210
Silicone[d]								
Trisiloxane	Dow Corning 704[e]	484	1.07	3×10^{-6}	275	220.5	1.72	220
	Dow Corning 705[e]	546	1.09	4×10^{-8}	275	215.9	1.76	250
Ether								
PPE	Santovac® 5[f,g]	447	1.2	6×10^{8}	350	205.8	1.84[h]	275
Fluoro-carbon								
PFPE[i]	Fomblin® Y-HVAC 25/9[j]	3400	1.90	3×10^{-7}	None	29.3	1	230
	Krytox® 1625[k]	3700	1.88	3×10^{-7}	None	41.8	1	230

[a] Edwards High Vacuum, Grand Island, NY.
[b] Balzers High Vacuum, Fürstentum, Liechtenstein.
[c] IVACO, Inc., Churchville, NY.
[d] Not recommended for use where electron beams could cause polymerization.
[e] Dow Corning, Inc., Midland, MI.
[f] Monsanto, Inc., St. Louis, MO.
[g] Excellent oxidation resistance.
[h] 4-ring either.
[i] Suitable for use where electron beams could cause polymerization.
[j] Montedison USA, New York, NY.
[k] du Pont and Co., Chemicals and Pigments Department, Wilmington, DE.
Source: Adapted from O'Hanlon [31].

Tetraphenyl silicone is extensively used in quick-cycled, unbaked systems because of its moderate cost, low backstreaming, thermal stability, and oxidation resistance. It has a freezing point of 20.5°C. Occasionally, it will be found frozen in an unheated storage area. Pentaphenyl silicone has improved stability and reduced vapor pressure and is used widely in systems that are baked to achieve the lowest ultimate pressures. The ultimate pressure in a diffusion pump charged with tetraphenyl silicone is dependent on purity, specifically, the quantity of low molecular weight impurity [33]. Pentaphenyl silicone is rapidly degraded by BCl_3 [34] and slowly degraded by CF_4 and CCl_4. Pentaphenyl ether and pentaphenyl silicone have extremely low vapor pressures. Explosions could not be induced in DC-705 or pentaphenyl ether in a system pressurized to 1/2 atm [35] with pure oxygen. When using pentaphenyl ether, it is absolutely necessary to restrict the cooling water flow so that the ejector stage operates at a wall temperature of 45–50°C. The ejector stage should be warm in all pumps to achieve adequate fluid degassing. However, if the ejector is too cold in a PPE-charged pump, a large fraction of the very viscous fluid will remain on the ejector and inlet walls. Some workers use a slightly larger fluid charge than recommended by the manufacturer to compensate for the fluid that resides on cool walls. PPE is suitable for use in mass spectrometers, leak detectors, residual gas analyzers, electron microscopes, and electron beam mask generation systems, because it does not form an insulating film. It is also the most stable diffusion pump fluid available for pumping corrosive gases; however, Lewis acids will degrade PPE in diffusion pumps [34].

PFPE fluid is suitable for some diffusion pump applications. From activation energy measurements, decomposition in a diffusion pump at 250°C has been estimated to be 0.009% (Krytox) and 0.14% (Fomblin) in 10 years [34]. Fomblin thermally decomposes at the C—C bond to yield equal molar amounts of C_3F_6O, COF_2, and CF_3COF [28,36]. The decomposition of Krytox was reported to be essentially the same, but with less COF_2 [35]. PFPE decomposes on electron or ion impact into low-molecular-weight radicals and therefore does not form a film; this makes it useful for diffusion-pumped electron beam systems. Residual gas analysis has shown the presence of high-molecular-weight fractions up to $M/e = 240$ advising against their use in heavy ion acceleration systems where exchange processes are dangerous [37]. PFPE is extremely stable for pumping all the usual reactive gases such as oxygen except for Lewis acids and fluorinated solvents. Since the boiler operates at temperatures over 200°C, PFPE, a Lewis base, reacts and decomposes into toxic vapors. Under no circumstances should diffusion pumps charged with PFPE be used to pump on BCl_3, AlF_3, and such. Pearson et al. [37] report on diffusion pumps charged with Fomblin Y-H VAC 18/8 used for pumping HF and UF_6. After prolonged use at pressures of 3 Pa, solids of UO_2F_2 and UF_4 deposited, and a dark sulfur-containing colloidal suspension was observed.

Mineral oils and polyphenyl ethers have been shown to be the only fluids stable in pumping tritium [17]. Polyphenylethers were found to be the preferred diffusion pump fluid for pumping tritium, whereas mineral oil was the choice for mechanical and turbo pump systems. PFPE was not satisfactory because of the large quantity of corrosive radiolysis products (HF, F, and COF_2) formed while pumping tritium.

The interior of a pump must be thoroughly cleaned before changing fluids. Hydrocarbon oils are easier to remove than silicones, but severely contaminated pumps that use either fluid may be cleaned successively in decahydronaphthalene, acetone, and ethanol. If the pump is relatively clean, acetone and alcohol are usually adequate. Polyphenyl ether is soluble in trichloroethylene and in 1,1,1-trichloroethane, but the latter is less toxic than the former. Pumps charged with PFPE are cleaned with a fluorinated solvent such as trichlorotrifluoroethane or perfluorooctane [38]. If the pump fluid level is low, it is good practice to drain completely and refill rather than add fluid. During operation the light fractions of broad molecular weight fluid are selectively removed, and the fluid's viscosity increases slowly with time. Gas bursting may be observed for several days following cleaning and changing fluid. Have patience while waiting for a newly charged pump to reach its ultimate pressure.

Pump boiler power may have to be changed if the fluid has thermal properties significantly different from those of the fluid for which the pump was designed. Most American pumps are designed to operate with tetra-phenyl siloxane. The heat necessary to maintain the boiler temperature depends upon the heat capacity C, and latent heat of vaporization h_ν, of the new fluid and also depends upon pump heat losses. Calculating their respective heat requirements may allow one to estimate boiler power differences. The heat required to vaporize 1 mL of fluid is $\rho[C(T_{\text{boiler}}-T_{\text{wall}})+h\nu]$. Latent heats and heat capacities of representative fluids are given in Table 17.2.

17.1.4 Reclamation

Procedures such as settling, filtering, adsorption, and distillation to remove contamination from pump fluids during reclamation. Costly fluids can be economically purified, while the cost of reclaiming inexpensive mineral oil is about the same as that of new fluid. Before considering reclaiming, one is advised to consult a firm specializing in reclaiming to determine how fluids should be segregated for shipment and how to specify the quality of the purified fluid. Cleaning procedures vary with the fluid and the type of contamination. Their cost will be dependent on degree of contamination, the fluid type, and the desired quality of the purified fluid. For example, very low vapor pressure color centers are costly to remove from a silicone fluid, but do not affect pump operation. Therefore, color specification increases reclaim cost. Technology and economics of pump fluid reclamation has been reviewed by Whitman [39].

17.2 Lubricants

Mechanical pumps with gear drives, bearings, seals, and vacuum system parts such as feedthroughs and bolts require lubrication. Moving parts may be lubricated in vacuum with liquid, grease, dry film, or, in certain cases, no lubricant—only dissimilar materials.

Lubricating systems have three components: the materials to be lubricated, the lubricant, and the environment [21]. These three components are interdependent, because surface adhesion and lubrication are not only characteristic of the material and the lubricant, but also interact with gas. Vacuum lubrication is unique because gas has been removed. Lubricants prevent moving surfaces from contact. Moving surfaces generate friction and produce heat, whereas wear destroys material and produces debris. In atmosphere, wear, not friction, is responsible for most problems. In a vacuum, friction becomes a problem as convection cooling decreases, lubricant evaporation rates increase, and reduced oxidation of metals hastens the onset of cold welding and alters the size distribution of wear particles [40,41].

Early papers have reviewed lubrication processes in air or vacuum [42–45]. In this section we review fluid rheology and techniques for vacuum lubrication.

17.2.1 Lubricant Properties

Fluid flow is a physical property of a lubricant that determines its usefulness in vacuum and atmospheric applications.

Full film, elastohydrodynamic, and boundary lubrication are three forms of lubrication that require fluids with different properties. The ratio of the lubricant film thickness to the surface roughness is important. Lubricant thickness depends on the absolute viscosity, η, the relative surface velocity, U, and the load, L. Figure 17.2, the Stribeck curve, relates these variables to the coefficient of friction f.

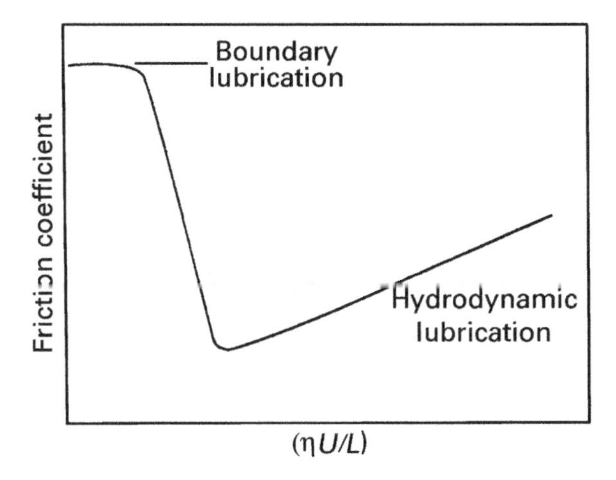

Fig. 17.2 Dependence of friction on viscosity, η, relative velocity, U, and load L.

A combination of high absolute viscosity, high relative speed, and low loading results in an oil film whose thickness is much greater than the roughness of either surface. The rapidly moving load rides on an oil wedge. This is called hydrodynamic or full-film lubrication. Viscosity is the most important lubricant property in this regime. Friction can be minimized for a given load or viscosity by the appropriate choice of viscosity.

At some value of low speed, low viscosity, and high loading, irregularities on the two surfaces will contact. This is the boundary lubrication region [41], which can be encountered in pump bearings and vanes during starting and stopping and between slowly sliding surfaces. The relation between friction coefficient, f and surface irregularities is [44]

$$f = s/p \tag{17.5}$$

s is the shear strength, and p is the yield pressure of the metal. Both quantities are related to the material's structure. The coefficient of friction can be reduced by decreasing the shear stress, increasing the yield pressure, or increasing the area over which the force is distributed. In the boundary region, the coefficient of friction is determined not only by the manner in which the adjacent surface peaks contact, but by the additives that affect the chemistry of these contacting surfaces.

The elastohydrodynamic region lies in between these two regimes. In this region oil undergoes a tremendous pressure increase. Its viscosity increases exponentially, and it behaves like a low-shear solid that can deform surface irregularities without contact.

In the following sections, we discuss absolute viscosity, kinematic viscosity, viscosity index, and techniques for their measurement.

17.2.1.1 Absolute Viscosity

The definition of viscosity was given in (2.21) and is repeated here in a different form.

$$\tau = \eta s' \tag{17.6}$$

Here t is the shear stress and s' is the rate of shear. The viscosity of a Newtonian fluid is independent of shear rate. Not all fluids behave in this manner. The viscosity of emulsions like hand cream, grease, and some synthetic fluids decreases with increasing shear rate. They are one type of non-Newtonian fluid. The decrease is small for the synthetic fluids we use, thus η will be considered a constant. The viscosity of grease decreases rapidly with shear rate until it approaches the viscosity of the oil from which it is formulated.

The viscosity of a liquid is predominantly a result of cohesive forces between the molecules. Since cohesion decreases with temperature, the viscosity of a liquid decreases on heating. In SI, dynamic viscosity, η has units of Pa-s. One mPa-s = 1 cP—a handy conversion to remember since the useful literature is tabulated in c.g.s. units. Research instruments measure absolute viscosity of liquids by timing the flow through a long capillary tube under constant head pressure. Such instruments are required for ASTM and ISO standards measurements. The volume flow rate through the tube (m^3/s) is related to the viscosity by the Poiseuille equation:

$$\frac{V}{t}=\frac{\pi d^4 \Delta P}{128\eta l} \quad (17.7)$$

ΔP *is* the pressure drop in the tube, d *and* l are the tube diameter and length, respectively. This equation neglects end effects and is reasonably accurate for long tubes. Engineering instruments attempt to measure absolute viscosity by measuring the torque on an immersed rotating spindle or a rotating cone adjacent to a flat plate. Because the flow is not laminar for low-viscosity liquids, corrections are needed to obtain absolute viscosity [45].

17.2.1.2 Kinematic Viscosity

Kinematic viscosity of a gas or liquid is absolute viscosity divided by density, $\nu = \eta/\rho$. In the c.g.s system, kinematic viscosity is expressed in units of Stokes. One Stoke = 1 cm^2/s. SI units are m^2/s, but data are usually labeled in mm^2/s because 1 cS = 1 mm^2/s. Kinematic viscosity, like diffusivity and permeability, is a *transport* property; observe that all three quantities have dimensions of L^2/T. Kinematic viscosity is measured directly in research instruments by timing the (Poiseuille) flow of a liquid through a long capillary under its own head [46]. Engineering instruments use short tubes or orifices in which the flow is not always laminar, so it is necessary to tabulate factors by which their readings can be converted to kinematic viscosity. Saybolt and Redwood instruments measure the times required for a known quantity of oil with a falling head to flow through short tubes of specific dimensions. The Engler instrument measures the ratio of the times required by equal volumes of oil and water to flow through a particular tube. The Saybolt (American), Redwood (English), and Engler (German) are the most common engineering instruments used for measuring the kinetic viscosity of oils. They are being replaced slowly by long tube instruments to meet SI standards. Appendix F.6 gives the factors for converting between various units.

The variation of kinetic viscosity with temperature for petroleum oils empirically fits the equation:

$$\log\left(\left(\nu+0.7\right)\right)=A+B\log T \quad (17.8)$$

Measurement of ν at two temperatures, 40 and 100°C, is adequate for interpolation and extrapolation down to the pour point when plotted according to Eq. (17.8). This curve also fits most synthetics except chlorosiloxanes, which show curvature when plotted in this empirical way.

17.2.1.3 Viscosity Index

V.I. is an empirical way of classifying how kinematic viscosity varies with temperature. It is an arbitrary *and* purely historical scheme. It recognizes the fact that the (high paraffin content) oils from the U.S. State of Pennsylvania have a uniformly *low* change in viscosity with temperature. These oils were arbitrarily assigned a V.I. = 100. Oils from the U.S. Gulf coast have a high naphthene content, and the greatest viscosity slope. They were assigned a V.I. = 0. V.I. is calculated from the 40 and 100°C viscosity values of the unknown oil, labeled U, and the tabulated viscosity values of oils of index 0, labeled L, and of index 100, labeled H. See Fig. 17.3. The unknown and the two standards must have the same viscosity at 100°C. The formulas for calculating V.I. are not given because they cannot be used without the tables in the ASTM standard [32]. Alternatively, V.I. may be obtained graphically from Appendix F.5. This plot was generated from the ASTM data. It greatly simplifies the procedure and yields sufficiently accurate results. The V.I. system was originally intended for use with mineral oils and was modified for use with synthetics. Some synthetics have a negative index, whereas other synthetics have indexes >100%. Lubricant fluids are described by three parameters: ν(40 °C), ν(100 °C), and V.I. However, only *two* of the three are needed for a complete description of an oil, as the three are interrelated. There is a practical use for this information: Resellers may choose one of three different pairs to disguise the fact that two differently

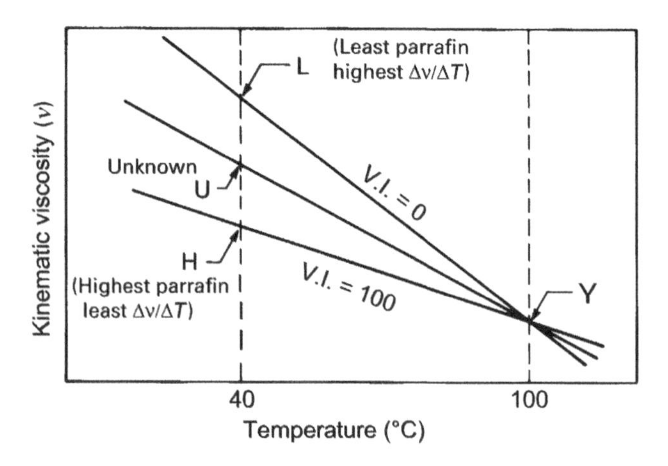

Fig. 17.3 Oil viscosity index.

branded fluids are identical. For example, a fluid may described by its {ν (100°C) and V.I.}, or its {ν (40°C) and ν (100°C)}, or its {V.I. and ν (40°C)}.

17.2.2 Selecting Lubricants

We can lubricate moving surfaces with liquids, such as petroleum oils, synthetic fluids, or greases, or we can choose from a variety of solid films such as silver, or molybdenum disulfide. The use of these materials in vacuum gives different results than when they are used in air. Each material has its own problems when ambient gas is removed.

17.2.2.1 Liquid

Petroleum and synthetic fluids are used to lubricate vacuum pumps and moving surfaces in vacuum. In the full-film region, fluid fills the spaces between moving surfaces and keeps them apart. The viscosity should be low enough to allow rotating or sliding motion at the lowest ambient temperature and remain sufficiently high at operating temperature (high V.I.). The viscosity required for full-film lubrication depends on the operational temperature range and the load. From Fig. 17.2 we see that the ratio of η/L should remain a constant. Therefore, high-viscosity fluids are required for heavy loads. Within a class of fluids, vapor pressure is usually inversely proportional to viscosity. Lubrication in the full-film regime is dependent on the ability of the fluid to adhere to the surface and form a film of adequate shear strength, which will support the bearing load.

Other necessary attributes are adhesion, stability, and heat conductivity. The oil must wet the surface and form a film, which is stable under high shear rate. Shear generates heat, which must be removed to prevent degradation and further viscosity decrease.

Oil adhesion is determined by the strength of the liquid–metal bond. Liquids with unsaturated bonds strongly adhere to or wet metal surfaces. Cohesion is a measure of intermolecular attraction. Oils that are very cohesive do not disperse or spread out rapidly on the surface of a metal. Certain highly cohesive oils such as polyphenyl ether adhere to metals strongly and also form a barrier that prevents creep. The outermost molecules align themselves in such a way that the exposed groups have a low liquid–metal adhesion. These fluids form a high contact angle with a metal surface and are termed autophobic. Fluids with good adhesion that do not have a resistance to creep will flow and wet a metal surface. Hydrocarbon oils have this property. Except for a fluid like penta-phenyltetramethyl trisiloxane, silicone fluids have high creep rates.

Sliding and rolling surfaces are easily lubricated with liquids for long lifetimes. The simplest liquid lubrication system is a wiped fluid coating. More complex systems can be designed to provide a continuous supply by migration or evaporation from a reservoir. If the fluid has been sufficiently outgassed, it can be used up to the temperature at which the vapor pressure of the fluid is intolerable. Vapor pressures of common vacuum pump fluids are given in Appendices F.2 and F.3. Polyphenyl ether is a common choice for lubricating hand-operated sliding and rotating surfaces in vacuum. The mechanisms used to supply the lubricant can also cause organic contamination of the vacuum chamber. Contamination can be reduced by choosing oil of sufficiently low vapor pressure or by enclosing the lubricated area with a shroud. A creep barrier will prevent lubricant from creeping along the shaft where it passes through the shroud. Polytetrafluoroethylene (PTFE) is a hydrocarbon creep barrier, while nylon is a fluorocarbon creep barrier. It may be necessary to cryogenically cool the shroud in the area near the region where the sliding or rotating shaft enters the chamber.

Rolling friction is encountered with ball or roller bearings. A rolling surface looks much like a sliding surface when viewed microscopically, and the effects on viscosity, adhesion, and heat conduction are the same. High-speed ball bearings such as those found on turbomolecular pumps require oil of lower viscosity than is used on slow-speed bearings.

Boundary lubrication requires fluids with unique properties, which are usually attained with additives. Sulfur, chlorine, and lead make effective boundary lubricants for steel-on-steel because they form low-shear-stress compounds that are wiped or etched from the high spots. These chemicals function as etchants. Phosphorus functions by forming iron phosphide at hot spots that plastically flows and fills in the asperities. Plastic deformation, like etching, causes the load to be redistributed over an increased area [47]. In both cases the shear strength is reduced, the yield pressure is increased, and the friction is reduced according to Eq. (17.5).

Oiliness, or lubricity, is an important property of boundary lubricants. It is an imprecise term and cannot be defined quantitatively. It refers to the ability of polar molecules to align themselves in double layers that slide easily over adjacent double layers. Animal fats have a high degree of oiliness, but they do not make acceptable hydrodynamic or boundary lubricants because they are not chemically or thermally stable. Some esters are stable and are used as oiliness additives.

17.2.2.2 Grease

Greases are made from a heavy petroleum distillate or, more commonly, a thickened petroleum or synthetic liquid. Petroleum oil can be vacuum distilled to yield wax-like, high-molecular-weight grease (petrolatum) which is uniform in composition. Grease also can be made by the process of gelling or thickening a liquid. Various compounds such as clay, esters, metal soaps, and powdered PTFE are used to increase the viscosity of oil to the consistency of grease. The oil is either physically entrapped in the thickener, adsorbed on it, or held in place by capillary action [21]. The typical starting material for a thickened grease is an oil whose

Table 17.3 Properties of Representative Vacuum Greases

Trade Name	Chemical Type	Temperature Range (°C)	P_V at 25 °C (Pa)	Sp. Gr. at (25 °C)	Penetration (mm)[a]
Celvacene[b] Light	Ester-thickened hydrocarbon	−40–90	10^{-4}	—	150
Apiezon[c] AP-100 Grease	PTFE-thickened hydrocarbon	10–30	10^{-8}	1.042	—
Apiezon[c] L Grease	Distilled hydrocarbon	10–30	10^{-8}	0.896	—
Dow Corning[d] High Vacuum	PTFE-thickened silicone	−40–260	10^{-7}	1.0	<260
Krytox[e] LVP L-10	PTFE-thickened fluorocarbon	−20–200	10^{-13}	1.94	280

[a] ASTM-D-21.
[b] Inland Vacuum Company, Churchville, NY.
[c] Edwards High Vacuum, Grand Island, NY.
[d] Dow Corning Co., Midland, MI.
[e] du Pont and Co., Chemicals and Pigments Department, Wilmington, DE.

room-temperature viscosity is about 500 mm^2/s. The room-temperature viscosity of a thickened grease is much higher than the base oil at low shear rates, but it drops by a factor of 30 at high shear rates.

Greases are characterized by their chemical type, method of thickening, vapor pressure, service temperature range, reactivity, and consistency. Characteristics of representative greases are shown in Table 17.3. The vapor pressure of grease increases with temperature, according to the properties of the oil and the thickener. Laurenson [48,49] has shown the vacuum performance of grease to be dependent on the vapor pressure of the grease and its method of manufacture. He demonstrated that the evaporation rate and quantity of a gelled grease made from hydrocarbon oil or silicone fluid of moderate vapor pressure depended on the total mass of the grease in vacuum, while the evaporation rate of a molecularly distilled grease depended only on the surface area. He attributed this to the fact that the molecularly distilled grease evaporated only from the surface, while the surface oil molecules of a filled grease evaporated from the large effective surface area of the gel. These molecules were replaced by migration from the bulk, as oil feeds a lamp wick. He could not experimentally observe the mass dependence of evaporation rate of filled greases made from very-low-vapor-pressure silicones or PFPE.

The ASTM penetration test [50] measures the depth in millimeters that a particular cone will penetrate grease in a known time. This is a measure of grease consistency. No. 2 grease with a penetration of 265–295 mm is typically used for low-speed bearings. Soft greases are needed for high-speed ball bearings.

Grease lubrication is used for long-life, low-maintenance applications. Grease is often used in low-speed ball bearings, rotary feedthroughs, and has been used in turbomolecular pump bearings [51]. Hydrocarbon grease has a characteristically low maximum service temperature. It is rarely much above room temperature for distilled greases, but is higher for filled greases, if the gelling agent is properly chosen. PFPE and silicone fluids have high maximum service temperatures. They are typically formulated from a mixture of fluid and powdered PTFE. Silicone fluids are poor steel-on-steel lubricants, and greases made from them are also poor lubricants unless thickened with a lubricant like PTFE. If a large reservoir of grease is needed, a distilled hydrocarbon, or filled grease made from very-low-vapor-pressure silicone or PFPE fluid, is recommended [49]. If a thin film is adequate, gelled grease may be used. PFPE greases are preferred to silicones where electron bombardment can cause fragmentation. The reaction between greases and elastomers is similar to that for fluids and elastomers shown in Appendix F.1. The gelling agent may cause an additional reaction. Manufacturers can provide information on the reactivity of greases containing proprietary additives.

Bolt greases can be used around, but not in vacuum systems. Such greases are formulated for the extreme boundary lubrication region. They contain extreme-pressure additives such as silver, copper, lead, or molybdenum disulfide. Base oils and additives used in these greases may have high vapor pressures.

17.2.2.3 Solid Film

Equation 17.1 shows that friction can be reduced in the boundary region by increasing the contact area and reducing the shear force of the lubricant. The important property of a solid film, or dry lubricant, is vapor pressure. Low vapor pressure reduces system contamination and allows operation at high temperatures. Solid-film lubricants are useful when loads are extremely high, speeds are low, surface temperatures extreme, and designs are simplified for maintenance-free long-life applications. Dry lubrication is limited by its finite thickness and the debris generated by its sacrificial removal. Dry lubrication failure often results from a local defect and therefore the lifetime of a dry lubricated system is not as predictable as that of a liquid lubricated system in which the lubricant evaporates or migrates in a uniform manner [43].

The desirable properties of a dry lubricant are low vapor pressure, low shear strength, and good adhesion to the base metal. Many solids have been used as dry lubricants. Among them are graphite, sulfides, and selenides of molybdenum and tungsten, gold, silver, and PTFE. It was originally thought the lubricating ability of graphite and other layered solids resulted from their loosely bound layered structure. The weak binding between layers was thought to allow sliding with low shear. This theory was found wanting when it was discovered that graphite was not a good lubricant in vacuum or at high altitude, but only in the presence of water vapor [52]. MoS_2 is a good lubricant in vacuum [53]. The difference in the lubricating ability of these two materials is a result of their differing structure. Graphite consists of a layered structure with a high interlayer binding energy. Contamination from water vapor lowers the binding energy and allows motion with low shear. Molybdenum disulfide films are oriented layers of S–Mo–S with a low binding energy between the adjacent sulfur layers, so contamination is not necessary for low-shear sliding [54].

Molybdenum disulfide and tungsten disulfide are the most widely used solid vacuum lubricants. Farr [55] has reviewed its structure and properties as they apply to lubrication. Mattey [56] described how these and other solid films have been used in space hardware. MoS_2 has a very low vapor pressure and can be applied by many techniques including sputtering and spraying. MoS_2 has been deposited by dc sputtering [57] and rf sputtering [58]. Sputter-deposited films give satisfactory lubrication but does generate debris [58]. MoS_2–graphite–sodium silicate coatings on steel have been shown to give good performance and long life [59]. Wear rates $<3\times10^{-18}$ $m^3/(N\text{-}m)$ were observed for spray-gun-coated parts; the wear rates were 100 times larger for MoS_2 films applied in an aerosol. Friction and wear of MoS_2-lubricated steel surfaces has been studied [60]. In the pressure range 10^5–10^4 Pa the friction and the wear decreased, because of the removal of water vapor, and remained constant in the pressure range 10^4–10 Pa. In the pressure range 10–0.1 Pa the friction continued to decrease, while the wear increased greatly. The friction decrease was attributed to the removal of oxygen. The wear increased because heat could not be dissipated in vacuum. Below 0.1 Pa no further change in wear or friction was observed. O_2 increased the friction by oxidizing MoS_2 to MoO_3 and MoO_2 [60,61].

Soft metals will also lubricate sliding surfaces. Pair hardness, ductility, and redeposition were shown to be important in designing a low-friction, long-life system [62]. The best gear pair was a silver-plated aluminum gear running against a MoS_2–graphite-sodium silicate-coated steel gear. Gold plating stainless steel bolts has proven to be an effective technique for preventing the galling of threads following system baking.

PTFE coatings have also been used for vacuum applications. PTFE transfers and re-coats [42]. The long PTFE molecules are oriented parallel to the direction of sliding [63].

Almost anything placed between two moving surfaces will reduce friction and wear. The environment affects the adhesion of a lubricant, its evaporation, oxidation, and intercrystalline forces. If we remove all foreign materials, similar metals will instantly cold weld. However, the cold welding of adjacent surfaces can be eliminated with materials of dissimilar lattice constants, such as sapphire balls in stainless steel races [43]. The friction coefficient of this system is higher than a lubricated system.

References

1 Thomson, G.W. and Douslin, D.R. "*Techniques of Chemistry*", in *Physical Methods of Chemistry, Part 5*, A. Weissberger and B.W. Rossiter, Eds., Vol. 1, Wiley, New York, 1971, p. 74.
2 Knudsen, M., *Ann. Physik* **28** (75), 999 (1909); **29**, 179 (1909).
3 Deville, J.P., Holland, L., and Laurenson, L., *1965 Transactions of the 3rd International Congress on Vacuum Science and Technology, Stuttgart, Germany, July 1965*, H. Adam, Ed., Pergamon, New York, 1965, p. 153.
4 Hickman, K.C.D., in *Transactions of the 8th National Vacuum Symposium of the American Vacuum Society, 1961*, L.E. Preuss, Ed., Pergamon Press, New York, 1962, p. 307.
5 Laurenson, L. and Troup, P., *J. Vac. Sci. Technol. A* **8**, 2817 (1990).
6 Crawley, D.J., Tolmie, E.D., and Huntress, A.R., *Transactions of the 9th National Vacuum Symposium Transactions of the Ninth National Vacuum Symposium of the American Vacuum Society, Oct. 31–Nov. 2, 1962, Los Angeles*, G. Bancroft, Ed., Macmillan, New York, 1962, p. 399.
7 Rettinghaus, G. and Huber, W.K., *J. Vac. Sci. Technol.* **9**, 416 (1962).
8 Dennis, N.T.M., Colwell, B.H., Laurenson, L., and Newton, J.R.H., *Vacuum* **28**, 551 (1978).
9 Huntress, A.R., Smith, A.L., Power, B.D., and Dennis, N.T.M., in *Transactions of the 4th. National Symposium. on Vacuum Technology*, W.G. Matheson, Ed., Pergamon, New York, 1957, p. 104.
10 ASTM D2715-92, *Standard Test Method for Volatilization Rates of Lubricants in Vacuum*, American Society for Testing Materials, Philadelphia, PA, 2007.
11 ASTM D-156, *Annual Book of ASTM Standards, Part 23*, American Society for Testing Materials, Philadelphia, PA, 1981, p. 111.
12 Zuidema, H.H., *The Performance of Lubricating Oils*, Reinhold, New York, 1959, p. 30.
13 ASTM D-92, *Annual Book of ASTM Standards, Part 23*, American Society for Testing Materials, Philadelphia, PA, 1981, p. 33.
14 J.M. Kuchta, *Summary of Ignition Properties of Jet Fuels and Other Aircraft Combustible Fluids*, U.S. Bureau of Mines, Pittsburgh Mining and Safety Research Center, AFAPL-TR-75-70, Air Force Aero Propulsion Laboratory, Wright-Patterson Air Force Base, Ohio, Sept. 1975, p. 16.
15 Kendall, B.R.F., *J. Vac. Sci. Technol.* **21**, 886 (1982).
16 Fomblin Pump Oils, Product Brochure, Solvay S.A, Brussels, Belgium.
17 P. Chastagner, *Selection of Fluids for Tritium Pumping Systems, 13th Annual Symp. on Applied Vacuum Science and Technology, Clearwater Beach, Florida, Feb. 6, 1984, DuPont*, Aiken, S. Carolina, (1984).

18 Cameron, A. and McEttles, C.M., *Basic Lubrication Theory*, 3rd ed., Horwood, Chichester, 1980, p. 33.

19 Burch, C.R., *Proc. R. Soc. A* **123**, 271 (1929).

20 Davy, J.R., *Industrial High Vacuum*, Pitman, London, 1951, p. 109.

21 Booser, E.R. "Lubrication and Lubricants", in *Kirk–Othmer Encyclopedia of Chemical Technology*, 3rd ed., 14 ed., M. Grayson and D. Eckroth, Eds., Wiley, New York, 1980, p. 484.

22 Zuidema, H.H., *The Performance of Lubricating Oils*, Reinhold, New York, 1959, p. 177.

23 Kreuz, K.L., *Lubrication* **56**, 77 (1970).

24 Fulker, M.J., Baker, M.A., and Laurenson, L., *Vacuum* **19**, 555 (1969).

25 Smith, A.L. and Saylor, J.C., *Vac. Symp. Trans.* **31** (1954).

26 Mahoney, C.L. and Barnum, E.R. "Polyphenol Ethers", in *Synthetic Lubricants*, R.C. Gunderson and A.W. Hart, Eds., Reinhold, New York, 1962, p. 402.

27 Holland, L., Laurenson, L., and Baker, P.N., *Vacuum* **22**, 315 (1972).

28 Sianisi, D., Zambeni, V., Fontanelli, R., and Binaghi, M., *Wear* **18**, 85 (1971).

29 Du Pont and Co., Krytox Pump Fluids, Product Brochure, DuPont and Co., Wilmington, Delaware.

30 N.D. Lawson, Perfluoroalkylpolyethers, Report No. A-70020, Du Pont, Wilmington, DE, Feb. 1970.

31 O'Hanlon, J.F., *J. Vac. Sci. Technol., A* **2**, 174 (1984).

32 ASTM D-2270, *Annual Book of ASTM Standards, Part 24*, American Society for Testing Materials, Philadelphia, PA, 1981, p. 277.

33 D. Petraitis, *Society of Vacuum Coaters*, 24th Annual Technical Conf., Dearborn, MI, 1981, page 73.

34 H.W. Lehmann, E. Heeb, and K. Frick, *Proceedings of the Third Symposium. on Plasma Processing*, **82**, No. 6, J. Dieleman, R.G. Frieser and G.S. Mathad, Ed, The Electrochemical Society, Pennington, NJ, (1982), p. 364.

35 Solbrig, C.W. and Jamison, W.E., *J. Vac. Sci. Technol.* **2**, 228 (1965).

36 R.K. Pearson, J.A. Happe, G.W. Barton, Jr., LLL Report UCID-19571, Lawrence Livermore National Laboratory, Livermore, CA, *Sept.* 27, 1982.

37 Luches, A. and Perrone, M.R., *J. Vac. Sci. Technol.* **13**, 1097 (1976).

38 L. Laurenson, *Ind. Res. Dev.*, Technical Publishing, New York, Nov. 1977, p. 61.

39 Whitman, C.B., *J. Vac. Sci. Technol., B* **5**, 255 (1987).

40 Miller, R., Cooper, D.W., Nagaraj, H.S., Owens, B.L., Peters, M.H., Wolfe, H.L., and Wu, J.J., *J. Vac. Sci. Technol., A* **6**, 2097–2102 (1988).

41 Tipei, N., *Theory of Lubrication*, Stanford University Press, 1962, p. 11.

42 Buckley, D.H., *Proc. 6th Intl. Vac. Congr. 1974, Jpn. J. Appl. Phys., Suppl. 2, Pt. 1*, 297, 1974.

43 Friebel, V.R. and Hinricks, J.T., *J. Vac. Sci. Technol.* **12**, 551 (1975).

44 Bowden, F.D. and Tabor, D., *The Friction and Lubrication of Solids, Part II*, Clarendon Press, 1964.

45 Van Wazer, J.R., Lyons, J.W., Kim, K.Y., and Colwell, R.E., *Viscosity and Flow Measurement*, Interscience, New York, 1963.

46 J.F. Swindells, R. Ullman and H. Mark, in *Technique of Organic Chemistry*, 3rd ed., A. Weissberger, ed., Vol. 1, Physical Methods of Organic Chemistry, Part 4, Interscience, New York, 1959, p. 689.

47 R.E. Hatton in: *Synthetic Lubricants*, R.C. Gunderson and A.W. Hart, Eds., Reinhold, New York, 1962, p. 402.

48 Laurenson, L., *Vacuum* **27**, 431 (1977).

49 Laurenson, L., *Vacuum* **30**, 275 (1980).

50 ASTM D-217, 1981, *Annual Book of ASTM Standards, Part 23*, American Society for Testing Materials, Philadelphia, 1981.

51 Osterstrom, G. and Knecht, T., *J. Vac. Sci. Technol.* **16**, 746 (1979).

52 Savage, R.H., *J. Appl. Phys.* **19**, 1 (1948).

53 Haltner, J., *Wear* **7**, 102 (1964).

54 Bryant, P.J., Gutshall, P.L., and Taylor, L.H., *Wear* **7**, 118 (1964).

55 Farr, J.P.G., *Wear* **35**, 1 (1975).

56 Mattey, R.A., *Lubr. Eng. (ASLE)* **34**, 79 (1978).

57 T. Spavins and J.S. Przybyszewski, *Deposition of Sputtered Molybdenum Disulfide Films and Friction Characteristics of Such Films in Vacuum*, NASA TDN-4269, December 1969.

58 Vest, C.E., *Lubr. Engng. (ASLE)* **34**, 31 (1978).

59 A. Thomas, The Friction and Wear Properties of Some Proprietary Molybdenum Disulphide Spray Lubricants in Sliding Contact with Steel. Report No. ESA CR(P) 1537, European Space Tribology Laboratory, Risley, England, January 1982.

60 Kurilov, G.V., *Sov. Mater. Sci.* **15**, 381 (1979).

61 Matsunaga, M. and Nakagawa, T., *Trans. ASLE* **19**, 216 (1976).

62 Kirby, R.E., Collet, G.J., and Garwin, E.L., in *Proc. 8th International. Vacuum. Congress, Vol. II*, J.P. Langeron and L. Maurice, Eds., IUVSTA, Cannes, France, 1980, p. 437.

63 Tabor, D., in *Microscopic Aspects of Adhesion and Lubrication*, J.M. Georges, Ed., *Tribology Series*, Vol. 7, Elsevier, Amsterdam, 1982.

Part V

Systems

Two-cylinder, piston-type, mechanical vacuum pump, made by Ferdinand Ducretet and Ernst Roger, Paris, 1920–1930. Donated to AVS National Office (New York, NY) by Mr. and Mrs. Donald M. Mattox. Permission from AVS, the Science and Technology Society of Materials, Interfaces and Processing.

One hundred years ago, a basic vacuum system, like that illustrated above, consisted of a pump, valve, gauge, and glass chamber. Today's systems incorporate these components, but they look different, are far more complex, and are constructed from materials that were not known at the time.

The following group of seven chapters focuses on three types of vacuum systems used in low and high technology industries: systems that will pump to the high

A Users Guide to Vacuum Technology, Fourth Edition. John F. O'Hanlon and Timothy A. Gessert.
© 2024 John Wiley & Sons, Inc. Published 2024 by John Wiley & Sons, Inc.

and ultrahigh vacuum ranges, as well as systems that will pump a large gas flow in the medium and low vacuum ranges. One way of categorizing systems is by their use, and how function emphasizes different facets of the technology.

Rough vacuum systems, described in Chapter 18, are used in low technology applications as well as for load locks in medium and high technology applications. One of the most challenging tasks facing vacuum system operator is the reduction of water vapor levels in the vacuum chamber. This is discussed here in some detail. The single chamber high vacuum systems discussed in Chapter 19 involve large process gas loads. Large gas loads, along with the need for line-of-sight motion from hearth to substrate, influence pump choices and chamber designs.

The ultraclean vacuum systems described in Chapter 20 are, for all practical purposes, ultrahigh vacuum systems whose processes operate in medium and high vacuum range. The achievement of ultraclean vacuum begins with the careful selection of materials, joining and cleaning techniques, and pumping technology. However, ultraclean vacuum cannot be sustained unless systems are maintained using ultraclean operational practices. Ultraclean vacuum systems are now commonly used in the manufacture of items such as densely populated, defect-free semiconductor chips, magnetic recording heads, and ultrathin laser diodes.

A discussion of ultraclean manufacturing would not be complete without including a detailed description of the environment in which the system is located. Entire sectors of modern high technology manufacturing operate within cleanrooms. Not only must all the liquids, gases, and solid materials used within these systems be contaminant-free, but so must the air in the facility; quality must be controlled in every facet of the process. A thorough discussion of contamination control is provided in Chapter 21, with the caveat that some mechanisms that generate or remediate contamination are discussed throughout the text, where they specifically relate to a material, gauge, pump, or procedure.

The systems described in Chapter 22 are used when high gas flows are required, such as those found in sputtering and sputter etching. Because the gas flow quantities and operating pressures are outside the normal high vacuum system operating range, flows are throttled or, in some cases, high vacuum pumps are replaced by mechanical pumps. These systems, along with some ultraclean systems, often form one of the modules in the multichambered systems described in Chapter 23. High-volume production processes, which require near-ultrahigh vacuum conditions, are performed in complex, multichambered systems. These systems are designed to maintain cleanliness during critical portions of a complex manufacturing process, while at the same time, allowing rapid product processing and transfer between steps.

Chapter 24 reviews the elements of leak detection. The leak detector was developed as a metrology tool to find leaks during construction and operation of components and systems. Today, it is an integral component of modern manufacturing plants and in research and development laboratories, where it has been cleverly adapted for efficient use.

18

Rough Vacuum Pumping

The initial vacuum pumping phase—atmosphere to crossover—is not a frequently studied subject; however, it is extremely important. If it is not done properly, it can affect system performance and product quality. No one initial pumping recipe fits all situations. The rough pumping time of a large accelerator is a short portion of its time under vacuum, but a significant part of its time for an architectural glass coater's cycle. If pumped too slowly, product throughput will be reduced; if too rapidly, water aerosols will entrap particles. If the process chamber is not cleaned, debris will accumulate and extend pumping time. Pumping to an unacceptably low pressure will generate backstreaming contamination from oil-sealed mechanical pumps, whereas crossover at an unacceptably high pressure will generate contamination from overloaded high vacuum pumps. How fast should a system be evacuated? and to what pressure? These and other issues must be considered carefully when specifying roughing pumps, configurations, and pumping cycles for large systems.

Generations ago, when systems were small and of similar size, these issues were not a concern; a one-size-fit-all solution worked when chambers, pipes, and pumps were similar. That history no longer applies.

Large system pumping time depends on pump types and pipe sizes, which in turn, affect system cost, return on investment, product cycle time, and quality. Crossover pressure is a variable that can be adjusted by the operator. Most importantly, it is a variable that can be changed easily and without consideration of its consequences. We analyze two topics in this chapter: rough pumping rate, and crossover pressure range. We show that these depend on system design and objectives.

A Users Guide to Vacuum Technology, Fourth Edition. John F. O'Hanlon and Timothy A. Gessert.
© 2024 John Wiley & Sons, Inc. Published 2024 by John Wiley & Sons, Inc.

18.1 Exhaust Rate

Economics and product quality are the two issues that determine the speed at which atmosphere is pumped. In general, we wish to manufacture quality product, quickly, with low capital investment. These requirements can conflict. By understanding important vacuum concepts, we can develop best operational practices.

18.1.1 Pump Size

Pumping from an initial pressure P_{init}, typically atmosphere, to crossover can be modeled by a chamber of volume V connected to a pump of speed S_p, through a conductance C, as depicted in Fig. 18.1. The chamber is assumed to have an internal outgassing of $Q/S = P_{ult}$ and remain approximately constant during the rough pumping cycle. The time dependence of the chamber pressure is then described by:

$$P_c(t) = (P_{init} - P_{ult})e^{-\frac{S_c t}{V}} + P_{ult} \quad (18.1)$$

The pumping speed at the chamber exit that is required to reach a desired chamber pressure P_c is dependent on the system volume V, the initial pressure P_{init}, the outgassing rate, and the time specified by the process or customer:

$$S_c = -\frac{V}{t}\ln\left(\frac{P_c - P_{ult}}{P_{init} - P_{ult}}\right) \quad (18.2)$$

Note that the assumed short-term ultimate pressure P_{ult} is the quotient of the outgassing rate and chamber speed that is being calculated $P_{ult} = Q_{og}/S_c$. One must

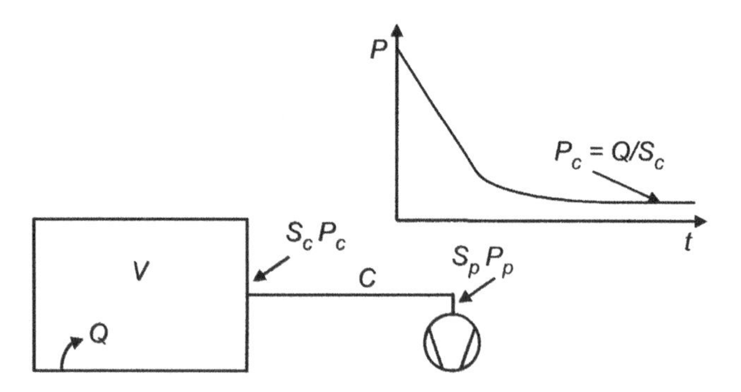

Fig. 18.1 Model for calculating the initial pumping time of a vacuum chamber from atmosphere to crossover. The rough pumping speed used to predict the pressure decrease is the speed measured at the chamber entrance.

either estimate this correctly or iterate to obtain the solution. For empty chambers, this will be a small quantity; however, when process chambers contain large gas loads, e.g., large quantity of paper or plastic, or an accumulation of debris from previous cycles, that term is significant. We desire to know the speed at the pump, and it can be obtained from:

$$\frac{1}{S_c} = \frac{1}{S_p} + \frac{1}{C} \tag{18.3}$$

Assuming we are pumping from atmosphere, the flow is viscous, and therefore C is pressure dependent. Because the conductance is pressure dependent, it will be large at atmospheric pressure and will decrease as the final pressure is reached. In practice, this means the pumping speed used in (18.1) is pressure-dependent and numerical methods are required to solve the problem exactly. One can obtain a simple approximate solution by assuming the conductance C to have a constant value—its smallest value—the value at the lowest pressure of interest. Using this value of conductance in (18.3) yields a *worst-case design* or the largest value of pump speed required. The actual exhaust time will be less than that calculated with (18.2) and (18.3). The speed of an oil-sealed vane pump can be considered to be constant for all practical purposes, as one does not use an oil-sealed pump in its compression-limiting region where the speed drops to zero. Large systems will have compound roughing pumps consisting of perhaps one or more vane pumps and one or more lobe blowers. Small systems may use a dry piston, scroll, screw, claw and lobe, or vane pump to rough pump from one atmosphere. The choice of pump depends on the desired system base pressure and product technology cleanliness level.

Speed of roughing is also a consideration. If the pumping time cannot be met economically or quickly with a normal roughing system, an alternative solution can be considered. One such special-purpose roughing system (for coating large area glass plates), uses one or two sequentially operated pressure dividing tanks configured to remove most of the atmospheric gas in an extremely short time. This design is analogous to that described in Section 14.4.1 for multiple sorption pumps. Pressure-division pumping makes use of Boyle's Law to exhaust a chamber quickly by suddenly opening a valve to a previously pumped ballast chamber. The chamber pressure drops suddenly by pressure division. After the sudden drop, the ballast tank is again isolated from the chamber and repumped slowly by a small pump during processing, so it is ready for the next roughing cycle. Two independent ballast tanks can be used if needed. Because particle-generating water aerosols are formed during rapid pumping, this technique cannot be used in processes in which particle formation is important; see the following section.

Load locks are an alternative way of reducing pumping time and maintaining clean process chambers. There are four elements of the load lock cycle. Product is first loaded in the lock; next, the lock is pumped. Third, it is transferred to the adjacent process or buffer chamber. And last, the isolated lock is vented to atmosphere. Load–unload locks and buffer chambers are now an integral part of clean and efficient system designs.

18.1.2 Aerosol Formation

Water aerosols are formed within a chamber when the pressure is reduced. Rapidly decreasing pressure causes water droplets to form because the temperature decreases. Supersaturated water vapor quickly condenses into droplets that grow. This happens most easily on micrometer- or nanometer-sized dust particles. A numerical simulation showed that water droplets grew to 45-μm diameter in 25 ms at 0.1 atm and to 10 μm in 25 ms at a 1 atm [1]. Growth ceases when the water vapor source becomes depleted.

In practice, rapid pumping produces large droplets that fall by gravity and collide with surfaces. After landing on a surface, the water droplet evaporates and leaves only its nucleus, consisting of dust or atmospheric materials, such as ammonium or sulfur salts [2]. For many years, rapid pumping was never identified as the source of contamination because the water evaporated so rapidly after arriving on a surface.

One may observe these transient aerosol clouds easily in a darkened room by rapidly opening the roughing valve while illuminating the chamber with a flashlight. This phenomenon has also been observed during crossover from roughing to high vacuum, but in much reduced concentration.

During *isothermal expansion*—equal temperature expansion—gas temperature does not change, because heat flow from the chamber walls by gas collisions keeps the temperature constant. During *adiabatic expansion*—expansion with no heat transfer—there is no heat flow from chamber walls to the gas, therefore, gas temperature must drop, because it does work on its surroundings. Over sixty years ago, operators of large space simulation chambers understood the gas did work on its surroundings, lost energy, and cooled [3]. They found that the gas in the center of an ideal large chamber could be as low as −115°C. In smaller systems, decreases to about −80°C have been observed. The temperature decrease in real systems depends on design. It can be near adiabatic, or near isothermal, or somewhere between. In all cases, the temperature will recover to ambient as the system comes to thermal equilibrium. A system with a spherical or cubical shape—that is, a system with a small surface-to-volume ratio, and a big pump—will behave more like an adiabatic system; its temperature will drop sharply and take a long time to return to ambient. A system containing many interior walls, with small, local volumes (large surface-to-volume ratio) and a small pump, will transfer heat rapidly to the gas during wall collisions; its gas temperature will drop only slightly and recover

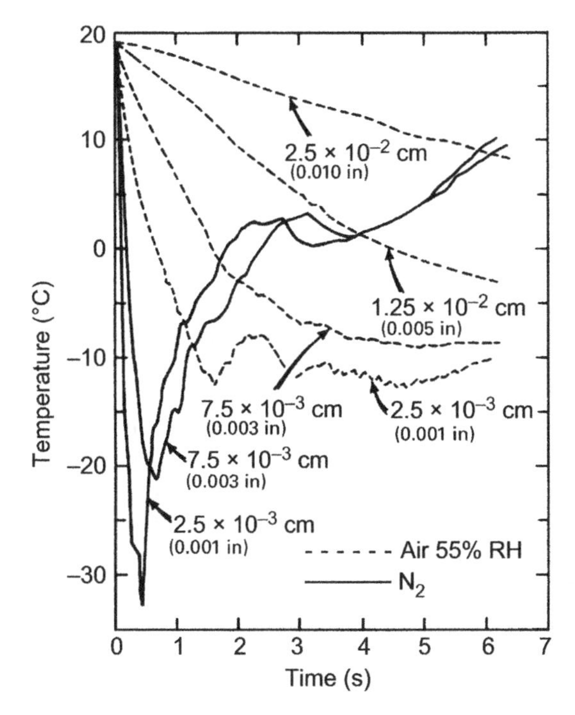

Fig. 18.2 Temperature versus time of fine thermocouples mounted in the center of a vacuum chamber and filled separately with dry nitrogen or air. Reproduced with permission from J.F. O'Hanlon and J-J. Shieh [4] Copyright 1991, AIP Publishing.

rapidly. Systems with high pumping speeds exhibit adiabatic behavior, whereas systems with low pumping speeds exhibit isothermal behavior.

Figure 18.2 illustrates the temperature changes recorded by an array of small-diameter thermocouples during the roughing of a 30-cm-diameter × 30-cm-high chamber. The thermocouples were located in the center of the volume. Either air of 55% relative humidity (RH) air or dry N_2 was separately pumped from atmospheric pressure. The temperature dropped adiabatically, then warmed to ambient temperature. The thermocouples consistently recorded the oscillations shown in Fig. 18.2, when surrounded by air. This was due to convection currents and the formation of insulating ice layers on the couples. Early measurements, using a laser particle counter in the roughing line, postulated the particles were condensed water [5]. This observation was soon confirmed [1,2,6].

During cooling, air becomes supersaturated, i.e., RH exceeds 100%. Supersaturation allows particles to nucleate and grow by one of three mechanisms. Table 18.1 summarizes the critical nucleation processes described by Zhao et al. [2,6]. The presence of sub-micrometer dust particles results in water

Table 18.1 Critical Saturation Ratio (S_c) for Three Condensation Processes and their Nuclei

Condensation Processes	Condensation Nuclei	S_c
Heterogeneous	≥0.1–0.002 μm	1–3
Condensation on ions (Singly charged)	Negative ions	4
	Positive ions	6
Homogeneous	Molecular clusters	3–8

Source: Reproduced with permission from Jun Zhao [6]. Copyright 1990, Jun Zhao.

condensation at just over 100% RH (critical saturation $S_c = 1$), whereas homogeneous condensation, i.e., water condensing on itself without a nucleating particle, requires a higher critical saturation of $S_c = 3–8$. These concepts were first recognized in 1880 by Sir John Aitken [7] and later used by Charles T. R. Wilson in the early 1900s in the development of his eponymous cloud chamber [8].

Temperature gradients and gravity redistribute water aerosols that transfer particles from the gas to the surface, so it is important to develop criteria by which supersaturation, and particle transport, can be avoided. A relationship between the initial RH, critical saturation S_c, and the degree of isothermal behavior, developed by Jun Zhao, was used to define a parameter he named the Zhao factor, Z [2,6]. To compute Z, one requires knowledge of the pumping speed, chamber volume, and total internal surface area chamber:

$$Z = \frac{\tau\omega}{\xi} \tag{18.4}$$

Under STP conditions, the rate of heat flow from the wall of an air-filled, stainless-steel vessel is $\omega = 0.0673$ m/s [6]. The vacuum chamber time constant is $\tau = V/S$ (s), and the chamber volume-to-surface ratio, $\xi = V/A$, has units of meters.

Figure 18.3 graphically displays the relationship between these variables. This graph can be used to design a pumping process that does not allow particles to become the nucleus of a water droplet that is then transported to a critical surface. To use this graph, one needs to know the initial RH of the air in the chamber and a nucleation mechanism. Typically, one assumes that water heterogeneously nucleates on ultrafine particles, that is, when $S_c = 1$ (see curve in Fig. 18.3). Next, one projects the intersection of the initial chamber RH across to the $S_c = 1$ curve, and down to the horizontal axis to determine the minimum Z. To prevent heterogeneous nucleation, Z must be greater than this minimum value, i.e., on the right-hand side of the $S_c = 1$ curve. If the chamber is known to be free of particles greater than 2-μm diameter, then negative ions could nucleate particles and one

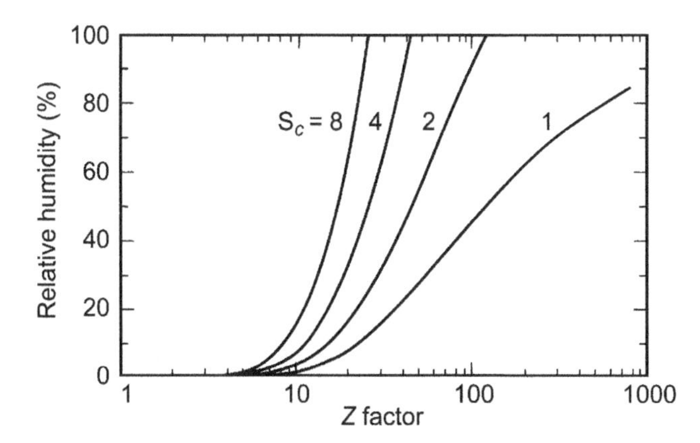

Fig. 18.3 Relationship between relative humidity, critical saturation ratio, and the degree to which the chamber is isothermal or adiabatic. An aerosol will not be formed if the intersection of RH and Z lies to the right of a given saturation curve. Reproduced with permission from Jun Zhao [6]. Copyright 1990, Jun Zhao.

should use the $S_c = 4$ curve. (The basis of the Wilson cloud chamber.) An example calculation using a model physical vapor deposition tool is illustrated in Fig. 18.4.

Zhao has provided us with a simple method for calculating the rough pumping speed that will prevent particles from becoming entrapped in water droplets during initial exhaust from atmosphere. Note that Eq. (18.4) contains the ratio τ/ξ. Since $\tau = V/S$ and $\xi = V/A$, this ratio, τ/ξ, reduces to A/S, eliminating the need to know chamber volume. Since most chamber designs are fixed, reducing the initial pumping speed—called soft roughing—is the easiest way to prevent particle entrapment in condensable vapors. It can be implemented with a choke in parallel with an (initially closed) roughing valve. At a reduced pressure, for example, 20,000 Pa (200 Torr), the main roughing valve can be opened.

Another technique eliminates condensation by measuring and calculating critical condensation parameters in real time and dynamically adjusting the pump's throttle valve [9]. This algorithm provides the shortest roughing time, as the roughing valve can be opened gradually opened.

If slow roughing adversely affects product throughput, other solutions may be considered. For example, one might purge the chamber with dry nitrogen to reduce the initial RH, and thereby the value of Z. Figure 18.4 shows the effect of four RH values on choked pumping speed. Alternatively, critical surfaces can also be located a few millimeters from a cover or shield plate, such that the *local volume* meets the Zhao criterion. It has been shown that closely stacked wafers can be exhausted at high speed without aerosol formation [4]. Systems designed for coating large glass sheets do not consider particle deposition, because product throughput takes precedence. It is not a concern in large space simulation chambers.

Aerosol Prevention Example

Example: Box coater with dimensions
$l = w = h = 1$ m.; volume $V = 1$ m^3
Chamber wall area = 6 m^2,
Internal fixture surface area = 6 m^2,
Total internal surface area = 12 m^2,
Roughing line diameter $d = 0.15$ m (~ 6").
Max. roughing pump speed at chamber,
$S = 1000$ cfm = 470 L/s = 0.47 m^3/s
Relative Humidity chamber air = 50%
From Eq. 18.4 we have:

$$Z = \frac{\tau\omega}{\xi} = \frac{V/S\,\omega}{V/A} = \frac{A\omega}{S} = \frac{(12\text{m}^2)(0.0673\text{ m/s})}{S}$$

To prevent water from nucleating on dust particles during initial roughing, the Zhao factor must be ≥ 110, as in Fig. 18.3.
Therefore:

$$S \leq \frac{(12)(0.0673)}{110} = 0.0073\text{ m}^3/s = 7.3\text{ L/s}$$

The pumping speed at chamber must be $S \leq 7.3$ L/s (at R.H. = 50%) to prevent aerosol formation during rough pumping and depositing dust on surfaces.

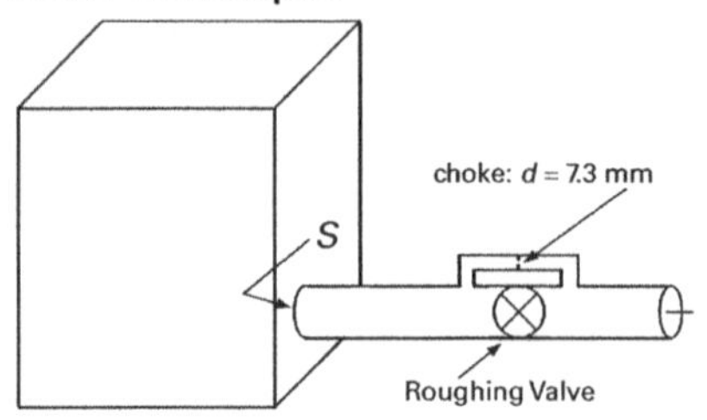

Initially pump through choke at 7.3 L/s, until pressure ~26,700 Pa (200 Torr).

Using Eq. (18.2) we calculate 3 min.
Then, open roughing valve

For four example RH values, the choked pumping speeds S, are:

RH (%)	S(L/s)	R	U_{EXIT} (m/s)
20	23	12,950	1.3
50	7.3	4110	0.41
73	2.1	1200	0.12
80	1	845	0.08

R = Reynolds Number in roughing line
U = ave. gas velocity in roughing line.

Fig. 18.4 An example calculation for determining maximum particle-free roughing speed.

18.2 Crossover

System crossover—the point at which the roughing phase ends and high vacuum phase commences—has been the subject of much oversimplification. The reason for this was already noted: in the early days, small systems with small roughing lines were common. Many texts, including early editions of this work, stated that one should not rough a chamber with an oil-sealed mechanical pump below a pressure of about 10–15 Pa (100–150 mTorr), lest backstreaming would contaminate the chamber. In reality, there are two criteria that must be met to determine an acceptable crossover pressure, and these are affected by system and pump size, and pump technology.

1) The crossover pressure must be *above* the pressure at which back-diffusing contaminants from the roughing pump could reach the chamber.
2) The pressure must be *below* the value at which the high vacuum pump would overload when connected to the chamber.

For the prototypical "small" vacuum system, the "100-mTorr rule" is valid, because at that pressure the chamber gas load is less than the overload condition of the typical high vacuum pump.

Widespread use of large vacuum processing systems in myriad manufacturing applications ranging from semiconductors, coated printing papers, secure documents, decorative plastics, and reflective glass, as well as the introduction of new types of roughing and high vacuum pumps, has rendered the "100-mTorr rule" obsolete.

Here we examine the two issues described above: rough pump contamination, and high vacuum pump overload. The physical principles that define overload differ for diffusion, turbomolecular, cryogenic, and ion pumps. First, let us examine roughing pump contamination.

18.2.1 Minimum Crossover Pressure

The minimum crossover pressure of an oil-sealed backing pump is limited by oil backstreaming, and this has been well-studied [10–13]. In one case, the backstreaming of oil through a 30-cm-long × 25-mm-diameter roughing line connected to a 4.5-m^3/h rotary vane mechanical pump was found to be small (~10^{-4} mg/min) at high pressures because of the viscous flushing action of the flowing air. At pressures below approximately 13 Pa (100 mTorr) the viscous flushing action was diminished until at a pressure of 1.3 Pa (10 mTorr), the backstreaming was 7×10^{-3} mg/min, or 70 times greater than at high pressures [11]. Experimental data for a 9-cm-diameter × 220-cm-long pipe attached to an oil-sealed mechanical pump are illustrated in Fig. 18.5.

The counterflow diffusion rate of the oil vapor is dependent on a number of variables including the gas flow rate, pressure, pipe diameter, mass of the exhaust gas, and, most neglected, the roughing pipe length. Horikoshi and Yamaguchi [12] quantified the back diffusion of molecular contaminants from an oil-sealed pump. They modeled the backstreamed concentration in the pipe as:

$$\frac{N_I(x)}{N_I(0)} = e^{-\frac{x}{L_D}} \tag{18.5}$$

$N_I(x)$ is the impurity concentration at a distance x from the pump entrance. $N_I(0)$ is adjacent to the pump entrance, and $N_I(x)$ is the concentration at a distance x from the pump entrance. L_D is the impurity back diffusion length [12] given by:

$$L_D = \frac{D_I\left[1+\left(\frac{M_I}{M_I+M_G}\right)^N\right]+ND_{I-G}}{N v_G\left[1-\left(\frac{M_I}{M_I+M_G}\right)^N\right]} \tag{18.6}$$

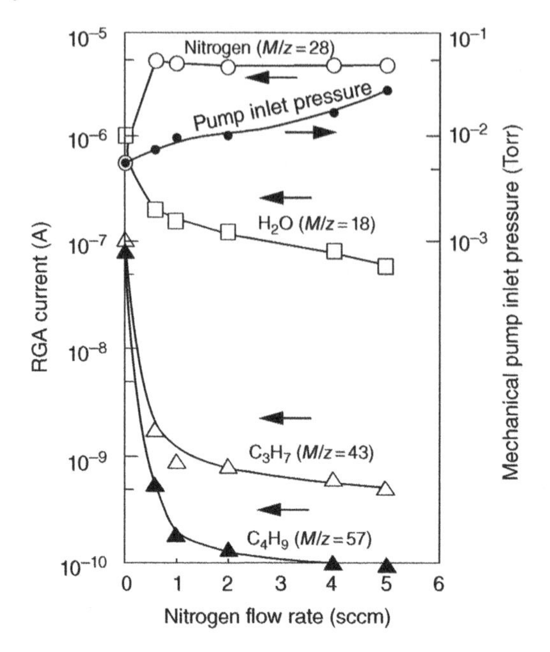

Fig. 18.5 Relation between partial pressures of backstreamed residual gases and the flow rate of nitrogen purge gas. Pipe length = 220 cm, pipe diameter = 9 cm. Reproduced with permission from Y. Tsutsumi et al. [13]. Copyright 1990, AIP Publishing.

D_I and $D_{I\text{-}G}$ are the diffusion constants of the impurity molecule, respectively, when its mean free path is greater than the pipe diameter Eq. (2.30) and when the carrier gas is in viscous flow, Eq. (2.35). $N = \lambda_{mol}/\lambda_{visc}$, and v_G is the average carrier gas velocity. Figure 18.6 sketches the fraction of a model impurity that would backstream in a 5.4-cm-diameter (2-in.) pipe connected to a 0.008 m^3/s (17 cfm) pipe as a function of its internal pressure and length. Argon was used as the model impurity, as it is close in mass to the $M/z = 39$ oil fragment; nitrogen was used as the carrier gas. These conditions and a pipe length of 0.5–1 m length represent those of the prototypical "small" vacuum system. The model demonstrates that such a system has very limited backstreaming of oil when the rough pumping is limited to pressures >10–15 Pa (100–150 mTorr). Note that this model does not account for surface migration. That can be reduced by lining a section of the pipe with a material such as Teflon, which is not wet by mineral oil.

Contamination from the roughing pump may be reduced by the use of liquid nitrogen [14] or ambient temperature trap and low-vapor-pressure oil. The liquid nitrogen trap with closely spaced surfaces is the most effective, but the most difficult to maintain and consequently infrequently used. Ambient

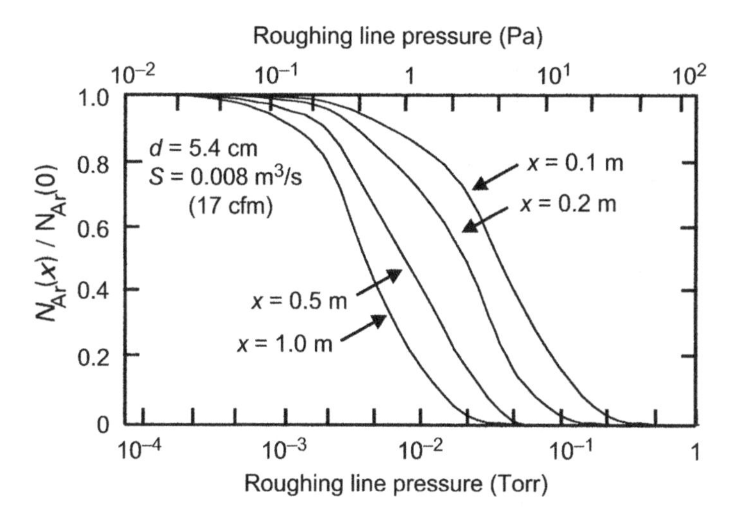

Fig. 18.6 Concentration of a model impurity, argon, counterflowing against a N_2 flow using a relation developed by Horikoshi and Yamaguchi [12].

temperature traps do not require constant refrigeration but eventually saturate. Zeolite [15,16], alumina [16–18], and bronze or copper wool have been used for this purpose. Water vapor will soon saturate a Zeolite trap [11,19]. Zeolite traps can remove more than 99% of the contamination [15], but they generate particles that can drift into valve seats and into the pump and hasten wear. Kendall [20] designed a thermoelectrically cooled (−40°C) zeolite trap, which included a bypass valve to avoid water vapor saturation during the initial portion of the pumping cycle. Catalytic traps [21], which operate on the principle of oxidation and reduction of copper oxide, have been developed [21,22]. Hydrocarbons were oxidized to CO_2 and H_2O in a heated catalyst. The catalyst was regenerated on exposure to air. All traps saturate. Experience has shown that traps are inadequately maintained in casual environments and provide a false sense of protection; however, they work well when used in production environments with scheduled maintenance.

Santeler [23] developed a gas purge technique to prevent mechanical pump vapors from contaminating a process chamber. His design is sketched in Fig. 18.7. Nitrogen or dry air is admitted to the roughing line at a point near the chamber isolation valve. The flow is adjusted so as to produce a pressure somewhat above that at which oil vapors could backstream, say, 10 Pa (0.1 Torr), for example, 54-mm-diameter pipe illustrated in Fig. 18.6. When crossover pressure is reached, the roughing valve is closed, and the preset flow prevents oil vapor backflow. Furthermore, it is impossible to pump the chamber below the set pressure.

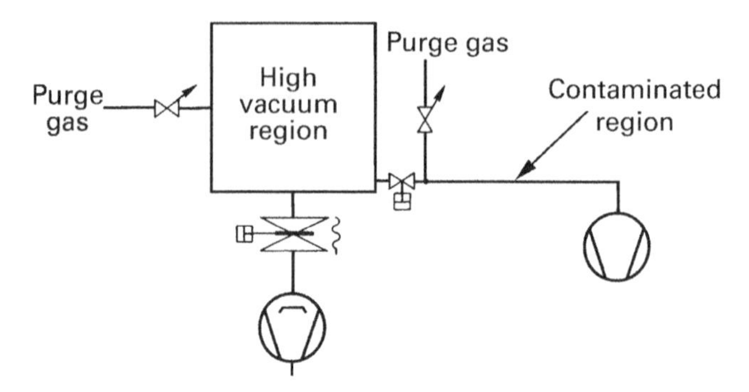

Fig. 18.7 Preventing hydrocarbon contamination from the roughing pump from reaching the process chamber by use of a purge gas. Adapted with permission from D. J. Santeler [23]. Copyright 1971, The American Vacuum Society.

Oil-free or "dry" roughing pumps do not backstream oil, but each has a useful lower pressure limit limited by its design and materials. For example, scroll pump users often add a 0.02 μm particle filter in the roughing line at the scroll pump entrance to prevent wear-generated particles from migrating to the process chamber.

18.2.2 Maximum Crossover Pressure

The maximum crossover pressures for momentum transfer pumps (diffusion and turbo) are determined by their *maximum gas load*, i.e., their overload condition. Capture pumps are distinct. Cryopumps have a maximum *impulsive heat load*, and sputter-ion pumps have a maximum *pressure* above which they will not operate. Let's look at these individually.

18.2.2.1 Diffusion

The maximum pressure at which a diffusion pump may be connected to a process chamber corresponds to the point where the *gas flow* from the chamber $Q = d(PV)/dt$, just equals the pump's maximum rated flow capacity. We noted in Chapter 12 that the rate at which energy is imparted to gas molecules is proportional to the electrical power delivered to the boiler. When the flow capacity of the pump is exceeded, the top jets will fail, because oil–gas scattering destroys the supersonic vapor stream first in the top jet. Continued increase of the inlet gas flow will cause the lower jets to fail in succession. The pressure–flow characteristics of a diffusion pump and roughing pump are shown in Fig. 18.8. Hablanian noted that the traditional speed versus pressure plot masks gas flow, and flow is the real independent variable [24]. A pressure versus flow plot visually reinforces the concept that diffusion (momentum transfer) pumps overload at a limiting gas flow. The sharpness of the flow–pressure behavior in the overload region is related

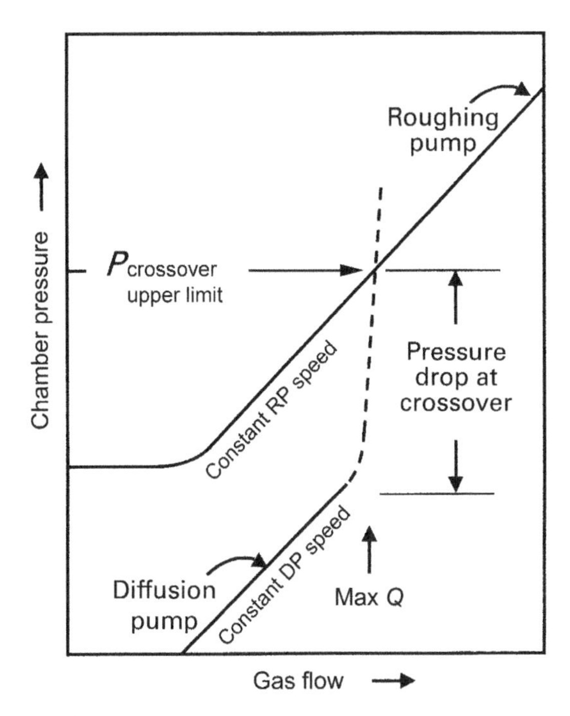

Fig. 18.8 Gas-flow-pressure characteristic of a diffusion pump and roughing pump. Adapted with permission from M.H. Hablanian [24]. Copyright 1992, AVS–The Science and Technology Society.

to the safety factor used in the jet design. The gradual transition from under to overloaded operation hides the fact that the top jet has failed and begins to backstream oil. It is best to crossover a bit below the maximum gas throughput value provided by the pump manufacturer.

The actual gas flow from the chamber can be measured with simple metrology tools. Figure 18.9 illustrates the necessary measurement procedure. First, one fills the chamber with its product load, substrate holders, deposition sources, and so on. Next, one pumps from atmosphere to a (first guess) crossover pressure and quickly closes the roughing system valve. Last, one calculates the gas load released in the chamber $Q = V\Delta P/\Delta t$, by measuring the rate of rise with a pressure gauge and stopwatch. Multiplying this by the system volume gives a gas load at this initial guess pressure. One then vents the chamber to atmosphere, to reestablish the appropriate initial conditions and repeats the process. This process is repeated for pressures above or below the first attempt until one arrives at a pressure where the loaded chamber outgassing rate is less than the maximum flow capacity of the high vacuum pump. Of course, the high vacuum valve is closed during these measurements.

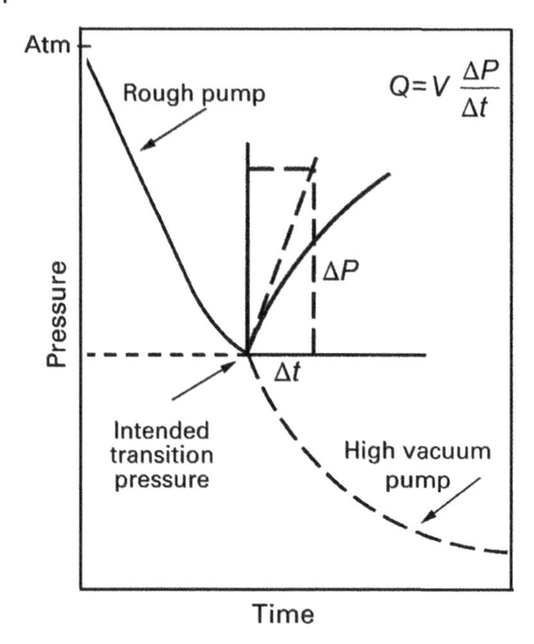

Fig. 18.9 Finding the proper crossover condition by performing a rate of rise measurement. Reproduced with permission from M.H. Hablanian [25], AIP Publishing.

If the net high vacuum pumping speed, and outgassing rates of chamber materials are known and assumed to remain constant during the roughing cycle, the maximum crossover pressure can be calculated to be:

$$P_{\text{crossover,max}} = \frac{Q_{\text{max}}}{S_{\text{net}}} \tag{18.7}$$

where S_{net} is the actual speed of the roughing system at the chamber entrance. When closing the roughing valve and opening the high vacuum valve, the pressure should first quickly drop by pressure division by the ratio of the speeds of roughing to high vacuum:

$$P_{\text{chamber}}(\text{after crossover}) = \frac{S_{\text{net}}}{S_{\text{high vacuum}}} P_{\text{chamber}}(\text{before crossover}) \tag{18.8}$$

If the pressure does not drop sufficiently after crossover, then the diffusion pump is still in its overload region [24]. In such a case, either a lower crossover pressure or a larger diffusion pump would be in order. This subject has been discussed in detail by Hablanian [26].

18.2.2.2 Turbo

Turbomolecular pumps like diffusion pumps, are momentum transfer pumps. They overload at a maximum *gas flow*. However, at overload they behave differently than a diffusion pump. Instead of the top jet failing, the rotational velocity of the blades decreases because the power to the drive motor is constant, and power is the product of rotational velocity and torque. Nesseldreher [27] has observed gross backstreaming of heavy oil fragments through turbomolecular pumps when their rotational velocity decreased to 40% of maximum. However, many pumps automatically remove drive power when $\omega < 80\% \ \omega_{max}$.

The safe maximum crossover pressure corresponds to the maximum *gas flow* at which the rotational velocity ω just begins to decrease. The speed decrease will not be sudden, because momentum transfer comes from blades whose stages rotate as one. The shape of the overload region is dependent on the staging ratio, or relative sizes of the turbo and roughing pump. Figure 18.10 illustrates this dependence. One can consider the manufacturer's stated value of Q_{max} (the point at which the speed begins to decrease) to be the maximum gas flow. The maximum pressure at crossover is then given by:

$$P_{crossover,max} = \frac{Q_{max}}{S_{roughing}} \tag{18.9}$$

Using the example in Fig. 18.10, one can see the effect of maximum crossover pressure on the roughing pump size. One must examine the pressure–speed performance characteristic of the roughing pump and ensure that its speed is retained at the lowest pressure—the actual crossover pressure—and not simply at the maximum crossover pressure.

Changing from an oil-sealed roughing pump to an oil-free roughing pump is of no value unless the replacement pump has adequate speed at the crossover pressure. For this reason, oil-free roughing pumps such as screw or scroll pumps are best suited for use with turbo-drag pumps and are not the best choice for use with conventional turbomolecular pumps.

Oil backstreaming in a turbomolecular pump at high values of inlet gas flow is a separate issue and is discussed in Chapter 22. Preventing foreline contamination from reaching the process chamber during power failure, when the rotational velocity decreases to zero, is described in Chapter 19.

18.2.2.3 Cryo

During crossover, the thermal load from the warm incoming gas will cause the temperature of the two refrigerator stages to increase. The thermal load is proportional to the heat content of the gas impulse. In Chapter 14, we described the temperature ranges of the two pumping stages. The first, or warm stage, operates

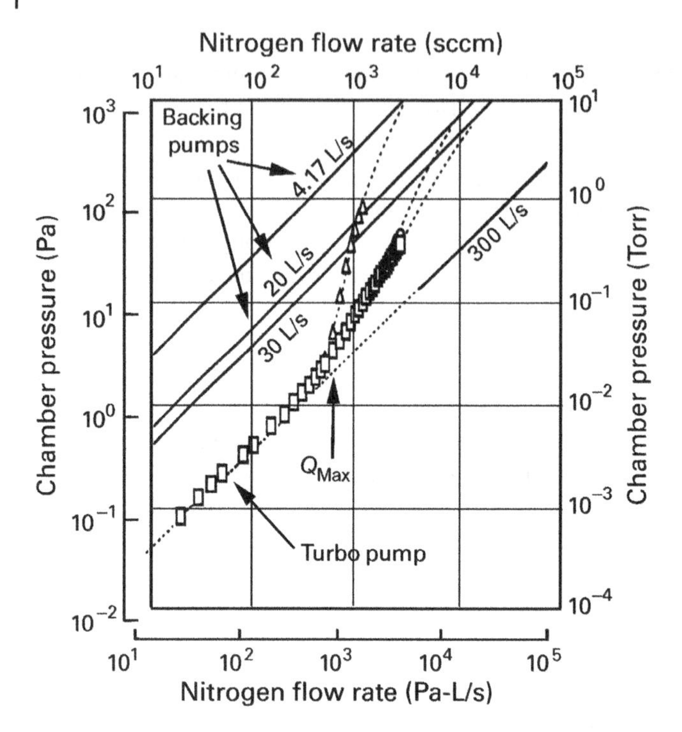

Fig. 18.10 Chamber pressure as a function of gas throughput in a 300-L/s turbomolecular pump, measured individually for backing pumps with speeds of 4.17, 20, and 30 L/s. Dotted lines represent measured data. Solid lines are constant pumping-speed lines. Reproduced with permission N. Konishi et al. [28] Copyright 19962, AIP Publishing.

in the range of 30–100 K, whereas the second, or cold stage, remains in the range of 10–20 K. The heat capacity of the second stage is less than that of the first stage; it pumps everything except water vapor. Its temperature increase is a sensitive measure of pump capacity. Heat loading is the major difference between a cryopump and a momentum transfer pump. Momentum transfer pumps overload at a critical *gas flow*. Cryopumps overload at a critical instantaneous *heat quantity*.

One recommended practice defines cryopump crossover as the maximum quantity of *nitrogen* $(PV)_{max}$ that can be condensed in a short time on the second stage of a newly regenerated cryopump, without the second-stage temperature exceeding 20 K [29]. The relation between this maximum *impulsive nitrogen heat load* sometimes called the gas burst rating and crossover pressure for small pump volumes is given by:

$$P_{\text{crossover,max}}\Big|_{\text{unloaded pump}} = \frac{(PV)_{\text{max}}}{V_{\text{chamber}}} \tag{18.10}$$

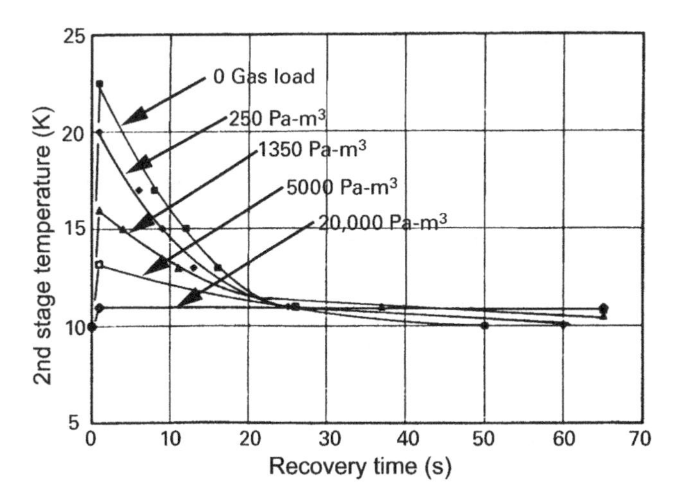

Fig. 18.11 Temperature of the cryopump second stage versus time for five different values of previously pumped nitrogen gas load on the second stage. In each case, the nitrogen gas impulse load was $PV = 20\,\text{Pa-m}^3$ (150 Torr-L or 200 mbar-L) of nitrogen. C. Juhnke et al. [30]/Adapted with permission from Elsevier.

Impulsive heat load should not be confused with gas flow. This definition applies to an "unloaded," or freshly regenerated pump. However, in normal service, the loading increases with time; eventually, the pump must be regenerated. Figure 18.11 illustrates the effect of gas (heat) loading on the second-stage temperature behavior during successive crossovers. As the loading increases, the peak temperature at the cryo surface decreases, but the time to recover to base temperature increases [30]. This increase in recovery time is a result of the insulating effect of the ice load. This interrelation between recovery time and accumulated gas load affects the design of cryo-pumped, rapid-cycle load locks. If the required load lock cycle time cannot be met, then either the gas load at crossover must be reduced (reduced crossover pressure) or the refrigeration capacity of the pump must be increased (increased pump size).

The actual crossover pressure must consider the prior gas loading when the system is cycled rapidly. The variable temperature recovery time of the cold stage is a consequence of the rate at which heat in the incoming gas can be removed and transported across the accumulated ice deposit.

Bentley [31] states a rule of thumb for the maximum gas load is $PV_i/W_2 \leq 4\,\text{Pa-m}^3/\text{W}$ (40 mbar-L/W or 30 Torr-L/W), where W_2 is the refrigeration capacity (Watts) of the second stage. If the gas burst is too large, water vapor in the gas burst could reach the adsorbent stage during viscous or transition flow and hinder the pumping of hydrogen and helium. If the second stage charcoal were saturated with helium, it would desorb, create a thermal short to the housing, and cause the pump to warm and cease pumping. Should that happen, the pump will need to be regenerated.

18.2.2.4 Sputter-Ion

Ion pump crossover must take place below a *maximum pressure*, typically ~10^{-3} Pa (10^{-5} Torr). Historically, small ion-pumped systems were rough-pumped with liquid-nitrogen-cooled sorption pumps, followed by pumping with a titanium-sublimation pump. Large and more modern systems are more typically roughed with an oil-free mechanical pump and turbo pump combination. As noted in Chapter 13, a plasma will form in an ion pump, if it is ignited at a pressure greater than ~10^{-3} Pa. The heat released by the plasma will then warm the electrodes and release adsorbed gas. This causes the pressure to increase additionally, resulting in thermal runaway and a pump that no longer functions.

In summary, the historical definition of crossover pressure is no longer valid, as modern systems use both larger chambers and pumps, as well as alternative technologies. The upper crossover-limiting criterion for sputter-ion pumps is a maximum value of gas pressure. The upper crossover criteria for turbo and diffusion pumps are maximum gas flow. For a cryopump, the upper pressure limit is related to its maximum heat impulse, and that depends on the heat quantity of the incoming gas. This is discussed in some detail in Chapter 22. Tables 18.2 and 18.3 summarize these issues for lower and upper crossover pressures. Any pressure within the system's range is acceptable. That said, one must remember that chamber debris will affect the gas load, which the high vacuum pump must remove. Table 18.4 reviews some rough pumping best practices that summarize this chapter.

Table 18.2 Lower Crossover Pressure Limits

Rough vacuum pump type	Lowest crossover pressure limited by
Oil sealed rotary	Oil backstreaming
Lobe blower	Oil backstreaming and staging ratio
Scroll	Ultimate pressure
Multistage claw or lobe	Ultimate pressure
Screw	Ultimate pressure
Multistage dry piston	Ultimate pressure
LN_2 sorption	Pump capacity, number of stages

Table 18.3 Upper Crossover Pressure Limits

High vacuum pump type	Highest crossover from roughing limited by
Diffusion	Maximum gas flow
Turbomolecular	Maximum gas flow
Cryogenic	Maximum heat impulse
Titanium sublimation	10^{-1} Pa (10^{-3} Torr) and Ti sublimation rate
Sputter-ion	10^{-2} Pa (10^{-4} Torr)

Table 18.4 Best Practices for Rough Pumping

Optimize rough pumping speed to reduce particle redistribution within chamber if it is important.
Clean debris from chamber often to minimize particle redistribution during venting to atmosphere.
Maintain roughing pumps according to manufacturer's schedule.
Change oil at required maintenance intervals to limit issues related to degraded or dirty fluid.
Be aware that change in process effluent or pump temperature can significantly shorten the replacement interval of mechanical pump fluids.
Fill pump with inert fluids when pumping oxygen.
Monitor base pressure of mechanical pump as a check on pump fluid contamination.
Use appropriate gas ballast to limit or reduce condensation of water and other vapors into rough pump fluids.
Monitor and add pump fluid more frequently when using gas ballast.
Set maximum diffusion and turbo pump crossover pressures based on pump's rated maximum foreline pressure, gas flow, and chamber gas load.
Set minimum diffusion and turbo pump crossover pressures considering also foreline trap design and type of mechanical pump.
Set maximum crossover of a cryopump by the pump's rated maximum heat load (also known as the Gas Bursting Rating).
Set maximum crossover pressure of a sputter-ion pump at or below 10^{-3} Pa (10^{-5} Torr).
Do not use fluid-sealed mechanical pumps to rough-pump a sputter-ion-pumped system.
Use dry mechanical pumps to exhaust clean chambers.
Use appropriate foreline traps to reduce backstreaming from fluid vapor of sealed pumps toward chamber or high vacuum pumps, and also to reduce fluid from diffusion pump entering mechanical pumps.
Install normally closed valves above all mechanical pumps to limit the chance of fluids or particles being transported toward the chamber during power outages.

(Continued)

Table 18.4 (Continued)

Use appropriate foreline filters or screens to limit particles and small debris from entering mechanical pumps and causing premature internal wear.
If process effluent gases can lead to chemistry that forms particles within fluid-sealed pumps, incorporate appropriate oil filtration system on mechanical pump. Also consider using a mechanical pump that is designed and constructed for corrosive, toxic, or particle-forming applications (often known as "chemical-series" mechanical pumps).
Use exhaust filters or mist coalescers to limit mechanical pump fluid loss to facility ventilation system.
Check condition of belts, shaft couplers, and related components on mechanical pumps often, maintain a supply of these wearable components, and replace them before component failure.
Observe the manufacturer's lobe blower upper pressure limit.
For high volume dry pumps, set and monitor various water flows and temperatures at manufacturer-recommended parameters for various locations of the pump. Check flows often in locations where other vacuum systems are added or removed from connection to process chilled water lines.

References

1 Wu, J.J., Cooper, D.W., and Miller, R.J., *J. Vac. Sci. Technol., A* **8**, 1961 (1990).

2 Zhao, J., Liu, B.Y.H., and Kuehn, T.H., *Solid State Technol.* **33**, 85 (1990).

3 Santeler, D.J., Jones, D.W., Holkeboer, D.H., and Pagano, F., *Vacuum Technology and Space Simulation*, NASA SP-105, National Aeronautics and Space Administration, Washington, DC, 1966, p. 80.

4 O'Hanlon, J.F. and Shieh, J.-J., *J. Vac. Sci. Technol., A* **9**, 2802 (1991).

5 Chen, D., Seidel, T., Belinski, S.E., and Hackwood, S., *J. Vac. Sci. Technol., A* **7** (3), 3105 (1989).

6 J. Zhao, *Thermodynamics and Particle Formation During Vacuum Pump-Down*, Ph.D. Dissertation, University of Minnesota, 1990.

7 Aitken, J., *Proc. R. Soc. Edinburgh* **11**, 1880–1881 (1880; *Trans. R. Soc. Edinburgh,* **30**, 1880; and *Nature,* **23**).

8 Wilson, C.T.R., *Proc. R. Soc.* **86A**, 285 (1911); *Proc. R. Soc.,* **87A**, 277 (1912).

9 Wu, J.J., Copper, D.W., Miller, R.J., and Stern, J.E., *Microcontamination* **8**, 27 (1990).

10 Jones, D.W. and Tsonis, C.A., *J. Vac. Sci. Technol.* **1**, 19 (1964).

11 Baker, M.A., Holland, L., and Stanton, D.A.G., *J. Vac. Sci. Technol.* **9**, 412 (1972).

12 Horikoshi, G. and Yamaguchi, H., *J. Vac. Soc. Jpn* **25**, 161 (1982).

13 Tsutsumi, Y., Ueda, S., Ikegawa, M., and Kobayashi, J., *J. Vac. Sci. Technol., A* **8**, 2764 (1989).

14 See for example:Danielson, P.M. and Mrazek, F.C., *J. Vac. Sci. Technol.* **6**, 423 (1969).

15 Fulker, M.J., *Vacuum* **18**, 445 (1968).
16 Craig, R.D., *Vacuum* **20**, 139 (1970).
17 Holland, L., *Vacuum* **21**, 45 (1971).
18 Baker, M.A. and Staniforth, G.H., *Vacuum* **18**, 17 (1968).
19 Singleton, J.H., *J. Phys. E.* **6**, 685 (1973).
20 Kendall, B.R.F., *Vacuum* **18**, 275 (1968).
21 T. Kraus, F.R.G. Pat. No. 1,022,349, Aug. 28, 1956.
22 Buhl, R., *Vak.-Tech.* **30**, 166 (1981).
23 Santeler, D.J., *J. Vac. Sci. Technol.* **8**, 299 (1971).
24 Hablanian, M.H., *J. Vac. Sci. Technol., A* **10**, 2629 (1992).
25 M. H. Hablanian, *J. Vac. Sci. Technol., A* **10**, 2629 (1992).
26 Hablanian, M.H., *High Vacuum Technology: A Practical Guide*, 2nd ed., Marcel Dekker, New York, 1997 Chapter 10.
27 Nesseldreher, W., *Vacuum* **26**, 281 (1976).
28 Konishi, N., Shibata, T., and Ohmi, T., *J. Vac. Sci. Technol., A* **14**, 2858 (1996).
29 PNEUROP: Vacuum Pumps, Acceptance Specifications, Refrigerator Cooled Cryopumps, Part 5, PNRASRCC/5. Frankfort (1989).
30 Juhnke, C., Klein, H.H., Schreck, S., Thimm, U., Häfner, H.U., Mattern-Klosson, M., and Mundinger, H.-J., *Vacuum* **44**, 717 (1993).
31 Bentley, P.D., *Vacuum* **30**, 145 (1980).

19

High Vacuum Systems

Single-chamber systems capable of producing a high vacuum environment are widely used. High vacuum pumping systems reviewed in this chapter encompass those that will produce a base pressure of an order of 10^{-6} Pa (10^{-8} Torr), and a process pressure $<10^{-4}$ Pa (10^{-6} Torr). The base pressure of such a system is determined by its pumping speed and chamber gas load. The pressure will be higher during the process because of gas evolved from the source and other heated surfaces. The pump combinations used on high vacuum chambers must have a high enough pumping speed not only to reach the base pressure, but also to maintain the process pressure. In addition, production systems should be able to reach this base pressure quickly. They must pump water efficiently, because it is the most difficult component of atmospheric air to pump.

In this chapter we discuss systems with these objectives in mind. We describe the operation of diffusion, turbo, ion, and cryo-pumped systems. Included is a procedure for starting, stopping, cycling, and fault protecting each system type. We describe problems commonly encountered when using these pumps and include general best practices as well as those for each pump type.

19.1 Diffusion-Pumped Systems

The layout of a small diffusion pump system is shown in Fig. 19.1. Several accessory items are shown as well. Not every system has or needs all these additional items; however, we show all here to discuss their location and operation. The diffusion pump, gate valve, mechanical pump, and liquid nitrogen cold trap over the diffusion pump are the basic components of a diffusion-pumped system.

In less critical applications, a water baffle can be used in place of a cold trap. Backstreaming rates of 5×10^{-6} (mg/cm^2)/min for DC-705 fluid over a 6-in.

A Users Guide to Vacuum Technology, Fourth Edition. John F. O'Hanlon and Timothy A. Gessert.
© 2024 John Wiley & Sons, Inc. Published 2024 by John Wiley & Sons, Inc.

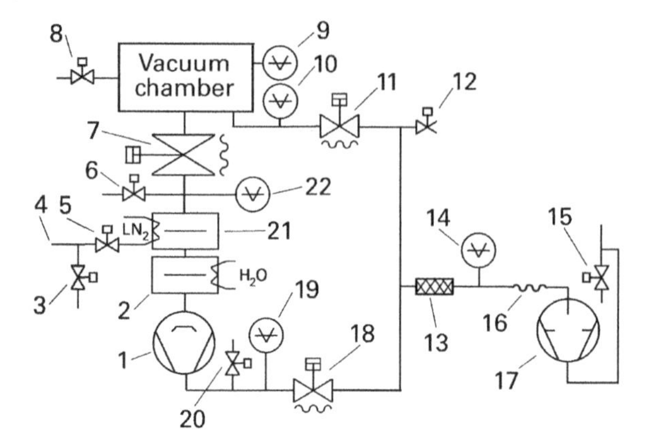

Fig. 19.1 Diffusion-pumped system: (1) Diffusion pump, (2) partial water baffle, (3) LN_2 vent valve, (4) LN_2 inlet, (5) LN_2 fill valve, (6) port for gas purging diffusion pump, (7) bellows-sealed high vacuum valve, (8) chamber bleed valve, (9) chamber ionization gauge, (10) chamber thermal conductivity gauge, (11) roughing valve, (12) mechanical pump vent, (13) roughing line trap, (14) roughing line thermal conductivity gauge, (15) sump for collecting condensable vapors, (16) bellows for vibration isolation, (17) mechanical pump, (18) foreline valve, (19) foreline thermal conductivity gauge, (20) leak testing port, (21) liquid nitrogen trap, and (22) diffusion pump ionization gauge.

diffusion pump and simple liquid nitrogen trap have been measured [1]. A value of 8×10^{-6} (mg/cm^2)/min above a 4-in. diffusion pump baffled with a chevron cooled to 15°C may be inferred from other data [2]. A cold-water baffle will provide adequate protection for many applications, provided the diffusion pump is charged with a pentaphenyl ether or pentaphenyl silicone fluid. As Singleton pointed out, these two fluids are most valuable in systems, which are trapped with a less-than-elegant trap [3]. The simplest way to reduce backstreaming in a low-cost system is to use a high-quality diffusion pump fluid.

In critical applications, a liquid nitrogen trap and a water- or conduction-cooled cap are used. The liquid nitrogen reservoir is a pump for water vapor and a trap for fluid fragments. A 6-in. diffusion pump and trap combination, like the one shown in Fig. 19.1 will have a net pumping speed of 1000 L/s for air at the chamber. A matching liquid nitrogen trap can pump water vapor at speeds up to 4000 L/s. Diffusion pumps, fluids, and traps of modern design are capable of reducing backstreaming rates to a level below that from other sources, mainly mechanical pump fluid contamination from the roughing pump.

The arrangement that is shown here for automatically filling the liquid nitrogen trap is convenient, when there is a long vacuum-jacketed supply line that must be chilled each time the trap needs filling. This does not eliminate the need for phase separation on large distribution systems. Without this arrangement the incoming liquid boils, and the gas quickly warms the trap and releases condensed vapors.

To avoid trap warming, a valve is added. When the low-level sensor is activated, the vent valve (#3 in Fig. 19.1) is opened, and the warm gas is vented to the atmosphere. A second sensor is placed in the supply line near the vent valve. When cooled by the liquid, it closes the vent valve and opens the fill valve. The controller then performs its normal function. A commercial controller with two-level sense elements may be modified to perform this function. If the trap is filled from a Dewar with a short length of tubing, this gas bypass operation is not required. Alternatively, we could use a triaxial delivery pipe in which gas bubbles are automatically vented [4]. If the liquid nitrogen is pressurized, its temperature will be above 77 K. Simple LN_2 traps containing a spheroidal reservoir are refilled when the trap is, say, half full. As the level drops, CO_2 is released from the warmer surface, and is repumped when the trap is cooled. At a temperature of 80 K, the vapor pressure of CO_2 is 7×10^{-7} Pa (5×10^{-8} Torr). As a result, ion gauge readings in the chamber will increase and decrease as the trap cools and warms. This effect is not seen with constant temperature traps in which the reservoir is located out of the pumping path and connected to the trapping surfaces at its coldest location (bottom of the reservoir).

Thermal conductivity gauges are located in the work chamber and in the foreline for control and fault protection, and in the roughing line to monitor mechanical pump blank-off pressure. One ionization gauge is located in or on the chamber, and a second is positioned between the cold trap and the gate valve. Many systems are designed and constructed without a gauge at this latter point; however, it is extremely useful. The most straightforward diagnostic measurement on a diffusion pump is its blank-off pressure, and that measurement can be made only with a gauge located below the main valve. Some larger traps are fabricated with an ion gauge port; gate valves with extra piping or ports on the diffusion pump side facilitate easy mounting of a lower ion gauge. In general, it is advisable to install the gate valve with its seal plate facing upward to keep the valve interior under vacuum at all times. This reduces the volume and surface area exposed to the atmosphere in each cycle and minimizes pumping time. All gauge tubes, both ion and thermal conductivity, should be positioned with their entrances facing downward or to the side to prevent accumulation of debris.

Some other items illustrated in Fig. 19.1 are not often needed or found. A leak-detection port located in the foreline provides the best sensitivity and speed of response for leak detecting the chamber, trap, and diffusion pump. A leak detector may be attached here while the system is operating. A water flow restrictor on the inlet side of the diffusion pump cooling coil will save water and increase the oil ejector stage temperature. Traps can be used in roughing lines and, when properly maintained, can reduce the transfer of mechanical pump oil to the chamber. In this system the trap is used as a roughing trap, to prevent mechanical pump fluid from reaching the chamber, and to prevent diffusion pump fluid from reaching the mechanical pump.

19.1.1 Operating Modes

Systems have three distinct operating modes: chamber pump and vent, system final power off, and system initial power on. Let us begin with the first phase. Assume the pumps are operating and that the chamber is at atmospheric pressure. The high vacuum and roughing valves are closed. During the chamber rough pumping cycle, the foreline valve is closed and the roughing valve is open. A thermal conductivity gauge and valve controller senses the rough-to-diffusion pump crossover pressure and operates the valves. The time constant of a 0–150-Pa thermocouple gauge is typically 2 s [5]. In a very small system, with a system roughing time constant less than the gauge time constant, the pressure will reach a level less than that set on the gauge. This problem is easily corrected by adjusting the set point to a pressure higher than actually desired or by using a gauge with a faster time constant. The fact that the gauge reading lags the system pressure can be used as a quick check for leaks or excessive chamber outgassing. When the gauge reaches the desired crossover pressure, the controller closes the roughing valve, and opens the foreline valve; a timer may delay the opening of the high vacuum valve. The thermal conductivity gauge will reach a pressure minimum that is below the set point and will drift upward at a rate determined by the real or virtual leak. If the pressure increases beyond the set point pressure before the timer opens the valve, the sequence may be programmed to abort. If the pressure is below the set point at the end of the timed interval, the high vacuum valve will be programmed to open. The diffusion pump and liquid nitrogen trap will then pump the system toward its base pressure. The time required to reach the base pressure is a function of diffusion pump speed, liquid nitrogen trap size, chamber volume, surface area, and chamber cleanliness. Water vapor is the main species being pumped, and a liquid nitrogen trap pumps water vapor efficiently. Diffusion pumps with expanded tops and oversized liquid nitrogen traps have high pumping speeds for water vapor.

Complete system shutdown—removal of all utilities—begins by closing the high vacuum valve and warming the liquid nitrogen trap, while the diffusion pump continues to remove the evolved gases and vapors. If the trap is allowed to equilibrate with its surroundings naturally, it will take 4–10 h, depending on trap design. This time can be shortened considerably by inserting a plastic tube in the liquid nitrogen reservoir and bubbling dry nitrogen gas through the liquid. When the liquid nitrogen trap temperature reaches 0°C, the power to the diffusion pump is removed. When the diffusion pump body cools to 50°C, the foreline valve is closed. The mechanical pump is then powered off and vented. Venting of the diffusion pump should be done with the valve located above the trap (#6 in Fig. 19.1). Valve 20 in Fig. 19.1, a leak testing port, should never be used to vent the pump. The diffusion pump should not be vented by opening the foreline valve and allowing air to flow from the roughing trap and pump, because this would transport mechanical pump oil vapor into the diffusion pump body. The pump should be cooled to <50°C and venting should proceed slowly. Last, cooling water in the diffusion pump should be turned off. If the pump interior is exposed to ambient air when cooled below the dew point, water vapor will condense and contaminate the pump fluid.

The last important phase, initial system starting, begins by flowing cooling water through the diffusion pump jacket and powering on the mechanical pump. After the gas in the mechanical pump piping is exhausted to crossover pressure, the foreline valve may be opened and the diffusion pump heater power supply activated. On a system operated by an automatic controller, a preset gauge performs this function. Diffusion pump heating times range from 15 min for a 4- or 6-in. pump to 45 min for a 35- or 48-in. diameter pump. When starting a diffusion pump, it is best to minimize the initial backstreaming transient and prevent it from contaminating the region above the trap, before the trap is cooled. This can be accomplished by precooling the trap, and pumping through crossover as quickly as is possible.

19.1.2 Operating Issues

The operations sketched in the preceding section did not deal in depth with some of the real and potential problems in diffusion-pumped systems—problems such as diffusion-pump fluid backstreaming, trap operation, and fault protection. Let us consider these here.

Backstreaming is a subject for which many vacuum system users have a biased view. Unfortunately, the common definition—transfer of diffusion pump fluid from the pump to the chamber—is for many the general definition. It is incorrect in diffusion pump systems, because mechanical pump oil can be the largest organic contaminant in the work chamber.

Diffusion pump fluid backstreaming is controlled by the diffusion pump, traps, baffles, and system operating procedures. Most importantly, the backstreaming from *both* sources must be reduced to a level that can be tolerated. Chapter 12 describes methods for reducing diffusion pump contamination. A high-quality diffusion pump fluid, coupled with a water-cooled or conduction-cooled cap, and a liquid nitrogen trap will minimize diffusion pump backstreaming.

The two critical system-operation concepts for minimizing diffusion-pumped system backstreaming are: crossing over to high vacuum pumping in the allowed pressure range and pumping through the crossover region as rapidly as is possible. A thorough and detailed discussion of these topics is given in Section 18.2.

The dominant gas load in all unbaked vacuum systems is water vapor. A liquid nitrogen-cooled surface will provide a high pumping speed (14.5 L-s^{-1}-cm^{2}) for this gas load. The vapor pressure of water at this temperature is ~10^{-19} Pa (10^{-21} Torr), and its the pumping speed is neither a function of the operating pressure nor sensitive to small changes in a trap's surface temperature. Several gases can sublime at low pressures and during trap temperature fluctuations. Carbon dioxide is partly condensed; its vapor pressure is 10^{-6} Pa (10^{-8} Torr) at 77 K. Methane and CO have high vapor pressures and are not condensed at 77 K. They are adsorbed to some small degree on liquid-nitrogen-cooled surfaces. Two-stage closed-cycle Freon-cooled refrigerators (−40°C) are quite useful for pumping water vapor.

Hengevoss and Huber [6] described the sublimation of gases from a liquid nitrogen trap as it slowly warmed. CH_4 sublimes first, followed by CO, then CO_2. Santeler [7] has shown that the sublimation of trapped CO_2 can be eliminated by delaying full cooling of the trap until the system pressure falls below 10^{-3} Pa or by momentarily heating the trap to 135–150 K after the system pressure falls below 10^{-6} Pa. Because CO_2 is partly condensed on the trap, it can cause problems. Siebert and Omori [8] have shown how certain liquid nitrogen cooling coil designs with temperature variations of only 1 K allow the release of enough CO_2 to cause total pressure variations of 20% in a system operating in the 10^{-4}-Pa (10^{-6} Torr) range.

In any vacuum system, faults may occur that could potentially affect equipment performance or product yield. A leak in the foreline, inadequate mechanical pump oil level, or a foreline valve failure could cause the fore pressure to exceed its critical value. When the fore pressure exceeds its critical value, pump fluid will backstream to the chamber. In addition, many diffusion pump system faults could cause harm to the pumping equipment. Either cooling water failure, or an inadequate diffusion-pump fluid level, will result in excessive fluid temperatures, accompanied by fluid decomposition. Decomposition products are both vapors, and tar-like substances that deposit on the pump's interior surfaces.

Diffusion-pumped systems should be equipped with sensors that will remove power from the system or place it in a standby condition in case of loss of utilities, a leak, loss of cryogen, or loss of pump fluid. Pneumatic and solenoid valves, which leave the system in a safe position during a utility failure, are used on automatically controlled systems. When the compressed air or electrical power fails, all valves except the roughing pump vent should close. The inbuilt roughing pump vent should open to prevent mechanical pump oil from pushing upward into the roughing line. A flow meter may be installed at the outlet (not the inlet!) of the diffusion pump cooling water line. The system controller uses the signal from the flow meter to close both high vacuum and foreline valves and to remove power to the diffusion pump. The use of a water filter on the inlet to the diffusion-pump cooling water line will prevent deposits from clogging the line and causing untimely repairs. A thermostatic switch mounted on the outside of the diffusion

Table 19.1 Best Practices for Diffusion-Pumped Systems

Check mechanical pump fluid level periodically. Replace when significantly discolored or is producing varnish on the sight glass.
Check mechanical pump belts and shaft connectors periodically.
Vent mechanical pump exhaust properly.
Operate roughing pump in the viscous flow regime to limit backstreaming.
Clean roughing traps periodically.
Measure diffusion pump heater power when system is new.
Maintain diffusion pump oil level and cooling water flow.
Maintain diffusion pump forepressure below its critical limit.
Maintain diffusion pump inlet flow below its overload limit.
Use inert diffusion pump fluid when pumping oxygen.
Maintain diffusion pump water flow rate and temperature properly.
Use water-flow switch to power off diffusion pump if water flow is interrupted.
Interlock diffusion pump shutoff to trip off at maximum foreline pressure.
Use a water filter between diffusion pump and liquid nitrogen trap
Turn off cooling water when diffusion pump is vented to atmosphere.
Keep LN_2 trap vent line free of obstructions.
Service oil-sealed turbo-pump bearings by product specification.
To prevent total bearing failure, remove turbopumps from service promptly if "bearing noise" is detected.
Maintain LN_2 trap under vacuum when cold or warming to ambient.

pump casing performs these tasks; a thermostatic switch mounted on the boiler can detect a low fluid level. The set points on thermocouple and Pirani gauge controllers are used to signal an automatic controller to close the high vacuum valve in case of a chamber leak, and to isolate and remove the power from a diffusion pump with excessively high forepressure. Many liquid nitrogen controllers have low-level sensors, which close the high vacuum valve when the cryogen is exhausted. If not, a thermoswitch can be attached to the trap vent line; the boil-off keeps the vent-line temperature below 0°C. With automatic control and scheduled maintenance, backstreaming can be minimized to a point at which it is a small source of organic contamination. Table 19.1 describes some diffusion-pumped system best practices.

19.2 Turbo-Pumped Systems

Turbomolecular-pumped systems are configured in two ways: with a gate valve, separate roughing line, and foreline, or as a system with no valves, where the chamber is roughed through the turbomolecular pump. Figure 19.2 illustrates a

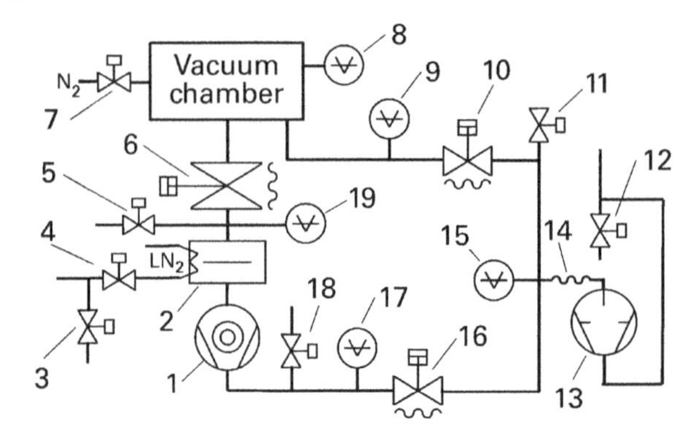

Fig. 19.2 Turbo pump system, with a separate roughing line: (1) Turbomolecular pump, (2) liquid nitrogen trap, (3) LN2 vent valve, (4) LN2 fill valve, (5) turbomolecular pump vent valve, (6) bellows-sealed high vacuum valve, (7) chamber bleed valve, (8) chamber ionization gauge, (9) chamber thermal conductivity gauge, (10) roughing valve, (11) mechanical pump vent valve, (12) sump for collecting condensable vapors, (13) mechanical pump, (14) bellows for vibration isolation, (15) mechanical pump thermal conductivity gauge, (16) foreline valve, (17) foreline thermal conductivity gauge, (18) leak testing port, and (19) turbomolecular pump ionization gauge.

system with a gate valve and a separate roughing line. This system is much like a conventional diffusion pump system. The entire roughing and foreline sections are identical to those found in a diffusion-pumped system. The criteria for sizing the forepump are similar to those of a diffusion pump. Diffusion pumps require a forepump whose speed will keep the forepressure below the critical forepressure at maximum throughput. Turbopumps require a forepump of sufficient capacity to keep the blades nearest the foreline in molecular flow or just in transition flow at maximum throughput. For classical turbo pumps without drag stages, the maximum steady-state inlet pressure P_{in} is typically 0.5–1.0 Pa (4–7 mTorr), and the maximum forepressure, P_f, is about 30 Pa (~260 mTorr). Because the throughput is constant, $P_{in}S_{in} = P_f S_f$, or:

$$\frac{P_f}{P_{in}} = \frac{S_{in}}{S_f} \tag{19.1}$$

Equation 19.1 states that the staging ratio or ratio of turbomolecular pump speed to forepump speed, S_{in}/S_f, should be in the range of 30–60. A more restrictive condition is placed on forepump size for turbomolecular pumps with low hydrogen compression ratios. A staging ratio of 20:1 is necessary for pumps with maximum compression ratios for $H_2 < 500$, in order to pump H_2 with the same speed as heavy gases.

A conventional liquid nitrogen trap is not required on a turbomolecular pump to stop bearing or mechanical-pump fluid backstreaming. The compression ratios for all but hydrogen, and helium, the lightest gases, are high enough so that none will backstream from the foreline side to the high vacuum side, provided that the pump is rotating at rated angular velocity. Oil backstreaming is totally eliminated in magnetically levitated turbomolecular-drag pumps backed by a dry roughing pump. A liquid nitrogen-cooled surface will not trap the small amount of hydrogen found in the chamber that results from compression limits. Liquid nitrogen traps are used to increase the system pumping speed for water vapor. Incorrect as it may seem on first consideration, the optimum place to locate this liquid nitrogen-cooled water vapor pump is directly over the throat of the turbomolecular pump. Alternatively, one could locate the liquid nitrogen-cooled surface within the chamber; when located here, it is called a Meissner trap.

Figure 19.3 depicts a system with no gate valve—the chamber is roughed directly through the pump. Because there is no roughing line, the problems associated with roughing line contamination and improper crossover are eliminated. Physically the system becomes simpler. No high vacuum valve is needed and only one ion gauge and thermal conductivity gauge are required. This system must be completely powered off each time the chamber is opened to the atmosphere. This makes the use of liquid nitrogen-cooled surfaces awkward because they must be warmed each time the system is stopped.

Roughing through the turbo pump places a size restriction on the mechanical pump, if the turbomolecular and mechanical pumps are to be started simultaneously. If the mechanical pump is small, it will not exhaust the chamber to the transition region before the turbomolecular pump reaches maximum rotational speed. When this happens, the motor over-current protection circuit will shut down the turbomolecular pump. A properly sized and operated mechanical pump will

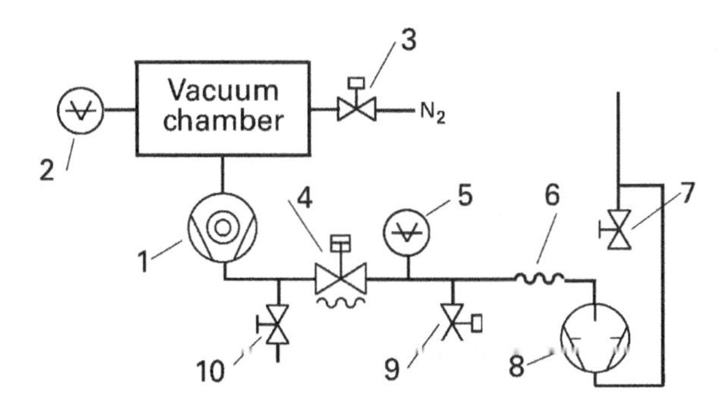

Fig. 19.3 Valveless turbo pump system: (1) Turbo pump, (2) chamber ionization gauge, (3) chamber pump and vent valve, (4) mechanical pump isolation valve, (5) foreline and mechanical pump thermal conductivity gauge, (6) bellows for vibration isolation, (7) sump for condensable vapors, (8) mechanical pump, (9) mechanical pump vent, and (10) leak testing port.

exhaust the chamber to 20–200 Pa (0.15–1.5 Torr) by the time the turbomolecular pump has reached, say, 80% of its rated rotational speed and prevents the backstreaming of the mechanical pump oil. Most turbomolecular pumps are designed with an acceleration time in the range of 5–10 min.

19.2.1 Operating Modes

The operation of the turbomolecular-pumped system with a separate roughing line, shown in Fig. 19.2, is much like that of the diffusion-pumped system described in Fig. 19.1. Before rough pumping commences, the high vacuum valve and the roughing valve are closed, and the foreline valve is open. Chamber pump-down begins by closing the foreline valve and opening the roughing valve. As in the diffusion or any other high vacuum system, the rough pumping hardware varies in complexity with chamber size. For 500-L/s or smaller turbomolecular pump, a two-stage rotary vane pump or a small dry pump is often used. For turbo pumps larger than 1000 L/s, a lobe pump backed by a rotary vane pump, or a larger dry pump is used. At crossover pressure, the roughing valve is closed, and the foreline and high vacuum valves are opened. If the turbo pump is crossed over at a flow less than the maximum throughput, the speed will drop quickly as given in Eq. (19.1). Here, as in diffusion-pumped systems, the dominant species is water vapor, and the pumping rate will be controlled by the speed of the liquid nitrogen trap. If no liquid nitrogen trap is used, this system will pump water vapor somewhat more slowly than an untrapped diffusion pump of the same speed. The large unbaked internal surface area of the pump adsorbs water during the early stages of the high vacuum pump-down cycle and then emits it at lower pressures.

System shutdown begins by closing the high vacuum valve and warming a liquid nitrogen trap if used, as in Section 19.1.1. When the trap has equilibrated, the foreline valve is closed and the power to the turbo pump motor is removed. The rotor will now decelerate. Typically, it should take 10 min or more for the rotor to come to a complete stop, but if that were to happen hydrocarbons from the foreline would rapidly diffuse to the region above the pump inlet. To prevent backstreaming of mechanical pump oil vapors and turbomolecular pump lubricating-oil vapors, the pump is vented with a reverse flow of dry gas. Argon, nitrogen, or dry air should be admitted at a point above the pump inlet or partway up the rotor stack when the rotor speed has decreased to approximately 50% of maximum rotational speed. The flow should continue until the pump is at atmospheric pressure. This can be properly accomplished by admitting gas through valve 5 (Fig. 19.2). Turbomolecular pumps should not be routinely loaded with atmospheric pressure gas pulses while running at rated speed. It does not improve bearing life. At any time after the foreline valve is closed, the mechanical pump system can be shut off and vented by valve 11 in Fig. 19.2, or its internal vent. The cooling water should be promptly shut off to prevent internal condensation. Tempering the water to just above the dew point will eliminate condensation that may form on outer portions of the pump body during normal operation.

System starting begins by initiating cooling water flow, opening the foreline valve, and simultaneously starting the mechanical and turbo pumps. After the pump accelerates to rated rotational speed, typically within 5–10 min, an optional liquid nitrogen trap may be filled. At this point, the chamber may be pumped as described in the preceding section.

Operation of the system with no gate valve, illustrated in Fig. 19.3, is considerably simpler. Operation begins by opening the cooling water and foreline valves and starting the mechanical and turbomolecular pumps simultaneously. If the rough pump has been chosen to make the chamber roughing cycle equal to the acceleration time, the system will exhaust the chamber to its base pressure without backstreaming foreline oil vapors.

This system is vented and stopped by closing the foreline valve, waiting for the rotor speed to decrease to 50% of maximum rotational speed, and admitting dry gas above the pump throat (valve 3 in Fig. 19.3). The vent valve is closed when the system reaches atmosphere, or it will pressurize the chamber. The rough pump is powered off, and cooling water flow is stopped.

19.2.2 Operating Issues

Some issues of turbo-pumped systems are unique, and some are shared with diffusion-pumped systems. All are easily solved.

Nearby amplifiers may pick up electrical or mechanical noise emanating from turbo-pump power supplies. Improper connection of Earth and neutral in three-phase supplies may also generate noise. Connecting the ground of each piece of equipment, including the pump, to the ground terminal on the most sensitive amplifier stage and then grounding to Earth can eliminate other electrical noise. This is most efficiently done with solid copper strips connected to a "quiet ground" that is isolated from the building power system ground. A quiet ground is constructed from wide, low-inductance copper strips connected to a copper rod driven deeply into Earth. Mechanical vibration may be reduced significantly by use of bellows between the pump and chamber and elastomeric mounts.

Turbo pumps must be protected against mechanical damage as well as against the loss of cooling water because the pump is a high-speed device with considerable stored energy. If a large solid particle enters the rotor or a bearing seizes, serious damage will be done to the pump. The expense of repair easily exceeds that of scraping the varnish from a diffusion pump. Such catastrophes need not and do not occur, if the most elementary precautions are taken. A splinter shield located at the pump throat adequately protects the rotors and stators from

Table 19.2 Best Practices for Turbo-Pumped Systems

Periodically change fluid in liquid-lubricated turbos; check levels.
Use adequately sized forepump.
Use vibration mounts to prevent vibration-induced damage.
Vent turbo pump from high vacuum side.
Maintain fore chamber of classical turbo in molecular flow.
Operate pump at its rated rotational speed.
Maintain cooling water flow at all times when pump is operating.
Use inlet screen to prevent objects from falling into pump inlet.

physical damage at some loss in pumping speed; some pumps are available with side entrance ports. Water cooling is used in oil-lubricated or grease-packed bearings. Proper cooling is necessary to extend bearing life. Most pumps contain a cooling water temperature sensor, so it is generally not necessary to add an external flow sensor. A water-flow restriction may be added to conserve cooling water. Any device with internal cooling water passages requires a clean water supply. Even though filters are used, it is advisable to reverse-flush the pump water lines once or twice a year to remove material that has passed through the filter. If the water supply is unreliable, use a local recirculating water cooler. No protection is needed for loss of liquid nitrogen because it does not serve a protective function.

Turbo pumps will give reliable trouble-free operation, if lubricating fluid is changed at recommended intervals and they are protected against cooling water failure, power failure, mechanical damage, and excessive torque. This protection is routinely provided. Table 19.2 describes some turbo pump best practices.

19.3 Sputter-Ion-Pumped Systems

The sputter-ion pump is the most common of all ion pumps; and is commonly used with a titanium sublimation pump and liquid nitrogen cryo baffle to form a high vacuum-pumping combination that is easy to operate and free of heavy hydrocarbon contamination. Although some ion pumped systems may still use historic sorption-pumped roughing modules, many systems use a turbomolecular pump combined with a dry mechanical pump to provide hydrocarbon-free roughing to sufficiently low pressure.

A typical, small, sputter-ion-pumped system is shown in Fig. 19.4. The TSP and cryo surface may be in the chamber on which sputter-ion pump modules are peripherally located or they may be in separate units as sketched here. Titanium is most effectively sublimed on a water, rather than a liquid nitrogen-cooled, surface in a system that is routinely vented to atmosphere. If liquid nitrogen cooling were used, the film on the

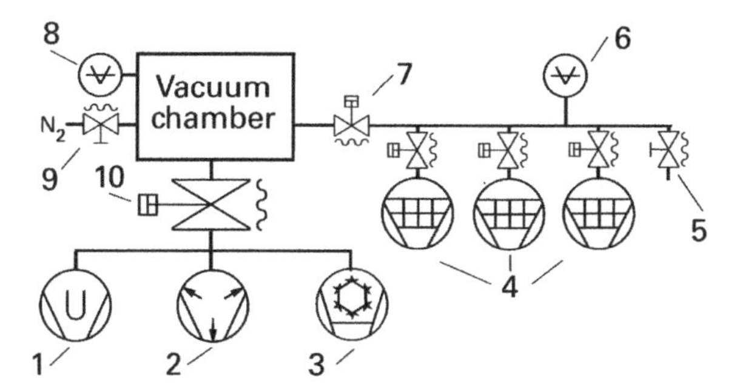

Fig. 19.4 Components of a small sputter-ion-pumped system: (1) Titanium sublimation pump, (2) sputter-ion pump, (3) liquid nitrogen-cooled array for pumping condensable gases, (4) sorption roughing pumps, (5) port for the attachment of a gas aspirator, carbon vane, or scroll roughing pump, (6) thermocouple or Pirani gauge, (7) roughing valve, (8) ionization gauge, (9) chamber release valve, and (10) high vacuum valve.

cooled surface would be composed of alternate layers of titanium compounds and water vapor. Because water vapor is the dominant condensable vapor in rapid-cycle systems, titanium flaking, and pressure bursts would frequently result. In a cycled system, it is best to separate the two functions and condense water vapor on a (valved) liquid nitrogen-cooled surface and titanium on a water-cooled surface.

A two- or three-stage sorption pump module is used to rough these systems, a gas aspirator, scroll, or carbon vane pump may be used to exhaust the chamber to about 15,000 Pa (110 Torr) before sorption pumping. Scroll pumps often require inlet particle filters to prevent backstreaming of particles to the chamber. The gas aspirator requires a high mass flow of nitrogen at high pressure and is noisier and less practical than a carbon vane or scroll pump. Neither pre-roughing pump is necessary, but their use do allow additional sorption cycles between sorption pump bakes. High-capacity modules equivalent to about 10–15 small pumps (10-cm diameter and 25-cm high) are available for roughing large volumes.

If the system is cycled frequently, a gate valve should be installed between the high vacuum pumps and the chamber to minimize operation of the ion pump at high pressures. An ion gauge is not needed on the pump side of the gate valve because the ion current serves as a de facto pressure gauge. A gas-release valve is provided for chamber release to atmosphere. Nitrogen is used for this function, because it is easily pumped by the sorption and ion pumps.

19.3.1 Operating Modes

For many small or academic sputter-ion-pumped chambers, the evacuation process still often begins with sorption pump chilling. Small commercial pumps will equilibrate in 15–30 min. If the sorption pumps have been saturated by prior use, they must be baked at a temperature of 250°C for at least 5 h before they are ready to chill. Pumping on a sorption pump with an oil-sealed mechanical pump is not a good way to speed its regeneration, because oil vapors would backstream and contaminate the sieve. This would permanently destroy the sorption pump sieve, as it cannot be removed by baking. The first stage of a two-stage pump manifold is used to rough to 1000 Pa (~10 Torr); the pump is then suddenly isolated from the manifold and the second stage is used to pump to 0.2–0.4 Pa (~200–400 mTorr). If a three-stage manifold is used, a pressure of 3000–5000 Pa (~30–50 Torr) is obtained in the first stage, 15 Pa (~150 mTorr) in the second, and 0.1 Pa (~1 mTorr) in the third. Rapidly cycled, staged pumping traps the neon that entered the first stage in viscous flow and also reduces the quantity of gas to be pumped by the last stage. Both effects reduce the ultimate pressure. At a pressure of 0.5 Pa (~5 mTorr) continuous sublimation may begin. The sputter-ion pump is started when the chamber pressure reaches 0.05 Pa (~5×10^{-4} Torr). The roughing line can be isolated from the system if the chamber and pump are clean. When the ion pump voltage increases to about 2000 V, its pumping speed will rapidly increase until it reaches its maximum value at about 10^{-3} Pa (~10^{-5} Torr). Below that pressure, continuous operation of the TSP is not necessary, because the sublimation rate of the titanium exceeds the gas flux. As the system pressure decreases, the interval between successive titanium depositions may be increased. When the layer is saturated, the pressure will rise as explained in Fig. 13.4. Timing circuits can control the sublimation time and interval between depositions.

A sputter-ion pump that has been exposed to atmosphere before its operation, will not pump gas as quickly as if it were clean and under vacuum. Operation of a system previously exposed to air begins with chilling the sorption pumps and water cooling the TSP surfaces. The sorption pumps and TSP are used to rough, as with a cycled operating system. When the sputter-ion pump is started, the system pressure will rise due to electrode outgassing. The solution is to continue pumping with the sorption and TSP until the outgassing load is reduced. If the outgassing is not reduced in a short time, power to the ion pump should be removed to avoid overheating. The outgassing should be reduced, after starting and stopping the sputter-ion pump power supply for a few 5-min cycles. When the pump reaches ~2000 V, the roughing line may be isolated from the system and normal operation resumed.

An ion pump is the easiest pump to stop. The high vacuum valve, if any, is closed and the power to the ion pump is removed. The entire system may remain under vacuum until it is needed again. The TSP cooling water should be disconnected if the system is to be vented to atmosphere to prevent condensation on the interior of the system. Table 19.3 reviews best practices for sputter-ion-pumped systems.

Table 19.3 Best Practices for Sputter-Ion-Pumped Systems

Start a sputter-ion pump below 10^{-3} Pa (10^{-5} Torr).
Operate sputter-ion pump below at pressures below 10^{-4} Pa
Isolate sputter-ion pump when venting chamber.
Keep sorption pump vent free from ice buildup.
Clean Ti flakes when replacing Ti filaments.
Use the power supply high current switch only when starting.
Remove polystyrene Dewars from sorption pumps while baking.
Sequentially operate multistage sorption pumps.
Adequately bake sorption pumps.

19.3.2 Operating Issues

One advantage of an ion pump is that no fault-protection equipment is needed to prevent damage from a utility failure. Loss of electrical power, cooling water, or liquid nitrogen will not harm the pump. Pumping will simply cease; gases will be desorbed from the walls and cryo baffle and the pressure will rise. If the pressure does not exceed 10^{-1} Pa (10^{-3} Torr), the pump can be restarted by applying power to the ion pump. However, if an internal cryo baffle is used, an overpressure safety valve needs to be installed.

Of the most severe ion pump concerns is the regurgitation of previously pumped gases, in particular hydrogen. A second concern is large gas loads, which ion pumps do not pump easily. Most of the pumping above 10^{-3} Pa (~10^{-5} Torr) is done by the TSP; at that pressure, filament life is short. Because of their slow pumping speeds, ion pumps are not suited for routine, rapid-cycle use.

19.4 Cryo-Pumped Systems

The layout of a typical cryo-pumped system driven by a helium gas refrigerator is sketched in Fig. 19.5. As in Fig. 19.1, more valves and auxiliary items are depicted than are necessary. The system requires no forepump, and mechanical pump operation is required only during roughing. After roughing, the rough pump may be turned off. A liquid nitrogen trap is not needed for the prevention of back-streaming from the cryogenic pump, but a cryo surface, cold-water baffle, or room-temperature baffle, none of which is shown in the sketch, may be necessary in the chamber to baffle the process heat load. These baffles also reduce the over-all system pumping speed. Most cryogenic pumps include a hydrogen vapor pressure or another gauge for monitoring second-stage temperature. The coarseness

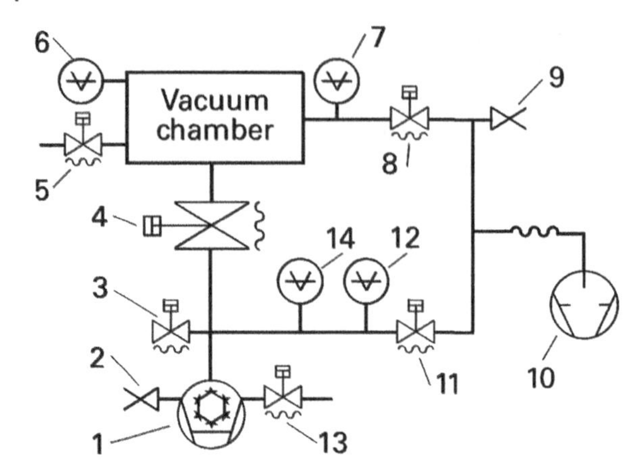

Fig. 19.5 Components of a helium gas refrigerator cryogenic pump system: (1) Cryo surfaces, (2) pressure relief valve, (3) flush gas inlet valve, (4) high vacuum valve, (5) chamber vent valve, (6) ion gauge, (7) thermal conductivity gauge, (8) roughing valve, (9) mechanical pump vent, (10) mechanical pump, (11) roughing valve, (12) thermal conductivity gauge, (13) vent valve, and (14) ion gauge.

of the gauge makes it difficult to read extremely low temperatures. A silicon diode or gold–germanium thermocouple is more accurate than a hydrogen vapor pressure bulb at cold-stage temperatures.

19.4.1 Operating Modes

A roughing pump evacuates the chamber until the crossover. See Section 18.2.2.3, *overload condition*. At that time, the roughing valve is closed, and the high vacuum valve is opened. The time required to reach the system base pressure is a function of the history of the cryopump, its radiation loading, and other traditional chamber characteristics. The chamber is cycled to atmosphere in the usual way by closing the high vacuum valve and venting with dry nitrogen gas.

System shutdown begins by closing the high vacuum valve, removing power to the compressor, and equilibrating any liquid nitrogen trap in the chamber or pump with nitrogen gas. An unpowered pump should be regenerated before being placed in operation.

19.4.2 Regeneration

The regeneration procedures, which are recommended or available within automatic controllers, make use of various techniques such as external heat, internal heat, pumping, and gas flushing. The object is to remove the captured gases from the pump after power has been removed. Not all procedures clean the sorbent and pumping surfaces properly. For example, removing the power and warming the refrigerated surfaces will allow water vapor to transfer from the warm stage to the cold stage, and eventually to form a puddle in the bottom of the pump housing. This happens because the warm stage reaches ambient before the cold stage. Cryopumps contain a safety relief, but desorbed vapors should not simply be discharged through it. Rather, the regeneration gas load should be pumped away in a mechanical pump using a controlled nitrogen gas flush, but not at pressures at which oil could backstream if oil-sealed pumps are used. Loading the charcoal with water vapor only increases the cleaning time. Procedures that advocate heating and uncontrolled mechanical pumping or flushing nitrogen through the pump are to be avoided, as are partial regeneration procedures in which argon may be transferred to the charcoal. Charcoal that is loaded with argon will not pump hydrogen at high speed.

One effective procedure consists of directing warm nitrogen purge gas into the cold stage and pumping the exhaust gases with a mechanical pump. The nitrogen gas flow quantity is adjusted to a large enough value to prevent the mechanical pumping line from ever going below a pressure that could cause oil backstreaming, if an oil-sealed pump were used. It is important to remove the heat from the nitrogen purge when the elements reach room temperature, or it will delay cooling [9]. It is important to direct the incoming warm gas flow to the first stage [10,11] or regeneration times will be considerably lengthened. A second effective procedure uses an electrical heater to warm the second-stage charcoal array to 100 K, while room-temperature nitrogen is passed through and exhausted by a pump. Upon reaching 100 K, heater power is removed, and it cools to its limiting temperature [12]. Preventing argon transfer to the charcoal from the second stage condensation array, or water vapor transfer from the first stage condensation array is key to any effective regeneration procedure, lest the helium and hydrogen pumping speeds become seriously compromised.

Ice deposits will not cause frequent regeneration in pumps used expressly at high vacuum. Frequent regeneration to remove large argon ice deposits is a concern for pumps used in sputtering systems. Regeneration is required when the power fails or dips for a brief time. Momentary loss of power is a serious concern, especially if the charcoal is saturated with helium. If power is lost for a short time, the helium will be released from the sorbent and will conduct a large amount of heat from the chamber walls to the pumping surfaces. It will serve no purpose to rough the pump to 20 Pa (150 mTorr) because that is not sufficient to prevent continued conductive heat transfer. A pump will thermally overload after a burst of helium and then require complete regeneration. If power is disconnected long enough for the first stage to release water vapor, the vapor will migrate to the charcoal and prevent it from pumping He or H_2, and regeneration will be required.

For these applications, experience has shown that the mean time between regeneration is determined by external events such as regional power failure or intentional system shut-down, and not by ice loading. Ice loading is an issue in sputtering systems and is discussed in Chapter 22.

19.4.3 Operating Issues

The most important issue is minimizing the first-stage heat load. In addition to the 300 K radiation from nearby chamber walls, the first stage is subject to thermal radiation from any source such as an electron beam hearth, heater lamp, or sputter cathode. Heat loads up to 100–150 W, which are possible in many processes, easily exceed the capacity of the ~40-W refrigerator stages. To reduce the incident flux on the first stage, some form of baffling is necessary. The simplest is a reflective, polished ambient temperature baffle. If that is insufficient, a cooled chevron array may be required. Water or liquid nitrogen may be required for baffle cooling. In many instances the manufacturer is unaware of the details of the process and so cannot provide the correct baffling. It is the user's responsibility to ensure that the process does not thermally overload the pump.

Loss of pumping by power failure or gas overload will not harm the pump. Loss of pumping will delay operation and may destroy a partially completed experiment, but it will not damage the pump. It is like leaving a refrigerator door open; food spoils! A malfunctioning overpressure relief is about the only way that damage can be done to the operator and pump.

Leaks can develop after regeneration. Charcoal dust that migrates to the O-ring seal in the overpressure relief valve is a common leak source. One may either install a 0.02–0.05-μm diameter particle filter before the valve, or clean the O-ring with a dry, dust-free wipe following each regeneration cycle.

Cryogenic pumps do not handle all gases equally well. The capacities for pumping helium and hydrogen are much less than for other gases. Some gases are easily pumped, but they present safety problems. Because cryo surfaces condense vapors, they can accumulate significant deposits, which, when warmed, could react with one another or air. Some combinations can ignite, so hot filament gauges should never be used in the pump. Venting flammable or hazardous gases does not lessen the danger of an explosion or reaction. As discussed in Chapter 14, safety overpressure relief valves are mandatory. Cryopump system safety hazards have been reviewed by Lessard [13]. High neutron or gamma radiation will subject standard polymeric parts to degradation, radiation resistant polymers avoid this problem. Safety releases with all-metal construction are recommended for such applications [14].

High-purity helium (99.9999%) is required to fill the compressors. Neon is the most common impurity in helium and may condense on the surface of the

Table 19.4 Best Practices for Cryo-Pumped Systems

Always operate a cryopump with a safety release valve.
Always keep exit of safety release free of obstructions.
Crossover at high pressure to avoid backstreaming.
Periodically change oil adsorber cartridge.
Regenerate completely with gas purging or vacuum pumping.
Clean safety release gasket to avoid carbon dust accumulation.
Regenerate after power failure.
Maintain correct He gas pressure; recharge with ultrapure He.
Isolate pumping arrays from heat sources with baffles.
Mount ion gauges only on chamber side of high vacuum valve connecting to cryopump.
Do not mount any potential ignition sources on cryopump side of high vacuum valve, or on any lines used during cryopump regeneration.

low temperature stage and lead to seal wear. Table 19.4 summarizes some cryopump best practices.

19.5 High Vacuum Chambers

All-metal high vacuum, corrosion-free chambers are usually fabricated from TIG welded 304- or 316-stainless steels using elastomer-sealed flanges. Systems with base pressure requirements of 10^{-2}–10^{-3} Pa (10^{-4}–10^{-5} Torr) are often fabricated from epoxy-painted, cold-rolled steel. Viton is the preferred gasket material for high-quality systems because of its low permeability. Buna-N is used when cost is a factor and silicone is used in certain high-temperature applications. Any metal, glass, or ceramic, whose outgassing rate and vapor pressure is adequately low can be used in the chamber, assuming that it is thermally and chemically compatible. The use of elastomers in the chamber should be approached with more caution. Extreme heat, as well as excited radicals from glow discharge cleaning, sputtering, or ion etching, will decompose materials like polytetrafluoroethylene. Whenever possible, dense alumina ceramic electrical insulators should be used. Valves with bellows and O-ring stem seals are found on high vacuum chambers. Bellows stem seals are used in locations such as the high vacuum line, foreline, and roughing line, where a vacuum exists on both sides of the seat. Both sides of this valve are leak-tight, but the stem side has a larger internal surface area than the seat side. Valves with O-ring stem seals have a high leak rate and are used only for applications like the chamber air release, where the seat side faces the vacuum. Elastomer hose roughing lines or vibration isolation sections should be avoided. Rubber hose deteriorates rapidly. Some general high vacuum system best practices are described in Table 19.5.

Table 19.5 Best Practices for General System Operation

Assemble interior fixtures with clean hand tools.
Clean hand tools with detergent, rinse with water, and then IPA. Store in clean cabinet.
Blow dry parts with dry air or dry nitrogen, and not air from an oil-sealed compressor.
Clean O-ring groves before installing O-rings.
Use only high-quality O-rings.
Clean O-rings with a detergent wash, alcohol rinse, and air bake at 40°C.
Use a minimum amount of grease on gaskets; wear cleanroom gloves when applying.
Tighten flange bolt pairs in diametric opposition, and in a sequence that applies uniform pressure to flange. Numbering bolt-hole sequence on large flanges is helpful.
Vent chamber slowly, and with clean, dry gas to avoid contamination.
Ground all chambers.
Ground or shield unused electrical feed-throughs.
Record initial pumping rate and rate of rise.
Record initial clean RGA spectrum for later comparison.
Handle internal fixturing with clean, particle-free gloves.
Warm auxiliary cold traps and pumps before venting to atmosphere.
Monitor LN_2 cold trap exhausts, to prevent ice from plugging vents.
Place protective covers over glass pressure gauges.
Wear hearing protection near loud pumps.
Keep a maintenance logbook for each system; enter each maintenance event.

19.5.1 Managing Water Vapor

Water vapor is a major problem in vacuum systems, and this is due to its binding energy; it is "sticky." Hydrogen bonds quite strongly to steel; its presence does not affect pumping time in a high vacuum region. Inert gases bond weakly to steel; they neither stick nor affect pumping time. Because the binding energy of water is between, it alone determines the time required to reach base pressure in an unbaked chamber. This is graphically depicted in two illustrations. The first, Fig. 19.6, demonstrates the relation between the exposed quantity of water and the deposited amount. These data show that the *adsorbed* quantity was directly proportional to the *exposure* quantity [15]. The second illustration, Fig. 4.5 of Chapter 4, illustrates the time required to pump away these individual water vapor loads. Together, the two figures validate something that we all know—increased atmospheric exposure time results in significantly increased pumping time.

The initial water vapor load can be reduced in a number of ways that will minimize the pumping time; some are included in Section 18.1. Best practices are illustrated in Table 19.6. Nitrogen purging and nitrogen venting are not the same; they have different meanings. Venting to atmosphere with nitrogen means admitting nitrogen gas after the pump valve is closed. Purging with nitrogen means flowing

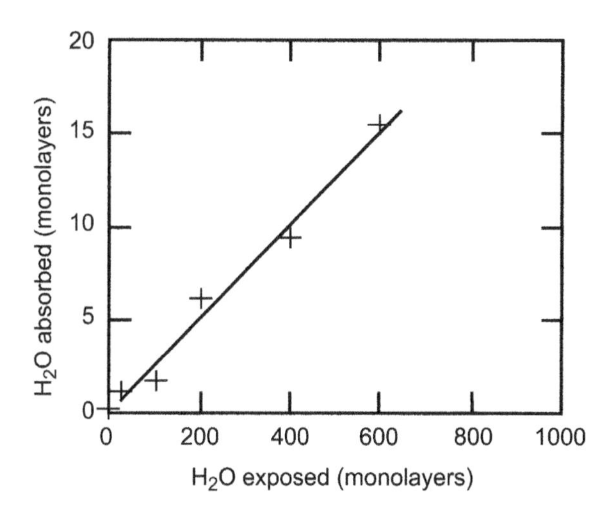

Fig. 19.6 Relation between water vapor exposure and water vapor adsorption on a stainless-steel surface. Adapted M. Li and H.F. Dylla [15].

Table 19.6 Best Practices for Managing Water Vapor

Expose the system to air for the minimum time necessary.
Flush the system with dry nitrogen when open, if useful.
Heat chamber surfaces when open to reduce water adsorption.
Add internal heat lamps to enhance water vapor desorption
Install load locks if possible.
Use multiple pumps on complex chambers.
Pump slowly, or flush with dry nitrogen to avoid aerosol formation.
Remove power to closed cycle water chillers when chamber is open.

nitrogen gas through the chamber to atmosphere. The effectiveness of a gas purge depends on geometry. Allowing nitrogen purge gas to flow through a chamber and exit via a small-diameter flange will reduce, but not eliminate, the back-diffusion of atmospheric water vapor. When open, most practical chambers expose large internal surface areas to atmosphere, so alternative means must be pursued. Heating the exterior walls to ~50°C reduces water adsorption. Reducing time that the chamber is exposed to air is the most effective procedure. One can minimize the initial water load by reducing the water vapor flux, exposure time, and adsorption probability.

Once the chamber doors are closed, the options are limited. The initial surface load of water must be removed to the level required to reach process pressure. Some processes allow for chambers to be modestly baked (~50°C), but some processes may have thermal limitations such as a Currie temperature. Baking reduces residence times and speeds pumping. Photon flux from quartz UV lamps, which is absorbed in near-surface layers, will desorb water. Desorption produced by the UV results from surface heating; the flux cannot illuminate all interior surfaces uniformly.

References

1 Hablanian, M.H., *J. Vac. Sci. Technol.* **6**, 265 (1969).
2 Holland, L., *Vacuum* **21**, 45 (1971).
3 Singleton, J.H., *J. Phys. E.* **6**, 685 (1973).
4 Vacuum Barrier Corp., Woburn, MA 01810.
5 Benson, J.M., *Transactions of the 8th National Vacuum Symposium of the American Vacuum Society, 1961*, Vol. 1, L. Preuss, Ed., Pergamon, New York, 1962, p. 489.
6 Hengevoss, J. and Huber, W.K., *Vacuum* **13**, 1 (1963).
7 Santeler, D.J., *J. Vac. Sci. Technol.* **8**, 299 (1971).
8 Seibert, J.F. and Omori, M., *J. Vac. Sci. Technol.* **14**, 1307 (1977).
9 Scholl, R.A., *Solid State Technol.* **187** (1983).
10 Bridwell, M. and Rodes, J.G., *J. Vac. Sci. Technol., A* **3**, 472 (1985).
11 Longsworth, R.A. and Bonney, G.E., *J. Vac. Sci. Technol.* **21**, 1022 (1982).
12 Häfner, H.-U., Klein, H.-H., and Timm, U., *Vacuum* **41**, 1840 (1990).
13 Lessard, P.H., *J. Vac. Sci. Technol., A* **8**, 2874 (1990).
14 K.M. Welch, "Capture Pumping Technology", Chapter 5 in: *Cryopumping*, 2nd ed., North Holland, 2001. p. 313.
15 Li, M. and Dylla, H.F., *J. Vac. Sci. Technol., A* **11**, 1702 (1993).

20

Ultraclean Vacuum Systems

Treatises on vacuum technology traditionally include a section entitled "ultrahigh vacuum." For those of us whose interest in vacuum dates to soon after the 1950s, that phrase indicated specific construction techniques, components, materials, and procedures such as those summarized in Table 1.1. The goal was the achievement of an ultrahigh vacuum (UHV) research environment. Today, this environment is required for myriad applications ranging from high-energy particle research to the high-volume manufacture of optical and magnetic films, semiconductor devices, as well as molecular beam epitaxy systems used to fabricate optoelectronic devices. Systems have grown in size and complexity and now include *in situ* metrology for real-time product analysis and process control. In parallel, the time to reach process conditions has been reduced for both technical and economic reasons.

The term "ultrahigh vacuum" is no longer adequate to describe the requirements demanded by these applications. The term "ultraclean vacuum" best describes the wide ranging and sophisticated requirements of modern high-end product manufacturing. Table 20.1, from ISO Standard 3529-1:2019(E), provides one definition of ultraclean vacuum. As it states, process conditions may take place in different vacuum ranges, but have a common issue of extremely high process gas purity that may include low particle density. Thus, UHV materials, construction techniques, and operating procedures become necessary for processes that operate in both the traditionally less demanding medium and the high vacuum ranges. In this sense, UHV is not an end unto itself but a pathway to ultraclean conditions.

Ultraclean vacuum processes require translational and rotational motion, electrical power and signals, cooling fluids, and optical instrumentation that function without adding contamination. Metrology is not simple; hot filaments can produce carbon monoxide, carbon dioxide, and methane. UHV pressure gauges and residual gas analyzers may provide incorrect information if mounted or operated improperly; misunderstanding of electron-stimulated desorption can result in

A Users Guide to Vacuum Technology, Fourth Edition. John F. O'Hanlon
and Timothy A. Gessert.
© 2024 John Wiley & Sons, Inc. Published 2024 by John Wiley & Sons, Inc.

Table 20.1 ISO Standards Definitions for Ultraclean Vacuum

Medium or high vacuum that requires special conditions for some gas species equivalent to UHV conditions.
Note 1: The requirements for the particular gas species (impurity) depend on the application.
Note 2: Hydrocarbons, CO, CO_2, and H_2O are typical impurity gases.
Note 3: The particular requirements may include specifications for low particle density.

Source: © ISO. This material is reproduced from ISO 3529-1:2019 with permission of the American National Standards Institute (ANSI) on behalf of the International Organization for Standarization. All rights reserved.

gauge inaccuracies. The need for distributed pumping to evacuate extremely long, narrow accelerator or beam chambers is well understood, and manufacturing systems have adopted these pumping techniques to minimize local contamination.

Some systems are mildly baked, whereas others, by virtue of their size, cannot be baked, but must be degassed by other methods. Some require base pressures in the range 10^{-8}–10^{-9}-Pa (10^{-10}–10^{-11}-Torr) before beginning processes that operate in the medium vacuum range. These chambers are constructed with UHV techniques including metal gaskets, and materials with low outgassing and permeation.

There are many sources and types of vacuum system contamination. Gases can desorb from the chamber walls or evolve from a pump. To place these sources in context, we begin with a modified version of a view first proposed by Santeler [1]. Figure 20.1 sketches a generalized vacuum system. The total pressure in the system from vacuum chamber sources is composed of the partial pressure of each gas or vapor desorbing from the walls; it is equal to the rate of desorption of that species divided by its pumping speed.

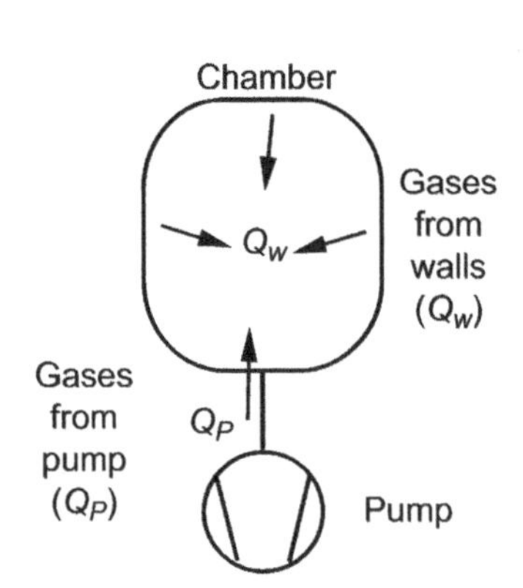

Fig. 20.1 High vacuum pumping system schematic.

$$P = \sum_i \frac{Q_i}{S_i} \tag{20.1}$$

Gases that back diffuse from the pump will contribute to the chamber background pressure.

$$P = \sum_i \frac{Q_{\text{pi}}}{S_i} \tag{20.2}$$

The total system pressure results from all sources. It is given by:

$$P = \sum_i \frac{Q_i}{S_i} + \frac{Q_{\text{pi}}}{S_i} \tag{20.3}$$

Santeler argues that the first term behaves normally, e.g., as pumping speed increases, the pressure contribution decreases, but the second term does not behave in a like manner. For example, more hydrogen would backstream from a larger sputter-ion pump. If the back-streamed gas load and the pumping speed were increased in proportion, the contaminant background pressure would not change. The result would be an effectively constant base pressure P_o, for a given pump. By this argument Eq. (20.3) may be simplified to read:

$$P = \sum_i \frac{Q_i}{S_i} + P_o \tag{20.4}$$

The production of ultraclean vacuum is therefore concerned with the choices of pumps, as well as chamber materials, components, and operating procedures. The contamination from the chamber may be reduced by the intelligent choice of UHV materials, fabrication techniques, and operating procedures; the gas load originating in the pump may be reduced by pump selection and processing procedures.

This chapter discusses the roles of pumps (turbo, ion, TSP, NEG, and cryo) and chambers (design, materials, and operation) in achieving ultraclean vacuum conditions. The chamber becomes important because it is easier to reduce chamber sources than to increase pumping speed. Pump choice is important because increasing S may not decrease P_o significantly as implied by Eq. (20.4). Pumps are selected by matching their performance with the application.

20.1 Ultraclean Pumps

Equation 20.4 indicates that a clean vacuum pump should reduce pressure by providing high pumping speed (low Q/S) and low backstreaming. In addition, the pump should not be a source of particles. Here, backstreaming is defined as the reverse flow of any gas, vapor, or material from the pump. To reduce the contaminant levels, rigorous adherence to exacting pumping procedures is necessary. For example, a cryogenic trap that has been accidentally warmed for only a few minutes will release condensed vapors. Although re-cooling will prevent further contamination, it will take time to repump the vapors that have been released, especially those that were marginally pumped on cryo surfaces.

Each pump type has its own characteristic class of gases and vapors that it will not pump and that it will generate as impurities. Each pump has its own characteristic particle source. The design and selection of pumps for ultraclean applications are done on an individual basis. There is often more than one "best" solution. The remainder of this section briefly reviews the selection and clean operation of some ultraclean configurations. Additional discussion is given in Section 2 of Chapter 21.

20.1.1 Dry Roughing Pumps

An oil-free roughing system is strongly preferred on ultraclean systems because oil filters and traps will not provide the required level of cleanliness for this class of system. Diaphragm, scroll, and multistage root pumps are excellent choices for oil-free roughing pumps. While diaphragm and scroll pumps can produce particle contamination, fast-closing valves, and 0.2-μm particle filters can be installed between the scroll pump and process chamber. If slightly lower pressure roughing is required, and a source of liquid nitrogen (LN_2) is readily available, the historic option of LN2-charged sorption pumps could be considered. In this case, a properly operated two-or three-stage Zeolite sorption pump system can cost-effectively rough a small chamber into the 10^{-2} Pa (10^{-4} Torr) range without fear of backstreaming particles or organic vapors.

20.1.2 Turbopumps

Although considered relatively clean HV and UHV pumps, turbopumps can contribute both hydrogen and lubricant vapors into the chamber. The hydrogen is primarily a result of the pump's limited compression ratio for this fast-moving molecule (of order 5000–15,000) and the hydrogen partial pressure in the foreline. Most modern turbopumps are designed with a series-connected molecular drag stage and will increase these pumps' hydrogen compression ratios. While turbopumps with magnetically levitated bearings eliminate the organic vapors present in pumps designed with oil-lubricated bearings, most modern pumps incorporate purging of the bearing regions to limit significantly limit contamination concerns from both vapors and particles. Finally, because turbopumps have a large internal surface area, cleanliness can be enhanced by baking one turbopump into an adjacent turbopump. This technique can reduce water vapor and other previously pumped vapors from migrating from the interior of the stator and rotor surfaces into the chamber.

20.1.3 Cryopumps

Closed-cycle helium gas cryogenic pumps are able to reach the UHV region required for rapid-cycle industrial processing systems. Cryogenic pumps can pump large but finite amounts of hydrogen, provided the cold-stage temperature remains at or below 20 K. Above 20 K, hydrogen will be released. Adsorption isotherms show the greatest capacity for pumping hydrogen at low temperatures, and this is a requirement for producing ultraclean conditions. Another important consideration in reaching very low pressures is the isolation of thermal and optical radiation from the cold stage. Baffles have been designed to maximize molecular transmission while minimizing optical transmission [2,3]. Indium thermal gaskets limit baking cryopumps to 70–100°C. Purging with warmed nitrogen is often used during cryopump regeneration, but turbopumps can be used during post regeneration to the same end.

Cryopumps use no high voltages and do not produce hydrocarbons or metal flakes. However, the displacer generates some vibration and the charcoal used on the cold stage sheds some dust and particles that can be transported to the chamber. Iwasa and Ito [4] developed a helium gas refrigerator pump using an Er_3Ni heat exchanger with which they were able to reach 3.6 K. Thus, they were able to produce a cryopump with no dust-producing charcoal. Cryopumps must be carefully damped for use with sensitive surface-analysis equipment such as SEM, ESCA, or SIMS. For UHV applications, it is mandatory that the Viton-sealed overpressure relief used in HV applications be replaced by its single-use, stainless-steel-foil counterpart. Cryopumps are often augmented by large area, LN_2 cryo arrays placed within the main chamber. These are designed specifically for pumping water vapor.

Demanding research applications may use all LN_2 and liquid helium pumps. They must be operated with care, so that vents do not become plugged. Liquid helium has become scarce and expensive; all-liquid pumps are no longer common. Those in use will have a gas helium recovery system to collect and compress the liquid boil-off. The recovered helium can them repurified.

20.1.4 Sputter-Ion, TSP, and NEG Pumps

Particle beamlines require distributed pumping. These have been designed using large areas of NEG pump material in combination with sputter-ion pumps [5]. The distributed NEG pumps remove all except the noble gases, which are then pumped by the sputter-ion pumps. Sputter-ion pumps require baking; they can be baked into the turbopumps used for system roughing. The main background gas present in a sputter-ion pumped system at low pressures is hydrogen, but small amounts of methane, ethane, and carbon monoxide are observed.

When used in combination with He closed cycle cryopumps, NEG pumps have been shown to improve pumping speed of an unbaked chamber and reduce the base pressure of a baked system [6]. NEG pumps have also been used in combination with turbopumps to produce enhanced pumping of hydrogen and water vapor [7]. Benvenuti et al. [8] have demonstrated that hydrogen, the dominant gas remaining at base pressure, could be pumped with higher speeds by depositing thin-film coatings of nonevaporable getter material on the system walls. Although distributed pumps are necessary to maintain high hydrogen pumping speed at low pressures, titanium sublimation pumps (TSP) have been considered [9,10]. Hydrogen pumping in TSPs is improved if the surfaces are liquid-nitrogen-cooled rather than water-cooled. Odaka and Ueda [11] have demonstrated that a 1000°C bake of the Ti–Mo TSP filaments removed large quantities of gaseous impurities. The water vapor load in an ultraclean system is much smaller than it is in an unbaked rapid-cycle system; therefore, the sublimed titanium film will not entrap enough water to make titanium flaking a problem. Momose, Saeki, and Ishimaru [12] identified TiO particles generated by ion pumps; they postulated that these particles were transported from insulating surfaces within the pump to the chamber by the nitrogen purge gas. In addition to their use in large distributed systems, getter pumps have been used as local pumps in sputter process chambers to pump impurity gasses selectively.

20.2 Ultraclean Chamber Materials and Components

AISI grades 304L and 316L stainless steel are often used to build ultraclean chambers. Stabilized grades such as 321 or 347 contain additives to reduce carbide precipitation; they are acceptable but expensive. All joints are made by Tungsten Inert Gas (TIG) welding or metal gasketed flanges; it is preferred that O-rings are not used anywhere in the high vacuum portion of the system, including the high vacuum connection to the pump or trap. However, O-rings are sometimes used for internal pass-through valves or on an outer door of a load lock. After the initial air content outdiffuses from O-rings within internal valves, they will no longer be a significant source of air contamination. However, they shed particles, and behave as sponges, releasing whatever process gas they were exposed to prior to valve closure.

Aluminum is increasingly being used for chambers and components. With air exposure, ordinary aluminum becomes covered with a porous, hydrated aluminum oxide over 10 nm thick. This layer becomes a source for water vapor and therefore is difficult to outgas. However, when processed by special extrusion, or by the EX-process, aluminum outgasses at a similar rate as stainless steel. The special extrusion process results in a dense aluminum oxide layer of approximately 3-nm thickness; it is achieved by extruding the aluminum in an atmosphere of argon and oxygen [13]. The surface finish on large aluminum components is realized by lathe machining in a chamber filled with argon and oxygen [14,15]; the resulting oxide is identical in thickness and density to that made by special extrusion. Outgassing rates of 10^{-11} Pa-m/s (10^{-14} Torr-L/s) are achieved with these treatments [16].

High-density alumina is a stable insulator, and it is stronger than quartz or machinable glass ceramic. The properties of this and other materials are discussed in Chapter 15. Metal-to-glass or metal-to-ceramic seals, discussed in Chapter 16, should not be subjected to temperatures higher than 400°C, and sealed copper-gasket flanges should not be baked at temperatures higher than 450°C. Unsealed flanges will tolerate higher temperatures. The flange and knife edge will suffer some loss of temper during an 800–950°C bake. The exact amount depends on the grade of stainless steel (grain growth retarders) and fabrication steps used in flange construction. Some commercial copper-gasket flanges are forged from grade 304, and some contain trace additives that will retard grain growth during heat treatment. Small grain size is necessary to prevent significant loss of hardness. Not all flanges are forged in the same manner or have the same properties; users with exacting requirements need exacting purchase specifications.

Parts should never be stored in plastic boxes or bags. Polymer storage bags contain plasticizers that readily outdiffuse. Sulfur from rubber bands that may be used with plastic bags will diffuse quickly through the storage bags [17]. Aluminum foil is used to cover open flange faces or to wrap cleaned parts until final assembly. However, one must take care to use foil that has been prepared for vacuum use will not contaminate the cleaned parts [18] (see discussion in Section 15.1.3). Inexpensive precleaned aluminum foil is commercially available.

Internal fixtures held together with screws will have gas trapped between flat surfaces and within and under screw threads. Slotted screws, which allow trapped gases to be vented, should be used within ultraclean chambers. Gold- or silver-plated bolts, used to secure copper gasket-sealed flanges, will prevent galling of threads. Thread galling can be a significant problem with systems that are frequently baked.

Connecting electrical power, instrumentation wiring, processing gases and liquids, and cooling fluids through an UHV wall is not trivial. There are few techniques for accomplishing these objectives. If one requires only linear motion the welded bellows, described in Fig. 16.23*a* allow motion while maintaining vacuum integrity with a solid metal wall. If rotary motion is required, the magnetic follower, described in Fig. 16.21 may be considered. If bake-out temperature is high, feedthrough demagnetization suggests switching to a "wobble stick" rotary feedthrough as described in Fig. 16.23*b*.

A solid stainless-steel interface separates vacuum from atmosphere. If cooling fluids, instrumentation, and electrical power must be provided to a rotating substrate holder, then a differentially pumped rotary motion feedthrough of the design illustrated in Fig. 16.19 is a solution. Magnetic fluid seals contain organic liquids that have finite vapor pressures and, as well, allow air permeation. Unfortunately, many cannot be baked at temperatures over 60°C. These characteristics often make these feedthrough types unacceptable in the lower pressure ranges of interest in ultraclean systems.

Ultraclean chambers are assembled within clean facilities using clean assembly techniques. After pretreatment and fabrication, components, wire, hardware, subassemblies, and all assembly tools are critically cleaned before being moved to the assembly cleanroom (see also Section 21.6). Clean tools remain best remain within clean assembly areas. After assembly and testing are complete, systems are double-wrapped for customer shipment.

20.3 Ultraclean System Pumping and Pressure Measurement

A generic pressure–time behavior during pumping is sketched in Fig. 4.10. The volume gas is removed first, followed by surface desorption, outdiffusion from the solid, and, last, permeation through solid walls. All of these processes, except initial volume gas removal, are greatly temperature dependent, because diffusion and desorption rates increase with temperature. Volume gas initially cools as it expands and produces aerosol particles that usually contain a dust particle in their center, as was described in Chapter 18.

The dominant gas desorbing from the interior surfaces in this portion of the pumping cycle is water vapor. Water vapor will bounce, re-adsorb, and bounce again during the process of being pumped. Eventually, the molecule will reach a pump entrance, where it has a probability of being pumped. Equation 4.16 estimates the residence time of a molecule. This relation predicts that the residence time is dependent on the ratio of chamber wall area to pump entrance area. If the pump entrance area is small (low pumping speed) compared to the total internal area of the system, the residence time will be long. Figures 15.2 and 15.4 show the outgassing rates, respectively, of a stainless-steel and an aluminum test chamber as they are pumped from atmosphere. These rates were shown to be approximately independent of surface treatment because water vapor was the dominant surface contaminant. Multiplying the vertical axis of either of these figures by the internal surface area and dividing by the pumping speed will yield a typical pumping time. Systems are designed to pump from atmosphere to 10^{-7}–10^{-8} Pa (10^{-9}–10^{-10} Torr) in a specified time, usually in the range 24–48 h. It is for this reason, load, and transfer locks are used.

Designers of long particle-beam accelerators learned quickly that distributed pumping is needed to reach suitably low pressures. Locating a pump at one end of a long, low-conductance tubular chamber would not provide adequate pumping for the entire duct. For that reason, beam systems are constructed in the form of continuously interconnected extruded tubes of cylindrical or oval cross-section. Typically, NEG pumping is provided continuously, and supplemented by locally placed sputter-ion pumps.

Designers of complex processing chambers with many baffles or internal shields can use the same concept in attempting to reach low pressures quickly.

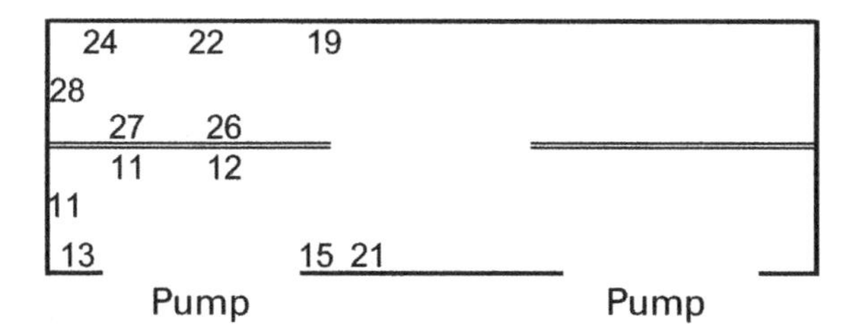

Fig. 20.2 Vacuum chamber containing arbitrary baffles and pumps. The numbers on the interior surface represent the number of bounces required by a molecule originating at that location before it reached a pump entrance.

Consider the diagram in Fig. 20.2. Here we have drawn a generic chamber of rectangular cross-section containing two pumps and two internal surfaces. These internal surfaces were added to simulate pumping in a complex chamber. The numbers located around the interior were generated by the Monte Carlo technique and represent the number of bounces required by a molecule to reach a pump starting from that location. The number of bounces is highly dependent on location and chamber geometry. If one were to install larger pumps at the existing locations, the number of bounces, and hence the time to remove molecules from the distant portion of the chamber, would be altered only marginally. Only by installing an additional pump or two in the upper chamber could one reduce the pumping time for molecules originating in the upper chamber. Distributed pumping is required to improve the ultimate pressure and time required to reach the base pressure of processing systems containing complex, conductance-limited pumping paths.

Surface preparation methods, treatments, and procedures used for reaching pressures in the 10^{-9} Pa (10^{-11}) Torr region have undergone major changes since UHV pressures were routinely attainable. The original procedures for Diversey cleaning chambers fabricated from 304 stainless steel, followed by a 450°C bake during each pumping cycle, are no longer used. Some very large systems cannot be baked. In large systems, various plasma discharge treatments such as those discussed in Chapter 15 are used to remove adsorbed gases and reduce the ultimate pressure. High temperature baking, if used, is done after chamber assembly, and pumping to prevent rediffusion of gases into the steel. After assembly, a 50°C bake has been considered by some users as suitable for UHV systems that may have trace amounts of surface organic contamination; anecdotal evidence stated that a higher temperature bake would polymerize any organic residues and prevent their removal. However, that is not an issue for surfaces that have been properly cleaned.

Unbaked areas of the system can dominate performance. The area of the system that cannot be baked depends on the pump and the efficiency with which interior surfaces can be baked. An ion pump can be completely baked. Turbomolecular and gas refrigerator cryopumps can be baked to 50°C with no problems. In some cases, interior surfaces may not completely bake. Consider again the illustration of Fig. 20.2. If the surfaces inside the chamber—simulated with a double line—could not reach the baking temperature, then molecular residence times would be significantly longer than on heated surfaces. Monte Carlo analyses of several chamber configurations showed that molecules bounce from heated and unheated surfaces in roughly the proportion of the heated-to-unheated surface area ratio. Since residence time is exponentially reduced with increasing temperature, complete baking of interior surfaces is desirable. UV lamps have been used to assist in this goal. However, they are most useful in applications where large areas can be illuminated.

It was commonly thought that CO and H_2 were the predominant residual gases found in all UHV systems. The data presented in Fig. 5.14 confirm that CO is only an artifact of hot filaments. Hydrogen is the dominant residual gas at UHV pressures. The source of this remaining hydrogen outgassing has been the subject of considerable research. Hydrogen is soluble in many metals. Bills [19] noted that there was insufficient atmospheric hydrogen to account for permeation through the bulk of a stainless-steel wall, but that other sources, such as the dissociation of water on the chamber exterior, could provide an adequate diffusion source. Redhead [20] has reviewed the status of hydrogen outgassing in UHV region. In Section 15.3.1 we discussed the diffusion model [21], which assumed hydrogen of one activation energy permeated by a diffusion-limited step with zero concentration at the vacuum interface. As Redhead noted, this predicted high outgassing rates well, but did not predict low levels of outgassing. Moore [22] applied an older recombination model that showed hydrogen atoms must recombine at the surface before desorption. Thus, the concentration of H at the vacuum surface cannot be zero. For low surface coverage the atoms must travel a distance before recombining and recombination becomes the limiting step. Moore solved this problem numerically by matching the diffusive flow of H from the near-surface concentration gradient with the observed outgassing rates by adjusting the surface recombination coefficient. This concept was used to model hydrogen outgassing from the after-shot dose in a fusion device [23].

A portion of the hydrogen dissolved in stainless steel is trapped at sites such as precipitates and in the Cr-rich phase of stainless steel [24]. These traps have activation energies ranging from 73 to 180 MJ/kg-mol. Redhead noted that baking at temperatures near 800°C depopulates a large number of hydrogen atoms trapped in these sites, some of which then occupy sites of lower activation energy. For this reason, baking only at a low temperature, say 50°C, may yield lower outgassing rates than by first vacuum firing at 800°C. Additionally, oxygen plays a role in reducing hydrogen outgassing rates. Early studies [25,26] showed that the outgassing rate of stainless steel decreased with increasing concentration of Cr in the oxide but was unchanged with increasing Fe content of the oxide. Thus, the oxide can become a hydrogen surface trap. These studies explained why the outgassing rate of stainless varied widely with surface treatment.

An experimental technique for reducing hydrogen outgassing makes use of a coating applied to the exterior surface of stainless steel [20]. A 5× reduction in outgassing was reportedly achieved from the reduction of the dissociation of water on the exterior stainless-steel surface. Hydrogen outgassing may not result solely from H within traps, but partly from interstitial permeation. Interstitial H has an activation energy of 56 MJ/kg-mol [25].

Interior chamber coatings of amorphous silicon and titanium nitride have been examined as methods of reducing hydrogen desorption from the interior of UHV/XHV chambers below that of baked stainless steel [27]. That study showed that 304 stainless steel coated with α-Si and mildly baked, achieved an outgassing rate comparable to that of uncoated stainless with a very high temperature bake. Such coatings might find use in analytical instruments.

The choice of gauges, where to mount them, and how to care for them during baking, are relevant issues for ultraclean systems. No gauge tube should be mounted with its entrance facing upward, lest it become a sink for debris. The gauge tube should be kept warmer than the system during cooling after a bake cycle. Gauges and RGA heads should be the last items cooled otherwise they become small capture pumps.

Perhaps the most critical issue is the choice of gauge. Hot filament gauges should be operated with emission currents of 10 mA so as to minimize ESD. As noted in Section 5.2.3, few commercial gauges were available for operation in this region, with the extractor gauge often being the preferred choice (see Fig. 5.13). Commercial availability of accurate UHV gauges is improving. The Stabil-Ion Gauge [28] has a quoted accuracy of ±4% accuracy at 3×10^{-9} Pa (2×10^{-11} Torr) with embedded chip [29]. The best "calibrated" cold cathode gauges are ±10% accuracy; however, their accuracy reduces to the standard cold cathode gauge accuracy of ±30% with use. Advanced gauges and RGAs incorporating the designs of Watanabe [30] for reducing ESD and electrode outgassing may be required. Much of the guidance discussed in this chapter is summarized in Table 20.2.

Table 20.2 Best Practices for Ultraclean Vacuum Systems

Construct systems from low-carbon steel and TIG welded joints.
Seal external joints only with metal gaskets; no elastomer gaskets.
Carefully *in situ* clean interior surfaces to reduce hydrogen desorption.
Use only dry mechanical or LN_2 sorption roughing pumps.
Use only turbomolecular, cryogenic, sputter-ion, or titanium sublimination, or non-evaporable getter high vacuum pumps.
Bake parts with threaded sockets, but not the screws, to prevent thread galling.
Use supplemental getter, or LN_2 cryo-panel pumping if needed.
Install metal knife-sealed overpressure relief on cryogenic pump
Measure base pressure with Extractor or Stabil-Ion Gauge
Install feedthroughs constructed with all-metal seals to allow high-temperature baking of vacuum system.
Design interior fixturing using very low outgassing, low permeability materials.
Install clean fixturing with clean tools in a clean assembly area.
Supply only ultrapure process gasses.
Access process chamber via load/unload locks.

References

1 Santeler, D.J., *J. Vac. Sci. Technol.* **8**, 299 (1971).
2 Benvenuti, C. and Blechschmidt, D., *Jpn. J. Appl. Phys. Suppl. 2* **13** (2-1), H. Kumagai, Ed., 77 (1974).Proceedings of the 6th International Vacuum Congress, Kyoto, Japan
3 Halama, H.J. and Aggus, J.R., *J. Vac. Sci. Technol.* **12**, 532 (1975).
4 Iwasa, Y. and Ito, S., *Vacuum* **47**, 675 (1996).
5 Benvenuti, C., *Nucl. Instr. Meth.* **205**, 391 (1983).
6 Giannantonio, R., Succi, M., and Solcia, C., *J. Vac. Sci. Technol., A* **15**, 187 (1997).
7 Pozzo, A., Boffito, C., and Mazza, F., *Vacuum* **47**, 783 (1996).
8 Benvenuti, C., Cazeneuve, J.M., Cicoira, F., Santana, A., Johanek, V., Ruzinov, V., and Fraxedas, J., *Vacuum* **53**, 219 (1999).
9 Benvenuti, C. and Hauer, C., *Le Vide, Suppl. 2* **201**, 199 (1980).
10 Rao, V., *Vacuum* **44**, 519 (1993).
11 Odaka, K. and Ueda, S., *Vacuum* **44**, 713 (1993).
12 Momose, T., Saeki, H., and Ishimaru, H., *Vacuum* **43**, 189 (1992).
13 Ishimaru, H., *J. Vac. Sci. Technol., A* **2**, 1170 (1984).
14 Chen, J.R., Narushima, K., Miyamoto, M., and Ishimaru, H., *J. Vac. Sci. Technol., A* **3**, 2200 (1985).
15 Miyamoto, M., Sumi, Y., Komaki, S., Narushima, K., and Ishimaru, H., *J. Vac. Sci. Technol., A* **4**, 2515 (1986).

16 Ishimaru, H., *J. Vac. Sci. Technol., A* **7**, 2439 (1989).
17 Sigmond, T., *Vacuum* **25**, 239 (1975).
18 Rao, V., *J. Vac. Sci. Technol., A* **11**, 1714 (1993).
19 Bills, D.G., *J. Vac. Sci. Technol.* **6**, 166 (1969).
20 Redhead, P.A. "*Intl. Workshop on Hydrogen in Mat's. Vac. Sys*", in *AIP Conf. Proc. 671, AIP Melville, NY*, G.R. Myneni and S. Chattopadhyay, Eds., 2003, p. 243.
21 Calder, R. and Lewin, G., *Br. J. Appl. Phys.* **18**, 1459 (1967).
22 Moore, B.C., *J. Vac. Sci. Technol., A* **13**, 545 (1995).
23 Akaishi, K., Nakasuga, M., and Funato, Y., *J. Vac. Sci. Technol., A* **20**, 848 (2002).
24 Yoshimura, T. and Ishikawa, Y., *J. Vac. Sci. Technol., A* **20**, 1450 (2002).
25 Ishikawa, Y., Yoshimura, T., and Arai, M., *Vacuum* **47**, 701 (1996).
26 Okuda, K. and Ueda, S., *Vacuum* **47**, 689 (1996).
27 Abdulla, M., Mamon, A., Elmustafa, A., Stutzman, M., Adderley, P., and Poelker, M., *J. Vac. Sci. Technol., A* **32** (2), 021604-1 (2014).
28 Arnold, P.C., Bills, D.G., Borenstein, M.D., and Borichevsky, S.C., *J. Vac. Sci. Technol., A* **12**, 580 (1994).
29 Fedchak, J.A. and Defibaugh, D.R., *J. Vac. Sci. Technol., A* **30**, 061601 (2012).
30 Watanabe, F., *J. Vac. Sci. Technol., A* **20**, 1222 (2002).

21

Controlling Contamination in Vacuum Systems

21.1 Defining Contamination in a Vacuum Environment

This chapter gives an overview of the origins and possible effects of contamination in vacuum systems. Many aspects will be presented in condensed form, and more detailed investigations will often be required to gain the necessary insight to understand how a particular vacuum contaminant is affecting a particular activity.

21.1.1 Establishing Control of Vacuum Contamination

Contamination control in vacuum technology typically begins with securing management support for a sufficient budget to develop and enact a plan. From that point, the plan will define specific contaminants, methods for their reduction, specialized training requirements, and ultimately, a method to measure success. Often, to address vacuum contamination issues, it is necessary to begin during product planning stages. For example, if vacuum components cannot be acquired from sufficiently clean sources, the user needs to develop and install component cleaning processes. The plan for success measurement could include direct measurements of contamination or improved product performance, or both.

Once corporate commitment and a contamination-control plan are established, the next critical step is getting "buy-in" from the people who will have to implement the plan on a daily basis. This can be a surprisingly complex task because it requires workers to develop an attitude to *combat something they cannot see.* Gaining sufficient buy-in of workers requires a deep understanding of contamination-related concerns downstream in process, in product development, or both. Workers will need to participate in unique training and develop new and specialized skills. Everyone involved in contamination-critical processing will

A Users Guide to Vacuum Technology, Fourth Edition. John F. O'Hanlon and Timothy A. Gessert.
© 2024 John Wiley & Sons, Inc. Published 2024 by John Wiley & Sons, Inc.

need to appreciate that the contamination control chain is only as strong as its weakest link. Contamination control requires commitment by all those involved.

21.1.2 Types of Vacuum Contamination

To a rough approximation, as shown in Fig. 21.1, one can assume that approximately *equal amounts* of contamination come from the following four areas: (1) process equipment, (2) the actual process, (3) chemicals or sources, and (4) personnel [1]. A closer inspection will often indicate that the largest source of these four areas relates to personnel. Vacuum contamination can further be divided into sources that are internal or external to the system. Internal sources can include particles and vapors originating from vacuum pumps, desorption of materials used to produce the vacuum components, process gases added to the system, and particles and vapors generated by the process. External sources include leakage, residuals from cleaning or packaging procedures, as well as vapor and particle contamination found in the local environment that may enter when the system is open or when product is introduced.

When defining the cleanliness of the vacuum environment, the most important question is what types of contamination are likely to cause problems? Vacuum-process contamination includes particles, gases, and films. Particles can exist as free-floating particles, or as particles bound to surfaces. Gas contamination can include atmospheric gases—residual or admitted through leaks or permeation through elastomers—and gas admitted or created during a vacuum process. Although certain atmospheric gases, such as water vapor, can form multilayers on internal surfaces that slowly desorb during a vacuum process, film contamination typically refers to materials that do not re-evaporate during process conditions. Often all of these sources are simultaneously present.

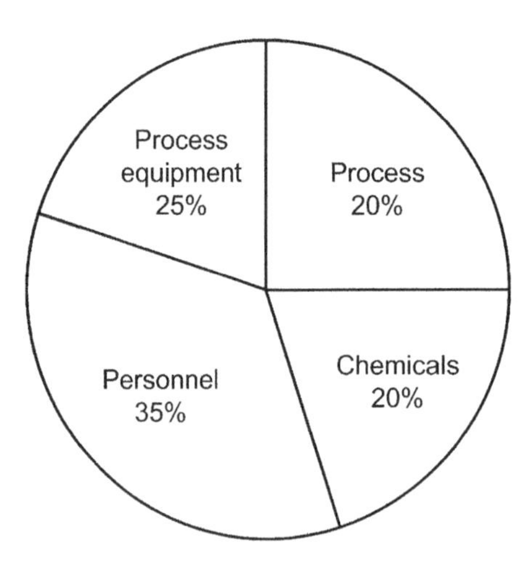

Fig. 21.1 Causes for contamination in the microelectronics industry. A.M. Dixon [1]/With permission of Elsevier.

Both the number of particles, their size, and size distribution are important. For particles in a gaseous environment, "Classification" (or "Class") is the term used to describe their number and size distribution. Note that "Classification" is also the term used to describe the particle cleanliness of clean room facilities that operate at atmospheric pressure. Historic definitions of clean room air quality in a volume, such as "Class 1000" and "Class 10," are no longer used.

In contrast to Classification, the term "Level" is used to describe the number and size distribution of particles on a *surface*. Although Class and Level will be defined differently, it is worth noting that these two contamination types are often related partly because airborne particles in a volume (defined by Class) can settle under gravity to become particles on surfaces (defined by Level).

The extent of gas contamination is defined by gas species, such as N_2, O_2, and H_2, and individual concentrations, expressed as partial pressure, percent, or parts per million (ppm). In vacuum technology, it is more typical to identify gas contamination by its partial pressure in units of, say, Pascal, mBar, or Torr. This is because the partial pressure of a particular gas will not change as *other gases* are admitted, removed, consumed, or evolved within the vacuum chamber—a manifestation of Dalton's Law discussed in Section 2.2.4. In contrast, if gas contamination is indicated by its relative concentration, it indicates a *ratio* of a gas contaminant to all other gases in the chamber. *Relative* concentration can change due to admittance or evolution of any *other* gas.

Film contamination is identified by either the thickness of the material, expressed in nm, or its surface mass density, typically expressed in $\mu g/cm^2$ of surface area. The composition of a film such as layout ink and pump fluid, and its vapor pressure is generally identified. Film contamination is considered to be material that will not become volatile under the pressure and temperatures expected during a vacuum process, and so it is often referred to as *nonvolatile residue* (NVR).

In addition to the above contamination categories, i.e., particles, gases, and films, there are contamination sources that can change their state. Water, as well as most other atmospheric gases, can exist as a gas, a film, or a solid, depending on gas and surface temperatures, and the particular chemistry of other gas-phase species. Another film contamination category consists of natural skin oils, bioaerosols, cutting fluids, or fractionated elements from vacuum-pump oils that can desorb from a surface to become relatively stable gas-phase species. A final example is biological species, e.g., mold, which can grow as relatively solid films on surface that later shed and become particles.

21.1.2.1 Particle Contamination

Particle contamination is often described as small bits of something that do not readily form a liquid or gas at typical process temperatures. Particle contamination has both equipment and environmental sources. Equipment sources include bits of abraded metal from the manufacturing process, and debris from elastomers, glasses, and ceramics that are generated during mechanical movement. Environmental sources can include soils, i.e., small bits of rock and sand mixed with partially decomposed organic matter, paint flakes as well as those generated from a wide range of biomaterials, such as flakes of human skin and hair, clothing fibers, plant pollen, plant decomposition, dust, smoke, and insects.

Volume Particle Contamination The number of naturally occurring particles of a certain size in a *given volume* tends to follow a power-law distribution: $y = ax^{-k}$. This natural distribution can often be approximated by a straight line in a log–log plot, as shown in Fig. 21.2 [2]. Specifically, for a volume with any given number of "larger" particles, the density of "smaller" particles increases logarithmically as the particle diameter decreases. For example, Fig. 21.2 shows relatively clean but unfiltered air to have ~100,000 particles ≥ 0.5-μm diameter/ft^3. However, the same volume of air will have ~300,000 particles ≥ 0.3-μm diameter (thus, 200,000 particles *between* 0.3 and 0.5-μm diameter), and ~750,000 particles ≥ 0.2-μm diameter (thus 450,000 particles *between* 0.2 and 0.3-μm diameter).

Earlier, we defined the terms "Classification" (or "Class") and "Level." One historic classification identification system was based on the number of 0.5-μm *diameter particles/ft^3*. This classification system originated in the 1960s when 0.5-μm diameter particles were the smallest particles that commercial particle counters could measure reliability to large concentrations [3]. Within this historic identification system, the air example above would be considered Class 100,000. Similarly, a Class 1000 environment would have ≤ 1000 particles/ft^3 ≥ 0.5-μm diameter, with increasingly larger numbers of smaller

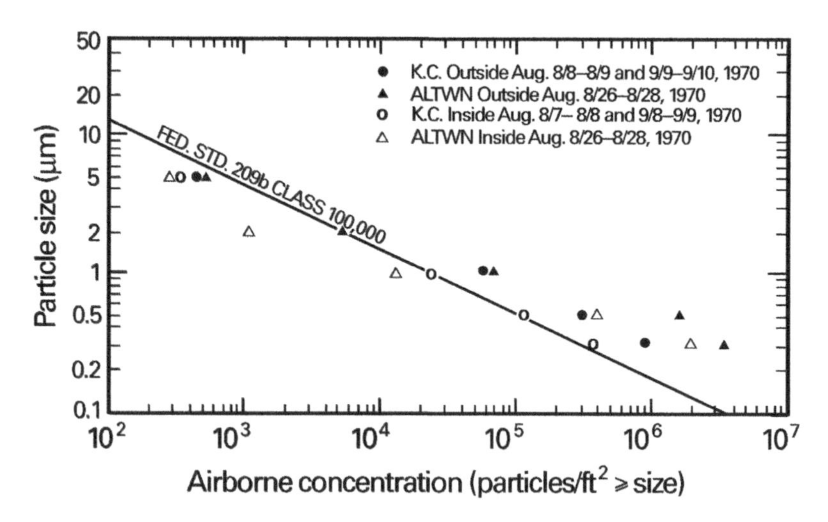

Fig. 21.2 Particle concentration of different sizes measured both inside and outside of manufacturing facilities, and at two geographic locations. P.W. Morrison and W.V. Pink [2]/ Reproduced with permission from John Wiley & Sons.

particles. This classification system was described in various editions of U.S. Federal Standard 209. The original version (209A) was released in 1963; the final version (209E) in 1992.

Although Federal Standard 209E remains widely referenced by vacuum technologists, it was canceled by the U.S. General Services Administration in 2001 and superseded by the International Standard ISO-14644-1 [4]. The ISO-14644-1 standard changed three critical parameters. First, because of increasing requirements of the microelectronic industry and improvements in particle counters, ISO-14644-1 references particles of ≥0.1-μm (100 nm) diameter instead of ≥0.5-μm diameter. Second, ISO-14644-1 uses the metric volume unit, m^3, not the English unit, ft^3, as its standard. Third, the *numerical identifier* in ISO-14644-1 is Log_{10} of the maximum number of allowed particles $\geq 0.1\,\mu m/m^3$. For example, ISO Class 5 ($\leq 10^5$ particles $\geq 0.1\,\mu m/m^3$) classification becomes the previous Federal Standard 209 for Class 100, after converting cubic feet to cubic meters ($35.2\,ft^3/m^3$). Table 21.1 describes the particle concentrations defined in ISO 14644-1 and adds relationship between the current ISO standard and the historic Federal Standard 209E.

Table 21.1 Maximum Number of Particles Allowed per m^3 ≥ Given Diameter, as Allowed in the Present ISO Standard 14644-1, and Correlated to Similar Concentrations per ft^3 in the Previous U.S. Federal Standard 209E

Maximum Number of Particles in Air (Particles per cubic meter)							
		Particle Size					
ISO Class	**Fed-Std 209E Class**	**≥0.1 μm**	**≥0.2 μm**	**≥0.3 μm**	**≥0.5 μm**	**≥1 μm**	**≥5 μm**
ISO 1		10	2				
ISO 2		100	24	10	4		
ISO 3	(Class 1)	1000	237	102	35	8	
ISO 4	(Class 10)	10,000	2370	1020	352	83	
ISO 5	(Class 100)	100,000	23,700	10,200	3520	832	29
ISO 6	(Class 1000)	1,000,000	237,000	102,000	35,200	8,320	293
ISO 7	(Class 10,000)				352,000	83,200	2930
ISO 8	(Class 100,000)				3,520,000	832,000	29,300

Note: Fed-Std 209(E) class definitions are taken from an obsolete U.S. Federal Standard and are shown here as an aid in comparing these to the current ISO definitions.
Source: © ISO. This material is reproduced from ISO 14644-1:2015 with permission of the American National Standards Institute (ANSI) on behalf of the International Organization for Standardization. All rights reserved.

Surface Particle Contamination: IEST Standard CC1246 Particle contamination on a surface is described by its "Level." The numerical identifier of the Level indicates that there is no more than 1 particle/ft^2 with a maximum diameter (in μm's) as indicated by the Level. For example, as shown in Table 21.2, Level 25 will have no more than 1 particle of 25-μm diameter/ft^2. Similar to volume *Classification*, the *Level* of a surface will also include allowed distribution of smaller particle sizes present on the surface. One historical surface cleanliness standard was the U.S. Military Standard 1246 (often called MilStd 1246) that was first introduced in 1962. It was revised 5 times by 1997, after which time the U.S. Army commissioned the Institute of Environmental Science of Technology (IEST) to adapt and

Table 21.2 Allowed Size and Distribution of Particles at Various Levels According to IEST-STD-CC1246E

Level	Particle Size (μm)	Count (#/0.1 m^2)	Count (#/Liter)
1	1	1	10
5	1	2.8	28
5	2	2.3	23
5	5	1.0	10
10	1	8.4	84
10	2	6.9	69
10	5	2.9	29
10	10	1	10
25	2	53.1	531
25	5	22.7	227
25	15	3.3	33
25	25	1	10
50	5	166	1660
50	15	24.6	246
50	25	7.2	72
50	50	1	10
...	...	...	...
1000	100	42,600	426,000
1000	250	1020	10,200

Note: The Level numeric identifier links to one particle of Level Number (in units of #/0.1m^2).
Source: Adapted from IEST-STD-1246D with permission from the Institute of Environmental Sciences and Technology (IEST), Schaumburg, IL USA.

maintain the standard for industrial use. The IEST standard underwent an additional five revisions, with the current IEST standard (2013) and identified as IEST-STD-CC1246E [5].

Examining Table 21.2, one notes that for Levels higher than Level 5, the Level identifies surface concentration of four different particle sizes. For example, a Level 10 surface must demonstrate ≤1.0 particles of >10-μm diameter/0.1 m^2, i.e., where the Level gets its name, *as well as* ≤2.9 particles >5-μm diameter/0.1 m^2, *and* ≤6.9 particles >2-μm diameter/0.1 m^2, *and* ≤8.4 particles >1-μm diameter/0.1 m^2. Although a maximum number of four different particle sizes for most Levels is sufficient for many vacuum applications, the standard provides Eq. (21.1). Using this equation, one may calculate similar guidance for higher Levels, and also the maximum number of allowed particles at any Level-of any particle diameter/0.1 m^2 [5,6].

$$\frac{\text{Particles} \geq \text{diameter}}{0.1\,\text{m}^2} = 10^{\left(0.926\times\left(\log_{10}^2(\text{Level})-\log_{10}^2\left(\text{diameter}[\mu\text{m}]\right)\right)\right)} \tag{21.1}$$

The above equation can be transposed into Eq. (21.2) to allow calculation of the Level when the number of particles per square foot ≥ a given diameter is known:

$$\text{Level} = 10^{\sqrt{\log_{10}^2\left(\text{diameter}[\mu\text{m}]\right)+\frac{\log_{10}\left(\frac{\text{Particles}}{0.1\,\text{m}^2}\geq\text{diameter}\right)}{0.926}}} \tag{21.2}$$

To use Eqs. (21.1) and (21.2) properly, we note that it is necessary to calculate the number of particles/0.1 m^2 at a given Level for two different particle diameters. The number of particles allowed for this Level between those two diameters will then be the number of the smaller particles with the number of larger particles subtracted [6].

Equations 21.1 and 21.2 allow the two-dimensional relationship shown in Fig. 21.3 to be calculated, and also show (for Level 50) the allowable number of particles of their sizes as indicated in Table 21.2. The figure also suggests which Levels will lead to visibly dirty surfaces (Level ~500 and higher), and which Levels will demonstrate low surface scattering, and therefore appear visually "clean" (Level ~100 and lower). In addition to surfaces, the IEST standard also provides guidance for the allowed maximum number and size of particles suspended in a fluid. This guidance originates from the historic requirements of the mid-1950s gyroscopic guidance systems. One example and related discussion of how to combine both volume and surface contamination standards is provided in Ref. [6]. In this example, the highest Levels (i.e., dirtiest components or facilities) are located far upstream in a large vacuum research process, while components closer to the heart of the vacuum process, and thus requiring increasingly cleaner specification, are assigned increasing lower Level designations.

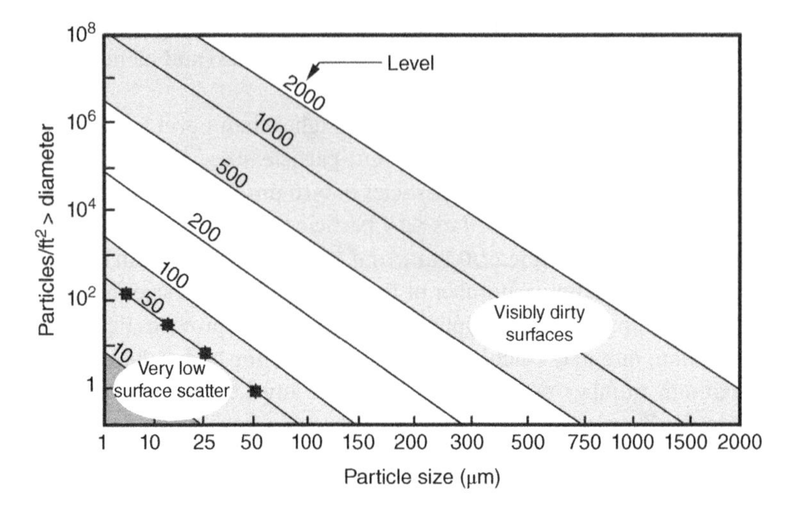

Fig. 21.3 Surface cleanliness chart derived from ISET-STD-CC1246. Black dots indicate the 4 particle sizes and concentrations allowed for Level 50. Adapted with permission from I.F. Stowers [6].

Particle Adhesion We have discussed the expected number and size distribution of particles located on a surface in a vacuum system, rather than in the gas phase. The mechanisms that would lead to the retention on these particles on a surface include gravity, electrostatic charging, molecular dipole attraction, and physical entrapment. Considering these mechanisms individually, *gravity* will tend to be the dominant settling and adhesion mechanism for particles ≥300 μm [7], at which point, a particle's weight will exceed most other attractive forces. Many of the typical particle sources listed above, such as soils, paint, and fabrics, will fall into this size range, as well as particles generated during system assembly, maintenance, or internal mechanical movement during processing. For these reasons, particle-sensitive surfaces generally should not be facing upwards or located near the bottom of the chamber. *Electrostatic Attraction* of particles occurs when the particle acquires a net electrical charge, with which it induces an opposite charge on a surface, and the opposite charges attract each other. Furthermore, the induced surface charge can attract and retain additional contamination from the surrounding environment. *Molecular Dipole Attraction* is similar to electrostatic attraction, but here the particle retains no net charge, but rather its charge is distributed to produce a dipole moment that can be attracted to or induce a surface charge, or both, resulting in attractive forces similar to electrostatic attraction. Finally, particles can become *physically entrapped* on a surface, and this mechanism becomes more pronounced as the surface roughness increases or the surface demonstrates significant porosity. Particle adhesion to a surface can also be influenced by adhesion-induced deformation [8]. In a later section, we describe how particle entrapment can be limited by proper machining, surface polishing of vacuum-exposed surfaces, or both.

21.1.2.2 Gas Contamination

Gas contamination in a vacuum system is described by its partial pressure in units of Pa, Torr, or mBar. Although a surface will generally not have a "gas contamination" specification ("gas" is defined in a volume), many gases and vapors can and do adsorb on surfaces in significant quantities [9]. It is crucial to understand that the vacuum chamber will always have a residual-gas contamination equal to the base pressure—regardless of the process pressure during operation. See Dalton's Law, Section 2.2.4. Therefore, control of gas contamination begins early in the vacuum system design process when the base-pressure specification is assigned. As one example, assume a vacuum chamber is evacuated to 1×10^{-6} Pa (~1×10^{-8} Torr), after which a sputtering process is performed at 1 Pa (~10 mTorr) using high purity "Research Grade" Ar. In this case, contamination from the 1×10^{-6} Pa base pressure of unknown gas would represent about 1 ppm (1 ppm = 1×10^{-6} Pa/1 Pa) of the Ar sputtering gas pressure, assuming the Ar gas was perfectly pure. However, 1 ppm is also the order of contamination that would be expected from *Research Grade* Ar gas. Therefore, in this example, the combined gas contamination during the sputtering process can be estimated to be *at least* 2×10^{-2} Pa, e.g., 1×10^{-6} Pa from the base pressure, and 1×10^{-6} Pa from the Ar. Because the contamination amounts from base pressure and the sputtering gas are equal in this example, this is often considered a reasonable system and process design. If gas contamination were needed to be reduced, contamination from both the system base pressure and the sputtering gas may need to be reduced. To reduce gas contamination from system base pressure, a vacuum-system design more consistent with UVH may need to be considered. See Chapter 20. To reduce contamination from the Ar process gas, inert gas purification with a NEG purifier could be considered. See Section 13.1.2. Generally, relatively few vacuum sputtering systems are designed to achieve a base pressure of 1×10^{-6} Pa. Base pressures of order 10^{-4} Pa are more typical. However, the above example shows that in a sputtering process with 1×10^{-4} Pa base pressure, the amount of gas contamination from the base pressure could be 100 times more than the contamination from the high purity process gas.

To consider how this amount of gas contamination may affect a product, consider further the above example of sputtering a film in high purity Ar gas. From Eq. (2.10), we know that at a pressure of 10^{-4} Pa, the time to form a monolayer (ML) of air at room temperature will be ~2.5 s. If our sputtering process demonstrated an effective deposition rate of 1 nm/s (= 600 Å/min, or ~5 mL/s for many materials), during the 2.5 s needed to form 1 mL of air, one would deposit 12.5 mL of sputtered film, i.e., 5 mL/s × 2.5 s = 12.5 mL. Therefore, assuming the sticking coefficient for air on the particular deposited film is ~1, one might expect the sputtered film to contain as much as ~8% air molecules (1 mL air distributed throughout 12.5 mL of sputtered film). See Section 13.1.1. Practically, this could mean that a sputtered film of a reactive material such as Ti could contain a very significant amount of air, even if the Ti sputtering target and the argon gas were both of very high purity. These air molecules would potentially form TiO_2 and/or TiN phases in the Ti film. Although these examples may suggest it would be nearly impossible to vacuum deposit a film that is relatively free of gas contamination, nature is often not quite that unkind. As shown in Table 13.2, chemically reactive surfaces such as Ti, have room-temperature sticking coefficients near 1 (~0.8). However, most surfaces are less chemically active than Ti, and arriving air molecules, mostly O_2 and N_2, have lower sticking coefficients and would be less likely to become entrained in the film. Note that sticking coefficients generally increase with the mass of gas-phase molecules [10], thus gas contamination by larger molecules such as hydrocarbons from vacuum pumps, is often a concern.

21.1.2.3 Film Contamination

Film contamination in a vacuum system usually refers to materials that become attached to the internal surfaces of a chamber. Some film types, such as water, may reside only temporally on the surfaces, and evaporate during preprocess or process evacuation, or with increased surface temperatures. These are referred to as volatile residues. In contrast, other films become more permanently affixed to the surface and are referred to as NVR. NVR is typically characterized by its mass per unit area or its thickness. NVR in vacuum systems is typically found in the form of continuous or discontinuous regions of deposited films, carbon, and sulfur-containing layers.

Organic contaminants also form NVR's. Similar to gas contamination, organic contamination can initially exist in the gas phase, where it is described by its partial pressure or percent contamination in ppm. Unlike particle contamination, vaporized organics are more likely to become films not only at the bottom of the vacuum chamber but also at its sides and top. Organic contaminants are often composed of large molecules each containing 10s-to-100s of Atomic Mass Units (AMU). However, most large organic molecules will still be much smaller than most particles. Large molecules require higher temperatures to volatile than small molecules, i.e., they have lower vapor pressures than smaller molecules. Therefore, organic molecules can remain attached to surfaces even if the vacuum chamber is subjected to elevated temperatures consistent with vacuum bake-out in the 250–400°C region.

As shown in Table 21.3, IESC-Standard-CC1246 identifies contamination Levels associated with maximum amounts of allowed NVR. Note that an NVR Level of less than A/10 indicates that the surface will generally have less than one atomic layer of NVR. Obviously, this layer of NVR will be discontinuous.

Table 21.3 Allowed Nonvolatile Residue (NVR) Levels as Permitted by IEST-STD-CC1246E

NVR Cleanliness Level[a,b]	Limit, NVR[a] (mg/ft^2 or μg/cm^2)	Limit, NVR[b] (μg/0.1 m^2)
A/100	0.01	10
A/50	0.02	20
A/20	0.05	50
A/10	0.1	100
A/5	0.2	200
A/2	0.5	500
A	1.0	1000
B	2.0	2000
C	3.0	3000
D	4.0	4000
E	5.0	5000

[a] Units used in some obsolete versions of Mil-STD1 provided for comparison.
[b] Units used in ISET-STD-1246D.
Source: Adapted from IEST-STD-1246D with permission from the Institute of Environmental Sciences and Technology (IEST), Schaumburg, IL USA.

21.2 Pump Contamination

Pumps are one of the most common sources of contamination in a vacuum system. Because vacuum systems incorporate different types of pumps, each system will have a different potential for pump contamination. We review these pump-contamination sources by describing the likely sources as the process vacuum level decreases from low/rough vacuum to UHV. Details of how specific contamination is produced from specific types of pumps are described in other chapters.

21.2.1 Low/Rough and Medium Vacuum Pump Contamination

The Low/Rough vacuum range, defined in Table 1.1, extends from atmospheric pressure to 100 Pa (atm – ~1 Torr), while the medium vacuum range extends from 100 Pa down to 0.1 Pa (~1–10^{-3} Torr). Mechanical and venturi pumps dominate these pressure ranges. Mechanical pumps can be further subdivided into subcategories fluid-sealed and dry pumps, while venturi pumps can be grouped by their source of flowing media, i.e., air, water, steam, N_2, Ar, etc. Contamination sources in venturi pumps will depend primarily on the flowing media and are not discussed here. Instead, we focus on contamination from mechanical pumps.

21.2.1.1 Fluid-Sealed Mechanical Pumps

Although fluid-sealed mechanical pumps are becoming less common in the microelectronics industry, they still comprise approximately half of all low/rough/medium vacuum pumps used in other vacuum industries [11], and thus an appreciation of their contamination issues remains relevant. There are two primary forms of contamination from fluid-sealed mechanical pumps. The first can occur if the pump is attached to an evacuated volume, and the pump stops operating. Atmospheric pressure will then slowly force *pump fluid* into the pump's internal regions, out the intake port, and toward the evacuated region via the roughing or forelines. System designs to limit this *fluid* contamination include a normally closed shut-off valve that closes on power failure. It is located between the pump and the roughing or forelines. Fluid contamination can be further limited by an "anti-suck-back" valve that is incorporated by pump vendors into most fluid-sealed mechanical pumps.

The second, and always present cause of contamination is backstreamed vapor from pump fluids. As discussed in Section 18.2.1, backstreaming occurs when pump fluid *vapors* migrate from the inlet of the pump toward the chamber. This backstreaming vapor can include the original "large" oil molecules of "new" fluid, as well as "smaller" molecules produced as new fluid cracks into smaller molecular fragments—a process called "fractionation"—and its extent depends on fluid operating temperatures and process effluents. Although backstreaming can contaminate a vacuum system with both large and small fluid molecules, smaller molecules have higher vapor pressure, so the proportion of fractionated molecules in the backstream is often higher. See Appendixes B.5 and B.6. As one example, contamination by hydrocarbon-based mechanical pump fluids is typically identified with residual gas analysis by clusters of peaks >40 AMU, where the clusters are separated by 14 AMU. Fourteen AMU represents an H–C–H radical. Long chains cleave in use or in a mass analyzer at these locations as illustrated in Fig. 9.5. Backstreaming can also include materials that become entrained or are formed from effluents, such as water and alcohols interacting with the pump fluid. Because different pump fluids and vacuum processes yield different levels of fractionation, fluid chemistries and gas entrainment, combinations of pump, pump fluid, and process effluent will yield the potential to contaminate a vacuum chamber in different degrees and compounds.

Although it should be assumed that backstreaming from fluid-sealed mechanical pumps is always present to some extent, it can be reduced by proper system design and operating procedures. Strategies to limit backstreaming often begin by appreciating that backstreaming of pump fluid vapor, or any gas, becomes increasingly likely when gas flow from the chamber changes from viscous to transition to molecular flow. As discussed in Section 18.2.1, backstreaming in a 2.5 cm dia. (~1 in.) tube 30 cm long was found to be relatively small when the tube pressure is >15 Pa (~0.15 Torr), because the line remains in viscous flow. However, backstreaming increases by a factor of 70 when the line is pumped to 1.3 Pa (~0.01 Torr). Although this example suggests that controlled flushing the roughing line during evacuation can limit backstreaming into a chamber, it should be noted that even at pressure >15 Pa, some fluid backstreaming will be present, as can be seen from Figs. 18.5 and 18.6. For vacuum systems that need to remain relatively free of hydrocarbon contamination, it is preferable to replace fluid-sealed pumps with dry pumps.

21.2.1.2 Dry Mechanical Pumps

Dry mechanical vacuum pumps do not use a sealing fluid in their evacuated regions, relying instead on close tolerance between moving and stationary components, or sealing with elastomeric materials. See also Sections 10.3–10.8. For these reasons, backstreaming of *pump fluid vapors* is eliminated as a major concern. A remaining contamination concern is that nearly all mechanical dry pumps contain low-pressure fluid lubricants or greases that are isolated from the evacuated region with elastomeric seals. These elastomeric seals can abrade or decompose, leading to particle and lubricant contamination, which can backstream into the process chamber in a similar way to that described for fluid-sealed pumps. The low/rough vacuum dry pumps discussed here are positive-displacement pumps, with ultimate base pressures of order 1 Pa (~0.01 Torr). Although the lower region of this pressure range is slightly higher than for many fluid-sealed vacuum pumps, fluid-sealed pumps are seldom operated for extended time in the 0.01–1 Pa (10^{-4}–10^{-1} Torr) range, primarily because of backstreaming concerns. Considering these points together, it becomes clear why many low/rough vacuum dry pumps can be direct replacements for many historic oil-sealed pump applications. The following discussion organizes dry pumps from the smallest to largest when the pump size is defined by its typical pumping speed range.

Some of the smallest dry mechanical vacuum pumps are the diaphragm pumps described in Section 10.7. Contamination from diaphragm pumps is primarily due to particles generated from either the materials or coatings of the bellows, used in lower performance pumps with base pressures ~100 Pa (~1 Torr) or from the elastomeric diaphragms that are used in higher performance pumps, with base pressures ~1 Pa (~0.01 Torr). Particle generation from bellows or diaphragm elements depends on the bellows and coating materials, elastomers, age of these components, and effluents being pumped. Elements are typically replaced after ~10,000 h (~1 year) of *continuous* operation, but earlier replacement may be warranted for certain process effluents or system designs. Slightly larger than diaphragm vacuum pumps are dry-piston pumps. Often all materials in these pumps that contact vacuum are fabricated from—or coated with—low-friction elastomeric materials such as PTFE. Because there is no liquid lubricant between the moving and stationary components, i.e., piston and cylinder, respectively, similar to diaphragm pumps, particle generation will depend on the particular materials, the effluent being pumped, and component age.

Larger-size scroll pumps are rapidly replacing many fluid-sealed pumps because of availability of similar performance ranges, ease of operation, and somewhat comparable initial costs. As described in Section 10.5, scroll-pump functionality combines both a very close tolerance between a moving and stationary surface, i.e., the sides of the scrolls, as well as elastomeric seal on the ends of the scrolls—the tip seals. The tip seal will abrade due to sliding motion, similar to the piston/cylinder components of the dry piston pump, leading to a potential for particle contamination. One often overlooked advantage of a scroll pump is that the motion of the orbiting scroll is often achieved by welded-steel bellows fitting. This means that particle generation and failure concerns of elastomeric rotary seals can be eliminated.

In a manner similar to backstreaming from fluid-sealed pumps, particles generated within diaphragm, piston, and scroll pumps can backstream toward the vacuum chamber when these lines are not in viscous flow. During power failure, particles residing within the evacuated regions of the pump can be driven by atmospheric pressure toward the vacuum chamber via the pump exhaust port. The potential for much of this particle contamination can be limited by proper use of gas purging and ballasting during pump operating as well as maintaining the roughing and forelines in viscous flow. Gas purging or ballasting of the pump not only limits condensation within the pump that can also accelerate corrosion within the pump but helps to sweep particles formed withing the pump out the exhaust. It should be assumed that all of the dry pumps discussed so far will exhaust particles, and therefore need to be properly connected to adequate facility ventilation, especially, in cleanroom environments. Finally, to limit backstreaming of particles during power failure, particle-sensitive systems generally configure these pumps with normally closed valves (close on power failure) near the intake port.

Multistage lobe pumps are a recently developed type of dry pumps that are available in sizes similar to scroll pumps, but which significantly limit particle generation within the pump. This pump is sometimes referred to as a micro-roots or a micro-lobe pump and is described in Section 10.4. Because these pumps have no elastomeric seals between their operational rotors and stators, there are no seals to abrade and form particles. Its lobes and castings are not in contact but have close tolerances. Some pumps' designs allow separate purging of the evacuated bearing and seal regions to limit potential contamination, and also to limit toxic or corrosive effluents from entering the bearing or gearbox regions. Multistage lobe pumps will generally include gas purging/ballast capability to limit condensation within the evacuated regions, and flush particles that may form in these regions due to effluent reactions. For all of these reasons, this pump type is often preferred for roughing or backing vacuum processes with low tolerance to particle contamination.

Moving up in pump size, the final category of dry mechanical pumps consists of claw and screw pumps. Although their pumping process is very different from each other, their potential contamination sources are similar. Multistage lobe, claw, and screw pump all contain counter-rotating shafts. Because their internal operating temperatures are often high, up to 350°C, the preferred material for claw and screw pump construction is low thermal expansion carbon steel. Screw and claw pumps are often used for pumping corrosive and particle-laden gases, because of their ability to sweep process particles through the pump or be cleaned with fluids, in the case of constant-pitch screw pumps. For these applications, the carbon steel components in the evacuated regions of claw and screw pumps are typically coated with a thick layer of an appropriate protective material such as PTFE. They are potential sources of particle contamination in the form of flaking or abrasion of the surface-protective materials, and corrosion products originating in regions where carbon steel has lost its protective coating and subsequently corroded. Unlike multistage lobe pumps, claw and screw pumps are configured with gearboxes containing a low vapor pressure lubricating fluid. Vapor or fluid leakage between these gearboxes and evacuated regions can lead to contamination, while decomposition of sealing materials can lead to particles. The processes for which these pump types are chosen often result in high (viscous) gas flow in the roughing or forelines that limits the reverse movement of potential vapors or particles toward the chamber. The increasing loss of the protective coatings and subsequent corrosion on the shafts will eventually cause not only increasing particle formation but loss of pump performance. This will necessitate frequent commercial disassembly, critical cleaning, reapplication of protective coatings, and rebuilding.

21.2.2 High and UHV Vacuum Pump Contamination

High vacuum pumps are nearly always operating in transition or molecular flow. Therefore, particles or vapors generated by their operation can lead to contamination in the process chamber via backstreaming. High vacuum pumps are either momentum transfer or capture. For fluid-sealed diffusion pumps, the primary contamination concern is from vapors from either the operating fluid or from gases that may have been previously pumped or have entered from the backing pump. For turbomolecular pumps, vapor contamination can originate from bearing lubrication or backstreaming from a mechanical pump, while particle generation can result from previously condensed process effluent being released from fast-spinning surfaces. For capture pumps, particles can be released from gettering or absorption surfaces, while vapors can be released from gettering, condensation, or sorption surfaces.

21.2.2.1 Diffusion Pumps

Sections 12.4 and 19.1 discussed how chamber contamination from diffusion pumps during steady-state operation related primarily to vapor backstreaming from fractionated pump fluid molecules [12]. Although these fragments can originate at several locations within the diffusion pump, good pump design and operation procedures will limit most of these sources. Important design attributes often include both a water-cooled baffle directly above the diffusion pump and a LN_2-cooled trap located between the water-cooled baffle and the high vacuum valve. See Fig. 19.1. In this design, heavy molecular-weight fragments from the pump vapor, e.g., C_8H_{10} and larger, are stopped by the water-cooled baffle and allowed to drip back into the pump. The water-cooled baffle also thermally insulates the LN_2 trap from the pump, slows LN_2 evaporation, and limits the depletion of heavy molecular weight pump fluid from the pump during long-term operation. Although intermediate-weight fragments, e.g., C_6H_6, will traverse the water-cooled baffle, their progress toward the chamber will be limited by cryo-condensation on the LN_2 trap. Very light fragments composed of molecules with one or two carbons, e.g., CH_4 (methane), C_2H_6/C_2H_4 (ethane/ethylene), CO, CO_2, etc., are not effectively stopped by a LN_2 trap. Although these gases and fragments can exist for various durations and condense in the chamber, and potentially be incorporated into depositing films, at room temperature, they will re-evaporate from surfaces, and have potential to return to diffusion pump for possible permanent evacuation [12].

Although the above discussion indicates a diffusion-pumped system can be designed so the chamber remains relatively free of contamination of most vapor contaminations, this is mostly true when the pump is operated in steady state. However, normal use of a diffusion-pumped system requires frequent nonsteady-state procedures, and this can lead to opportunities for significant contamination. The most typical of these is evacuating the chamber from atmospheric pressure with a mechanical pump to a sufficiently low pressure, and then switching chamber pumping from the mechanical pump to the diffusion pump, i.e., during "crossover." Potential causes of vapor contamination during crossover are described in Sections 18.2 and 19.1. This contamination can come from either mechanical- or diffusion-pump vapors or both, and they can enter the chamber through the roughing line, or the high vacuum pump via its foreline. To limit these sources, the following procedures provide initial guidance: (A) stop roughing the chamber at the proper crossover pressure; (B) during crossover, assure that the outlet pressure of the diffusion pump remains "safely" less than its critical value—for the highest gas throughput expected. A typical safety factor would be 0.5 of the diffusion pump's "critical" foreline pressure; and (C) use proper traps and valve-opening times to assure that crossover occurs as rapidly as possible. This last guidance is perhaps the most difficult process parameter to optimize because it will depend on the pump, chamber size, baffle and trap configuration, and high vacuum valves operational characteristics. The goal during crossover is to limit the brief time that the top stages of the diffusion pump are in a condition of high-pressure "overload." In overload, the reduced mean free path of vapor molecules ejected from the pump's vapor jets potentially enables these molecules to diffuse into the chamber. The effect of this situation is generally limited by use of proper baffles, traps, and can be additionally limited by controlled opening of the high vacuum valve.

21.2.2.2 Turbo- and Turbo-Drag Pumps

Conditions that lead to contamination from turbomolecular and turbomolecular-drag pumps were reviewed in Section 19.2.2. Although turbo and drag pumps have far fewer contamination concerns than diffusion pumps, a similarity exists when a turbo-pumped system uses a fluid-sealed mechanical pump for roughing and backing. In this case, backstreaming mechanical-pump vapors can enter the chamber through either the roughing or forelines. The presently preferred method to limit both of these concerns is to use of a dry mechanical pump for both roughing and backing described previously. If instead a fluid-sealed mechanical pump is used, roughing the chamber no lower than the required crossover pressure for the turbo will limit chamber contamination from the roughing line. Additionally, assuring that the critical foreline pressure of the turbo, or drag pump is not exceeded is not only critical for maintaining sufficient rotation speed but also maintaining sufficient compression near the exhaust stages to avoid backstreaming. Specifically, rotation speed of the pump should be maintained at greater than ~40% of its full rotation speed, especially during any transient or high-throughput conditions, e.g., crossover or high-flow processing, to maintain sufficient compression ratio at both the inlet, e.g., turbo, or outlet, i.e., drag ends of the pump.

The main difference between residual gases in a turbo- or drag-pumped system and a diffusion-pumped system will be the significantly lower pumping speed for the lightest gas, hydrogen, in the turbo-pumped system as described in Section 11.3 and Fig. 11.4, compared to a diffusion-pumped system, described in Section 12.2 and Fig. 12.4. Although helium will also have a relatively low pumping speed in a turbopump, the residual gas in even a properly baked UHV system pumped with a turbomolecular pump will be 90–95% hydrogen [13]. This low pumping speed for helium is the basis of the important "counter flow" design using turbo-pumped leak detectors that are described in Chapter 24. Helium outgassing from the pump or chamber materials is low compared to hydrogen. It is important to know that hydrogen contributes a significant portion of the partial pressure because even small amounts of hydrogen can be very chemically active in many vacuum-synthesis processes. Therefore, vacuum technologists should remain aware that vacuum processes may produce very different (and sometimes better) results when using a turbo-pumped system versus a diffusion-pumped system. This is not necessarily because of reduced fluid-vapor contamination compared to a diffusion-pumped chamber, but because of the increased residual hydrogen "contamination" in the chamber of a turbo-pumped system.

The final potential source of turbopump contamination is rotor bearings and their potential lubricants. Most historic turbopumps were manufactured with oil-lubricated bearings located near the exhaust end of the turbopump. In these designs, potential vapor contamination from the lubricant could be limited by using the previously mentioned guidance of maintaining proper rotation speed and sufficiently low fore-pump pressure. Although turbopumps with oil-lubricated bearings remain available, because of their high reliability and lifetime, when properly maintained, more recent designs often include greased, or ceramic, bearings or magnetically levitated bearings using ceramic bearings as an emergency backup. These designs allow for advantages such as choice of turbopump mounting orientation, ease of maintenance, and the potential for very long-term operation with some bearing designs. Although the magnetically levitated bearing designs have the potential to be free of both fluid-lubricant vapors, grease vapors, or ceramic dust, most modern bearing designs include allowance for gas purging of the bearing region, which can further limit the potential for contamination during operation, as well as during start-up and shut down procedures [13].

21.2.2.3 Cryopumps

Operation and contamination concerns of modern cryopumps are reviewed in Sections 14.4.2 and 19.4.3. Particle concerns result primarily from the activated carbon used as the cryosorbing media for pumping H_2, He, and Ne. See Figs. 14.12–14.13. The activated carbon is glued to the metal surfaces of the (colder) second stage "array," and over time the vibration from the closed cycle He expander can dislodge small carbon particles. Gravity will cause most of these liberated carbon particles to fall down toward the bottom of the pump, where they can fall onto the sealing surface of the pressure-relief valve, leading to post-regeneration leaks. However, cryogenic pumps can operate in any orientation, and in some of these orientations, the liberated carbon particles may fall out of the cryopump and into other critical surfaces or samples within the chamber. Furthermore, similar to all particles generated in a vacuum environment, fluid mechanics calculations considering diffusion and gravity indicate that small particles of order 300–500 μm have potential to *diffuse* to more remote locations within the system. This type of particle transport can be further influenced by electrostatic charging.

A properly regenerated cryogenic pump has a high pumping speed for most gases, and very high pumping speed for water vapor (~3× higher pumping speed for water, relative to N_2, O_2, Ar, or H_2). However, as the activated carbon cryosorption material reaches its capacity for H_2, He, or Ne, the pumping speeds for these three gases decreases significantly. The effect is a vacuum process that is sensitive to hydrogen can produce very different results before and after a regeneration. This guidance suggests two things: (1) the vacuum technologist should be mindful to record in the system log when a cryogenic pump is regenerated; and (2) if vacuum processing results are different before versus after regeneration, the first thing to realize is that the process may be more sensitive to hydrogen than may have been previously appreciated. Indeed, many important material-science discoveries can be linked to this type of informed observation.

21.2.2.4 Sputter-Ion and Titanium-Sublimination Pumps

As summarized in Chapters 13 and 20, Ti-sublimination and sputter-ion pumps are often used together in UHV systems. Both pumping techniques are hydrocarbon-free, have no mechanically moving parts, and produce no vibration. When operated properly, the pumps can achieve very low pressure in the 10^{-9} Pa (~10^{-11} Torr) range, and so are often chosen for vacuum characterization techniques that require either very long mean-free paths, and/or long ML formation times, but do not generate significant throughput into the pump. Most vacuum characterization processes can account for foreign particle contamination, and careful particle analysis can determine their origin. The evaporation of Ti or Ta followed by many durations of gettering reactive gas species, e.g., O_2, N_2, etc., will often lead to the formation of thin oxide and nitride layers that can generate very small particles. Furthermore, because of the close proximity of these gettering surfaces to plasmas within the ion pump, these small particles can become charged, and distribute to more remote regions within the vacuum chamber. For these reasons, sputter-ion and Ti-sublimination pumps are generally positioned well below any critical surfaces.

The main residual gases from ion- and Ti-sublimination pumped chambers will include the inert gases, i.e., primarily Ar, Ne, and He, but also methane and ethane. The amount of inert gas will depend on specific type of ion-pump design used and its ability to "bury" and retain, and thus pump inert gases. See Section 13.2. If a Ti-sublimination pump or a non-evaporative getter pump is used alone, it will have no ability to pump inert gas, and therefore Ar, Ne, and He will constitute most of the residual gas at low pressures. Methane and ethane will also generally be present in an ion- and Ti-sublimination-pumped chamber. These gases form by high-temperature reaction of hydrogen with carbon impurities entrained in Ti or W filaments, i.e., from the Ti sublimination pump sublimination sources, or the W-filamented ion gauges, respectively. Unfortunately, methane and ethane are only marginally pumped on Ti gettering surface—even if the surface is maintained at liquid-nitrogen temperatures. Furthermore, any methane that has been gettered onto the Ti gettering surface will tend to be readily displaced by other typical gases. See Table 13.1.

21.3 Evacuation Contamination

In addition to contamination from the pumps used to evacuate a chamber, there are other sources of contamination linked directly to the evacuation process. Here, we describe sources and methods for mitigating their effects.

21.3.1 Particle Sources

The two sources of chamber contamination are airborne debris and surface contamination. Airborne substances, as noted elsewhere, contain a whole host of contaminants, many of which are common to the atmosphere we breathe, whereas others are locally unique. Some airborne particles are water soluble; others are not. Most particles are composed of inorganic or organic materials; some are of biological origin. In all cases, their sizes and densities vary. Figure 21.2 presents examples of airborne particle densities and size distributions. All particles are potential sources of concern, but the level of concern is quite product dependent. Any solid particle can obscure light and produce a photolithographic defect, and some particles may chemically react with a critical surface.

Surface contamination is found on internal fixtures and chamber walls. Loosely bound surface particles can dislodge and move to a critical surface due to surface vibration. Particles that were not removed from these surfaces during cleaning remain bound, and some become so firmly embedded that they cannot be removed. Particles that can be dislodged by vibration are a potential contamination source. Motion, as simple as closing a chamber door, banging on a wall, starting a pump, moving a wafer holder, or rotating a shaft, can dislodge particles.

Airborne and surface contaminants are too small to be observed by eye, but they can be collected and analyzed. Atmospheric aerosols are typically collected on a dry 0.45-μm-pore-diameter filter using a small pump to draw an air sample through the media. Surface particles can be collected with a small piece of dry filter media; however, those that are deeply embedded will not be collected. Capturing surface contaminants on wet filter media is not advisable, as water-soluble compounds will be dissolved. Both surface and airborne samples can be analyzed with either a high-power optical microscope or an SEM. One must first coat nonconducting samples with a ML of conducting carbon or gold. Some particles can be identified by shape; other particles require EDX analysis while in the SEM. Walter McCrone's encyclopedic work contains myriad photographs of a broad range of particulate substances [14], and Hinds [15] describes methods for sample collection and measurement of aerosol particles.

21.3.2 Remediation Methods

To reduce vibration-induced contamination, pumps can be mounted on elastomeric pads, and their piping connected to chambers via metal bellows. Doors can be opened and closed carefully, motion feedthroughs and valves can be programmed to start and stop softly, and as mentioned elsewhere, critical surfaces can sometimes be designed to face downward to avoid collecting debris.

Historically, turbulence during rapid evacuation was assumed to be the root cause of surface particulate contamination. Turbulence was observed over 75 years ago by C. Hayashi of Ulvac Corp. (C. Hayashi, personal communication, October 21, 2008), who considered the swirling motion of water mist in a glass bell jar during initial evacuation to be proof that turbulence produced contamination. In that same era, vacuum metalizing operators quite accidentally discovered they could eliminate the dull appearance of deposited aluminum films by manually opening the roughing pump valve slowly. In 1962, Ames discovered that particle contamination of vacuum-deposited thin films could be prevented by evacuating the chamber from atmosphere "at a significantly controlled, slow rate" [16]. In the ensuing decades, attempts such as those by Ho [17] and Bowling and Larrabee [18], were made to connect contamination and turbulent flow; however, no statistically valid correlations were ever found. In 1987, based on *in situ* laser particle counter measurements, Borden [19], postulated that water vapor, not turbulence, was the origin of roughing contamination. Following that observation, Chen et al [20] observed particle counts 1000× larger than measured in the initial cleanroom air and attributed these to moisture condensing on ultrafine particles that were initially smaller than the counter's detection limit.

Soon after, Zhou [21] and Wu et al. [22] provided the physical understanding of heterogeneous condensation of water on airborne particulates during roughing. Both gravity and the resulting swirling motion of the locally cold air mass transported these airborne contaminants to clean surfaces, where the water evaporated and obscured both their origin and mode of transport. The details of Zhou's method were described in Section 18.1.2. The discrepancy between the historical assumption of isothermal laminar flow (**R** <1200), and current understanding of near-adiabatic cooling is illustrated in the example calculation given in Fig. 18.4. If one were to assume that laminar flow was necessary to prevent particle deposition, one would simply restrict the roughing speed to 2.1 L/s, independent of relative humidity (RH) or interior chamber surface area. In this example, that speed would *not* prevent heterogeneous nucleation in air of RH >73%, and for lower values of RH, it would significantly lengthen roughing time.

Zhou's method is used on ultraclean processing systems. Pumping speeds are reduced just enough to keep the relative humidity below 100%, and at that speed, roughing line flow may be turbulent for most values of RH. The vacuum technologist cannot adjust chamber volume, or internal surface area, but can control the initial relative humidity by partial venting with a dry gas and control the roughing pump speed with a throttle valve.

21.4 Venting Contamination

Rapid venting with high-velocity air does produce significant particulate contamination and in proportion to flow, as observed by Chen et al. [20]. Flow velocities can be reduced with a low-pressure throttle valve, a flow meter, or with a sintered metal diffuser. Laminar flow conditions can be determined using relations found in Section 3.3.2 or in Roth [23]. During venting, water vapor cannot condense on fine particles, because the gas temperature increases.

Vent gas, supply line pressure regulators, valves, and vent lines are all potential contamination sources. Atmospheric air is dirty. Most large vacuum process facilities supply nitrogen gas from a large LN_2 Dewar that is usually located outside the facility. Although the N_2 gas from this LN_2 source is clean and dry, the contamination level of the supply lines, regulators, and valves may be less certain. Furthermore, even if the N_2 distribution system is well-designed and pre-cleaned, maintenance activities can compromise its contamination level. For these reasons, processes with critical vapor or particle specifications should be configured with point-of-use filters, typically 0.05 μm, to remove both a water-vapor and particles.

Sample loading and maintenance operations are both very large sources of chamber contamination. For a vacuum chamber that must be opened for sample loading or maintenance, the amount of water vapor that will absorb on the inside chamber walls, and the number of particles that can enter the chamber, will depend on the amount of time the chamber is opened, and the relative humidity of the environment to which the chamber is exposed. One obvious method to limit both of these contamination sources is to place the loading port, or load lock of the vacuum system within a humidity-controlled cleanroom. Although this configuration will not eliminate contamination sources, the amount of contamination entering the chamber will be much less subject to seasonal humidity variations or other activities occurring within or outside the process facility that could vary volumetric particle density near the system or the samples.

Although very high purity vent gas, cleanroom-type environments for sample loading, and load locks, are certainly preferred options, the cost of these facilities and designs can be prohibitive. Less costly options can include incorporating a chamber "lid lock" that allows samples to be exchanged through a smaller opening while dry gas purging maintains the lid opening in viscous flow. This technique can limit both particles and water vapor from entering the system. Other techniques include performing a pump–purge–pump cycle, while purging with hot gas, purging with a heavier gas like argon, or briefly heating the chamber "cooling water" before final pumping, or combinations of the above. All these techniques quicky reduce the amount of water vapor that has physiosorbed on the inside of the chamber during sample loading or maintenance. Depending on the particular vacuum process to be performed, other reduction techniques include initiating a preprocess glow discharge, admitting UV radiation, and generating ozone within the chamber [24].

21.5 Internal Components, Mechanisms, and Bearings

Components that go into a chamber often contribute both vapor and particle contamination and we describe methods used to assess this contamination. Section 21.9 describes cleaning procedures used to reduce this contamination before the components are placed in the system.

It can be surprisingly difficult to measure the contamination originating from a particular component when used within a vacuum system. Part of this is because components often contain many different materials, and each material type may respond differently under differing process conditions. Although it is relatively easy to measure contamination in a controlled-air environment, air-based measurements may not accurately indicate the extent of particle generation in vacuum. Some reasons for this include the effects of atmospheric moisture on bearing and sliding surface lubrication, reducing particle generation, and altering the electrostatic environment of the component. Contamination generated during component operation may be different from that generated during evacuation or venting.

One procedure used to measure in-vacuo particle generation from components is described in [25]. Dry-filtered N_2 gas is admitted into either the volume of a component or into a chamber containing the component to be tested. Once the N_2 gas achieves a designated pressure, the N_2 inflow ceases, and the chamber is evacuated through a particle counter that records the number and size distribution of particles released during evacuation. The process can be repeated to investigate how particle generation changes—and often reduces—with subsequent evacuations. Although this method can provide significant insight, particles will pass through the detector primarily while viscous-flow conditions exist in the exhaust line. Particle generation during conditions of molecular flow will not be detected. Similar methods have been developed that use particle counters under continuous-flow conditions. Rather than measuring particles generated *during* (real or simulated) vacuum-processing steps, other methods measure the number of particles, or their impact on products *after* various vacuum-process steps, or both. In one such study, the number and effect of particle generation from an electron-beam deposition process are detailed [26]. As shown in Fig. 21.4, using automated particle counting equipment on Si wafers, particle generation occurring during several steps of the process sequence was isolated. Less costly versions of this type of analysis can include capturing particles on clean filter paper and analyzing at ~50× with an optical microscope. However, if more advanced microscopy analysis is added, e.g., SEM with EDX, the shape, and composition of different particles can be revealed. This can suggest the origin of the particles, and thereby indicate where contamination reduction efforts may be most effective.

Internal rotary or linear motion devices can generate significant particle contamination. Motion is often required for sample exchange or to enhance deposit uniformity, or both. The motors that produce this movement are typically located outside the vacuum environment, and motion transfer is through vacuum feedthroughs. In addition to potential contamination from linear- or rotary-motion feedthroughs described in Section 16.3.4, transferring motion within the chamber typically requires chain, polyurethane belt, screw drives, and associated bearings; each of these can generate particles. Figure 21.5 compares particle

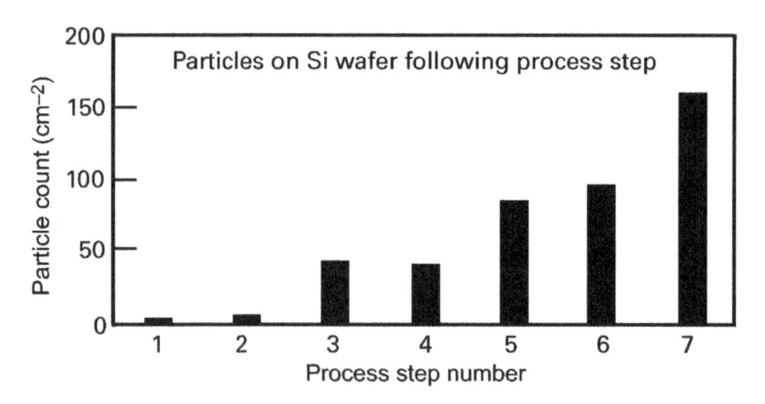

Fig. 21.4 Particles 1–6 μm^2 on Si wafers added by indicated process step. Step 1: Wafer chemically cleaned and spun dry. Step 2: Wafer transferred to load lock. Step 3: Evacuation and soft venting of load lock. Step 4: Transport from load lock to deposition chamber. Step 5: Energizing evaporation source with shutter closed. Step 6: Evaporation without rotation. Step 7: Evaporation with rotation. Numerical data from R.A. Bowling and G.B. Larrabee [18]. Reproduced with permission of Gessert Consulting, LLC.

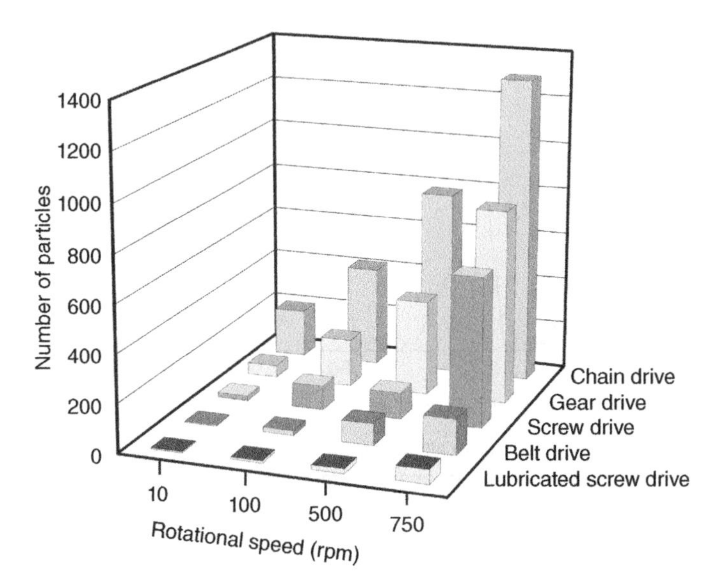

Fig. 21.5 Comparison of particles generated from various internal drive processes. Particles were collected on filter paper and viewed with a 50× optical microscope. Data from H. Patton, LLNL. Reproduced with permission of Gessert Consulting, LLC.

generation from several different internal motion mechanisms, as well as its dependence on rotation speed.

Typical internal system fasteners include screws, nuts, bolts, and pins. As indicated in Section 20.2, vapors can desorb from threaded fastener surfaces, or from trapped volumes. To minimize these sources, the use of well-cleaned and vented fasteners is required. Unlike components that demonstrate relative movement *during* vacuum processes, fasteners primarily generate particle contamination when tightened or loosened. However, these particles will accumulate in the chamber during maintenance activities, and potentially be redistributed during evacuation and venting, as we discussed in Sections 21.3–21.4; these particle generation sources should also be minimized. This is typically accomplished by using fasteners that are commercially produced with roll-formed, rather than cut threads, followed by critical cleaning, and then coating with materials such as Ag, Au, MoS_2, or WS_2. These materials act as a solid lubricant and reduce the onset of thread galling and subsequent particle formation [27]. Additional vapor and particle reduction can be achieved by designing components with captured pins for alignment and structure, thus allowing one to minimize the number of threaded fasteners used only to hold these pins in place.

Valves are the final example of internal particle-generating components. In addition to vapors and gases released from elastomer seals during valve closure

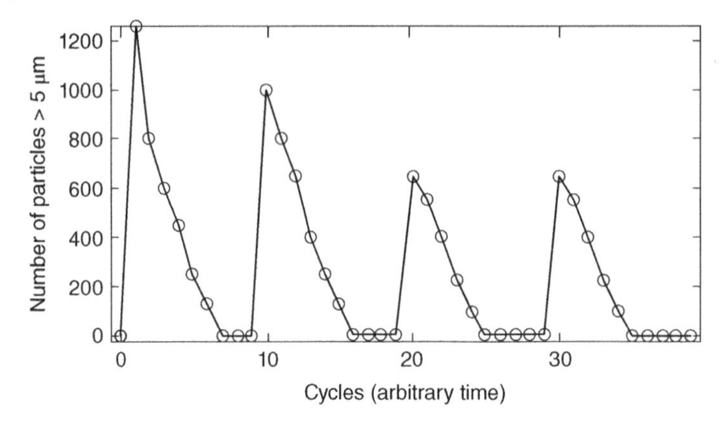

Fig. 21.6 Particle generation resulting from HEPA-filtered air flowing through a representative gate valve. Measurements performed at atmospheric pressure and as a function of increasing valve cycling. Data from H. Patton, LLNL. Reproduced with permission of Gessert Consulting, LLC.

depicted in Fig. 16.9, particles are also released. Measurements of particle release during valve closing have revealed significant contamination. Although valve designs and orientations will produce varying amounts, compositions, and sizes of particles, the origin of these particles is generally a combination of abrasion of elastomers, rollers and bearing surfaces used in the activation mechanism, and residual materials from previous processes. Figure 21.6 characterizes particle generation of a representative gate valve. Although the number of particles can diminish with subsequent valve activations, even after many valve cycles, related particle generation can remain significant. A similar example is reported in reference [28].

21.6 Machining Contamination

Most surfaces exposed to a vacuum environment will require machining such as cutting, milling, turning, grinding, and polishing. Because of the local forces and temperatures generated during these processes, all of these processes have significant potential to leave materials deposited, attached, and entrained on the surfaces. Although the following overview will focus on metals, similar concerns relate to other typical vacuum materials such as ceramics, glasses, and some machinable elastomers.

21.6.1 Cutting, Milling, and Turning

Most metal vacuum chambers and components will be fabricated from much larger pieces of "stock." Machining often begins with mechanical or laser cutting to separate a smaller "workpiece" from the stock, followed by milling, turning, boring, and drilling to produce precision dimensions. All of these processes require machining with a metal "tool" made of material harder than the workpiece to slowly remove small amounts of metal, producing significant local heat that can microscopically deform or alter near-surface structure. For most machining processes, much of the heat is removed by flowing a "cutting fluid" between the tool and the workpiece. Although the fluid cools both, thus helping maintain material properties of each, the heat can cause the cutting fluid to break down, leaving deposits on the surface that can be difficult to remove. Further, the cutting process has the potential to press both the tool and the workpiece cutting fluid residuals and previously removed materials into the grains and subsurface regions of the polycrystalline metal being machined. For all of these reasons, fabrication of contamination-sensitive vacuum chambers and components require step-by-step identification and control of the manufacturing process to control gas, film, and particle contamination. For example, not only must the cutting fluid meet environmental and machining requirements needed for the product, but the eventual cleaning process must be chosen to remove residuals of each particular cutting fluid.

21.6.2 Grinding and Polishing

The surface roughness of a vacuum component is critical for several reasons. First, the surface roughness will influence the actual area available for gas adsorption/desorption, and so will affect surface outgassing. Second, the surface roughness will influence the extent to which particles will adhere to the surface, and so will affect how easily the surface can be cleaned. Finally, burnishing and rolling processes can yield a surface that may appear very flat and smooth, but these processes can produce a surface that will trap gases and particles, as illustrated in Fig. 21.7*a*. For this reason, the preferred machined surface for vacuum use is formed with sharp tooling as shown in Fig. 21.7*b*. Following machining, surfaces exposed to vacuum typically undergo grinding and perhaps polishing. Because *mechanical* grinding or polishing processes can embed particles into the near surface that could be later released as contamination, *chemical- and electro-polishing* are preferred for vacuum use.

Electropolishing typically removes ~10–15 μm of material from the surface. This has the added benefit of removing any subsurface contamination, such as residual particles or burrs from machining. Electropolishing will typically reduce the surface roughness of as received or machined stock, but the amount of reduction will depend on pre-electropolishing surface condition. The roughness of electropolished stainless steel or aluminum for vacuum applications is often specified as a Number 4–16 finish (±4–16 μ-in. average surface roughness).

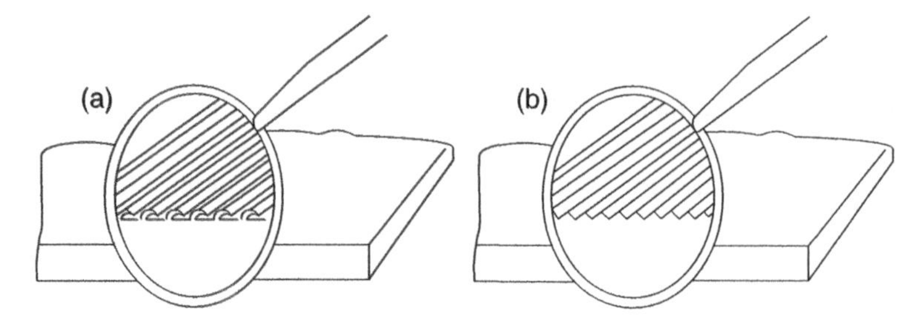

Fig. 21.7 (a) Surface that may appear smooth but that has potential for gas, particles, and cutting fluids to be trapped in subsurface regions. (b) Preferred machined surface for vacuum components that have been formed with sharp tooling. Reproduced with permission of Gessert Consulting, LLC.

While electropolished surfaces are viewed as the least likely to capture and hold particles and reduce water adsorption, their effect of reducing or increasing out-diffusion of hydrogen from the near-surface bulk continues to be debated.

21.6.3 Welding

Welding is widely used to join pieces of machined metals together for vacuum chambers and components. Because welding involves the rapid melting and refreezing together of metal alloys, alloy separation can occur in the weld zone. This can produce different surface compositions in the weld zone, and lead to potential for contamination. For example, most stainless steels are alloys of at least Fe, Cr, and Ni. See Chapters 15, 16, and Appendix C.8.

Depending on the alloy and its carbon concentration, the welding process can produce a weld zone surface containing elemental Fe, also referred to a *free* Fe. This free Fe will corrode much more aggressively than the surface of non-welded regions yielding iron oxide particle contamination. Low contamination vacuum welding requires that the parts to be welded and any welding filler material, be critically cleaned both before and after the welding process. Pre-weld cleaning must include removal of all dirt, oils, layout materials, and cutting fluids. Post-weld cleaning typically includes chemical etching to remove discoloration associated with welding ("heat scale"), followed by electropolishing to remove any free Fe and Ni that may have phase separated in the weld zone. An added benefit of electropolishing, once free Fe and Ni are removed from the surface, is the creation of a highly corrosion-resistant region that is enriched with Cr in the form of a passivating oxide. Grain boundary corrosion resulting from improper cooling of the weld zone was depicted in Fig. 16.2.

21.7 Process-Related Sources

Most vacuum processes will contaminate a vacuum environment to some extent. This is especially true of deposition and etching processes. In these processes, and in addition to the formation of particular deposits from the deposition or etching process, the vacuum chamber often undergoes significant heating which can liberate gases or particles that are weakly bound to surfaces. Furthermore, in etching and sputtering processes, ions are generated that can impart considerable energy onto surface (10s–100s of electron volts), and these ions can dislodge not only weakly, but more strongly bound species from the surface. Various types of chemical vapor deposition processes can not only form particles on surfaces but in the gas phase. Processes such as "plasma ashing" (turning films such as photoresist into ash) can liberate copious amounts of residue. Although the wide range of vacuum processes makes it impossible to address this topic in detail, the following provides general guidance for many vacuum processes.

21.7.1 Deposition Sources

One often overlooked source of contamination in a vacuum-deposition process is the actual evaporation or the sputtering source, i.e., the target itself. The purity of an evaporation material or sputtering target is stated by vendors and later reported in the literature. For vacuum deposition sources, purity is typically between 99.9% (3 9s) and 99.99999% (7 9s) purity. While low purity materials are often suitable for optical or protective coatings, the highest purity materials are required for active semiconductor films, e.g., sources for Si, Ge, III-V, or II-VI materials used in molecular beam epitaxy, most vacuum source material requirements fall somewhere in between these purity endpoints.

However, vacuum technologists should be aware of some key caveats in even these high-purity specifications. First, the purity specification from a vendor is based only on the impurities *tested for*, not necessarily for all elements that could be present. This means that the vacuum technologist needs to review and carefully consider which particular elements have been measured for each particular source used, and often more importantly, which elements *have not* been looked for. Additionally, purity of metals is often specified on a "metals basis," meaning that purity is based primarily on other metals that may be present. However, in many vacuum deposition processes, oxygen, or nitrogen, i.e., nonmetals, often become entrained in the source during either manufacture or storage. These elements may not be indicated in the purity analysis, but will still be released during the deposition process, yielding the potential to form oxide and/or nitride phases in a film. Although calculations of the ML formation time Eq. (2.10) using the base pressure and deposition rate can provide an estimate of purity of the final film, outgassing "contaminants" from the source can complicate this estimate. Contamination from source outgassing is typically limited by proper source "soaking" prior to deposition that allows for outgassed contaminants to be pumped or gettered before the deposition commences, that is before the shutter is opened.

Contamination from sputtering targets requires even more consideration. In this case, the "purity" of the sputtering target is often based on the purity of the starting chemical constituents and *not the final target*. Target manufacture typically includes many process steps before a target of precise composition, density, dimensions, and required bonding is produced. Each of these steps can add contamination that are generally not be reflected in the target's purity specification. When the purity of the as-deposited and bonded target needs to be known, it is prudent to also produce smaller "witness targets" using all the same process steps as used to produce the full-size target, and then subjecting these witness targets to necessary compositional analysis. Next, and unlike thermal evaporation, the sputtering process is much "cooler" than thermal evaporation, thus sputtering does not provide enough thermal soaking to remove entrained gases. Instead, contamination control in sputtering processes depends more critically on understanding what particular pre-sputtering conditions are required for a particular target to yield required film reproducibility, and also how exposure of the target to non-standard conditions, e.g., a standard atmosphere during maintenance, will alter required pre-sputtering conditioning.

Sputtering generally requires an inert gas, and sometimes a reactive gas, as used in reactive sputtering to form compounds. The impurities contained in these gases constitute additional sources of contamination.

21.7.2 Leak Detection

As described in Chapter 24, nearly all vacuum technologists will need to leak-detect components or systems. Contamination during leak detection can be significant and therefore needs careful consideration for particle- or vapor-sensitive processes. Many vacuum technologists believe that a modern mass spectrometer leak detector (MSLD) that uses dry pumping technologies, such as turbo-drag high vacuum pumps backed by dry mechanical pumps, presents little risk of contamination. Certainly, MSLD with dry pumping has significantly reduced contamination concerns compared to historic MSLD's that used fluid pumps such as diffusion and oil-sealed rotary vane, we will show that potential contamination opportunities remain, even when using modern leak detectors.

Two common leak detection methods are often used before turning to an MSLD: the rate-of-rise test, and the alcohol penetration test. During a pressure rate-of-rise test, the vacuum system is evacuated to a low pressure, followed by closing the high vacuum valve connecting the chamber to the pump. If the "normal" outgassing rate of the chamber is known and reproducible, the presence of an unintended leak will cause the chamber pressure to increase faster than if a leak were not present. Although this procedure, described in Chapter 24, sounds easy, it is often complicated, because few vacuum technologists actually know the "normal" outgassing of their chamber, and how this outgassing may be altered by process or seasonal changes. However, from a contamination point of view, a chamber leak allows water vapor, atmospheric and other gases, and particles to enter.

In the alcohol-penetration leak-checking procedure, the chamber is pumped to low pressure, and while pumping, a *small amount* of liquid alcohol, such as methanol or isopropyl alcohol, is dropped near a suspected leak. If a leak exists, the liquid alcohol will be initially drawn into the leak, expand considerably, and the measured pressure in the evacuated component will increase significantly. For small leaks, the pressure is often monitored with an ion gauge. Alcohol is replacing air entering via the leak, and it can reside in the system as a contaminant for long times, especially if the chamber is cooled. Be aware that the alcohol will temporarily freeze in small leaks, and appear to seal the leak, until the site is heated or warmed to ambient.

Historic MSLDs with fluid-charged pumps are still widely used in contaminant-tolerant vacuum applications. Hence, proper fluid selection and use of LN_2 traps are required to limit contamination from entering both the mass spectrometer and the component under test. In contrast, Fig. 24.2 shows a modern MSLD configured with dry pumps will eliminate contamination due to pump fluid vapors. Scroll pumps, used on most modern leak detectors, have the potential to produce particle contamination. Some modern MSLDs can operate in either the *normal-flow or counter-flow modes*. When operated in counter flow, the vacuum region being leak checked is directly connected to the input of the scroll pump; the evacuated system is connected to the foreline between the turbo and the scroll pump as shown in Fig. 24.2. In this configuration, particles from the scroll pump can backstream into the chamber.

Bellows tubing, typically stainless steel, is used to connect a leak detector to either the foreline or chamber of the system under test. Two tubes—a "dirty" and a "clean" tube should be used, carefully labeled and not interchanged to avoid contamination. Even a dry-pumped vacuum system's foreline will be more contaminated than the chamber, and this foreline contamination will contaminate the "dirty" connection tube, and if the two are inadvertently interchanged, will later contaminate a clean chamber or component.

Finally, venting of any clean vacuum region following use of a MSLD should be carefully considered. If venting is performed through a MSLD, the vent supply of the MSLD is often not high-quality N_2, but only ambient air. If venting through the MSLD, any previous contamination brought into the leak detector from a chamber or foreline, has the potential to be transported from the MSLD into the clean vacuum component when venting through the leak detector in viscous flow. Preferred practice when venting a clean vacuum component following leak detection is to first close a valve between the MSLD and the clean component, vent the clean component with clean N_2, and then vent the leak detector separately through its own venting port.

21.8 Lubrication Contamination

Operations within a vacuum chamber often require some type of relative motion of surfaces or bearings [29–31]. When these surfaces are moving without lubrication, cold welding can eventually lead to surface damage [32]. In air environments, cold welding is significantly reduced by a combination of convective cooling of the surfaces, while surface friction is reduced by environmental water vapor acting as a lubricant. However, in a vacuum system, neither convective cooling nor water vapor is available, and so cold welding (galling) occurs quickly. Here we discuss various options for lubrication within a vacuum system as well as contamination concerns associated with these options.

21.8.1 Liquid Lubricants

Wet lubricants and their associated contamination concerns when used in mechanical pumps are discussed in Section 21.2.1.1. Although some of these low vapor pressure fluids could be considered for certain other lubrication uses in a vacuum system, wet lubricants for vacuum technology are more often in the form of a grease [33]. A grease is produced by combining a low vapor pressure vacuum fluid "base gel" with larger "thickener" particles, bonding the gel to the thickener by dipole attraction. The larger particles resulting from the gel + thickener combination are generally referred to as a "grease." The chemistry of the particular combination largely determines which vacuum application the grease may be suited for. This is not only because the size of the molecules and bonding strength will affect its vapor pressure, but also because it will affect the lubricity and range of operating temperature. Furthermore, if the grease experiences use conditions that separate the gel from the thickener, the separate components may still provide some lubrication. In these cases, the lubricity and vapor pressure of the individual component can become critically important.

The vapor pressure of grease is typically specified at 25°C, and characteristics of representative grease types are given in Table 17.3. The extent of vacuum grease evaporation can be estimated from the vapor pressure and the exposed area, this rate will represent the number of molecules being admitted as possible contamination into the adjacent vacuum environment and/or nearby surfaces. The vapor pressure of many vacuum greases is very low, in the range 10^{-7}–10^{-13} Pa (10^{-9}–10^{-15} Torr), and the amount of surface exposure of the grease is usually small. Therefore, the total amount of vaporous grease-related contamination is generally low and can represent a lower contamination source than would result from the particles generated when a bearing or sliding surface is not lubricated.

Creep, or lubricant surface migration, is a property of most liquid lubricants including grease. Although creep can be detrimental in many vacuum applications, it can be a benefit if constrained by the use of barrier films. It can replenish itself after being squeezed out from a contacted area—by creeping back into the same area before another contact is made.

21.8.2 Solid Lubricants

Advantages of dry lubricants include cleanliness, wide range of use temperatures, and surface friction that is largely independent of temperature. However, unlike a wet lubricant that can replenish itself through creep, dry lubricants function by sacrificial action, and the dry-lubricated component will generally fail once the lubricant has been completely consumed. This sacrificial action will also generate particles that must be considered in the design and placement of any dry-lubricated surfaces. Unlike wet or other lubricants that can be applied in the field, many dry lubricants are commercially fabricated, and include thin films of soft metals, e.g., Ag, Au, In, and Pb, on harder surfaces. Harder anti-gall coatings for specialized applications are also available and can include Ni, Cr, TiN, or hard anodized layers. In addition to reducing friction and improving wear resistance, the solid coating can also enhance the appearance, hardness, electrical conductivity, and corrosion resistance of a component. Application processes for these materials include electrodeposition, vacuum evaporation, sputtering, and ion plating. The effectiveness of the coating is typically further enhanced by well-considered surface preparation techniques, such as electropolishing.

Dissimilar materials are another solid lubrication option, that is, to uses fastener components fabricated from dissimilar materials. Although this is similar to the options described above with films of soft or hard metals, an entire fastener can be produced from a different type of metal. For the case of chambers or components fabricated from 304 or 316 stainless steel, fasteners are manufactured from the stainless-steel alloy known as Nitronic 60®, an alloy that includes manganese, silicon, some nitrogen. While the corrosion resistance of this alloy is between that of 304 and 316 stainless steels, its yield strength and resistance to galling are superior under many conditions [34]. Another option is to install Heli-Coil® inserts of Nitronic 60 alloy into the female treads in 304 or 316 stainless-steel chambers or components. In this way, the typical stainless-steel fastener can be used, but particle reduction results without the use of wet- or dry-coated fasteners.

21.8.3 Lamellar, Polymer, and Suspension Lubricants

These lubricants could be viewed as between liquid and dry lubricants. Lamellar materials include MoS_2, $MoSe_2$, and WS_2 that form a layered structure where the interlayer bonding is much weaker than the intralayer bonding, i.e., material layers slide relative to each other [29]. These materials have vapor pressures on the order of 10^{-12} Pa (10^{-14} Torr) but unlike graphite that requires water vapor to be a lubricant, they achieve their lowest friction and highest wear resistance in the absence of absorbed water [31]. In some cases, MoS_2 powders are applied directly to components by techniques such as burnishing or vacuum sputtering. In other processes, the lamellar powders are mixed with a binder material that provides enhanced adhesion of the lubricant to the surface [29]. It has also been found that the use of a polytetrafluoroethylene (PTFE) base layer can provide a 10× increase in wear life over a MoS_2 layer alone [32]. An often-used field application process is to suspend the lamellar particles in a solvent carrier and apply it in a semiliquid form. The solvent eventually evaporates, leaving a loosely bound dry coating. Typical polymer coatings include PTFE and polyimide. In addition to low vapor pressure and low friction, PTFE has the added advantage that it is readily transferred by rubbing against any clean metallic surface.

Regardless of the material or application technique, and their low vapor pressures, these coatings can liberate particles from the surfaces, and their use requires careful consideration for particle-sensitive vacuum processes. A final lubrication option is a commercial process known as Dicronite®. This process forms a lubricant film by spraying high-pressure, room-temperature WS_2 onto dry surfaces yielding a high lubricity and adherent lamella structure that is ~0.5 μm thickness. Although the process can be cost-effective for many vacuum applications, reports suggest the reproducibility and the durability of the coating have not yet been found to match that of sputtered MoS_2 films [35].

21.9 Vacuum System and Component Cleaning

A critical step in system contamination control is assuring adequate cleanliness of a new chamber and its components. Although this may sound simple, many of the procedures used to fabricate components involve processes that can add contamination onto surfaces, trap contamination into near-surface layers, or add gaseous contamination. Here we describe general approaches to cleaning that are typically used for new vacuum chambers and components, and as well suggest possible cleaning procedures for systems that may have been in service for long times.

21.9.1 Designing a Cleaning Process

There are four major categories of contamination to consider when a vacuum component cleaning process is designed. These are: (1) Surface-film contamination; (2) Subsurface contamination; (3) Particle contamination; and (4) Gas-phase contamination. To address each contamination type properly, both the final cleanliness specification and the measurements to assess the extent of cleanliness need to be identified and understood.

Surface-film contamination typically includes layout fluids that may have been used *prior to* machining, cutting fluids used *during* machining, residuals from surface-finishing processes, greases, oils, dirt, and such from post-fabrication handling and packaging. While film contamination can be visible on polished flat surfaces, it will also be present (but not as visible) on less polished or “unseen” machined surfaces, such as internal threads. To remove film contamination, it is desirable to know the composition of the films and identify particular cleaning agents and processes that will remove each known contaminant [36]. When the extent and type of contamination are not known, often a combination of both organic and inorganic contamination is assumed. Organic film contamination, e.g., oils, greases, elastomer residues, and biomaterials, are often removed by soaking and rinsing with a sequence of organic solvents such as acetone, methanol, and isopropyl alcohol that will dissolve the film into solution. These are typically followed by a final soak or high-pressure spray with clean deionized water. Although removing organic film contamination with an appropriate solvent is important, it is not very effective at removing inorganics. For many metal surfaces, inorganic contamination will include films from manufacture or transport of the stock materials containing sodium and sulfur compounds, and residuals from machining processes such as layout or cutting fluids or both. These residuals are typically removed using dilute aqueous detergent solution. In this case, the detergent is selected based not only on contaminant and final requirements of the component but also on the cleaning method. For example, while a soaking process may be effective with a detergent that produces foam, high-pressure spraying or mechanical washing process often benefits from a low-foam detergents. [36].

Some origins of *subsurface contamination* in stainless steels were briefly discussed in Section 21.6. For the case of stainless steel, this type of contamination can result from processes such as rolling and cutting of stock at the foundry with carbon steel tooling. Other subsurface contamination can result from machining and/or surface-finishing of workpieces. Because subsurface contamination can be firmly bonded to the near-surface region or trapped by surface pores or irregularities, cleaning with organics or aqueous detergents will remove primarily only the weakly bound contaminations. To remove the more firmly bonded contamination, the surface layers are generally removed by either sequential chemical etching processes or electropolishing. For chemical etching, and following removal of surface films, an etch is generally chosen that will oxidize the metal surface, leaving a relatively thick surface oxide (often called "smut") that can appear dull. This oxide is then removed by a second, preferential etch that will remove the oxide smut. This typically improves the surface appearance. Generally, sequential process steps of oxidization, oxide removal, and rinsing with deionized water must be conducted to produce a surface that is both free of subsurface contamination and is also visually uniform and appealing. However, this type of surface etching can produce a surface that remains contaminated with particles, and so the etching is often completed with high-pressure spray washing.

Surface particle contamination was historically addressed with cleaning solutions and processes that have been largely phased out due to health or environmental concerns [37]. Instead, it has been found that an environmentally friendly, final spray clean with high-pressure deionized water is very effective in removing film or subsurface contamination or both, yielding surfaces with very low particle contamination [38]. Table 21.4 summarizes the main results [39], while additional guidance is available in reference [40].

Finally, because component or system assembly generally has to occur in an environment containing at least atmospheric oxygen and nitrogen, *gas-phase contamination* is generally addressed only after the vacuum chamber has been evacuated. As described earlier, gases that are adsorbed onto surfaces will eventually be released and potentially become contaminant gases in the chamber volume. In vacuum processes, the amount of outgassing from the surfaces can be compared to other sources of contaminant gas, such as impurities in process gases. If the amount of surface gas is dominating the gas contamination and affecting process or product quality, procedures such as thermal baking of the surface, possibly with purging using nonreactive gases, are typically performed. See Section 4.3.2.

21.10 Review of Clean Room Environments for Vacuum Systems

In many of the suggestions and examples discussed in this chapter, contamination within a vacuum chamber is assumed to derive from either the pumps, condensation of atmospheric water vapor, process gases, mechanisms, materials, or lubricants. However, the environment surrounding the vacuum system is another major contamination source. This environment will typically contain gases, vapors, and particles that enter a vacuum chamber during assembly,

Table 21.4 Efficiency of Removing Aluminum Oxide Particles from Optical Surfaces Using Various Cleaning Processes

	Spray Pressure	Particle-Removal Efficiency		Lowest Achieved Particle Conc. for Particles >5 μm
		>5 μm	>1 μm	
Cleaning Process	**MPa (psi)**	**(%)**	**(%)**	**(#/cm^2)**
Scrubbing and dragging with lens tissue using ethanol and acetone	—	99.6–99.98		2–40
Spray with liquid solvent, for 5–30 s	0.35 (50)	97	3	1500
Fluorocarbon (Freon TF®)	7 (1000)	99.7–99.9	81	10–35
Fluorocarbon (Freon TF®) Water	17 (2500)	98–99.5	—	2–60
Ultrasonic agitation of Freon TF® for 1–2 min. duration	—	24–92	1	9000–70000
Sequential Cleaning Process: (i) Ultrasonic agitated TWD-602, (ii) Freon TF® pressure spray, (iii) vapor degrease, (iv) immersion in boiling solvent, (v) ultrasonic agitated rinse	—	92	—	4000
Compressed gas jet for 10 s duration (Texwipe Micro-duster®)	—	50–61	—	5000–5800
Vapor degrease in Freon TF®	—	11–28	—	65000–80000

Source: Adapted from "Comparison of cleaning methods used for removing particulate aluminum oxide from polished metal and glass surfaces," *Energy and Technology Review, Lawrence Livermore Laboratory Defense Programs, Contamination Control Technology*, I.F. Stowers and H.G. Patton, October 1979, p. 6. Used with permission of Lawrence Livermore National Laboratory *Note*: High-pressure water spray is as, or more effective than historically used, but environmentally unfriendly, fluorocarbon-based processes.

maintenance, and the routine operations of opening and introducing materials. For these reasons, operating a vacuum system to achieve low contamination generally requires that sample loading and unloading occur within a clean-room environment. The remainder of the system can be located in a less clean maintenance area.

21.10.1 The Cleanroom Environment

Figure 21.2 and Table 21.1 describe the number of particles that are present in typical room-air environments, how the number of particles tends to increase with decreasing particle size, and how historic and modern cleanroom *Classifications* are determined. The goal of the cleanroom environment is to reduce the number of airborne particles so that they don't become surface particles (as defined by its *Level*). Particle reduction in air is accomplished by the use of high-efficiency particle air (HEPA), or ultra-high efficiency particle air (ULPA) filters. HEPA filters remove 99.99% of airborne particles to 0.3 μm, whereas ULPA filters are 99.999% efficient to 0.1–0.3 μm. The operational lifetime of HEPA filters is significantly increased by proper use and maintenance of appropriate prefilters that are located upstream from the HEPA filters. In a cleanroom, the HEPA filter units are typically positioned in the ceiling so that the HEPA-filtered air is directed downward toward the cleanroom floor. See Fig. 21.8. In some designs, the filtered air passes through a perforated floor after which it is directed through a return plenum through a heating-ventilation-air-conditioning (HVAC) region, and then back through the HEPA filter units for reuse in the cleanroom, Fig. 21.8*a*. In other designs with less stringent requirements, the filtered air may be directed sideways when it reaches a solid floor and returned to the HEPA filters via air-return vents located on the lower portion of the walls, near the floor, as sketched in Fig. 21.8*b*. Regardless of the design, the air velocity of the descending clean air within the cleanroom is controlled to be ~1.5–2.0 ft/s. This air velocity exceeds the settling rate of the largest particles, while the air exchange rate is high enough to assure that particles generated by people or equipment are quickly removed from within the volume of the cleanroom. Typical room air exchange rates are ~1 air exchange/10 s. The basics of HEPA filter materials, construction, design, placement, use, and maintenance are discussed in many references [1–3].

21.10.2 Using Vacuum Systems in a Cleanroom Environment

Modern high-end cleanroom design includes options for incorporating advanced vacuum processing systems. The design goal is for sample loading and unloading to take place in the cleanroom environment, while routine maintenance activities are performed in an adjacent area. Although the particular approach to vacuum system location in a cleanroom depends on the type of vacuum processes and product being produced, most designs benefit from well-designed *service chase* areas.

In some designs, service chases are in non-clean areas that are completely separated from the cleanroom environment. In this design, all penetrations for equipment service lines between the cleanroom and service chase must be well sealed to prevent particles from entering the cleanroom. Also, the relatively dirty chase

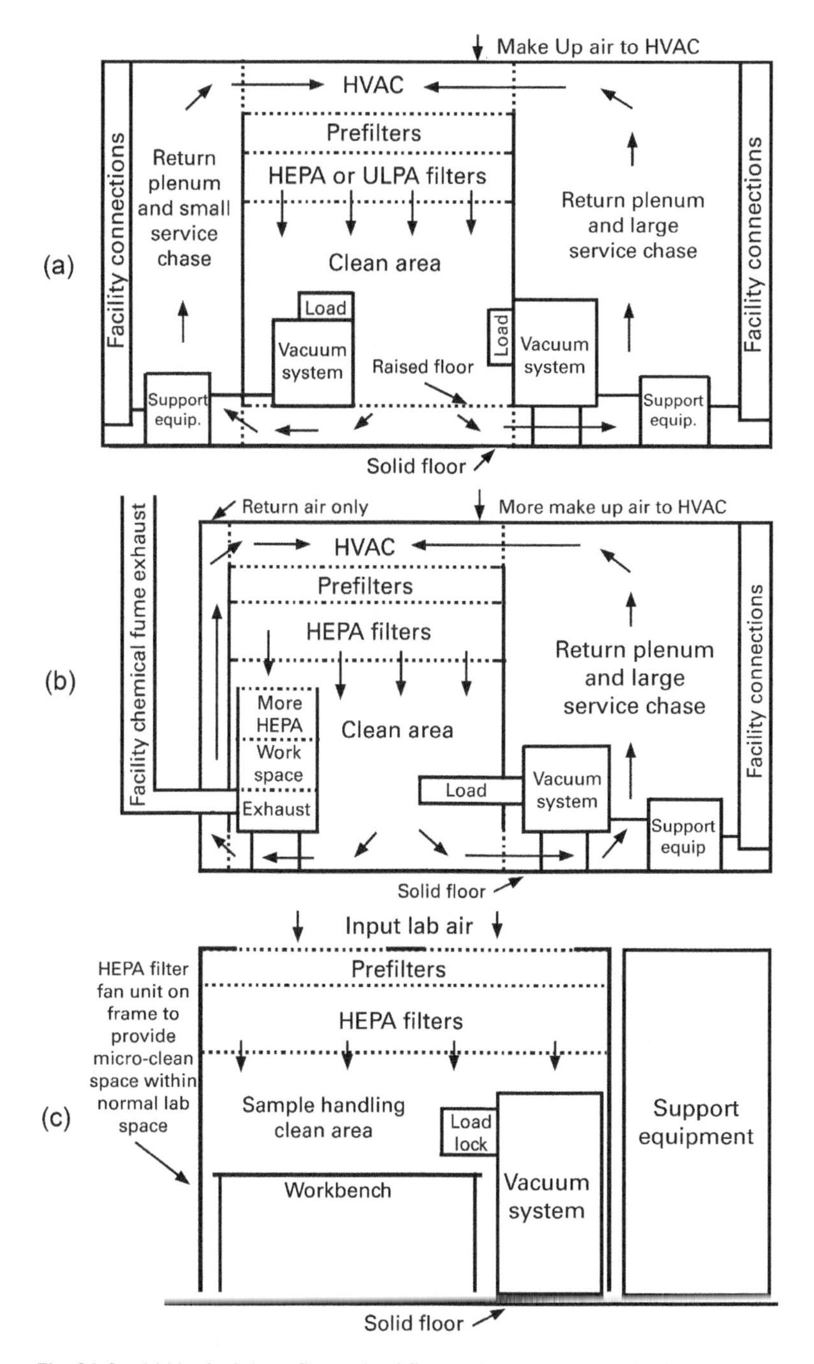

Fig. 21.8 (a) Vertical downflow, raised floor exit cleanroom with air return plenums containing tools and utility connections; (b) vertical downflow, sidewall exit cleanroom with air returns and tools in service chases; (c) Portable micro-clean room providing a locally clean space for a workstation and a single tool. Reproduced with permission of Gessert Consulting, LLC.

cannot be accessed directly from the cleanroom, or while wearing cleanroom garments. In other designs, the service chase also provides parts of the air-return plenums for the cleanroom. Some examples of this are shown in Figs. 21.8*a* and *b*. Benefits of this design include bringing service connections between system cleanroom and service chase support equipment in the service chase, through relatively large and unsealed openings. This is because air moving through the large openings is flowing at sufficient velocity to limit particles moving from the chase to the cleanroom. However, because the return plenum or chase area will typically be dirtier than the cleanroom, it is still generally not recommended to enter the chase directly from the cleanroom or entering this area with cleanroom garments. This limitation can make troubleshooting and maintenance of the vacuum systems located in the cleanroom difficult. To allow the chase to be entered directly from the clean room, service chases can be designed to be at (or near to) the same cleanroom classification as the main cleanroom area. In this case, the service chase requires its own HEPA filtration, the air-return plenum cannot be part of the service chase, and the possibly dirtier, chase must be maintained at a lower pressure than the main cleanroom, i.e., air should flow only from the main cleanroom into the service chase. Finally, because the support equipment in the chase may produce more heat and contamination than typically produced in the main cleanroom, in this design the return plenums, HVAC, and HEPA filters may need to be separated from those of the main cleanroom.

Regardless of the design, most equipment connections to building facilities, such as cooling water, process, vent gases, and electrical, will be made in the service chase. If the main system is located within the clean area, Fig. 21.8(*a*–left), it is often expedient to locate vacuum system support equipment such as mechanical pumps, effluent treatment, etc., in the chase. As the cleanliness classification of the clean room increases, i.e., the specified particle concentration reduces, many cleanroom designs will enlarge the service chases to house not only vacuum system support equipment, but the majority of the vacuum system–leaving only the chamber door or the load-lock door protruding into the clean room, Fig. 21.8(*a*–right). In this way, both routine and nonroutine maintenance associated with the vacuum system can be conducted in the relatively non-clean service chase area, often without the need for gowning or using cleanroom-compatible tools and procedures. To assist this design, many vacuum-system suppliers can produce nominally identical vacuum systems configured for either wall mounting in a cleanroom or use in a non-clean environment. In addition to a well-designed service chase acting as a return plenum, locations in a clean room are often not appropriate or needed for a service chase, yet balancing airflow requirements in the cleanroom still requires using these areas for return plenums. In this case, the return plenums are sized for the return airflow expected in that part of the cleanroom.

As shown in Fig. 21.8(*b*–left), one typical application of this cleanroom design uses wall space for clean-air benches or cleanroom-compatible chemical fume hoods. In the case of cleanroom benches or fume hoods, it is important to consider how both air flow from the bench or hood may impact flow from the cleanroom's HEPA filters. One significant benefit of HEPA-filters benches and hoods in a cleanroom is that the classification of the worksurface within the bench or hood will generally be increased by at least one classification rating, i.e., using a HEPA-filtered hood in a cleanroom will increase the workspace classification from ISO 6 to ISO 5, i.e., historic Class 1000 to at least Class 100.

Although operation of a vacuum system within a cleanroom facility is generally expected to provide the lowest potential for environmental contamination of the chamber and product, often cleanroom-like environments are desired where a cleanroom is not available. In these situations, a free-standing HEPA-filtered *fan unit*, producing a local "micro-clean space," shown in Fig. 21.8*c*, is a cost-effective solution. The HEPA-filtered fan units are positioned so that the environment where sample loading and unloading occurs becomes a much cleaner space than its immediately surroundings. If samples can be brought into the cleaner sample handling area in a "clean state," its contamination will be limited during loading and unloading. Furthermore, micro-clean areas can often be designed to include space for preprocess sample preparation, i.e., preload sample cleaning, or space for low-contamination, post-process sample storage.

Much of the guidance discussed in this chapter is summarized in Table 21.5.

Table 21.5 Best Practices for Contamination Control

For contamination sensitive vacuum process, establish a contamination control program that specifies maximum contamination levels in volumes and surfaces
Assume that residuals from the base pressure will always be present in the vacuum chamber and can be incorporated into films and/or onto surfaces.
Use process gases that are of sufficient purity to limit gaseous and particle contamination within specified levels.
Be aware of specific contamination concerns in low-vacuum and high-vacuum pumps.
Establish initial evacuation rates to avoid heterogenous nucleation of water vapor and subsequent redistribution of particles with the chamber.
Establish final evacuation (crossover) pressure to maintain roughing line in viscous flow—to limit backstreaming of vapors and particles from the roughing pump.
Establish crossover procedures to limit vapors or particles entering the chamber from the pump, i.e., considering the maximum inlet pump pressure, throughput, and maximum foreline pressure.

(*Continued*)

Table 21.5 (Continued)

Establish process procedures that limit particle generation from internal components.
Locate the vacuum system to limit vibration-induced particles within the chamber.
Establish venting with appropriately clean vent gas and "slow venting" procedures to limit particle redistribution within the chamber.
Use appropriate materials and manufacturing processes to produce surfaces that produce a minimum amount of film, particle, and vapor contamination, and are easily cleaned.
Be aware that chemical and physical sources used in film deposition can be significant source of process contamination.
Be aware that leak detection can produce significant sources of chamber contamination.
Be aware that lubrication of internal components can limit but also produce various types of process contamination.
Use of appropriate material cleaning procedures can significantly reduce contamination from typical manufacturing or handling processes used for vacuum materials.
Incorporation of appropriate clean rooms or areas can significantly reduce vacuum process contamination and/or contamination reproducibility.

References

1 Dixon, A.M. "Guidelines for Clean Room Management and Discipline", in *Handbook of Contamination Control in Microelectronics*, D.L. Tolliver, Ed., Noyes, Park Ridge, NJ, 1988, p. 136.

2 Morrison, P.W. and Pink, W.V. "Fundamentals of Environment Control," Chapter 10", in *Environmental Control in Electronic Manufacturing*, P.W. Morrison, Ed., Van Nostrand Reinhold, New York, 1973, pp. 244–245.

3 M.N. Kozicki, S.A. Hoenig, and P.J. Robinson, *Cleanrooms, Facilities and Practices*, Van Nostrand Reinhold, New York, 1991, pp. 28–29.

4 ISO 14644-1:2015 "*Cleanrooms and Associated Controlled Environments, Part 1: Classification of Air Cleanliness by Particle Concentration*", Second Ed., *International Standards Organization*, Geneva, Switzerland, 2015.

5 IEST-STD-CC1246(E)2013, *Product Cleanliness Levels–Applications, Requirements, and Determination, Institute of Environmental Sciences and Technology* (IEST, Arlington Heights, IL), Available through American National Standards Institute (ANSI), Washington, DC, 2013.

6 I.F. Stowers, *SPIE Conference on Optical Manufacturing and Testing III*, SPIE, **3782**, 1999, p. 525.

7 The reason gravity will tend to be the dominant settling and adhesion mechanism for particles ≥300 μm is as follows: The terminal velocities of large particles (Reynolds No. > 1) are determined by inertial forces, and small particles (Reynolds No. < 0.4) by viscous forces. 300-μm-diameter is in the middle of this transition.

8 Rimai, D.S., Quesnel, D.J., and Busnaina, A.A., *Colloids Surf., A* **165**, 3–10 (2000).

9 R.A Outlaw and H.G Thompkins, "*Ultrahigh Vacuum Design and Practice*", Chapter 9, *Outgassing*, AVS Education Committee Book Series, Vol. **3**, F.R. Shepherd, Ed., pp. 88–121, AVS – Science and Technology of Materials, Interfaces, and Processing, 2009.

10 Ferrario, B. "Getters and Getter Pumps", Chapter 5", in *Foundations of Vacuum Science and Technology*, J.M. Lafferty, Ed., Wiley, New York, 1998, p. 281.

11 K. Pulidindi and A. Prakash, "Vacuum Pump Market Size", *Report ID: GMI2667*, Global Market Insights, Selbyville, DE, Nov. 2019.

12 Hablanian, M.H., *High Vacuum Technology*, 2nd ed., Marcel Dekker, Inc., New York, 1997, pp. 235–267.

13 Suurmeijer, B., Mulder, T., and Verhoven, J., *Vacuum Science and Technology*– English Translation, The High Tech Institute and Settels Savenije Van Amelsvoort, Eindhoven, The Netherlands, 2020, pp. 285–307.

14 McCrone, W.C., *The Particle Atlas: a Photographic Reference for the Microscopical Identification of Particulate Substances*, Ann Arbor Science Publishers, Ann Arbor, 1967.

15 Hinds, W.C., *Aerosol Technology*, Wiley, New York, 1999, Chapters 10 and 20.

16 Ames, I., Gendron, M.F., and Seki, H., *Transactions of the 9th National Vacuum Symposium*, Macmillan, New York, 1962, p. 133.

17 Hoh, P.D., *J. Vac. Sci. Technol. A* **2** (2), 198 (1984).

18 Bowling, R.A. and Larrabee, G.B., *Microcontamination Conference and Exposition*, Cannon Communications, Santa Monica, CA, 1986, p. 161.

19 Borden, P.G. and Baron, Y., *Water Aerosol Formation During Pump-Down in State-of-the-Art Process Equipment*, High Yield Technology, Mountain View, CA, 1987.

20 Chen, D., Seidel, T., Belinski, S., and Hackwood, S., *J. Vac. Sci. Technol., A* **7** (5), 3105 (1989).

21 J. Zhao, *Thermodynamics and Particle Formation During Vacuum Pump-Down*, Ph.D. Dissertation, University of Minnesota, 1990.

22 Wu, J.J., Cooper, D.W., and Miller, R.J., *J. Vac. Sci. Technol., A* **8**, 1961 (1990).

23 Roth, A., *Vacuum Technology*, North Holland, Amsterdam, 1990, pp. 62–65.

24 Vig, J.R., *J. Vac. Sci. Technol., A* **3**, 1027 (1985).

25 Dylla, F., Biallas, G., Dillon-Townes, L.A., Feldl, E., Myneni, G.R., Parkinson, J., Preble, J., Siggins, T., Williams, S., and Wiseman, M., *J. Vac. Sci. Technol., A* **17**, 2113 (1999).

26 Pindoria, G., Houghton, R.F., Hopkinson, M., Whall, T., Kubiak, R.A.A., and Parker, E.H.C., *J. Vac. Sci. Technol. B* **8**, 21 (1990).

27 MacGill, R.A., Castro, R.A., Yao, X.Y., and Brown, I.G., *J. Vac. Sci. Technol. A* **12** (2), 601 (1994).

28 L. Lilje, "Controlling Particulates and Dust in Vacuum Systems", *Proc. 2017 CERN-Accelerator-School Course (CAS) on Vacuum for Particle Accelerators*, Glumslov, Sweden, p. 351, 2017. https://arxiv.org/abs/2006.02820.

29 Lince, J.R., *Lubricants* **8** (74), 1 (2020).

30 Roberts, E.W. and Todd, M.J., *Wear* **136**, 157 (1990).

31 Roberts, E.W., *J. Phys. D Appl. Phys.* **45**, 503001–503017 (2012).

32 Friebel, V.R. and Hinricks, J.T., *J. Vac. Sci. Technol.* **12** (1), 551 (1975).

33 W.R. Jones and M.J. Jansen, Tribology for Space Applications, *Proc. Inst. Mech. Engineers. Part J, J. Engineering Tribology*, **222**, (8), 997, (2008)

34 Hsu, K.L., Ahn, T.M., and Rigney, D.A., *Wear* **60**, 13 (1980).

35 Anderson, M.J., Cropper, M., and Roberts, E.W. "The Tribological Characteristics of Dicronite", in *Proc. 12th Euro. Space Mechanism and Tribology Symp. (ESMATS)* Liverpool, UK, 19-21, Sept. 2007, Publication ESA SP-653, Aug. 2007.

36 McLaughlin, M.C. and Zisman, A.S., *The Aqueous Cleaning Handbook: A Guide to Critical-Cleaning Procedures, Techniques, and Validation*, 4th ed., Alconox, Inc., Technical Communications, 2015.

37 J.D. Shoemaker, Meltzer, M., Miscovich, D., Montoya, D., Goodrich, P. and Blycker, G., *Cleaning Up our Act: Alternatives for Hazardous Solvents Used in Cleaning*, Lawrence Livermore National Lab. Report UCRL-ID-115831, January 1994.

38 Stowers, I.F. and Patton, H.G. "Techniques for Removing Contaminants from Optical Surfaces", in *Surface Contamination*, K.L. Mittal, Ed., *Genesis, Detection, and Control*, Vol. *I*, Plenum Press, 1979, p. 341.

39 I.F. Stowers and H.G. Patton, *Energy and Technology Review, Lawrence Livermore Laboratory Defense Programs, Contamination Control Technology*, Lawrence Livermore Laboratory, October, 1979, p. 6.

40 Sasaki, Y.T., *J. Vac. Sci. Technol., A* **9** (3), 2025 (1991).

22

High Flow Systems

Not all thin-film deposition processes require high or ultrahigh vacuum. Some of the most interesting processes take place in the medium and low vacuum range and require a high gas flow. The pressure–speed ranges for some processes are shown in Fig. 22.1.

Sputter deposition is done in the 0.5–10-Pa (10^{-3}–10^{-1}-Torr) range. For certain materials, sputtering is the preferred deposition technique. Various plasma processes are performed in the range of 5–500 Pa (0.01–5 Torr). Plasma-deposited films are formed from the reaction of chemical vapors in the glow discharge. Plasma etching and reactive-ion etching are commonly used to pattern thin films. Polymer films are formed from the glow discharge polymerization of a monomer such as styrene. Plasma etching is a simple isotropic chemical etching process that uses chemically active neutrals in the discharge; for example, the plasma decomposes CF_4 and creates fluorine atoms that react with a silicon surface. The pump removes the volatile product SiF_4. Reactive ion etching is a directional process used for patterning thin-film structures. Its directionality is due to high-energy ions that are accelerated through the plasma sheath toward the surface, where they enhance the reactivity of the chemically active species with the unmasked portions of the film. Because ion-stimulated reaction proceeds at a rate many times faster than physical plasma etching, the thin film etches downward much faster than laterally. Vertical etching permits finely resolved lines. Reactive ion etching can be performed over the entire range of pressures used for sputtering and plasma etching. Any differences attributed to pressure are differences in nomenclature rather than theory. Low-pressure chemical vapor deposition (LPCVD) and reduced pressure epitaxy (RPE) are thermal processes that take place at low pressures. The thermal energy is typically provided by induction heating. LPCVD, which is done in the 10–100-Pa range, has attracted wide attention. The high diffusivity of the thermally active species at low pressure improves

A Users Guide to Vacuum Technology, Fourth Edition. John F. O'Hanlon
and Timothy A. Gessert.
© 2024 John Wiley & Sons, Inc. Published 2024 by John Wiley & Sons, Inc.

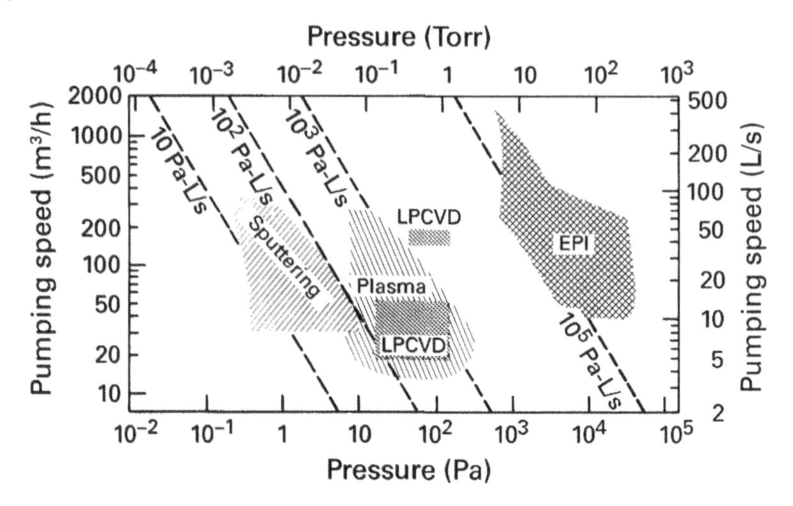

Fig. 22.1 Pressure–speed ranges for some thin-film growth, deposition, and etching processes that require medium to low vacuum and gas flow. Reproduced with permission from Applied Materials Inc.

the transport of the vapor throughout the reactor and allows the growth of uniform films on large wafers. RPE takes place in the 500–10^5-Pa (4–760 Torr) pressure range. Epitaxial films grown at reduced pressure are higher in quality and have less auto doping than films grown at atmospheric pressure.

Many processes performed in the medium or low vacuum range also require the use of vapors that are toxic, hazardous, or corrosive. Special precautions must be taken in the design, operation, and maintenance of these systems to ensure operator safety and equipment protection. Kumagai [1] describes a generalized approach to handling hazardous gases that both updates and adds content to the earlier AVS Recommended Practice for Pumping Hazardous Gases [2].

Thin-film deposition and etching processes span a pressure and gas flow range that far exceeds the capability of any one pump. The useful pressure for each process is dictated by the process's physics. Sputtering cannot commence until the pressure is high enough to initiate a self-sustained glow discharge, but the pressure must be low enough for the sputtered material to reach the anode without suffering a large number of gas collisions. Gas flows ranging from 10 to 10^6 Pa-L/s (~0.1–10^4 Torr-L/s) are needed for a different purpose in each process. In some processes, the high flow dilutes or replenishes the reactant species and simultaneously flushes away the products of reaction and other impurities; in others, it mainly serves to flush away impurities.

This chapter describes un-throttled mechanically pumped systems and throttled high vacuum systems. Mechanically pumped systems are used for ion etching, plasma deposition, LPCVD, and RPE. Throttled high vacuum pumps are used mainly for sputtering and ion etching.

22.1 Mechanically Pumped Systems

The pressure–speed operating ranges of rotary mechanical pumps, Roots-claw pump combinations, turbo-drag pumps, and throttled high vacuum pumps differ. Two-stage rotary vane pumps can operate in the sputtering pressure range with less than maximum pumping speed; however, at low pressures, they cause backstreaming of oil vapor. Rotary vane pumps are economical to speeds of 200–300 m^3/h and provide effective pumping for clean processes in the region delineated by speeds lower than this value and pressures greater than subject to backstreaming. This region is bounded by curve A in Fig. 22.2. Oil-free screw pumps are useful for dusty processes, as they do not contaminate and are little affected by corrosion or particulate contamination. Screw pumps have replaced rotary mechanical pumps for many applications.

For speeds greater than 200–300 m^3/h, a lobe pump backed by a rotary pump may be used. Again, oil backstreaming limits the lowest pressure of operation. At sufficiently low pressures the roughing line would be in free molecular flow and would allow mechanical pump oil to back diffuse to the Roots pump outlet, creep around the interior surfaces, and enter the process chamber. Oil contamination may be eliminated by bleeding gas into the roughing line to maintain the Roots

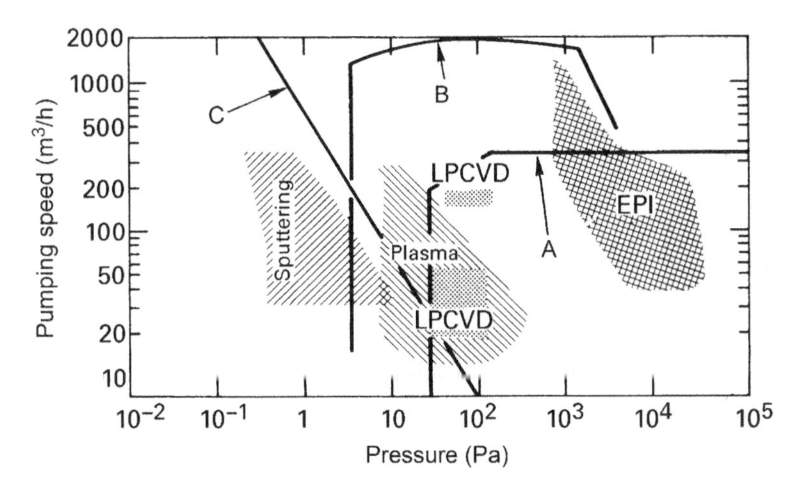

Fig. 22.2 Useful pressure–speed ranges for some pumping systems: (A) rotary mechanical pumps; (B) Roots pump backed by a rotary pump; and (C) throttled high vacuum pump. Reproduced with permission from Applied Materials Inc.

outlet at suitably high pressure. The Roots pump has an upper-pressure limit of ~1000 Pa (~10 Torr), which yields the useful operating region outlined by B in Fig. 22.2.

Dry pumps using Roots-claw combinations can pump from atmospheric pressure to process pressure without concern for oil contamination. A speed–pressure characteristic for one combination pump is sketched in Fig. 10.9. These pumps have proven to be valuable in systems that generate large quantities of dust, such as those producing silicon dioxide from the synthesis of silane and oxygen. The pulsating pressure observed in the exhaust of some dry mechanical pumps can result in enhanced counter-diffusion along the walls of the exhaust pipe. Figure 22.3 illustrates the vortices generated along the pipe walls by the pulsating pressure in an LPCVD pump. Air flows toward the pump during the low-pressure half-cycle in the core. During that time, it becomes entrapped in the vortices, as their layers rotate in opposite directions. On the next half cycle, when the pressure

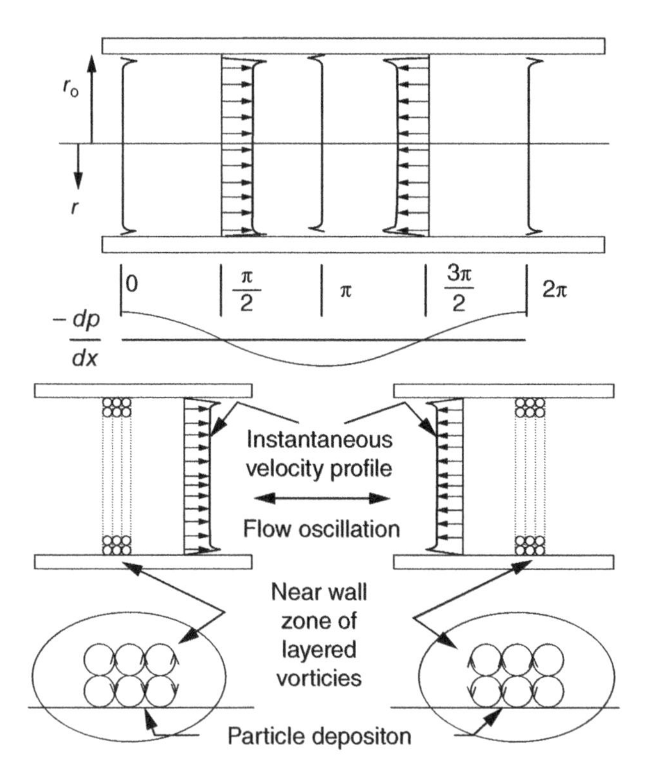

Fig. 22.3 Pressure pulsation within the exhaust of a dry pump. Air can counter-diffuse by alternately flowing in the core toward the pump during the negative half-cycle and remaining within the vortices during the positive half-cycle. R.A. Abreu et al. [3]/ Reproduced with permission from AIP Publishing.

pulse is outward, toward the exit, the air in the vortices moves closer to the pump. Counter diffusion results from this process of alternately moving toward the pump in one half-cycle, and "hiding" in the vortex currents during the other half-cycle [3]. This enhanced back-diffusion phenomenon is known as Richardson's annular effect [3,4].

22.2 Throttled High Vacuum Systems

The pressure range encompassed by sputtering and other plasma processes is above the operating range of all high vacuum pumps. They can be used with a throttle valve between the pump and the chamber. The throttle valve allows gas to flow from the process chamber to the pump while keeping the pressure in the pump below its maximum operating pressure.

Turbo and cryopumps will maintain chambers at sputtering pressures when the inlet to each is throttled to a pressure below its respective critical inlet pressure. Figure 22.2, curve C, outlines the upper throughput limit of a typical small, throttled high vacuum pump. This section discusses process chambers as well as the configuration and operation of turbo and cryogenic pumps for sputtering applications. Ion pumps are not considered because of their inability to handle high gas loads; diffusion pumps are no longer used on high-purity etching or deposition systems.

22.2.1 Chamber Designs

As we noted in Chapter 21, cleanliness and contamination-free needs force some systems to be designed and constructed according to UHV practice. To reduce overall contamination in the vacuum product to the part-per-billion (ppb) level, the ratio of background or chamber contamination to process gas pressure must be kept extremely small. Entrance and exit load locks are used to isolate the environment, and gases of high purity will be delivered to the system through gas distribution systems that will be leak tight at the UHV level. Many of the gases used in these processes are corrosive, explosive, or poisonous. Designs that focus on UHV construction techniques result in clean systems with reduced maintenance. For example, reduction of moisture levels to the ppb level is required to reduce corrosion from a reaction with an anhydrous etch gas such as chlorine.

Residual gases pose a greater problem in a sputtering system than in a high vacuum evaporation system, because of enhanced plasma desorption of wall impurities. Electron- and ion-impact desorption efficiently release gases from the chamber walls. They are more effective than a mild bake. If hydrogen is not removed from an argon discharge, the sputtering rate will be reduced [5] and hydrogen will become incorporated into the film [6]. Sputtering discharges with argon or other noble gases can be kept clean by operating the discharge in a static mode with selective pumping, in theory, or by permitting a large argon flow through the chamber during sputtering. A static discharge is maintained by exhausting the chamber and refilling it with argon to the operating pressure while other gases are selectively removed by an auxiliary pump located within the chamber. The ideal auxiliary pump does not exist. TSPs generate some methane and do not pump noble gas impurities. NEG pumps have found applications, especially for pumping hydrogen. For these reasons, static discharges are not used for sputtering.

Viscous flushing functions only if there is significant gas flow through the active sputtering region and chamber, and if the arrival rate of impurities from the gas source is much less than the desorption rate from the chamber walls. Lamont [7] has pointed out that high throughput alone does not guarantee adequate flushing; it is necessary that the gas stream velocity be large in the region in which the cleaning action is desired. Contamination originating from within the chamber can be reduced with high gas flow. In the high flow limit, the lowest possible level of contamination attainable is that of the source gas. In critical applications, the source gas is scrubbed by passing it through a titanium sublimation or a nonevaporable getter pump. Because of this, there is little point in flushing a small chamber at a rate greater than a few hundred Pa-L/s (a few Torr-L/s) with the purest available source gas [8]. Both the source gas cleanliness and the gas flow rate are important to the maintenance of conditions suitable for deposition of pure films.

Contamination-free sputtering requires high vacuum pumping to a suitable base pressure [9] followed by pre-sputter cleaning with the discharge operating and with the shutter covering the samples on which a film is to be deposited. The flushing time of a typical small system (<1 s) implies that a system could be cleaned by pumping to the process pressure and initiating a glow discharge without pumping to high vacuum [10]. Unfortunately, the glow does not clean all surfaces adequately, nor do surfaces outgas that rapidly. The continued evolution of water vapor from surfaces not exposed to the glow can cause oxygen or hydrogen contamination of deposited films. Most importantly, the only routine way to check for minute leaks in the system without the use of an RGA is to pump to the same low base pressure each time before opening the leak to argon flow. For this reason, *in situ*, partial pressure analysis is now common. Pre-sputtering cleans by several techniques. If the sputtered material is a getter, it may be allowed to deposit on the chamber walls where it serves as an effective getter pump [11]. The sputtered material also covers adsorbed gases, whereas the discharge cleans surfaces exposed to the glow.

Base pressure and the length of pre-sputtering time are process, equipment, and material dependent. No general observations can be made; for example, d'Heurle [12] showed that a 10-min pre-sputter cleaning was sufficient for aluminum, while a minimum of 1-h cleaning time for molybdenum [13]. Some film

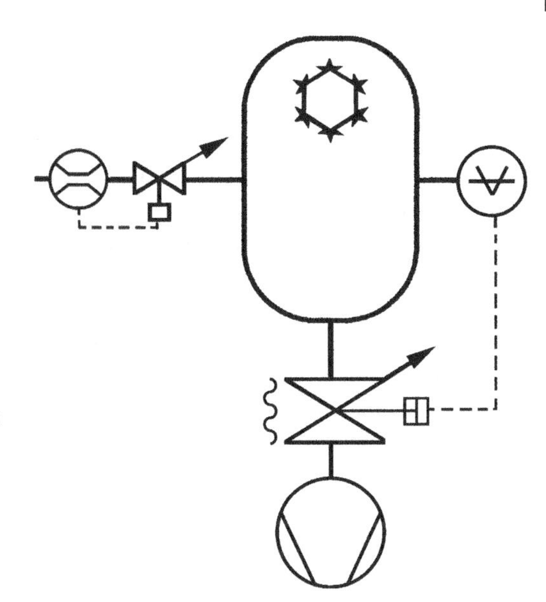

Fig. 22.4 Generic reactive processing chamber illustrating *independent* control of the gas flow and the chamber pressure. The gas flow is regulated by a thermal mass flow controller, while a capacitance manometer measures and maintains the chamber pressure by controlling the throttle valve located between the chamber and the cryo or turbopump. A cryo panel suitable for pumping water is shown located within the processing chamber. This is often done to improve the water pumping speed and the time required to reach the needed base pressure.

properties are so dependent on deposition purity that the base pressure and minimum pre-sputtering time are of crucial importance in the repeatable fabrication of uniform, high-quality films. In addition, the hydrogen background concentration can affect film purity and that is not always reported. The cleanest systems use entrance, transfer, cleaning, and exit locks to maintain purity in the process chamber and clean incoming substrates.

Precise process control must include *independent* control of both chamber pressure *and* process gas flow. Figure 22.4 illustrates a design that allows independent control of both variables. Note that *pressure* is controlled by the *throttle valve*, whereas process *gas flow* is controlled by a *mass flow controller*. If we were to fix the throttle valve setting, we would change the pressure when changing the gas flow. However, changing two variables (gas flow and pressure) simultaneously masks the process chemistry, because chemical reactions vary with reactant quantities *and* reaction rates. In the early days of reactive ion etching, this lack of understanding made data comparison between researchers impossible, because etch rates are controlled by both pressure and reactant quantity. Gas flow and chamber pressure must be *independently controlled* in order to achieve a reproducible, meaningful process.

22.2.2 Turbo Pumped

A turbo-pumped system suitable for high-flow applications needs to attain an adequate base pressure as well as exhaust a high gas flow at medium vacuum pressures. The two requirements are different but not in conflict. In Chapter 20 we discussed the requirements of a turbopump for good high vacuum pumping: high compression ratio for light gases, high pumping speed for all gases, and an auxiliary cryo surface for increased water pumping speed. The selection of a pump for a particular application was then made on the basis of pumping speed and compression ratio. Data that describe compression ratios for individual gases are most often available for zero gas flow. These data cannot be used to predict pump performance under gas flow. Limited high-flow data are available, and some results are presented here.

Figure 22.5 depicts the pumping speed for hydrogen and the argon throughput as a function of argon gas pressure. The argon throughput is the product of the argon pumping speed and the inlet pressure. For this pump, the hydrogen speed remained constant for argon inlet pressures ≤0.9 Pa (7 mTorr), above which it dropped precipitously. The sudden drop in hydrogen pumping speed corresponded to a similar sharp decrease in the rotational speed of the rotor blades in this pump. The exact pressure at which the rotational speed begins to decrease is a function of the pumping speed of the forepump. See Fig. 18.10. As the inlet pressure increased, the gas flow in the blades closest to the forepump changes from molecular to transitional and then to viscous. Near the onset of viscous flow, the added frictional drag requires more torque from the motor; the constant power motor responds by reducing rotational speed, i.e., power = speed × torque. The knee of curve A in Fig. 22.5 will move slightly to the right for a forepump larger than the 35-m^3/h pump used here and to the left for a smaller forepump. It is important to keep the pump running at full rotational velocity to maintain adequate hydrogen pumping speed. The blades will run at full velocity as long as the inlet pressure is suitably throttled below the pump's maximum rated flow.

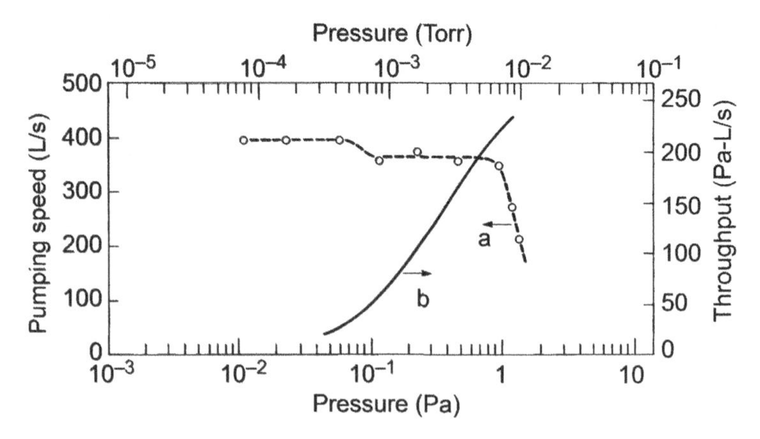

Fig. 22.5 Hydrogen pumping speed (a), and argon throughput (b), as a function of argon inlet pressure in a Balzers 400-L/s turbomolecular pump backed by a 35-m^3/h rotary vane pump. J.F. O'Hanlon [14]/Reproduced with permission from AIP Publishing.

Visser [8] has measured the methane compression ratio in a turbopump during argon flow. His results, shown in Fig. 22.6, quantified the increase in methane compression ratio with increasing argon flow.

Konishi, Shibata, and Ohmi [15] have shown that the staging ratio (ratio of turbo to backing pump speed) strongly affects the partial pressure of contaminant gases that originated in the backing pump. Figure 22.7 illustrates the partial pressure of helium challenge gas as measured in the process chamber for a range of nitrogen process gas flow backing pump sizes. A minimum in the contaminant gas (a maximum in its compression) was observed. In molecular flow, the compression ratio for helium is the same as stated by the manufacturer for the condition of no gas flow. As the pump inlet pressure increases, the flow enters the transition region and begins to sweep the helium toward the fore chamber. Eventually, the pressure becomes high enough so that the blades are in viscous flow and the pump's speed decreases. As soon as the rotor velocity begins to decrease, the helium compression ratio decreases. Figure 22.7 demonstrates that process contamination may be reduced by using a large

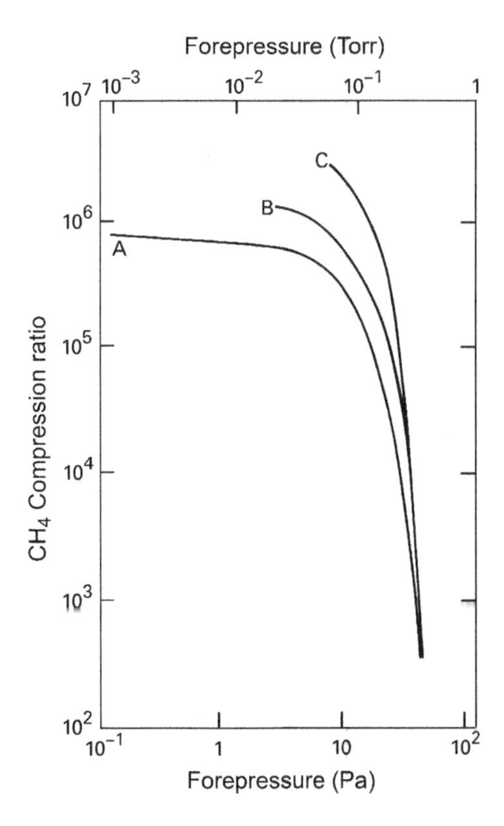

Fig. 22.6 Methane compression ratio as a function of argon gas flow in a Balzers 250 L/s turbomolecular pump. (A) Q(Ar) = 0, (B) Q(Ar) = 2 Pa-L/s, (C) Q(Ar) = 8 Pa-L/s. J. Visser, [8]/Reproduced with permission from Materials Research Corp.

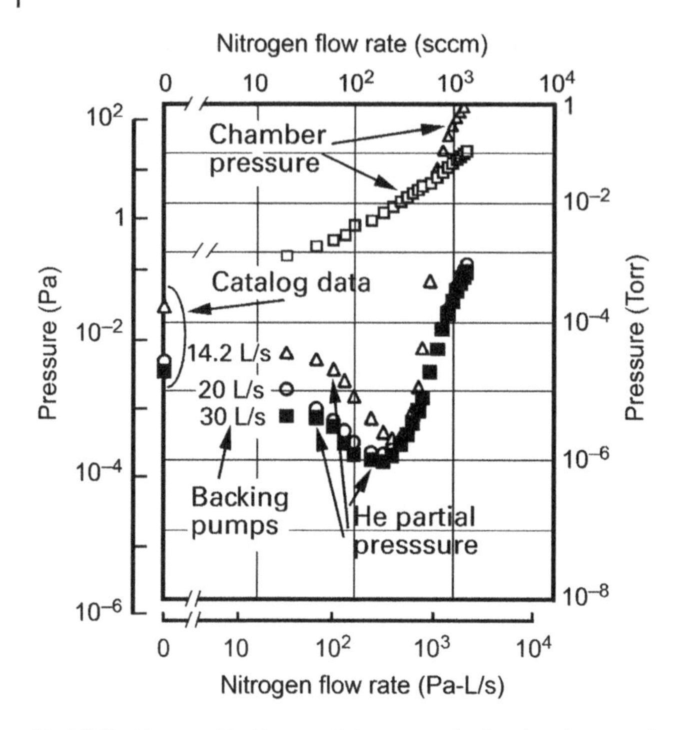

Fig. 22.7 Measured helium partial pressure in the chamber as a function of nitrogen gas flow rate for backing pumps having three different speeds. Helium gas challenge was 200 sccm. Catalog data were calculated from the compression ratios of the TMP. N. Konishi et al. [15]/Reproduced with permission from AIP Publishing.

backing pump and throttling the turbopump at least enough to maintain maximum rotor velocity. As noted elsewhere, some processes may be sensitive to or unknowingly dependent on small hydrogen partial pressures leading to unpredictable results.

Turbopumps, especially those with a molecular drag stage, are well-suited for this use, if throttled to keep the blades running at full velocity and backed by an adequately large mechanical pump. Today's clean systems use dry mechanical pumps to back a turbo-drag pump forming a completely oil-free system. When operated correctly, such systems will have a high pumping speed and a high compression ratio for light gases. The turbopump thus serves as a one-way baffle for light gases and hydrocarbons, whereas most of the compression is done by the dry mechanical pump. The turbopump is not appropriate for pumping on high-pressure plasma polymer deposition systems because material may deposit on and unbalance the rotors, and lead to pump destruction [16].

22.2.3 Cryo Pumped

The cryopump should be capable of both evacuating the chamber to an adequate base pressure and pumping a large gas flow. Chamber outgassing, and temperature and history of the cold stage, determine the base pressure. In Chapter 14, we discussed how a balance between the refrigeration capacity and heat loading determines the pump's temperature. Temperature was not the only factor that determined the pumping speed of a gas. It is also a function of the species and quantity of previously sorbed gasses.

It was also observed that the heat load carried to the pumping surfaces by the incoming gases under high vacuum conditions (low gas throughput) was insignificant in comparison to the radiant flux. If nitrogen were pumped with a typical two-stage cryopump, the time to deposit a 1-mm-thick condensed layer of solid nitrogen would be about 10^4 h at a pressure of 10^{-5} Pa (10^{-7} Torr) [17]. Therefore, neither the heat load of the incoming gas nor the resulting solid deposit is a major concern in the high vacuum region. This is not true when large quantities of gas are being pumped. As the gas flow to a cryopump is increased, its pumping speed changes. Figure 22.8 sketches the pressure dependence of the pumping speed over several flow regions. In the free molecular flow region, the pumping speed is constant. At somewhat higher pressures the speed increases due to the increased conductance as the gas enters the transition flow region. Under some circumstances, this flow will reach a maximum value (choked or critical flow) that is characteristic of the sonic velocity of the gas. At higher pressures, the heat conductivity of the gas becomes large and heat from the walls of the chamber will

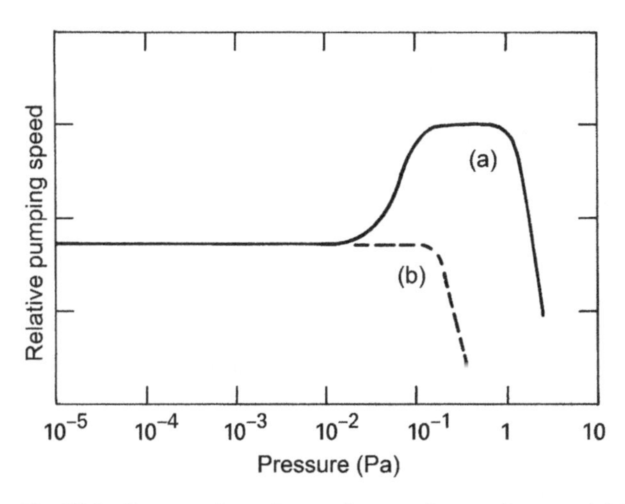

Fig. 22.8 Pressure dependence of cryogenic pumping speed. (a) Free surface, (b) practical baffled pump.

flow to the cooled surfaces by gas transfer. As these surfaces warm, the sticking coefficients decrease and pumping ceases. This behavior has been observed by Dawson and Haygood [18] for CO_2 and by Bland [19] for water vapor. Loss of pumping speed in a practical pump usually occurs at about 0.2–0.4 Pa (1.5–3 mTorr), where the heat loads exceed the capacity of the refrigerator. The pumping speed of a pump in the high-pressure region is sketched in Fig. 22.8*b*.

At high throughput, the gas carries most of the heat to the cryo surface. An argon flow of 180 Pa-L/s (1.35 Torr-L/s or 1.8 mbar-L/s) corresponds to 1-W incident heat flux on a 20 K surface. For gases such as nitrogen, which have a heat of condensation greater than argon, the heat flux will be proportionately larger. The majority of this heat flux will be absorbed by the cold stage. As an example, consider a two-stage cryopump with surfaces at 20 and 80 K, respectively, in which the argon first collides with the 80 K baffle, passes through the baffle, and finally sticks on the 20 K surface. For each kg-mol of argon that flows into the pump, the expander must remove a total of 13,862 kJ. This value is obtained by taking the difference in enthalpy between 300 and 20 K. See Table 22.1. If all of the argon were cooled to 80 K on impact with the 80 K baffle, a total of 4580 kJ/(kg-mol) would be removed. This value corresponds to 33% of the total heat that is removed during the pumping process. In practice, the gas is not cooled to 80 K on impact with the first stage, because the accommodation coefficient is not unity. If, for example, the warm stage had an efficiency of 0.5 for cooling the argon, then 85% of the total heat of the gas would remain to be removed by the 20 K stage. The cold stage removes at least 67% of the heat from argon because its vapor pressure is so low, and its heat of condensation is so large. Table 22.1 gives the approximate total

Table 22.1 Enthalpy of Gases Frequently Pumped at High Flow Rates

	Total Enthalpy[a] kJ/(kg-mole)		
Gas or Vapor	**300 K**	**80 K**	**20 K**
Ar	13,950	9370	**88**
N_2	15,580	9190	**134**
CCl_4	49,750	**2950**	738
CF_4	26,000	16,000	**670**
CF_3H	30,750	**2,678**	670
CF_3Cl	31,300	**20,230**	678

[a] Approximate total enthalpy at room temperature and nominal temperatures of 1st and 2nd cryopump stages. Gas and vapor enthalpies are given in bold for cryo surface on which they solidify.

enthalpy at the temperatures of the first and second stages of a cryopump for several gases of interest. The maximum throughput for each gas is determined by the maximum power that can be removed by the stage at which the gas or vapor condenses. The maximum throughput is gas dependent. CF_4 has a maximum throughput of half the value quoted by the manufacturer for argon. Several other gases used for reactive ion etching deposit on the 80 K stage and can completely close the baffles and quickly render the pump useless for high vacuum pumping unless it is first regenerated.

At high gas flow, the heat absorbed by the expander will be proportional to the total gas throughput, until the pressure is large enough for heat conduction from the walls to be appreciable. Heat conduction from the walls begins at a Knudsen number of ~1 and reaches its maximum or high-pressure value at $Kn = 0.01$. This heat load is added to the heat load due to heat of condensation but not in a linear manner. The largest amount of heat flowing by gas conduction goes to the 80 K pumping stage. This added warm stage load reduces the refrigeration capacity of the cold stage. For high gas flow applications, a cryogenic pump should be designed to minimize the heat load from the walls so that the expander may be most efficiently used to remove the heat of condensation of the incoming gas. For a typical pump, the Knudsen number for air at 0.2 Pa (1 mTorr) is about 0.3. At this value the heat capacity is a substantial fraction of the high-pressure value, permitting several Watts to flow from the chamber walls to the warm stage.

In some high gas flow situations, continued pumping of hydrogen is important but does not always happen. Pumps are designed so that the inner surface of the cold stage is baffled from the warm stage and covered with a layer of activated charcoal. This surface will pump hydrogen, neon, and helium effectively if it is adequately cooled. At 10 K, there is adequate sorption capacity for hydrogen on activated charcoal or molecular sieve. Argon will also be pumped on this surface. To avoid or reduce the argon pumping on the inner surface, it is baffled from the remainder of the system. If the baffle is optically dense for argon, the hydrogen pumping speed will be reduced, and if the baffle is open, the argon will readily pass through, condense on the inner surface, and cover the sorbent with solid argon. All pump designs are a compromise between these two concerns. In a high-flow application, however, the sorbent surface will become coated with argon rather quickly, regardless of the nature of the baffling. Once the sorbent becomes coated, cryotrapping is the only mechanism by which hydrogen can be pumped. Hengevoss [20] has shown that the cryotrapping of hydrogen in argon is strongly temperature dependent. It is nil in solid argon for temperatures greater than 20 K. The value of argon throughput, which will keep the second-stage temperature below 20 K, may be below the stated maximum value for a particular pump.

The constraints placed on a cryopump system for high gas flow are considerably different than those placed on pumps used for high vacuum. These systems require a throttle valve to keep the pressure in the cryopump below a value of 0.2–0.4 Pa (1.5–3 mTorr). The cryopumps must be able to remove the conductive heat load entering the chamber from the walls. This heat load may be removed by use of a high-capacity, two-stage gas refrigerator.

Regeneration in a sputter or etch system will obviously be more frequent than in a system used for high vacuum. Automatic controllers are available for performing this function on idle time. If a cryopump is used at the design-limit argon throughput, the cold stage will be heated to a temperature at which it will not pump hydrogen or helium; either supplemental pumping or throughput reduction will be necessary to pump these gases. Furthermore, the pump's maximum throughput is dependent on the vapor pressure, specific heat, and heat of sublimation of the gas or vapor being pumped.

Cryogenic pumps also suffer from the phenomenon of irreversible overloading. Pump warming can be triggered by a gas burst when the pump is operating near its throughput limit. It will then cease pumping and require regeneration, because the rate at which heat from a gas burst enters the pump is greater than the rate at which it can be removed by the refrigerator.

Cryogenic pumping of toxic or explosive gases presents serious safety concerns. If the pump were to suddenly warm, a large quantity of gas could be emptied into the exhaust system. If this gas were toxic, it could overload the exhaust scrubber. If the pump were condensing an explosive gas, operation of an ionization gauge in the pump body during release of the gas would present a serious problem. The ion gauge tube, and any other gauges that could act as ignition sources, should be located outside the pump body and interlocked so that its filament cannot be operated when gas is being released. A prudent operator would not choose cryogenic pumping for certain gases. These concerns should be understood before selecting pumps for sputter or etch processes. Lessard [21] describes the explosion hazards from hydrogen and oxygen, produced by the dissociation of water and entrapped in the cold stage. A potential hydrogen detonation is shown to be a function of the quantity of hydrogen available, which in turn is dependent on the charcoal quantity and the design of the cryopump baffles and louvers. Pump manufacturers limit this risk by limiting the quantity of charcoal and thus limiting the quantity of hydrogen that can be adsorbed in the charcoal. Ozone was shown to be a potential risk factor for a small number of system operators. It was concluded that certain sputter processes using oxygen could generate O_3 that could be trapped on the cold array. During regeneration, it is possible for the ozone to drip from the array in liquid form and detonate from the liquid state with a resulting loud pop or bang [21]. As indicated above, potential ignition source such as certain gauges should be located on the chamber side of the high-vacuum valve to limit the chance of ozone ignition during regeneration. Much of the guidance discussed in this chapter is summarized in Table 22.2.

Table 22.2 Best Practices for High Gas Flow Systems

Set chamber pressure independently of other variables using main pump throttle valve.
Set process gas flow independently of other variables using mass flow controller.
Determine mass flow rates of all incoming gasses with mass flow controllers.
Regenerate cryopumps frequently in high-flow processes, e.g., sputtering and etching.
Be aware that process differences observed before and after regeneration are often due to reduced hydrogen in the base pressure after regeneration.
Vent cryopump gases during regeneration to building exhaust if they may contain hazardous components.
Do not use cryopumps for process that have potential to produce ozone.
Use appropriate sound shielding for high volume vacuum pumps.
Operate high volume vacuum pumps with proper gas purging and at sufficient temperatures to limit condensation of process effluent within the pump.
Consider use of temperature-controlled jackets on fore lines and exhaust lines to limit and control condensation within these lines.
Clean screw, lobe, claw, reciprocating piston, and other dry pumps frequently when pumping corrosive and particulate-generating gases and vapors.
Use proper low vapor pressure lubricating fluids in gearboxes. Operate these gearboxes with vendor-recommended temperature ranges by control of cooling water flow and temperatures. Check the condition of gearbox lubricant often for proper color and viscosity.
Establish schedule for periodic factory rebuilding of high volume vacuum pumps, especially if effluent contains corrosive or condensing gases.
Carefully maintain process cooling water flow and temperatures to vendor specifications within various regions of high volume vacuum pumps.
Install O-rings manufactured from inert elastomers.
Remove chamber debris and clean sputter shields frequently.
Install proper effluent treatment equipment and processes on exhaust lines to limit release of toxic, flammable, corrosive, or particle-laden exhaust gases.

References

1 Kumagai, H.Y., *J. Vac. Sci. Technol., A* **8**, 2865 (1990).

2 O'Hanlon, J.F. and Fraser, D.B., *J. Vac. Sci. Technol., A* **6**, 1226 (1988).

3 Abreu, R.A., Troup, A.D., and Sahm, M.K., *J. Vac. Sci. Technol., B* **12**, 2763 (1994).

4 Richardson, E.G. and Tyler, E., *Proc. Phys Soc. (London)* **42**, 1 (1929).

5 Stern, E. and Caswell, H.L., *J. Vac. Sci. Technol.* **4**, 128 (1967).

6 Cuomo, J.J., Leary, P.A., Yu, D., Reuter, W., and Frisch, M., *J. Vac. Sci. Technol.* **16**, 299 (1979).

7 Lamont, L.T., *J. Vac. Sci. Technol.* **10**, 251 (1973).

8 Visser, J., *Trans. Conference and School on Elements and Techniques and Applications of Sputtering*, Brighton, November 7–9, 1971, p. 105.

9 Maissel, L.I. "The Deposition of Thin Films by Cathode Sputtering", in *Physics of Thin Films*, G. Hass and R.E. Thun, Eds., Vol. **3**, Academic, New York, 1966, Chapter 5, p. 106.
10 Shirn, G.A. and Patterson, W.L., *J. Vac. Sci. Technol.* **7**, 453 (1970).
11 Theuerer, H.C. and Hauser, J.J., *Appl. Phys.* **35**, 554 (1964).
12 d'Heurle, F.M., *Metall. Trans.* **1**, 625 (1970).
13 Blachman, G., *Metall. Trans.* **2**, 699 (1971).
14 O'Hanlon, J.F., *J. Vac. Sci. Technol.* **16**, 724 (1979).
15 Konishi, N., Shibata, T., and Ohmi, T., *J. Vac. Sci. Technol., A* **14**, 2958 (1996).
16 J. Vossen, Dry Etching Seminar, New England Combined Chapter and National Thin Film Division of AVS., October 10–11, 1978, Danvers, MA.
17 Davey, G., *Vacuum* **26**, 17 (1976).
18 Dawson, J.P. and Haygood, J.D., *Cryogenics* **5**, 57 (1965).
19 Bland, M.E., *Cryogenics* **15**, 639 (1975).
20 Hengevoss, J., *J. Vac. Sci. Technol.* **6**, 58 (1969).
21 Lessard, P.H., *J. Vac. Sci. Technol., A* **8**, 2874 (1990).

23

Multichambered Systems

Multiple chamber systems are economically important in the manufacture of many common consumer products, both low- and high-tech. Coated paper for inkjet and laser printers, computer chips, coated plastic films for food, and anti-static packaging are a few of the items that we encounter daily. Other materials fabricated in multichambered systems are used to form component parts of products we may not recognize—products such as retail sales security tags, secure paper currency, thermally efficient home, commercial, and automobile window glass, solar cells, image recorders, and active matrix display panels.

The underlying structure, or substrate, on which films are deposited and patterned, can be either flexible (for example, a roll of plastic film) or rigid (for example, a glass or metal plate or a silicon wafer). In the case of rigid substrates, we can further choose between "in-line" (serial) processing or "cluster" (random) processing. In an in-line system, the products march sequentially through the serial chambers in lockstep. In a cluster, tool the process sequence for cleaning, etching, deposition, or annealing is chosen by the design of the particular device or product. The same equipment can produce multiple products by altering the process and its sequence.

The choice of equipment depends on many factors. The product volume, size, allowed defect levels, product yield, and profit margin all influence the cost of manufacturing and the type of system that works best. Low-cost consumer products need to be coated rapidly and cheaply; this requires high product throughput in systems with low capital investment. Some high-tech products are exceedingly sensitive to cross-contamination from a prior process step; these systems require careful isolation between process steps. No single system design fits all needs. In this chapter, we review some common designs in order that we understand why products are manufactured in specific systems and how these systems are configured. We review machines for handling flexible

A Users Guide to Vacuum Technology, Fourth Edition. John F. O'Hanlon
and Timothy A. Gessert.
© 2024 John Wiley & Sons, Inc. Published 2024 by John Wiley & Sons, Inc.

and rigid substrates, as well as discuss multichamber pressure reduction techniques for sampling atmospheric pressure gases.

23.1 Flexible Substrates

A large number of consumer products that are manufactured in sheet form require some form of thin-film coating. Many polymer films for electrostatic and other packaging, decorative coating, anti-theft and security devices, videotape, capacitors, and heat-reflecting and magnetic storage applications, as well as paper for laser and inkjet printing, are economically coated in vacuum deposition systems. These coating processes are only economical, if the material can be deposited at high speeds on large rolls of material. The machines designed for this purpose are called web coaters. The process is known as semicontinuous coating.

The dominant outgassing from the feedstock rolls comes from the ends of the rolls, typically of order 1 m. long and 0.5–1-m diameter. The gas load, primarily water vapor, largely originates in the end faces of the feed rolls. Depending on the nature of the material, the total gas load from the feed roll, and chamber walls could be as large as 7000–10,000 Pa-L/s (50–70 Torr-L/s). A reasonable deposition pressure for a typical process might be ~10^{-2} Pa (10^{-4} Torr). The necessary pumping speed for a single chamber system would therefore be >10^5 L/s. Clearly, it would be uneconomical to coat material in a system with such a high capital investment. The equation $S = Q/P$, tells us that low pumping speed is required to remove gas at high pressure. Thus, a two-chamber design in which the feed and winding roll chamber is maintained at a high pressure and in which the deposition chamber is maintained at a low pressure will be more economical than a single chamber. Figure 23.1*a* illustrates a two-chamber design. The feedstock is loaded on the upper left spindle. It is threaded around the central drum, then back to the winding roll, where it is stored on the winding roll. Although not shown in Fig. 23.1*a*, web coaters often contain a glow discharge cleaning apparatus between the feedstock roll and the first slit. This system is designed to remove large amounts of water vapor from the feed material before coating. The inner slits are adjacent to the central drum separate the dividing partition and central drum on both sides and both ends of the roll. The entire chamber is constructed in two parts: one part, bounded by the outer slits, contains the rear system wall, deposition tooling, and pumping system connections. The central section containing the three rolls and inner slits can be removed to facilitate loading and unloading feedstock and finished product.

Two separate pumping systems are now required: one to pump the winding chamber and one to pump the deposition chamber. However, the sizes of the pumps are significantly less than those required for a single chamber design.

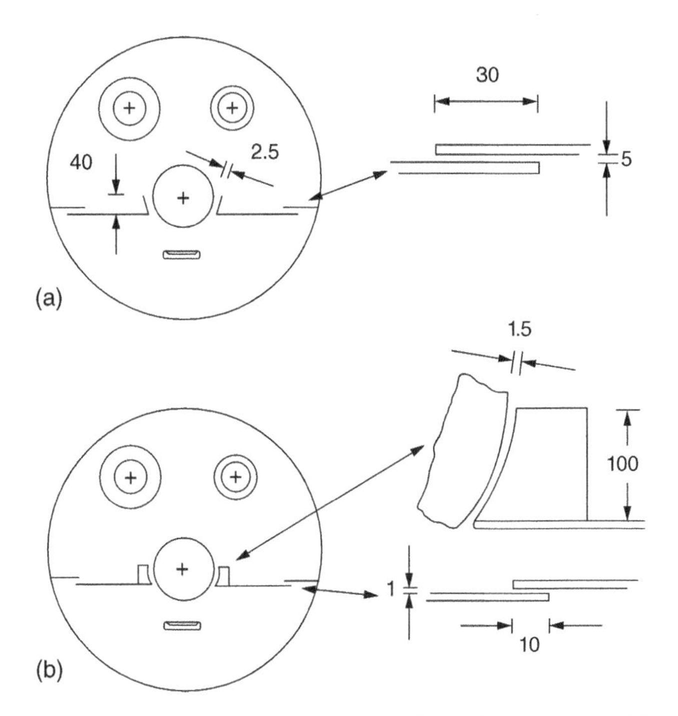

Fig. 23.1 Schematic representation of slit configurations for different two-chamber vacuum web coaters. (a) basic slits; (b) improved slit design to reduce gas flow to deposition chamber. J. Kieser et al. [1]/Reproduced with permission of Elsevier.

The deposition chamber pump needs to exhaust gas from the deposition process (typically sputter deposition), the outgassing from the deposition chamber walls, and the gas load from the winding chamber that flows through the inner and outer slits. Sophisticated inner slits with reduced spacing have been designed to reduce the gas flow to the deposition chamber; one such design is shown in Fig. 23.1*b*.

The gas flow from the winding chamber to the deposition chamber is primarily water and possibly organic vapors. Demands for reducing the contamination in the deposition chamber necessitate further design changes in the system, shown in Fig. 23.2. In the coating system depicted here a third chamber is used to remove gas at a pressure intermediate to the winding and deposition chambers. With three chambers, one can tolerate increased throughput through the slits, smaller pumps, or lower levels of contamination in the deposition chamber. Pressure in the deposition chamber is the primary web coater design issue [1]; however, capital cost and cleanliness are very important.

Air-to-air coating—a configuration in which product moves from air through a system and back to air, without encountering a valve—is economical for coating materials such as long rolls of wire or sheet stock [1]. Multiple chambers isolated

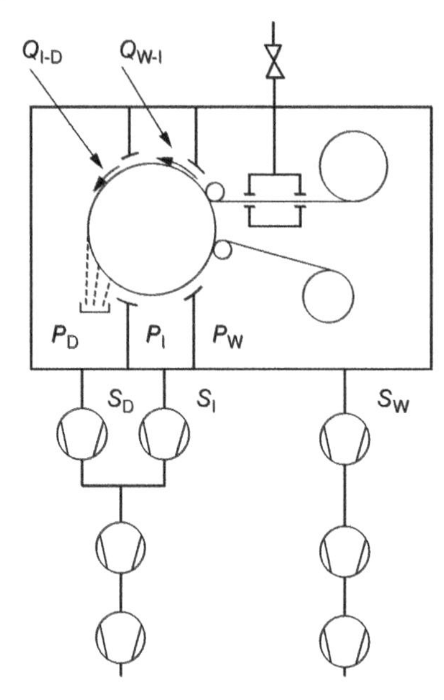

Fig. 23.2 Schematic representation of an advanced vacuum web coater containing an intermediate chamber. The gas flow between the chambers is indicated. A glow discharge unit for pretreatment of the web is shown in the winding chamber. J. Kieser et al. [1]/ Reproduced with permission of Elsevier.

by narrow slits (sheet stock) or closely spaced tubes (wire stock) are used to reduce sequentially the pressure to the central process chambers where the material is cleaned and coated, followed by a reverse sequence of exit isolation chambers.

A novel design for multilayer deposition on flexible substrates, Fig. 23.3 was developed by Madan [2]. In this design, the flexible substrate can be transported between a feed roll and winding roll during reactive sputter deposition. The feedstock can be moved in either direction between left or right rolls. Both rolls are contained within a removable cassette.

Electrical properties of photovoltaic cells and thin-film-transistor displays deposited on flexible substrates, for which this equipment was designed, are extremely sensitive to film deposition process contamination that originated from a previous step. To minimize contamination between two process steps, the cassette, containing either thin stainless steel or a thin plastic

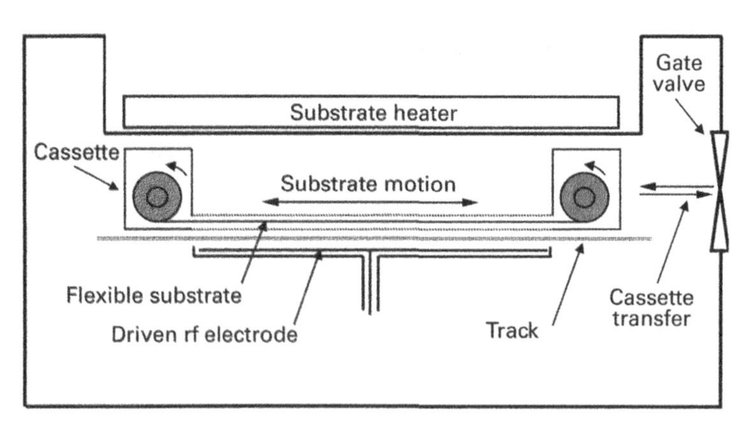

Fig. 23.3 Schematic view of a transportable cassette for holding a rolled, thin, flexible substrate. This cluster-tool configuration has one gate valve. Alternatively, gate valves may be configured at both ends for use in an inline system. Reproduced with permission of MVS Systems, Inc [2].

substrate, can be removed and transferred between deposition chambers. This design would represent a hybrid inline-cluster system.

23.2 Rigid Substrates

Thin, flexible substrates in roll form are ideally suited for coating in semicontinuous deposition systems. Rigid substrates cannot be handled in the same fashion. Rigid substrates can be coated in either inline systems, or in systems designed in the form of a cluster. In this section we give examples of both forms of system design. The design of choice depends on the size of the substrate, capital equipment investment, process flexibility, tolerable cross-contamination levels, product throughput, and cost and value added. For extremely high volume, high throughput manufacturing, large inline designs appear to be the most efficient. Cluster designs appear to be most efficient for devices that would require extensive process isolation.

23.2.1 Inline Systems

Inline systems contain integrated atmospheric- or vacuum-based pre-cleaning zones, because operator handling constitutes a large source of contamination. Figure 23.4 illustrates a system containing an atmospheric steam cleaning chamber followed by vacuum drying chambers to eliminate the water vapor load prior

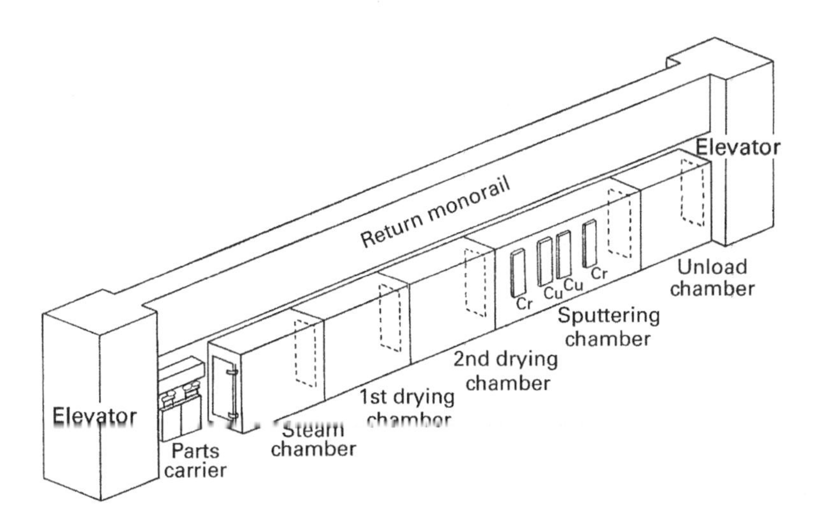

Fig. 23.4 An inline system designed for the deposition of chrome–copper–chrome metallurgy on ceramic substrates. The substrates are steam cleaned and dried before metal deposition. T. Rogelstadt and G. Matarese [3]/Reproduced with permission from AIP Publishing.

to entering the sputter deposition chamber. This system was designed to deposit multilayer Cr–Cu–Cr films on ceramic substrates. The substrates were mounted on carriers, which were used for transport through the cleaning, drying, and deposition chambers. In this system, argon was used as the sputtering gas. No reactive depositions were required; therefore, no isolation chambers were used to separate the individual metal depositions. Adequate cross-contamination reduction could be maintained without intermediate isolation zones, or without slit valves between each metal deposition. The number of series magnetron stages and transport speed can be adjusted to control each layer's thickness. This is an example of a straightforward inline system that incorporates pre-cleaning external to vacuum before formation of the metal layers by physical vapor deposition.

When different gas mixtures are required for each film layer, gas cross-contamination can be a problem. Early designs for fabricating thin film solar cells used one deposition chamber to deposit an entire *p–i–n* structure. First a B_2H_6-doped SiH_4 gas mixture was introduced to deposit the *p*-doped silicon layer. Next the intrinsic layer was formed by pyrolysis of SiH_4. Finally, phosphene was added to the silane to form the *n*-doped silicon layer. Residual diborane caused boron contamination of the intrinsic layer and reduced the sharpness of the *p–i* interface. Residual phosphorous contamination from the last stage also contaminated the boron-doped layer in the next device.

Another design iteration featured separate chambers for each layer deposition. This configuration used an entrance load lock, separate deposition chambers, and an exit lock. Each chamber is isolated with a slit valve. However, cross-contamination still occurred while the slit valve was open [4].

The next-in-line design modification used isolation chambers between process chambers to further reduce cross-contamination from the levels of chambers only separated by only a valve. Figure 23.5 illustrates one such design. Isolation chambers are used between each process and between entrance and exit locks and processes. Load locks use soft roughing to prevent aerosol formation. The isolation chamber between the entrance lock and first process helps remove water vapor from incoming substrates. Quartz lamp heaters as well as cryo panels increase the

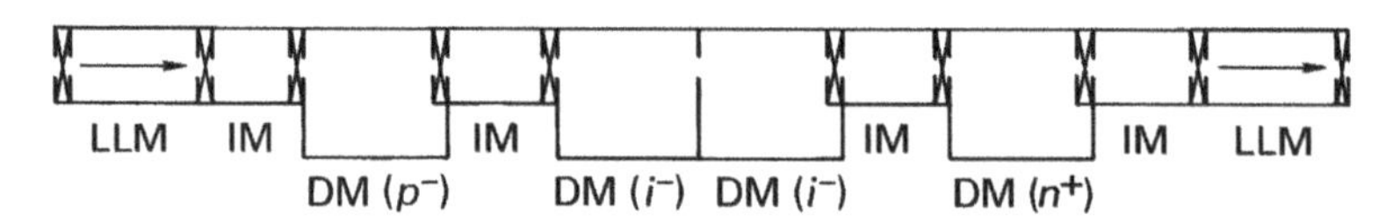

Fig. 23.5 Inline system for the deposition of *p–i–n* solar cells. This system contained isolation modules located between the individual deposition chambers. The isolation modules reduce the cross-contamination between deposition chambers. (LLM = load lock module; IM = isolation module; DM = deposition module.) Adjacent chambers are isolated with gate valves. A. Madan [5]/Reproduced with permission from SPIE.

rate at which water vapor is removed. The number of deposition modules can be altered to meet the relative film thickness requirements. Isolation chambers reduce cross-contamination, but proper electrode design and shielding are necessary to reduce particle contamination due to sputter desorption from chamber fittings and walls. Modified opening and closing of slit valves is required to prevent particle generation caused by suddenly opening and closing slit valves.

Inline systems are ideally suited for coating large architectural-size glass plates. An example system is illustrated in Fig. 23.6. The system is comprised of several zones. First, the plates are washed and then dried. Great care must be taken in the drying process to eliminate as much water vapor as possible. Water will remain, and that must be pumped from the glass surface before it is coated. This is done in the vacuum entry zone. The plates are loaded into the entry lock and pumped rapidly by opening the valve connecting it to a large, evacuated ballast tank, at which time its pressure drops suddenly by pressure division. After the pressure divider step, the ballast tank is isolated and slowly repumped for the next cycle. While this is happening the entry lock is evacuated to about 13 Pa (100 mTorr) with roughing pumps. Following this, the valve connecting the load lock to the holding chamber (10^{-3} Pa) is opened. The entry holding chamber and load lock are relatively sized so that both chambers assume a pressure less than the overload condition of the holding chamber diffusion pumps. The valve between the load lock and holding chamber is now

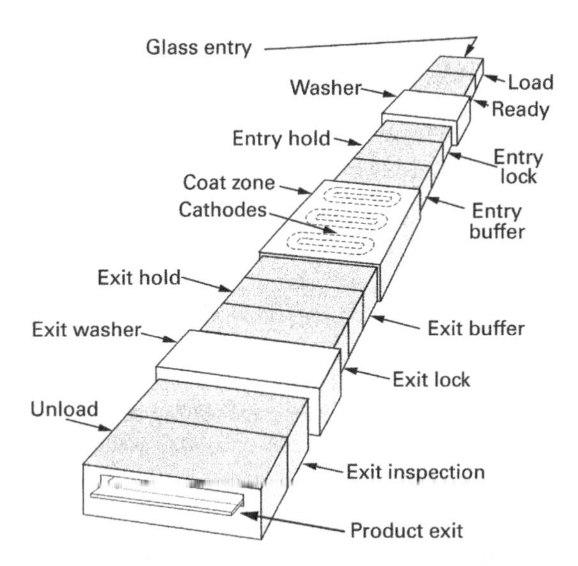

Fig. 23.6 Large double-ended, continuous flow production glass coating system. This system is comprised of load, ready, pre-clean, coat, post-clean, inspection, an unload sections. Russell J. Hill and Steven J. Nadel [6]/Reproduced with permission from von Ardenne GmbH.

closed and the system is pumped for a time necessary to remove the remaining water vapor. Dry air could be used to flush the load lock to reduce the initial water vapor load. Cryo pumping installed in both load lock and entry holding chambers will further reduce water vapor partial pressures before the plates are transported to the entry buffer where they await coating. The transport system in the buffer chamber is designed to increase the transport speed of the plates to the speed required within the coating chambers. Note that this design does not use isolation valves to prevent cross-contamination between adjacent sputter deposition targets. Some sputter targets deposit nitrides using argon–nitrogen mixtures and other targets deposit oxides using argon–oxygen mixtures. These gases cannot enter adjacent deposition modules without affecting the optical properties of the deposited films. Isolation is a necessity, but isolation valves cannot be used in a high-speed, continuous transport design. Instead, an isolation scheme similar to the one described in Fig. 23.7 is used. The process zones in this system design are pumped with four process pumps to handle the large gas flow needed for sputtering. The two isolation zones are each evacuated with a separate pump. Each isolation pump removes the gas load from the process zone to which it is adjacent. Closely spaced plates form isolation tunnels that limit the gas flow from the two process zones. Ideally, the pressure in the two isolation chambers is low enough that the net gas flow through center tunnel in each group of three isolation tunnels should be insignificant. The gap between the top spacer and the upper surface of the glass plates can be adjusted to compensate for the differing incoming product thickness, and thus minimize the gas losses at the isolation slits. The remaining vacuum portion of this system

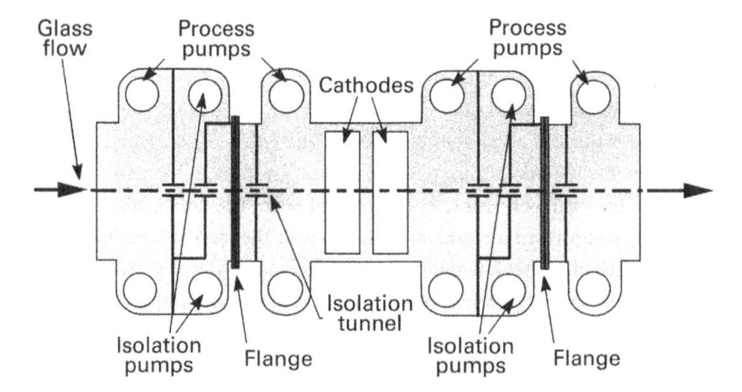

Fig. 23.7 Detail (vertical cross section) showing the method of gas isolation between adjacent process zones in one continuous flow production glass coating system. Each process zone is pumped with four process pumps, whereas each isolation zone is pumped with two pumps. Each isolation pump removes the gas load from the process zone to which it is adjacent. Isolation tunnels limit the gas flow between process zones. Russell J. Hill and Steven J. Nadel [6]/Reproduced with permission from von Ardenne GmbH.

consists of the exit chambers. Non-vacuum cleaning, inspection, and unloading steps complete the system.

Inline systems are extremely efficient for coating high volumes of large substrates. Deposition chambers can be reconfigured with different cathodes or gases to deposit a range of films with differing solar properties. High product volumes make this an efficient low-cost process.

23.2.2 Cluster Systems

Inline systems are not required to achieve isolation and high product quality. Cluster system designs with a central isolation chamber containing wafer transport hardware, can be viewed in one sense as serial processing wrapped around an isolation chamber. Its single isolation chamber has a distinct cost advantage over large inline systems; however, cluster tools cannot handle the same product throughput or massive substrate sizes as inline systems. Cluster system designs do have the advantage of random access to process modules. Thus, one could change a layer sequence without refitting or changing a process chamber. One example of a single cluster tool is illustrated in Fig. 23.8. This system contains one load lock

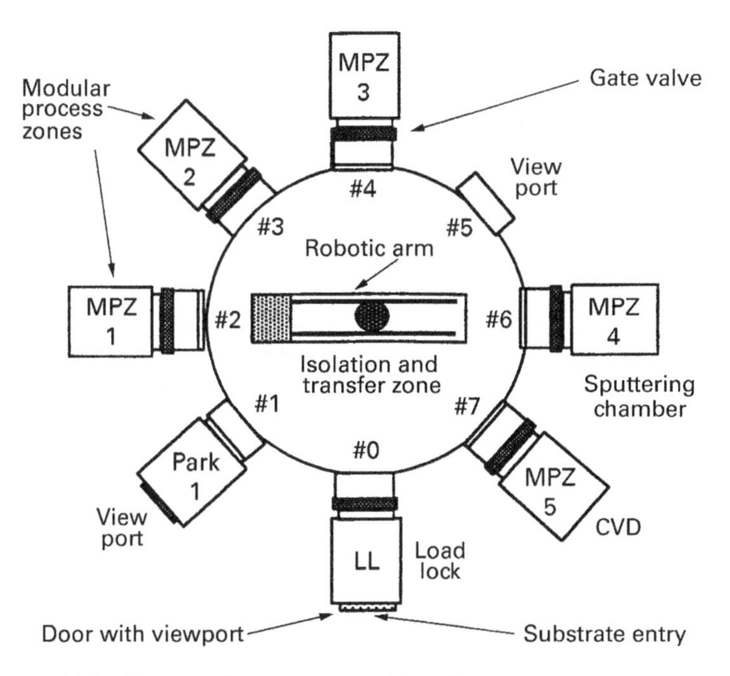

Fig. 23.8 Cluster tool system designed for silicon device fabrication. This system contains a central transfer/isolation chamber, one load–unload chamber and one auxiliary chamber. The auxiliary chamber can be a buffer or holding chamber. Reproduced with permission of MVS Systems, Inc. [2].

through which all substrates are loaded and unloaded. To allow flexibility in substrate movement, one chamber is used as a buffer to park or store substrates. Each chamber contains a specific process such as reactive sputtering or chemical vapor deposition. The robotic arm is a required feature of all cluster tools, and it is used to transfer substrates between process chambers. If needed, a second cluster tool could be connected to the first and provide a greater number of process zones and simultaneous depositions.

The second cluster tool example is depicted in Fig. 23.9. This chamber contains separate exit and entrance load locks. The dual robotic arm, which can access two chambers simultaneously, permits high product throughput. Silicon wafers, for which this system was designed, are loaded externally in ultraclean, isolated cassettes. A separate handling system transfers the incoming 300-mm-diameter wafers in the cassette to or from their respective unload or load lock. The load and unload chambers contain separate pumping systems capable of pumping each lock to ~10^{-3} Pa (10^{-5} Torr). The transfer chamber can be pumped to a pressure of 10^{-5} Pa (10^{-7} Torr) during product transfer. Each process chamber is isolated from the transfer chamber with a slit gate valve.

A second design of increased complexity, shown in Fig. 23.10, uses two transfer chambers. However, the transfer chambers are not identical and are not designed simply for increased product throughput. In this system, the load/unload locks are pumped to a pressure of 10^{-3} Pa, transfer chamber #1 is pumped to a pressure of 10^{-5} Pa, (10^{-7} Torr) whereas the right-hand #2 transfer chamber is pumped to

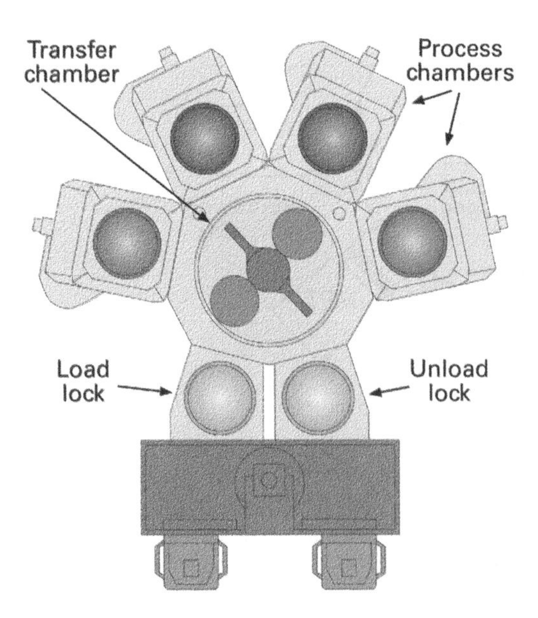

Fig. 23.9 The Dielectric Etch eMax™ Centura® cluster tool is designed for processing silicon microelectronic devices. This system contains a central transfer chamber and dual load locks—one used for product load and the second used for product unload. Reproduced with permission from Applied Materials Inc.

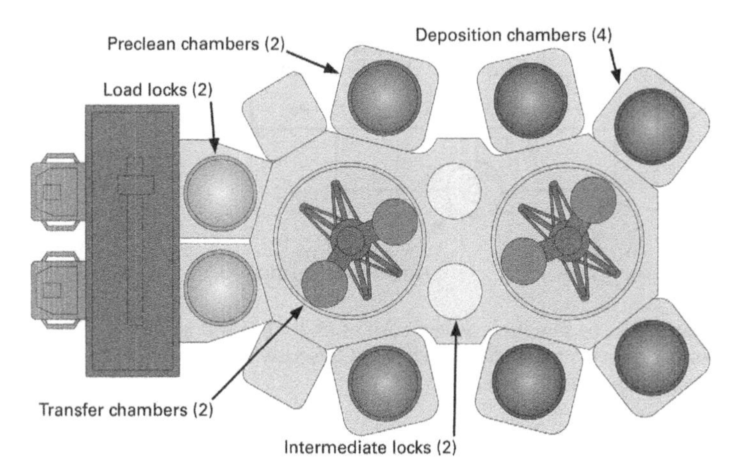

Fig. 23.10 Endura® xP cluster tool system for deposition of metal films. This system contains two transfer chambers and dual load locks. The transfer chambers communicate by means of intermediate locks—one for incoming product and one for outgoing product. Reproduced with permission from Applied Materials Inc.

a pressure of 10^{-6} Pa (10^{-8} Torr). Product wafers are transferred between the transfer chambers through intermediate locks. The intermediate or isolation locks are not pumped but assume the pressure of the chamber to which they are connected. The purpose of this dual-transfer chamber design is to isolate critical processes from atmospheric contamination. In a single-transfer-chamber system, cross-contamination from the load lock to the transfer chamber can enter the process chamber. Since the limiting pressure is no less than 10^{-5} Pa, residual water vapor or oxygen desorbing from surfaces can back diffuse into adjacent process chambers. Precleaning, which does not demand ultraclean vacuum pressures, is done in the left side. Moving a wafer from transfer chamber #1 to the transfer chamber #2 further reduces the cross-contamination from atmosphere, allowing transfer chamber #2 to attain a reduced base pressure. In turn, this reduces the residual cross-contamination to the critical metal deposition processes. These chambers can be pumped to pressures of order 10^{-7} Pa (10^{-9} Torr) allowing ultraclean metal films to be deposited by physical vapor deposition.

Each of the multichambered systems described in this chapter had a different design goal. Some were designed for extremely high product throughput, some for flexibility, and others for high volume manufacturing in an ultraclean environment. Each was designed with the goal of reduced capital cost and reduced unit cost for a specific product. Innovative designers will continue to create new designs that are based on new materials and a thorough understanding of fundamental vacuum principles.

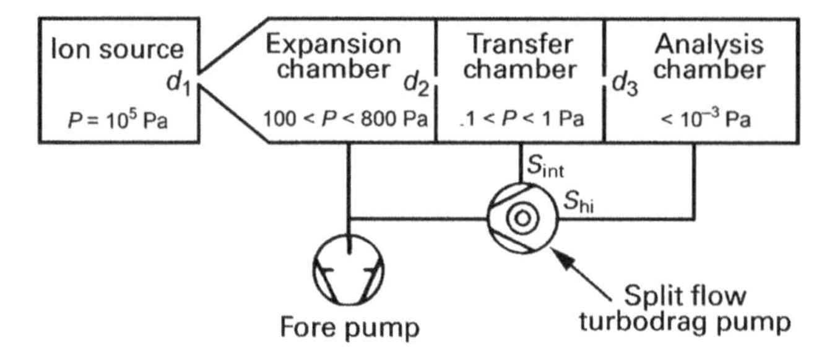

Fig. 23.11 Schematic of a gas sampling system such as might be used on an atmospheric pressure ionization mass spectrometer. O. Ganschow and A. Conrad [7]/ Reproduced with permission from Elsevier.

23.3 Analytical Instruments

Instrumentation systems such as atmospheric pressure ionization mass spectrometers (APIMS) and gas chromatograph–mass spectrometers (GC–MS) require multichambered pumping. In each case, the incoming gas source is at atmospheric pressure. The analyzer chamber cannot operate at this pressure because the mean free path in that chamber must be long. One must take care in specifying these systems—they are ultraclean systems. Although they operate in several pressure regions, their purpose is to detect trace impurities, thus, one should *never* use elastomeric gaskets in their construction. The resulting permeation would corrupt their data and severely limit their use. In the APIMS, ionization is done in a corona discharge at one atmosphere; however, the mass analysis must be performed at a pressure of order $\leq 10^{-3}$ Pa in order to avoid gas–gas collisions. Typical pressure-reducing systems use 2–3 gas expansion chambers. Since a small pump can remove gas rather efficiently at high pressures, such a gas sampling system is cost-effective, especially when combined with the efficiency of a split-flow turbo-drag pump and single forepump. An example of such a sampling system is illustrated in Fig. 23.11.

References

1 Keiser, J., Schwarz, W., and Wagner, W., *Thin Solid Films* **119**, 217 (1984).
2 A. Madan, Semiconductor Vacuum Deposition System and Method having a Reel-to-Reel Substrate Cassette, U.S. Pat. 625,408 B1 (issued July 10, 2001), MV Systems, Golden, CO 80433.
3 Rogelstadt, T. and Matarese, G., *J. Vac. Sci. Technol. A* **3**, 516 (1985).

4 Kuwano, Y., Ohnishi, M., Tsuda, S., Nakashima, Y., and Nakamura, N., *Jpn. J. Appl. Phys.* **21**, 413 (1982).
5 A. Madan. The Society of Photo-Optical Instrumentation Engineers, SPIE, 706, 72, 1986.
6 Hill, R.J. and Nadel, S.J., *Coated Glass Applications and Markets*, Von Ardenne Coating Technology, Fairfield, CA, 1999, p. 21.
7 Ganschow, O. and Conrad, A., *Vacuum* **47**, 757–762 (1996).

24

Leak Detection

At some time, we will be confronted with a system that does not behave normally—behavior that could be the result of a component malfunction, initial outgassing, or a leak. On another occasion, we may need to qualify a new system. When and how to hunt for leaks are two important and useful skills. This discussion will focus on how to find leaks, mainly by using a mass spectrometer tuned to helium. That said, simpler techniques for finding large leaks are included.

The decision of whether to search for leaks is as important as the method. Each new component or subassembly should be routinely leak-tested after welding or brazing. However, it is premature to leak test a new system during its first operation, because its performance does not meet the user's expectations. New systems often pump slowly, due to outgassing from fixtures, seals, and new fluids or lubricants. The patient operator will usually wait a few days before criticizing the base pressure. At that time, it would be useful to take an RGA scan of the system background or perform a leak test.

Searching for a leak in an established system is reasonably straightforward, especially when a history of the system is known. A well-documented log, such as is commonly used in development and production, assists the operator in determining the cause of poor performance. The log should contain important information: pressure versus pumping time, base pressure, rate of rise, maintenance history, and best of all, a background RGA scan taken under known, good performance conditions.

The most sensitive leak checking procedures are done with a helium-based mass spectrometer leak detector (MSLD) or residual gas analyzer (RGA). The MSLD contains a mass detector permanently tuned to ^{4}He ($M/z = 4$), and a portable, self-contained pumping system. Some modern mass-spectrometer-based units can also be tuned to ^{3}He and H_2. The first all-metal leak detector was designed by Nier for the Manhattan project [1,2] and had an ultimate sensitivity

A Users Guide to Vacuum Technology, Fourth Edition. John F. O'Hanlon and Timothy A. Gessert.
© 2024 John Wiley & Sons, Inc. Published 2024 by John Wiley & Sons, Inc.

of 1×10^{-4} Pa-L/s (1×10^{-6} sccs) [1]. By contrast, the most sensitive modern instruments using this technology have sensitivities of 10^{-10} Pa-L/s (10^{-12} sccs).

This chapter focuses on the principles of two popular MSLD instruments: the classical forward flow leak detector and the counter flow detector. MSLD mass filtering and detection fundamentals are described in Chapters 8 and 9. We begin by describing the instrument's most important parameters: sensitivity and response time. We conclude with descriptions of operating procedures, useful leak hunting hints, and discussion of a helium alternative: a semiconductor-based sensor for detecting H_2 in a dilute H_2–N_2 tracer gas.

24.1 Mass Spectrometer Leak Detectors

MSLDs, using either magnetic sector or quadrupole filters, detect helium in one of two configurations: forward flow and counter flow. Each is connected to a test object that is probed externally with helium. Helium permeating a leak then flows to mass analyzer where its partial pressure is measured. The MSLD inlet may also be connected to a capillary tube or a bypass element, which can sample or "sniff" helium emanating from a leak in a pressurized test object. Modern helium-based leak detectors use small, efficient quadrupole mass analyzers; however, some commercial instruments use magnetic sector analyzers, as only a small permanent magnet is needed for this narrow mass range. This detector is less costly than a quadrupole.

24.1.1 Forward Flow

The classical forward flow leak detector is sketched in Fig. 24.1. It consists of a small high vacuum pump—historically a diffusion pump, but now a turbomolecular pump—a magnetic sector mass filter tuned to $M/z = 4$, a cold trap, a separate mechanical pump, and an arrangement of valves in a manifold for exhausting the test sample. Some instruments have two mechanical pumps: one for backing the high vacuum pump, and a separate roughing pump for exhausting rather leaky test objects for long times. The test object, which can range from a small component to a large system, is connected to the inlet port, and evacuated. If the object is small, the internal roughing pump will suffice; however, if the volume is large, the system pump will be needed. After reaching a sufficiently low pressure, the test object is connected to the inlet of the leak detector. Helium sprayed around a leak will find its way into the high vacuum pump, through the cold trap, and into the mass analyzer. A second roughing pump is useful when long times are required to exhaust a test object with large leaks and to prevent excessively high turbo forepressure. The classical leak detector design requires a liquid nitrogen trap to prevent diffusion pump fluid

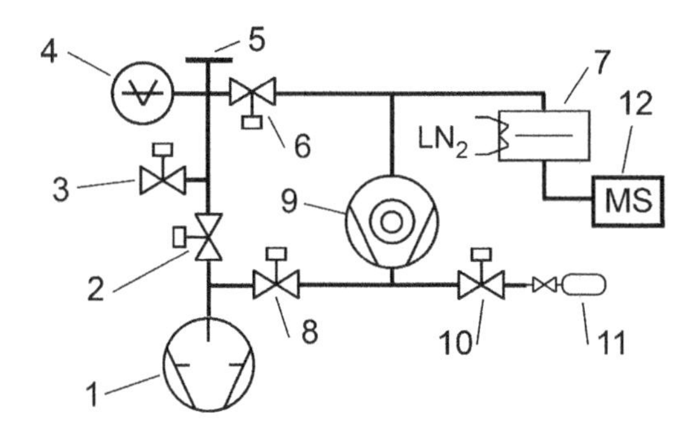

Fig. 24.1 Vacuum circuit of a classical forward flow leak detector. (1) Rotary vane pump, (2) Test port roughing valve, (3) Test port vent valve, (4) Pressure gauge, (5) Test port, (6) Inlet valve, (7) Liquid nitrogen trap, (8) Foreline valve, (9) Turbo pump, (10) Calibration valve, (11) Calibrated helium leak, and (12) Magnetic sector mass analyzer tuned to $M/e = 4$. W.G. Bley [3]/Reproduced with permission from Elsevier.

from reaching the mass analyzer and to reduce the partial pressure of water and other condensable vapors. Silicone diffusion pump fluids are not recommended for this application, as they decompose and form an insulating deposit on the deflection electrodes. An insulating deposit would severely alter the electric field near the deflection electrodes. An LN_2 cold trap is still required with a turbopump to remove condensable vapors.

24.1.2 Counter flow

The counter flow leak detector is illustrated in Fig. 24.2. The object under test is connected to the *foreline* of the high vacuum pump; however, the mass filter remains connected to the high vacuum pump inlet. The high vacuum pump used for this application must have a high compression ratio for atmospheric air, but a low compression ratio for helium. This can be accomplished with either a diffusion pump or a turbo pump. Recall that the compression of a diffusion pump jet is exponentially proportional to the oil velocity and the compression of a turbo pump blade is exponentially proportional to the blade tip velocity. Low helium compression can be achieved in a diffusion pump by reducing the number of stages, its heater power, and its height-to-diameter ratio. In a similar manner the rotational speed, and number of blade rows can be decreased, or blade angles can be increased. In this manner, high compression for all gases, except helium and hydrogen, maintains high vacuum in the mass analyzer, but their low compression allows them to backstream through the turbo or diffusion pump where it is detected at their inlet side.

By isolating the mass filter from the test sample, liquid nitrogen is no longer needed, and the instrument becomes truly portable. The counter flow concept is the most significant advance in leak detector in its seven-decade history.

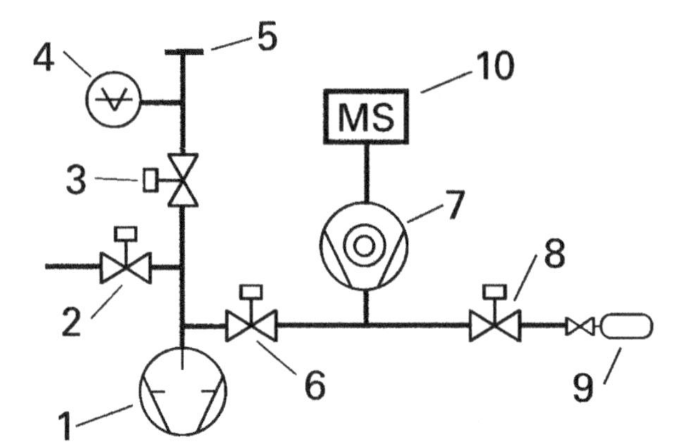

Fig. 24.2 Counter flow leak detector configuration. (1) Dry rotary vane pump; (2) Leak detector vent valve; (3) Leak detector connection valve; (4) Pressure gauge; (5) Connection port to an individual component, a system foreline, or its chamber vent; (6) Inlet valve, (7) Turbo-drag pump, (8) Calibration valve, (9) Calibrated helium leak, and (10) Magnetic sector or quadrupole mass analyzer tuned to $M/e = 4$, and additionally, in some models, $M/e = 2$ and 3. W.G. Bley [3]/Reproduced with permission from Elsevier.

24.2 Performance

Leak detector performance is primarily dependent on sensitivity and response time, but its overall performance cannot be determined without knowledge of the system under test.

24.2.1 Sensitivity

Both the RGA and the MSLD are sensitive to a threshold partial pressure of the tracer gas. In the best case, the minimum detectable partial pressure is the absolute sensitivity of the instrument above the background noise. In a typical operating system a residual background tracer gas pressure exists because it is regurgitated from a pump, back-diffuses through a pump, is released from permeable materials, or is desorbed from a cold trap surface. Residual helium is also present in the atmosphere at a level of about 5 ppm and may be present in significantly higher levels in facilities that routinely fill helium gas cylinders, perform extensive leak testing, or use helium for other purposes in close proximity. This residual gas pressure may be greater than the ultimate detectable pressure of the instrument and therefore reduce its useful sensitivity.

The helium leak flux Q_L is related to the helium pressure in the mass spectrometer by $Q_L = P_L S_L$. The sensitivity s of the forward flow leak detector has been defined [3] as:

$$s = \frac{P_L}{Q_L} = \frac{1}{S_L} \tag{24.1}$$

The variables in Eq. (24.1) are those for helium. In a similar manner, the sensitivity of the counter flow leak detector was shown to be [3]:

$$s = \frac{1}{\left(K_{\text{high vac}} \cdot S_{\text{fore pump}}\right)} \tag{24.2}$$

where the compression, K, is the compression ratio for helium in the high vacuum pump, and S is the helium forepump speed.

As the maximum inlet pressure is increased, the flow to either the forward flow or counter flow detector must be throttled, or the pressure in the mass analyzer will rise above its operating range. Figure 24.3 illustrates how the sensitivity decreases above the throttling pressures for the forward flow and counter flow units. This figure shows the counter flow unit to be more sensitive than the forward flow unit at high sampling pressures, or leak rates. The forward flow unit is more sensitive at the lowest possible leak rates. Chew noted that helium retention

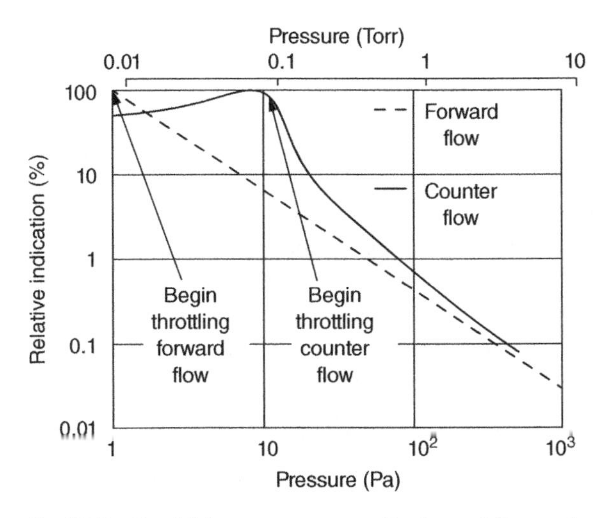

Fig. 24.3 Sensitivity versus pressure for forward flow and counter flow leak detectors connected to a pumped vacuum system. Throttling of the counter flow leak detector begins at 10 Pa (75 mTorr). Throttling of the forward flow unit begins at 1 Pa (7.5 mTorr) even when LN_2 is used. W.G. Bley [3]/Reproduced with permission from Elsevier.

in commercial mechanical pumps varied widely [4]; for some pumps, pump venting or gas ballast was necessary to remove dissolved helium.

From Eq. (24.1) one can see that the maximum detectable sensitivity is reached at zero helium pumping speed. Closing the valve between the system pump and chamber allows the leaking tracer gas to accumulate in the chamber. This is easily accomplished when leak detecting with an RGA and with MSLD units that are equipped with a valve between the ionizer and self-contained pump. A constant leak will cause the helium partial pressure to increase linearly with time at zero pumping speed. After time t_1 a detector will measure partial pressure P_1, and the leak flux will be given by $Q = P_1 V / t_1$. This is called the accumulation technique. Small volumes may also be tested effectively by isolating them completely from the MSLD. After time t, a connecting valve is opened, and the collected quantity of helium is allowed to flow to the MSLD. The minimum leak flux detectable by an RGA or MSLD in normal operation is ~10^{-9}–10^{-10} Pa-L/s (~10^{-11}–10^{-13} Torr-L/s). The accumulation technique has been used to find leaks as small as 10^{-11}–10^{-13} Pa-L/s (~10^{-13}–10^{-15} Torr-L/s) [5,6]. In practice, high vacuum system leaks $<10^{-8}$ Pa-L/s (10^{-10} Torr-L/s) are uncommon [7], as they tend to hydrolyze shut with atmospheric exposure.

24.2.2 Response Time

The maximum sensitivity of the leak detector can be realized only if the tracer gas has time to reach its steady-state value. For a system of volume V evacuated by a pump of speed S, the pressure change due to a sudden application of a tracer gas to a leak is given by:

$$P_{\text{tracer}}(t) = P_{\text{tracer}}(0)\left(1 - e^{St/V}\right) \tag{24.3}$$

where $P_{\text{tracer}}(t)$ is the background pressure of the tracer gas. At time zero, the pressure of the tracer gas in the system is $P_{\text{tracer}}(0)$. The pressure slowly builds to a steady-state value. Sixty-three percent of the steady-state pressure is reached in a time equal to the system time constant V/S; 5 time constants are required to reach 99% of the response. This means that a 100-L system pumped by a leak detector with a speed of 5 L/s will require application of the tracer gas for 20–100 s to reach maximum sensitivity. This is the case when an MSLD is connected to a chamber that has been isolated from the high vacuum pump by a valve. See Fig. 24.4*a*. Placing pumps in parallel, as illustrated in Fig. 24.4*b*, can reduce the time constant, but not without loss of sensitivity. If the leak detector can handle the gas load, the fast time constant can be retained in turbomolecular and diffusion pumps without loss of sensitivity by placing the leak detector in the foreline and isolating the mechanical pump from the system by a valve. See Fig. 24.4*c*.

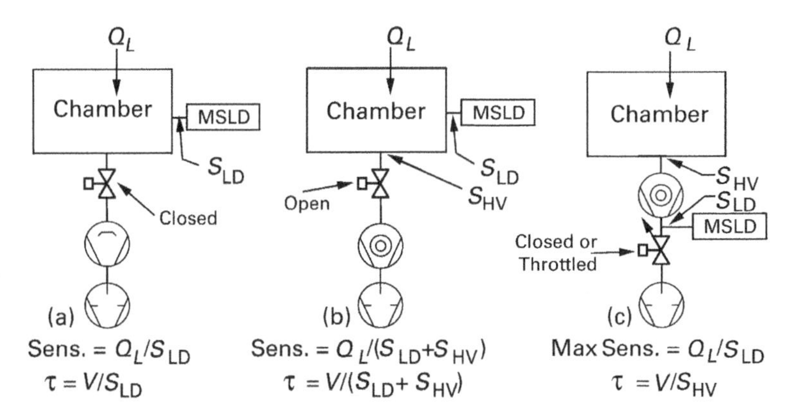

Fig. 24.4 Three techniques for using an MSLD to detect leaks in a vacuum chamber pumped by a diffusion or turbomolecular pump. Methods *a* and *c* have a high sensitivity; methods *b* and *c* have a fast response time. The sensitivity of method *c* is maximum when the throttle valve is closed.

This technique may be used if the gas flow is too large for the leak detector by allowing the leak detector to pump at its maximum flow rate, while pumping the remainder of the gas with the forepump. The fast time constant is retained, and the sensitivity is reduced only by the ratio of forepump-to-leak-detector flow.

Leak detection procedures for ion or cryogenic pumped systems differ because they are capture pumps. An ion pump may be helium-leak checked by momentarily removing power to the pump. A cryogenic pump cannot be shut down for leak checking because the evolved gas load will overload the pump in the MSLD. If the cold stage temperature is increased to 20 K, helium pumping by the sorbent bed will drop to zero, and the leak detector can be operated at its maximum sensitivity. If the cold stage is not warmed during leak detection, helium will accumulate and will desorb the next time the pump is exposed to a chamber gas load. A sufficient quantity of desorbed helium will thermally short the pump. As we discussed in Chapter 19, no amount of pumping will help; complete regeneration is required. Moraw and Prasol [8] described optimum conditions for leak detecting large space chambers.

24.2.3 Testing Pressurized Chambers

Capillary "sniffing" and bypass sampling are two methods for testing pressurized containers. It is best to test a system under conditions of normal use. Therefore, gas regulator fittings and gas distribution lines are often tested by internally pressurizing with helium and sniffing for helium escaping to atmosphere. The capillary probe is simply a long probe containing very-small-diameter tube with a

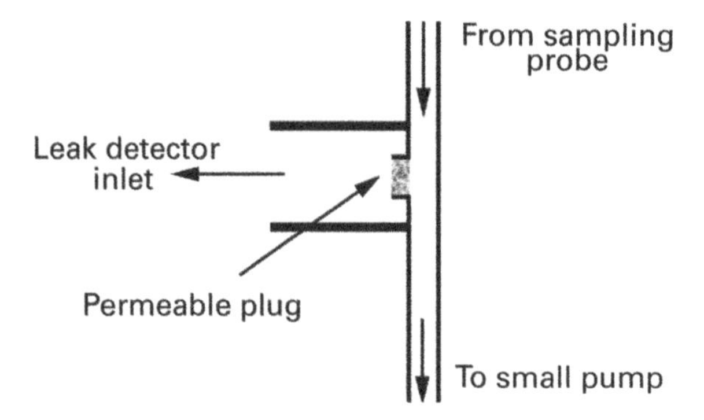

Fig. 24.5 A bypass sampling technique for sniffing He from internally pressurized vessels.

porous plug that limits the flow to ~10^{-3} sccs (10^{-1} Pa-L/s). The end of the probe is connected to the inlet of the MSLD. Molecular flow through the porous plug enhances the helium flow by a factor of ~2. See Eq. (2.20). The volume of the tube is sufficiently small to have a response time of ~2 s. A bypass sampler operates on an entirely different principle. The sampler, illustrated in Fig. 24.5, consists of a long tube connected at one end to a pump with a flow capacity of 1–2 sccs. A small fraction of this flow is pumped into the MSLD through a porous bypass connection. Including the mass separating ability of the porous plug, ~0.1% of the helium is sensed by the MSLD. The tube diameter is designed for a time constant of a few seconds.

24.2.4 Calibration

MSLDs require calibration. Minimum specified leak rates for systems and components use an MSLD tuned to ^{4}He to validate performance. Performance can be characterized by either a minimum detectable leak or a minimum detectable concentration ratio or both. Commercial instruments include a standard leak with a stated numerical value whose value will remain within a limiting range, say 10%, typically for a period of one year. Standard leaks are available in very fine values for determining the maximum sensitivity of an instrument, as well as for higher ranges for quantifying coarse leaks and concentration ratios.

Figures 24.1 and 24.2 show the respective locations of a standard leak cell in conventional and counter flow leak detectors. It is important for the leak to be in the same location as the test gas signal and be the only source of helium. Calibration is done with the detector's sample connection port isolated from any potential leak. MSLD instruments need to be powered on and operational for a sufficient time, typically a half-hour, for its amplifiers and mass filter circuits to stabilize.

The ISO Standard for leak detector calibration [9] allows for two standard leaks: a small leak for determining the minimum detectable leak, and a large, variable leak for determining minimum detectable concentration ratio of He:Air. The value of the small leak should be near that of the instrument's specified limit.

Internal calibration routines allow the leak detector to stabilize, and then open the valve to the calibrated leak. Again, time must be allowed for any excess concentration that accumulated in the small volume between the leak capillary and its external valve to be removed by the instrument's internal pump. The calibration procedure will reset the digital or analog indicator.

When leak detectors are used in a throttled mode, the system must be recalibrated with a larger standard leak. The background value of atmospheric helium is about 5 ppm, about 0.5 Pa (~4 mTorr), and this will limit the minimum detectable leak. When testing pressurized vessels with leak detectors in their "sniffer" or capillary bypass modes, one needs to remember that atmospheric background helium is part of the signal being analyzed. The ISO standard [9] describes detailed methods for calibrating MSLDs.

24.3 Leak Hunting Techniques

It is not possible to provide a complete list of useful hints, but the following should prove useful. Before beginning, it is invaluable to have a thorough understanding of the system under test, including its volume and pumping speed at chamber and foreline. These are required to calculate the system time constant, $\tau = V/S$, and response time.

A system that cannot be pumped below the operating range of its roughing pump has either a gross leak or a malfunctioning pump. Begin by hunting for large leaks. After these have been eliminated, close all the valves in the system and observe the pressure in the mechanical pump with a thermal conductivity gauge. If valveless, blank the mechanical pump with a flange containing only a thermocouple gauge and verify its base pressure. Next, if the pump is operating properly, sections of the foreline and roughing line can be isolated and pumped individually to discover a potentially leaky section.

If the system pumps to the high vacuum range, but cannot reach its usual base pressure, there may be a fine leak. There may not be a leak, rather, a faulty high vacuum pump, a leak on the pump side of the gate valve, a contaminated gauge, or considerable outgassing. If a leak has not been detected, close the high vacuum gate valve, and observe the downstream pressure. A low pressure on the pump side of the gate valve tells us the pump is operating properly. If the blank-off pressure is high, the pump or valve could have a

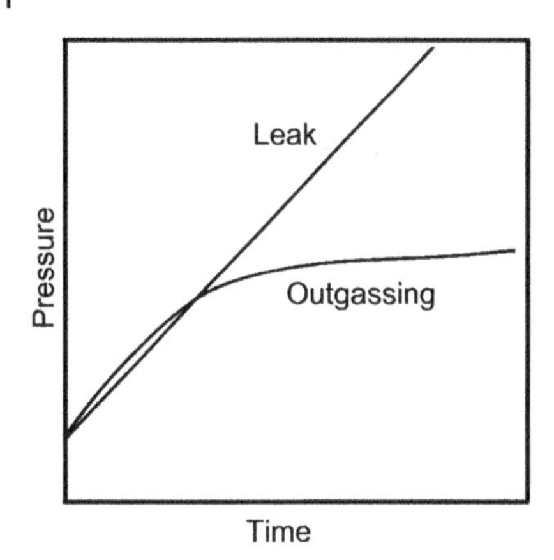

Fig. 24.6 Response of a sealed chamber to a leak and to outgassing from interior walls. N.G. Wilson and L.C. Beavis [10]/ Reproduced with permission from American Vacuum Society.

leak. A leak detector can be attached to the foreline of a diffusion or turbomolecular pump, or to a flange adjacent to a cryogenic or ion pump. A valve in the foreline, shown in Fig. 24.4c, allows the leak detector to be attached without removing power from the pumps. When specifying a new system, it is prudent to consider a port with valve at this location.

After the system has been verified to be free of leaks from the mechanical pump through the top of the high vacuum pump, one may open the chamber valve. Poor chamber pressure may result from outgassing, external leaks, or internal leaks. One simple way to distinguish between leaks and outgassing is to examine the plot of the system pressure versus time after closing the high vacuum valve. See Fig. 24.6. A molecular leak causes a linear increase in pressure with time. Outgassing causes the pressure to rise to a steady-state value that is determined by the vapor pressures of the desorbing species. If the system contains an RGA, a quick scan distinguishes between poor performance due to an atmospheric leak or outgassing—although outgassing of water vapor may be difficult to distinguish from a leak in an interior water line.

Figure 24.7 illustrates the time required for helium to permeate a gasket at room temperature and 80°C. There are two messages in this figure: do not attempt to leak check the system during baking; and helium background from a loaded O-rings will not decrease until it has out diffused and pumped. A coffee break may be required before proceeding. Helium may on occasion cause other difficulties. In differentially pumped O-rings, helium was observed to collect in sidewall pockets of an O-ring groove, where it remained until the flange was loosened [11].

Interior leaks in liquid nitrogen traps are quite difficult to locate. Leaks have been observed when LN_2 has been added to the trap or suspected when large signals at $M/z = 14$ and 28 are observed with an RGA. The trap will have to be removed from the system and leak checked with an MSLD to pinpoint the leak. Leaks at cryogenic temperatures can seal and not be visible at room temperature, and this frustrates repair. In some instances, it was found necessary to reweld all seams.

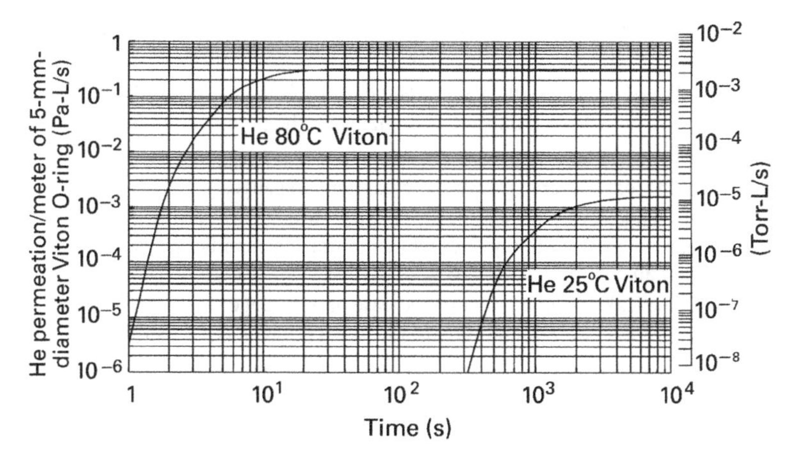

Fig. 24.7 Calculated helium permeation at 25 and 80°C through a Viton gasket initially containing no helium. The steady-state permeation rates were experimentally measured; the solubilities were derived from the measured time to reach equilibrium at room temperature.

Sometimes the source of a very large peak at $M/z = 28$ is not obvious; a possible source is an internal leak in a nitrogen gas supply valve, if used to vent chamber.

RGA's provide more information than an MSLD; this allows us to solve problems quickly. Air leaks and water outgassing are easily differentiated. Air leaks are discerned by the presence of oxygen at $M/z = 32$, except in large TSP systems in which the oxygen pumping speed is so large that the oxygen signal will not be seen and in baked UHV systems, where it is adsorbed on stainless steel walls. Outgassing and water line leaks each produce a large peak at $M/e = 18$, but they can be distinguished by rate of rise. The RGA has an added advantage of using tracer gases other than helium.

It is important to restate that helium background partial pressure in the area near leak testing can negatively affect results. For example, operations such as helium cylinder filling and repurification, produce high local helium background partial pressures. High helium background pressure reduces sensitivity and obscures small leaks. This is a serious problem, if the instrument's visual indicator is set to zero before beginning a search.

The MSLD has been universally accepted for finding small leaks in vacuum components and systems. The instrument is most valuable when it has internal calibration standards (accuracy), valves for throttling its internal pumping speed (accumulation technique), and inlet flow (foreline sampling technique). Table 24.1 provides some best practices for leak hunting. Excellent leak detecting suggestions are provided in the AVS [10], Pfeiffer [12], and Leybold [13] handbooks.

Table 24.1 Best Practices for Leak Detection

Best Practices for Leak Detection
Assume that MSLD or RGA is working, using a standard He leak before leak detection.
Spray helium around a large leak and listen for a change in the pump's pitch.
Spray alcohol on large leak, observe pressure increase on TC gauge.
Spray alcohol on a fine leak, observe pressure increase on ion gauge.
To determine if internal leaks are present in supply lines (water, CDA, N_2, etc.) pressurize then depressurize the supply lines while monitoring chamber pressure at lowest pressure attainable.
Spray helium on a fine leak, observe pressure decrease on ion gauge.
Spray helium on a sorption pump roughing line, observe a large pressure increase.
Use only non-silicone pump fluids and non-silicone-lubricants in MSLD or RGA.
Check new components and sub-assemblies first, before leak checking entire system.
Check welds and seals first, then face side of flanges.
Begin external leak check with helium at system top, and then proceed downward.
Spray small amounts of helium to reduce saturation and isolate other leak sites.
Wrap plastic around nearby locations and flush with nitrogen, when testing a site.
Seal one leak with alcohol (it will freeze) while testing nearby leak site, after testing, remove frozen alcohol gently with a heat gun.
Bake flanges to remove grease that has plugged small leaks.
Wait 20 min. for background helium pressure to decrease when O-rings become saturated with helium from testing. (See Fig. 24.7)
Drain interior water-cooling lines, flush them with warm nitrogen to dry interior and remove ice that may have formed in leaks after line was drained, and then leak test.
Bubble helium gas through liquid nitrogen traps to test for interior trap leaks.
Search for leaks in an ion-pumped system using oxygen tracer gas.
Search for leaks in a TSP system using argon tracer gas.
Regenerate cryopumps after completing leak testing with helium or hydrogen.
Power off all high voltage supplies before leak testing near any HV feedthroughs.
Test ultrapure gas delivery lines and connectors with sampling probes.
Test gas supply lines individually for internal valve leaks.
Cover tested system with plastic bag, and then spray helium inside bag for final check.

24.4 Leak Detecting with Hydrogen Tracer Gas

Helium has been the tracer gas of choice for over three decades, but it is a rare gas that has become increasingly more expensive. When used in large quantities, costly helium recovery, and repurification systems are used to reduce its loss. Hydrogen will pass through smaller leaks than helium because it is lighter and smaller in diameter. However, it is explosive, and for that reason, it was never used. Forming gas—a mixture of 5% H_2 and 95% N_2—is a low-cost, nonflammable, and safe alternative to helium. The hydrogen in Forming gas can be detected with a mass spectrometer tuned to H_2 or in a commercial leak detector, by a solid-state device whose surface is sensitive to the hydrogen concentration.

Commercial MSLDs capable of detecting hydrogen have hydrogen sensitivities of 5×10^{-10} sccs in vacuum and 5×10^{-8} sccs (sniffer), compared with helium sensitivities of 5×10^{-12} sccs (vacuum) and 5×10^{-8} sccs (sniffer). Hydrogen is a common background signal in an MSLD, and its resulting signal-to-noise ratio (i.e., minimum detectability concentration ratio) limits its ability to detect extremely small leaks. With a change in switch position, the same MSLD can use Forming gas as a tracer for coarse applications, or helium for applications demanding maximum sensitivity.

Efficient semiconductor-based sensors have been developed with sensitivities of 5×10^{-7} sccs. Semiconductor-based hydrogen sensors are not as sensitive as magnetic sector or quadrupole mass analyzers but are excellent for applications that do not require the sensitivity required for detecting extremely small high vacuum system leaks. The expense of initially leak testing very large chambers—a task that would consume considerable helium—can be reduced considerably by using forming gas and either an RGA or a leak detector having a solid-state hydrogen sensor. Solid-state hydrogen leak detectors are the low-cost choice for leak checking products such as air conditioners, transformers, water lines, medical and food packages, etc. These products do not need the 10^{-12} sccs (10^{-12} Torr-L/s) capability of an MSLD.

References

1 Nerkin, A., *J. Vac. Sci. Technol., A* **9**, 2036 (1991).
2 Nier, A.O., Stevens, C.M., Hustrulid, A., and Abbott, T.A., *J. Appl. Phys.* **18**, 30 (1947).
3 Bley, W.G., *Vacuum* **44**, 627 (1993).
4 Chew, A.D., *Vacuum* **53**, 243 (1999).
5 O'Hanlon, J.F., Park, K.C., Reisman, A., Havreluk, R., and Cahill, J.G., *IBM J. Res. Dev.* **22**, 613 (1978).
6 Watanabe, F. and Ishimaru, H., *J. Vac. Sci. Technol., A* **8**, 2795 (1990).
7 Beavis, L.C., *Vacuum* **20**, 233 (1970).
8 Moraw, M. and Prasol, H., *Vacuum* **28**, 63 (1978).
9 ISO 3530:1979 (R2021), *Vacuum Technology—Mass-Spectrometer-Type Leak-Detector Calibration*, The International Standards Organization, Geneva, Switzerland, Revised, 2021.

10 Wilson, N.G. and Bevis, L.C., in *Handbook of Vacuum Leak Detection*, W.R. Bottoms, Ed., *AVS Monograph #M2, 2nd Printing*, Education Committee, American Vacuum Society, New York, 1988.

11 Johnson, M.L., Manos, D.M., and Provost, T., *J. Vac. Sci. Technol., A* **15**, 763 (1997).

12 *Leak Detection Handbook*, Pfeiffer Vacuum GmbH, Asslar, Germany, 2013.

13 Rottländer, H., Umrath, W., and Voss, G., *Fundamentals of Leak Detection*, Leybold, GmbH, Köln, Germany, 2016.

Part VI

Appendices

A Users Guide to Vacuum Technology, Fourth Edition. John F. O'Hanlon
and Timothy A. Gessert.
© 2024 John Wiley & Sons, Inc. Published 2024 by John Wiley & Sons, Inc.

Appendix A

Units and Constants

A.1 Physical Constants

k	Boltzmann's constant	1.3804×10^{-23} J/K
m_e	Rest mass of electron	9.108×10^{-31} kg
m_p	Rest mass of proton	1.672×10^{-27} kg
N_o	Avogadro's number	6.02252×10^{26}/(kg-mol)
R	Gas constant	8314.3 J-(kg-mol)$^{-1}$-K^{-1}
V_o	Normal specific volume of an ideal gas	22.4136 m^3/(kg-mol)
σ	Stefan–Boltzmann constant	5.67×10^{-8} J-s^{-1}-m^{-2}-K^{-4}

A.2 SI Base Units

SI Base Quantity	Unit	Symbol
Length	Meter	m
Mass	Kilogram	kg
Time	Second	s
Electric current	Ampere	A
Thermodynamic temperature	Kelvin	K
Amount of substance	Mole	mol
Luminous Intensity	Candela	cd

A Users Guide to Vacuum Technology, Fourth Edition. John F. O'Hanlon and Timothy A. Gessert.
© 2024 John Wiley & Sons, Inc. Published 2024 by John Wiley & Sons, Inc.

A.3 Conversion Factors

Conventional unit	→ multiply by →	to get SI unit
Mass		
lb	0.45359	kg
Length		
Ångstrom	1.0×10^{-10}	m
Nanometer	1.0×10^{-9}	m
micrometer	1.0×10^{-6}	m
mil	0.00254	cm
inch	0.0254	m
foot	0.3048	m
Area		
ft^2	0.0929	m^2
$in.^2$	6.452	cm^2
ft^2	929.03	cm^2
Volume		
cm^3	0.001	L
$in.^3$	0.0164	L
gallon (US)	3.7879	L
ft^3	28.3	L
L	1000.0	cm^3
Pressure		
micrometer (Hg)	0.13332	Pa
N/m^2	1.0	Pa
millibar	100.	Pa
Torr	133.32	Pa
inches of Hg	3386.33	Pa
lb/in^2	6895.3	Pa
bar	100,000	Pa
atmosphere	101,323.2	Pa
Conductance or pumping speed		
L/h	0.000277	L/s
L/s	0.001	m^3/s
L/min	0.0166	L/s

(Continued)

Conventional unit	→ multiply by →	to get SI unit
m^3/h	0.2778	L/s
ft^3/min	0.4719	L/s
ft^3/min	1.6987	m^3/h
Outgassing rate		
$(Pa\text{-}L)/(m^2\text{-}s)$	0.001	Pa-m/s
$(mbar\text{-}L)/(m^2\text{-}s)$	0.1	Pa-m/s
$(Pa\text{-}m^3)/(m^2\text{-}s)$	1.0	Pa-m/s
$\mu L/(cm^2\text{-}s)$	1.33	Pa-m/s
$(Torr\text{-}L)/(cm^2\text{-}s)$	1333.2	Pa-m/s
Gas flow		
molecules/s (at 0°C)	4×10^{-18}	Pa-L/s
Pa-L/s	0.001	$Pa\text{-}m^3/s$
Torr-L/s	0.133	$Pa\text{-}m^3/s$
micron-L/s	0.13332	Pa-L/s
standard cc's/min (sccm)	1.69	Pa-L/s
Pa-L/s	3.6	$Pa\text{-}m^3/h$
atm-cc/s	101.323	Pa-L/s
standard cc's/s (sccs)	101.323	Pa-L/s
Torr-L/s	133.32	Pa-L/s
standard liter/min (slm)	1689	Pa/L/s
(kg-mole)/s (at 0°C)	2.48×10^{9}	Pa-L/s
Gas permeation constant		
cm^3 gas (STP)-cm thickness/ (cm^2 area-s-atmosphere)	0.0001	m^2/s
cm^3 gas (STP)-cm thickness/ (cm^2 area-s-Torr)	0.076	m^2/s
Dynamic viscosity		
newton-s/m^2	1	Pa-s
poise	10	Pa-s
Kinematic viscosity		
centistoke	1	mm^2/s
Diffusion constant		
cm^2/s	0.0001	m^2/s
$furlong^2/fortnight$	0.0334	m^2/s

(Continued)

(Continued)

Conventional unit	→ multiply by →	to get SI unit
Heat conductivity		
Watt-cm^{-1}-K^{-1}	100	J-s^{-1}-m^{-1}-K^{-1}
Specific heat		
J-kg^{-1}-K^{-1}	M	J-$(kg\text{-}mol)^{-1}$-K^{-1}
cal-$(g\text{-}mole)^{-1}$-K^{-1}	4184	J-$(kg\text{-}mol)^{-1}$-K^{-1}
BTU-lb^{-1}-$°F^{-1}$	4184M	J-$(kg\text{-}mol)^{-1}$-K^{-1}
Heat capacity		
J/kg	M	J-$(kg\text{-}mol)^{-1}$
cal-$(g\text{-}mole)^{-1}$	4184	J-$(kg\text{-}mol)^{-1}$
BTU/lb	2325.9M	J-$(kg\text{-}mol)^{-1}$
Energy, work, or quantity of heat		
ft-lb	1.356	J
kW-h	3.6	MJ
BTU	1055	J
kcal	4184	J
to get conventional unit	← divide by ←	SI unit

Appendix B

Gas Properties

B.1 Mean Free Paths of Gasses as a Function of Pressure at $T = 25°C$

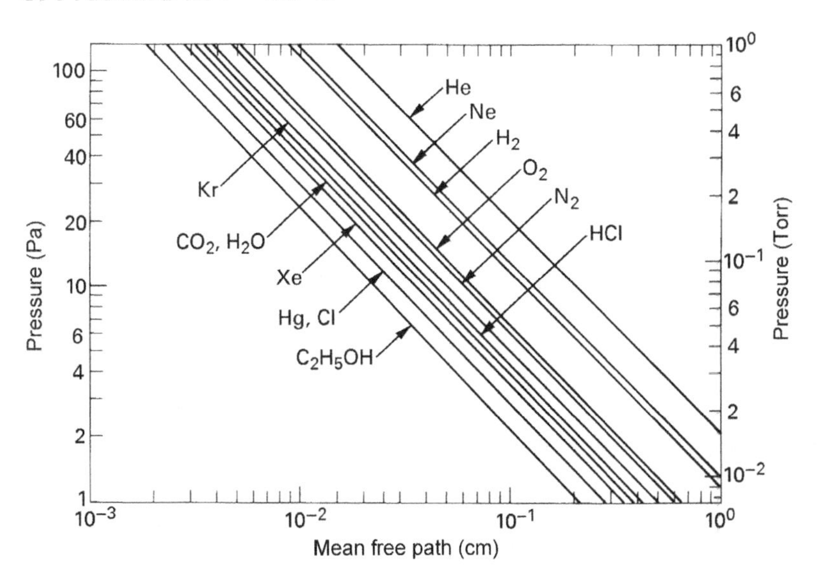

Source: Reproduced with permission from Vacuum Technology, p. 505, A. Guthrie. Copyright 1963, John Wiley & Sons, New York, 1963.

A Users Guide to Vacuum Technology, Fourth Edition. John F. O'Hanlon and Timothy A. Gessert.
© 2024 John Wiley & Sons, Inc. Published 2024 by John Wiley & Sons, Inc.

B.2 Physical Properties of Gasses and Vapors at $T = 0°C$

Gas	Symbol	MW[a]	Molecular Diameter[b] (nm)	Average Velocity[c] ($m\text{-}s^{-1}$)	Thermal Cond.[a,d] ($mJ\text{-}s^{-1}\text{-}K^{-1}$)	Dynamic Viscosity[a,d] (μPa-s)	Diffusion in Air[d,e] ($10^{-6} m^2\text{-}s^{-1}$)
Helium	He	4.003	0.218	1197.0	142.0	18.6	58.12
Neon	Ne	20.183	0.259	533.0	45.5	29.73	27.63
Argon	Ar	39.948	0.364	379.0	16.6	20.96	17.09
Krypton	Kr	83.8	0.416	262.0	6.81[f]	23.27	13.17
Xenon	Xe	131.3	0.485	209.0	4.50[g]	21.9	10.60
Hydrogen	H_2	2.016	0.274	1687.0	173.0	8.35	63.4[a]
Nitrogen	N_2	28.0134	0.375	453.0	24.0	16.58	18.02
Air		28.966	0.372	445.0	24.0	17.08	18.01
Oxygen	O_2	31.998	0.361	424.0	24.5	18.9	17.8[a]
Hydrogen chloride	HC1	36.46	0.446	397.0	12.76	14.25[h]	14.11
Water vapor	H_2O	18.0153	0.46	564.0	24.1[i]	12.55[i]	23.9[a,j]
Hydrogen sulfide	H_2S	34.08	0.47[k]	412.0	12.9	11.66	14.62[k]
Nitric oxide	NO	30.01	0.372[k]	437.0	23.8	17.8	19.3[k]
Nitrous oxide	N_2O	44.01	0.47[k]	361.0	15.2	13.5	13.84[k]
Ammonia	NH_3	17.03	0.443	581.0	21.9	9.18	17.44
Carbon monoxide	CO	28.01	0.312[a]	453.0	23.0	16.6	21.49
Carbon dioxide	CO_2	44.01	0.459	361.0	14.58	13.9	13.9[a]
Methane	CH_4	16.4	0.414	592.0	30.6	10.26	18.98
Ethylene	C_2H_4	28.05	0.495	452.0	17.7	9.07	13.37
Ethane	C_2H_6	30.07	0.53	437.0	16.8	8.48	12.14

[a] Reproduced with permission from Handbook of Chemistry and Physics, 58th ed., R. C. Weast, Ed. Copyright 1977, Chemical Rubber Co., CRC Press, West Palm Beach, FL.
[b] Reprinted with permission from Kinetic Theory of Gases, E. H. Kennard, p. 149. Copyright 1938, McGraw-Hill, New York.
[c] Calculated from Eq. (2.2).
[d] At atmospheric pressure.
[e] Calculated from Eq. (2.28).
[f] $T = 210\,K$, Reprinted with permission from *Cryogenic and Industrial Gases*, May/June 1975, p. 62. Copyright 1975, Thomas Publishing Co., Cleveland, OH.
[g] Footnote *e*, $T = 240\,K$.
[h] $T = 18°C$.
[i] $T = 100°C$.
[j] $T = 8°C$.
[k] Calculated from viscosity data.

B.3 Cryogenic Properties of Gases

Property	Units	He	H_2	Ne	N_2	Ar	O_2	Xe	CF_4
nbp liq.[a]	K	4.125	20.27	27.22	77.35	87.29	90.16	164.83	145.16
mp (1 Atm)[b]	K		14.01	24.49	63.29	83.95	54.75	161.25	123.16
Density of liquid, nbp[a]	kg/m^3	124.8	70.87	1208	810.0	1410	1140	3058.0	1962
Volume of liquid, nbp	$(m^3/kg) \times 10^{-3}$	8.01	14.1	0.83	1.24	0.709	0.877	0.327	0.597
Volume of gas at 273 K[a]	m^3/kg	5.602	11.12	1.11	0.79	0.554	0.698	0.169	0.274
Ratio V^g_{273}/V^l_{nbp}		699.4	788.7	1337	637.5	781.5	796.3	516.8	458.3
Heat of vaporization, nbp[b]	kJ/(kg-mol)	95.8	911.0	1740	5580	6502	6812.	12640.	12,000
Heat of fusion, mp[b]	kJ/(kg-mol)	16.75	118.0	338.0	714.0	1120.0	438.0	1812.0	699.0
Spec. heat, c_p^{vap}, 300 K[b]	kJ-(kg-mol)$^{-1}$-K^{-1}	20.94[c]	28.63	20.85	29.08	20.89	29.45	20.85	62.23

[a] Reproduced with permission from Cryogenic and Industrial Gases, May/June 1975. Copyright 1975, Thomas Publishing Co., Cleveland, OH.
nbp is "normal boiling point." As used in this table, 1 atm is defined to be 101,326 Pa.
mp is "melting point."
[b] Reprinted with permission from *Handbook of Chemistry and Physics*, 58th ed., R. C. Weast, Ed. Copyright 1977, Chemical Rubber Co., CRC Press, West Palm Beach, FL.
[c] At $T = -180°C$.

B.4 Gas Conductance and Flow Formulas

Note: In these formulas, pressure is in Pascal, volume in m^3, length in m, pumping speed in m^3/s, gas flow in Pa-m^3/s, velocity in m/s, and particle density in m^{-3}. Temperature is given in Kelvins unless otherwise stated.

Characteristic Numbers

Knudsen's number

$$Kn = \frac{\lambda}{d}$$

$$Kn = \frac{6.6}{P\ (Pa) d(mm)} \quad \text{(air, 22°C)}$$

Reynolds' number

$$\mathbf{R} = \frac{4m}{\pi kT\eta}\frac{Q}{d}$$

$$\mathbf{R} = 8.41\times10^{-4}\frac{Q(\text{Pa-L/s})}{d}\ (\text{air, } 22°\text{C})$$

Mach number

$$\mathbf{U} = \frac{U}{U_{\text{sound}}} = \frac{4Q}{\pi d^2 P U_{\text{sound}}}$$

Langhaar's number

$$l_e = 0.0568 d\mathbf{R}$$

Quantities from Kinetic Theory

Most probable velocity

$$v_p = (2kT/m)^{1/2}$$

Average velocity

$$v = 463 \text{ m/s} \quad (\text{air, } 22°\text{C})$$

RMS velocity

$$v_{\text{rms}} = (3kT/m)^{1/2}$$

Mean free path, one component gas

$$\lambda = \frac{1}{2^{1/2}\pi d_o^2 n}$$

$$\lambda(\text{cm}) = \frac{0.67}{P(\text{Pa})} \quad \text{or} \quad \lambda(\text{cm}) = \frac{0.005}{P(\text{Torr})} \ (\text{air, } 22°\text{C})$$

Mean free path of gas a in a mixture of gases a and b

$$\lambda_a = \frac{1}{\left[2^{1/2}\pi n_a d_a^2 + \left(1+\frac{v_b^2}{v_a^2}\right)^{1/2} n_b \frac{\pi}{4}(d_a+d_b)^2\right]}$$

Particle flux

$$\Gamma = n\left(\frac{kT}{2\pi m}\right)^{1/2}$$

Monolayer formation time

$$t_{ml} = \frac{1}{\Gamma d_o^2} = \frac{4}{nvd_o^2}$$

Ideal gas law

$$P = nkT$$

$$\frac{P_1V_1}{T_1} = \frac{P_2V_2}{T_2}$$

Specific heat ratio

$$\gamma \sim 1.4 \text{ (diatomic gas)}$$

$$\gamma \sim 1.667 \text{ (monatomic gas)}$$

$$\gamma \sim 1.333 \text{ (triatomic gas)}$$

Viscosity at normal pressures

$$\eta = \frac{0.499(4mkT)^{1/2}}{\pi^{3/2} d_o^2}$$

Viscosity at reduced pressures (free molecular)

$$\eta_{fm}(\text{Pa-s}) = \left(\frac{Pmv}{4kT}\right)$$

Heat conductivity at normal pressures

$$K = \frac{1}{4}(9\gamma - 5)\eta c_v$$

Diffusion constant, gas 1 in gas 2

$$D_{12} = \frac{8\left(\frac{2kT}{\pi}\right)^{1/2}\left(\frac{1}{m_1} + \frac{1}{m_2}\right)^{1/2}}{3\pi(n_1 + n_2)(d_{o1} + d_{o2})^2}$$

Diffusion constant, self-diffusion

$$D_{11} = \frac{4}{3\pi n d_o^2}\left(\frac{kT}{\pi m}\right)^{1/2}$$

Diffusion constant, molecular (Knudsen), pipe of radius r

$$D = \frac{2}{3} r v$$

Speed of sound in a gas

$$\mathbf{U}(\mathrm{m/s}) = v\left(\frac{\pi\gamma}{8}\right)^{1/2}$$

Conductance

$$C = \frac{Q}{(P_1 - P_2)}$$

Pumping speed

$$S = \frac{Q}{P}$$

Flow Regimes

Turbulent flow	$\mathbf{R} > 2200$
Choked flow	$\mathbf{U} = 1$
Viscous flow	$\mathbf{R} < 1200$, and $\mathrm{Kn} < 0.01$
Poiseuille flow	$\mathbf{U} < 1/3$, $\mathbf{R} < 1200$, $\mathrm{Kn} < 0.01$, and $l_e \ll l$
Molecular flow	$\mathrm{Kn} > 1$

Gas Flow Formulas

Continuum flow, thin aperture, any gas

$$Q = AP_1C'\left(\frac{2\gamma}{\gamma - 1}\frac{kT}{m}\right)^{1/2}\left(\frac{P_2}{P_1}\right)^{1/\gamma}\left[1 - \left(\frac{P_2}{P_1}\right)^{(\gamma-1)/\gamma}\right]^{1/2}$$

$$\text{for } 1 > P_2 / P_1 \geq (2/\gamma + 1))^{\gamma/(\gamma-1)}$$

Choked flow limit, thin aperture, any gas

$$Q = AP_1C'\left(\frac{kT}{m}\frac{2\gamma}{\gamma+1}\right)^{1/2}\left(\frac{2}{\gamma+1}\right)^{1/(\gamma-1)}$$

$$\text{for } P_2/P_1 \le (2/(\gamma+1))^{\gamma/(\gamma-1)}$$

*Continuum flow, thin aperture, air 22°*C

$$Q(\text{Pa-m}^3/\text{s}) = 766AP_1C'\left(\frac{P_2}{P_1}\right)^{0.714}\left[1-\left(\frac{P_2}{P_1}\right)^{0.286}\right]^{1/2}$$

$$\text{for } 1 > P_2/P_1 \ge 0.52$$

*Choked flow limit, thin aperture, air 22°*C

$$Q(\text{Pa-m}^3/\text{s}) = 200P_1AC'$$

$$\text{for air at } 22^\circ\text{C, when } P_2/P_1 \le 0.52$$

Viscous flow, long circular tube (Poiseuille)

$$Q = \frac{\pi d^4}{128\eta l}\frac{(P_1-P_2)}{2}(P_1-P_2)$$

Molecular flow:

$$Q(\text{Pa-m}^3/s) = a\frac{v}{4}A(P_1-P_2)$$

where a is the transmission coefficient and A is the entrance area.

Useful transmission coefficients

Five transmission coefficients for shapes of common interest are given here. Others are found in Fig's. 3.5–3.12.

1) *Very thin aperture, length $l \ll$ diameter d*

   $$a = 1$$

2) *Round pipe; any length l, radius r*
   Use a from Table 3.1 or Fig. 3.5, or use the following expression given by A. S. Berman, *J. Appl. Phys.*, **36**, 3356 (1965). $a = (K_1 - K_2)$. K_1 and K_2 are given

by the following formulas with the reduced length $L = l/r$, where l and r are the duct length and radius, respectively.

$$K_1 = 1 + \frac{L^2}{4} - \left(\frac{L}{4}\right)\left[L^2 + 4\right]^{1/2}$$

$$K_2 = \frac{[(8 - L^2)(L^2 + 4)^{1/2} + L^3 - 16]^{1/2}}{72L(L^2 + 4)^{1/2} - 288\ln\,[L + (L^2 + 4)^{1/2}] + 288\ln 2}$$

3) *Rectangular pipe, any width-to-thickness ratio (b/h), length l*
   Use a from Fig. 3.8, or use the following expression calculated from G. L. Saksaganskii, *Molecular Flow in Complex Vacuum Systems*, Gordon and Breach Science Publishers, London, 1988. p. 17.

$$a = c\frac{hb}{l(h + b)}$$

The coefficient c is given by

b/h	0.05	0.1	0.2	0.4	0.6	0.8	1.0
c	1.52	1.44	1.33	1.18	1.12	1.1	1.0

4) *Rectangular pipe; thin, slit-like (thickness h ≪ width b), any length l*
   Use a from Table 3.2, or use the following expression given by A. S. Berman, *J. Appl. Phys.*, **36**, 3356 (1965), and erratum **37**, 4509 (1966). $a = (K_1 - K_2)$. K_1 and K_2 are given by the following formulas with the reduced length L, given by $L = l/h$ (length-to-slit spacing).

$$K_1 = \frac{1}{2}\left(1 + (1 + L)^2 - L\right)$$

$$K_2 = \frac{3/2[L - \ln(L + (L^2 + 1)^{1/2})]^2}{L^3 + 3L^2 + 4 - (L^2 + 4)(L^2 + 1)^{1/2}}$$

5) *Annular cylindrical pipe, length l, inner radius r, and outer radius r_o*
   Use a from Fig. 3.6 or use the following formula given by A. S. Berman, *J. Appl. Phys.*, **40**, 4991 (1969).

$$a = \left\{1 + L\left[1/2 - A\tan^{-1}\left(\frac{L}{B}\right)\right]\right\}^{-1}$$

$L = l/(r_o - r_i)$. A and B are given in the formulas below. In these formulas $\sigma = r_i/r_o$ and has a range $0 < \sigma < 0.9$ and L has a range $0 \le L \le 100$.

$$A = \frac{(0.0741 - 0.014\sigma - 0.037\sigma^2)}{(1 - 0.918\sigma + 0.05\sigma^2)}$$

and

$$B = \frac{(5.825 - 2.86\sigma - 1.45\sigma^2)}{(1 + 0.56\sigma - 1.28\sigma^2)}$$

Combining conductance of objects in molecular flow:

1) *Parallel*

$$C_T = C_1 + C_2 + C_3 + \ldots$$

2) *Series, isolated*

$$\frac{1}{C_T} = \frac{1}{C_1} + \frac{1}{C_2} + \frac{1}{C_3} + \ldots$$

3) *Series, not isolated, equal entrance and exit areas (Oatley)*

$$\frac{1-a}{a} = \frac{1-a_1}{a_1} + \frac{1-a_2}{a_2} + \frac{1-a_3}{a_3} + \ldots$$

4) *Series, not isolated, unequal entrance and exit areas (Haefer)*

$$\frac{1}{A_1}\left(\frac{1-a_{1\to n}}{a_{1\to n}}\right) = \sum_1^n \frac{1}{A_i}\left(\frac{1-a_i}{a_i}\right) + \sum_1^{n-1}\left(\frac{1}{A_{i+1}} - \frac{1}{A_i}\right)\delta_{i,i+1}$$

where $\delta_{i,i+1} = 1$ for $A_{i+1} < A_i$, and $\delta_{i,i+1} = 0$ for $A_{i+1} \geq A_i$

Transition conductance, Knudsen's method

$$C = \frac{Q}{(P_2 - P_1)}$$

$$Q = Q_{\text{viscous}} + Z'Q_{\text{molecular}}$$

where

$$Z' = \frac{1 + 2.507\left(\frac{d}{2\lambda}\right)}{1 + 3.095\left(\frac{d}{2\lambda}\right)}$$

B.5 Vapor Pressure Curves of Common Gases

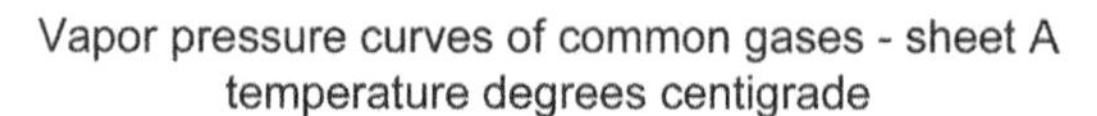

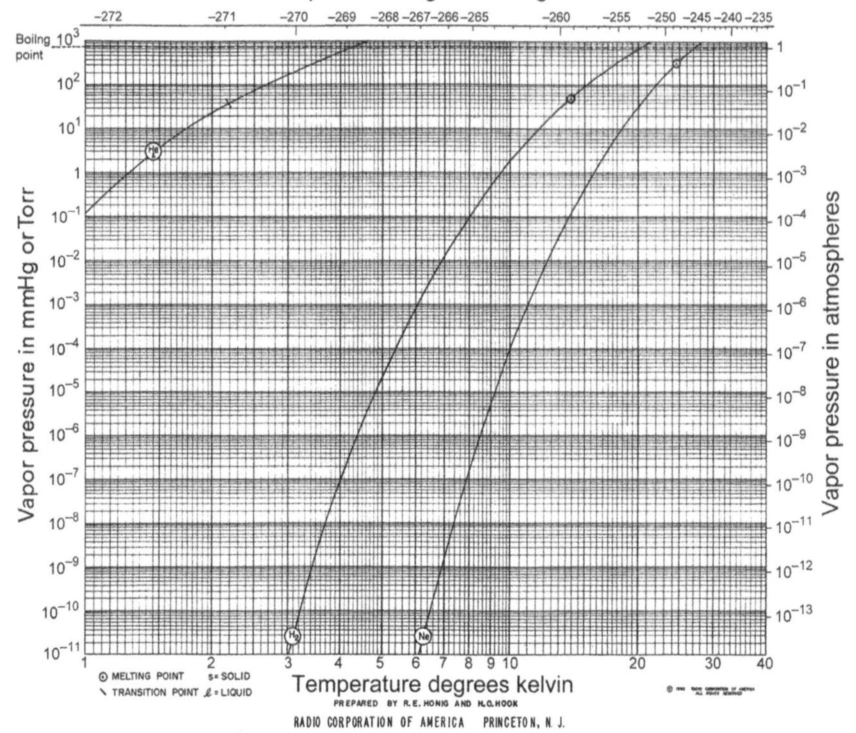

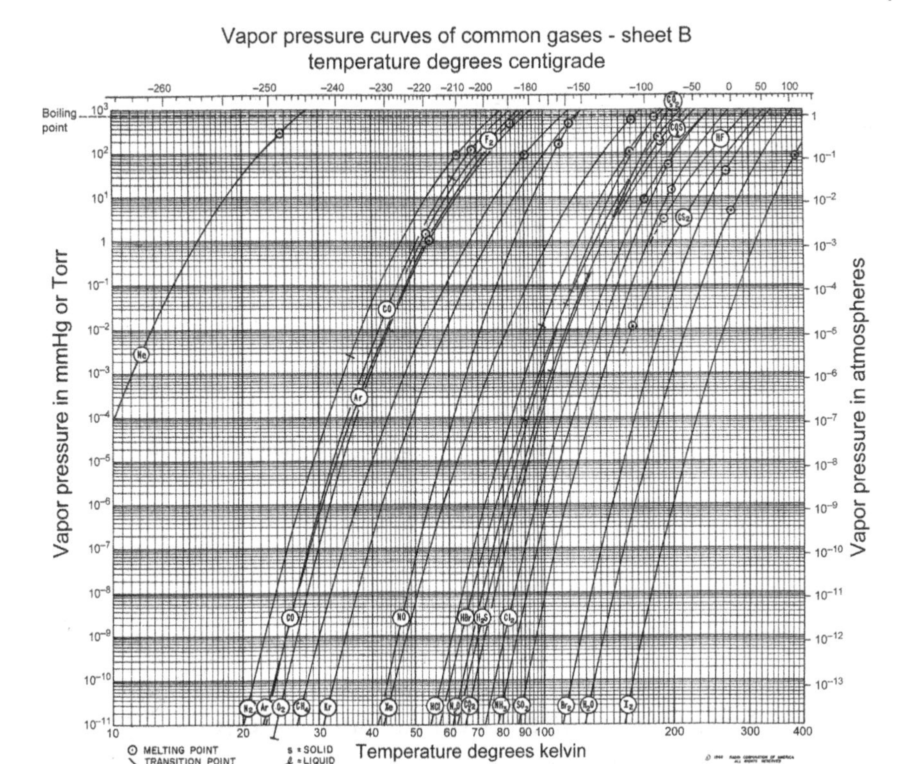

Source: Reproduced with permission from RCA Review, 21, p. 360, Sept. 1960, Vapor Pressure Data for Some Common Gases, by R. E. Honig and H. O. Hook. Copyright 1960, RCA Corporation.

B.6 Appearance of Discharges in Gases and Vapors at Low Pressures

Gas	Negative Glow	Positive Column
Argon	Blue	Violet
Carbon tetrachloride	Light green	Whitish green
Carbon monoxide	Greenish white	White
Carbon dioxide	Blue	White
C_2H_5OH	—	Whitish

(*Continued*)

(Continued)

Gas	Negative Glow	Positive Column
Cadmium	Red	Greenish blue
Hydrogen	Light blue	Pink
Mercury	Whitish yellow	Blue–green
Potassium	Green	Green
Krypton	Violet	Yellow pink
Air	Blue	Reddish
Nitrogen	Blue	Red–yellow
Sodium	Whitish	Yellow
Oxygen	Yellowish white	Lemon yellow with pink core
Thallium	Green	Green
Xenon	Pale blue	Blue–violet

Source: Reproduced with permission from: Materials for High Vacuum Technology, Vol. 3, p. 393, W. Espe. Copyright 1968, Pergamon Press.

B.7 DC Breakdown Voltages for Air and Helium Between Flat Parallel Plates

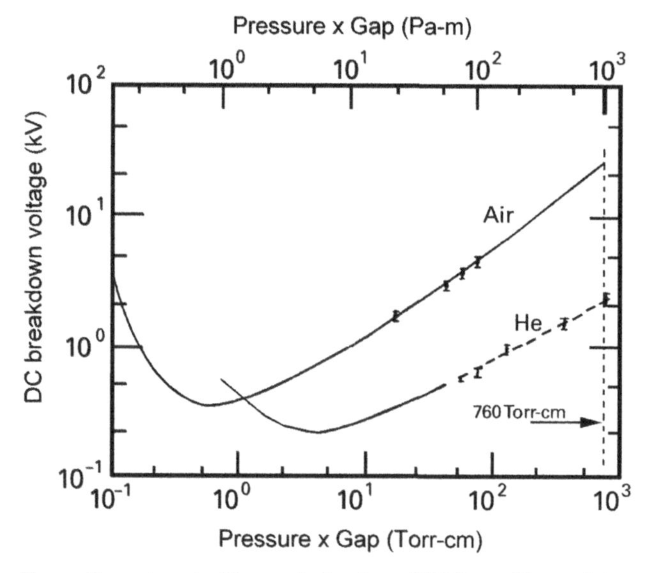

Source: Reproduced with permission from *IEEE Trans. Plasma Science*, **44**, 12, p. 3246, L. Babish and T. Loïko, Copyright 2016, IEEE.

B.8 Particle Settling Velocities in Air

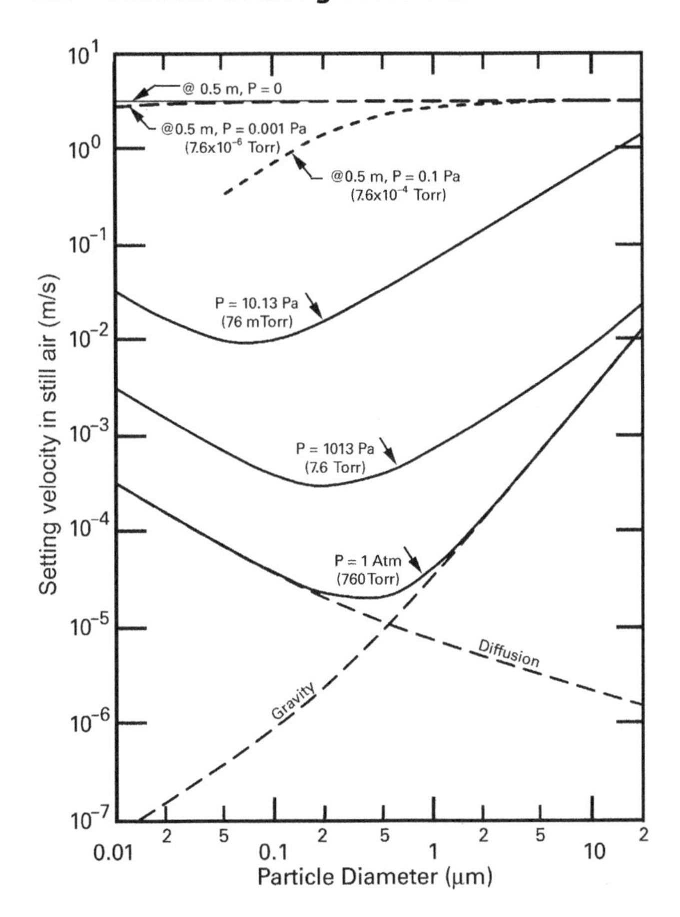

Fig. B.8 Calculated particle settling velocities in air for particles of density 1000-kg/m^3 (water) at various pressures. Diffusive velocities were calculated from $x = (Dt)^{1/2}$ at a diffusion time of 1 s. Gravitational settling velocities were calculated from Stokes Law. Both diffusion constant and viscous drag force was corrected for slip using relations given in W. C. Hinds, *Aerosol Technology*, Wiley, New York, 1999, Chapters 3 and 7. For comparison, particle velocities after ½-m fall at 0.001 Pa (7.6×10^{-6} Torr), 0.1 Pa (7.6×10^{-4} Torr), and UHV are shown.

Appendix C

Material Properties

C.1 Outgassing Rates of Vacuum-Baked Metals

Material	Treatment	q (10^{-11} Pa-m/s)
Aluminum[a]	15 h at 250°C	53.0
Aluminum[b]	20 h at 100°C	5.3
6061 Aluminum[c]	Glow discharge +200°C bake	1.3
Copper[b]	20 h at 100°C	146.0
Copper (OHFC)[d]	24 h at 100°C	2.90
	24 h at 250°C	0.181
Cr (0.5%)–Cu (99.5%) alloy[d]	24 h at 100°C	0.375
	24 h at 250°C	0.357
304 Stainless steel[a]	30 h at 250°C	400.0
Stainless steel[e]	2 h at 850/900°C vacuum furnace	27.0
316L Stainless steel[f]	2 h at 800°C vacuum furnace	46.0
U15C Stainless steel[g]	3 h vacuum furnace 1000°C + 25 h *in situ* vacuum bake at 360°C	2.1

[a] J. R. Young, *J. Vac. Sci. Technol.*, **6**, 398, (1969)
[b] G. Moraw, *Vacuum*, **24**, 125 (1974)
[c] H. J. Halama and J. C. Herrera, *J. Vac. Sci. Technol.*, **13**, 463 (1976)
[d] Y. Koyatsu, H. Miki and F. Watanabe, *Vacuum*, **47**, 709 (1996)
[e] R. L. Samuel, *Vacuum*, **20**, 295 (1970)
[f] R. Nuvolone, *J. Vac. Sci. Technol.*, **14**, 1210 (1977)
[g] R. Calder and G. Lewin, *Br. J. Appl. Phys.*, **18**, 1459 (1967)
Source: Adapted with permission from *Vacuum*, **25**, p. 347, R. J., Elsey, Copyright 1975, Pergamon Press.

A Users Guide to Vacuum Technology, Fourth Edition. John F. O'Hanlon and Timothy A. Gessert.
© 2024 John Wiley & Sons, Inc. Published 2024 by John Wiley & Sons, Inc.

C.2 Outgassing Rates of Unbaked Metals

Material	q_1 (10^{-7} Pa-m/s)	α_1	q_{10} (10^{-7} Pa-m/s)	α_{10}
Aluminum (fresh)[a]	84.0	1.0	8.0	1.0
Aluminum (degassed 24 h)[a]	55.2	3.2	4.08	0.9
Aluminum (3 h in air)[a]	88.6	1.9	6.33	0.9
Aluminum (fresh)[a]	82.6	1.0	4.33	0.9
Aluminum (anodized 2-μm pores)[a]	3679.0	0.9	429.0	0.9
Aluminum (bright-rolled)[b]	—	—	100.0	1.0
Duraluminum[b]	2266.0	0.75	467.0	0.75
Brass (waveguide)[b]	5332.0	2.0	133.0	1.2
Copper (fresh)[a]	533.0	1.0	55.3	1.0
Copper (mechanically polished)[a]	46.7	1.0	4.75	1.0
Copper, OHFC (fresh)[a]	251.0	1.3	16.8	1.3
Copper, OHFC (mechanically polished)[a]	25.0	1.1	2.17	1.1
Copper, OHFC (20°C)[c]	—	—	0.408	—
Chrome (0.5%)–Copper, OFE (99.5%)[c]	—	—	0.102	—
Gold (wire fresh)[a]	2105.0	2.1	6.8	1.0
Mild steel[b]	7200.0	1.0	667.0	1.0
Mild steel (slightly rusty)[b]	8000.0	3.1	173.0	1.0
Mild steel (chromium-plated polished)[b]	133.0	1.0	12.0	—
Mild steel (aluminum spray coated)[b]	800.0	0.75	133.0	0.75
Steel (chromium-plated fresh)[a]	94.0	1.0	7.7	1.0
Steel (chromium-plated polished)[a]	121.0	1.0	10.7	1.0
Steel (nickel-plated fresh)[a]	56.5	0.9	6.6	0.9
Steel (nickel-plated)[a]	368.0	1.1	3.11	1.1
Steel (chemically nickel-plated fresh)[a]	111.0	1.0	9.4	1.0
Steel (chemically nickel-plated polished)[a]	69.6	1.0	6.13	1.0
Steel (descaled)[a]	4093.0	0.6	3933.0	0.7
Molybdenum[a]	69.0	1.0	4.89	1.0
Stainless steel EN58B (AISI 321)[b]	—	—	19.0	1.6
Stainless steel 19/9/1-electropolished[d]	—	—	2.7	—

(Continued)

Material	q_1 (10^{-7} Pa-m/s)	α_1	q_{10} (10^{-7} Pa-m/s)	α_{10}
-vapor degreased[d]	—	—	1.3	—
-Diversey cleaned[d]	—	—	4.0	—
Stainless steel[b]	2333.0	1.1	280.0	0.75
Stainless steel[b]	1200.0	0.7	267.0	0.75
Stainless steel ICN 472 (fresh)[a]	180.0	0.9	19.6	0.9
Stainless steel ICN 472 (sanded)[a]	110.0	1.2	13.9	0.8
Stainless steel NS22S (mech. polished)[a]	22.8	0.5	6.1	0.7
Stainless steel NS22S (electropolished)[a]	57.0	1.0	5.7	1.0
Stainless steel[a]	192.0	1.3	18.0	1.9
Zinc[a]	2946.0	1.4	429.0	0.8
Titanium[a]	150.0	0.6	24.5	1.1
Titanium[a]	53.0	1.0	4.91	1.0

For all values of q, $q_n = qt^{-\alpha_n}$, where n is in hours. See Fig. 4.6 for a detailed explanation.
[a] A. Schram, *Le Vide*, No. 103, 55 (1963),
[b] B. B. Dayton, *Trans. 6th Natl. Vac. Symp. (1959)*, Pergamon Press, New York, 1960, p. 101,
[c] Y. Koyatsu, H. Miki, and F. Watanabe, *Vacuum*, **47**, 709 (1996),
[d] R. S. Barton and R. P. Govier, *Proc. 4th Int. Vac. Congr. (1968)*, Institute of Physics and the Physical Society, London, 1969, p. 775, and *Vacuum*, **20**, 1 (1970).
Source: Reproduced with permission from Vacuum, 25, p 347, R. J. Elsey. Copyright 1975, Pergamon Press.

C.3 Outgassing Rates of Ceramics and Glasses

Material	q_1 (10^{-7} Pa-m/s)	α_1	q_{10} (10^{-7} Pa-m/s)	α_{10}
Steatite[a]	1200.0	1.0	127.0	—
Pyrophyllite[b]	2667.0	1.0	267.0	—
Pyrex (fresh)[c]	98.0	1.1	7.3	—
Pyrex (1 month in air)[c]	15.5	0.9	2.1	—

For all values of q, $q_n = qt^{-\alpha_n}$, where n is in hours. See Fig. 4.6 for a detailed explanation.
[a] R. Geller, *Le Vide*, No. 13, 71 (1958).
[b] R. Jaeckel and F. Schittko, quoted by Elsey.
[c] B. B. Dayton, *Trans. 6th Natl. Symp. Vac. Technol. (1959)*, Pergamon Press, New York, 1960, p. 101
Source: Reproduced with permission from Vacuum, 25, p. 347, R. J. Elsey. Copyright 1975, Pergamon Press.

C.4 Outgassing Rates of Elastomers

Material	q_1 (10^{-5} Pa-m/s)	α_1	q_4 (10^{-5} Pa-m/s)	α_4
Butyl DR41[a]	200.0	0.68	53.0	0.64
Neoprene[a]	4000.0	0.4	2400.0	0.4
Perbunan[a]	467.0	0.3	293.0	0.5
Silicone[b]	930.0	—	267.0	—
Viton A (fresh)[c]	152.0	0.8	—	—
Viton A (bake 12 h at 200°C)[d]	—	—	0.027[e]	—
Polyimide (bake 12 h at 300°C)[d]	—	—	0.005[e]	—

For all values of q, $q_n = qt^{-\alpha_n}$, where n is in hours. See Fig. 4.6 for a detailed explanation.
[a] J. Blears, E. J. Greer, and J. Nightengale, *Adv. Vac. Sci. Technol.*, Vol. 2., E. Thomas, Ed., Pergamon Press, 1960, p. 473.
[b] D. J. Santeler, et al., *Vacuum Technology and Space Simulation*, NASA SP-105, National Aeronautics and Space Administration, Washington, DC, 1966, p. 219.
[c] A. Schram, *Le Vide*, No. 103, 55 (1963).
[d] P. Hait, *Vacuum*, **17**, 547 (1967).
[e] Pumping time is 12 h.
Source: Adapted with permission from *Vacuum*, **25**, p. 347, R. J. Elsey. Copyright 1975, Pergamon Press.

C.5 Permeability of Polymeric Materials

	Permeability (10^{-12} m^2/s)					
Material	Nitrogen	Oxygen	Hydrogen	Helium	Water Vapor	Carbon Dioxide
PTFE	2.5[a]	4.78[b](25)	20.0[a]	570.0[a]	23.6[b] (25)	—
Perspex	—	—	2.7[a]	5.7[a]	—	—
Nylon 31	—	—	0.13[c]	0.3[c]	—	—
Neoprene CS2368B	0.21[a]	1.5[a]	8.2[a]	7.9[a]	—	—
Viton	0.5[d] (25)	1.14[b] (25)	8.7[d] (25)	15[b] (25)	218[b] (25)	2.3[e] (25)
Kapton	0.032[c]	0.1[c]	1.2[c]	2.1[c]	—	0.2[c]

(Continued)

	Permeability (10^{-12} m^2/s)					
Material	**Nitrogen**	**Oxygen**	**Hydrogen**	**Helium**	**Water Vapor**	**Carbon Dioxide**
Buna-S	4.8[f] (30)	—	—	—	—	940.0[f] (30)
Perbunan	0.8[f]	—	—	—	—	23.0[f] (30)
Delrin	—	48.0[f]	—	—	17.0[f]	92.7[f]
Kel-F	0.99[f] (30)	0.46[f] (30)	—	—	0.22[b] (25)	—
Vespel	—	0.076[b] (25)	—	—	31.2[e] (25)	—
Polyvinylidene chloride (Saran)	0.0007[f]	0.004[f]	—	—	1.06[f]	0.022[f]
Mylar	0.018[d]	8.5[d]	—	1.7[g]	99[d]	0.37[d]
Silicone, dimethyl	210[g]	450[g]	495[g]	263[g]	28500[g]	2030[g]
Kalrez-dry	—	3.11[e] (25)	—	—	125[e] (25)	—

All measurements made at 23°C unless otherwise noted in parentheses after the value.
[a] Reprinted with permission from *Vacuum*, **25**, p. 469, G. F. Weston. Copyright 1975, Pergamon Press. Data derived by Weston from measurements made by Barton reported by J. R. Bailey in *Handbook of Vacuum Physics*, 3, Part 4, Pergamon Press, Oxford, 1964.
[b] Data taken from C. Ma, et al, *Journal of the IES*, March/April 1995. p. 43, using tracer gas measurements in an ultrapure nitrogen gas stream.
[c] Reprinted with permission from *J. Vac. Sci. Technol.*, **10**, p. 543, W. G. Perkins. Copyright 1973, The American Vacuum Society.
[d] Data taken from M. Mapes, H. C. Hseuk, and W. S. Tiang, *J. Vac. Sci. Technol. A*, **12**, 1699 (1994).
[e] Data taken from L. Laurenson and N. T. M. Dennis, *J. Vac. Sci. Technol. A*, **3**, 1701 (1985). Data at 25°C, from plot in Fig. 16.8, left.
[f] Reprinted with permission from *Vacuum Science and Space Simulation*, D. J. Santeler et al., NASA SP-105, National Aeronautics and Space Administration, Washington, DC, 1966, p. 216.
[g] Data taken from *Introduction to Mass Spectroscopy*, Varian Associates, Palo Alto, 1980, p. 49.

C.6 Vapor Pressure Curves of the Solid and Liquid Elements (Sheet A)

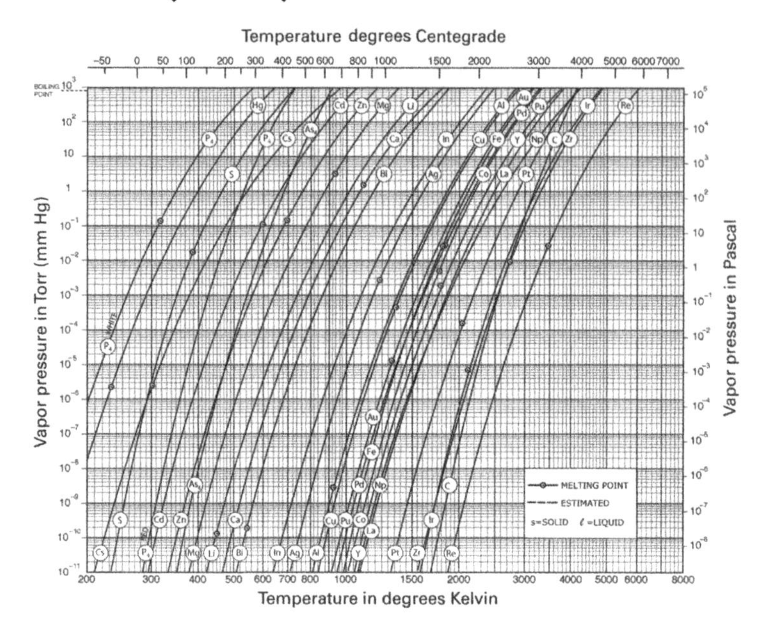

Vapor Pressure Curves of the Solid and Liquid Elements (Sheet B)

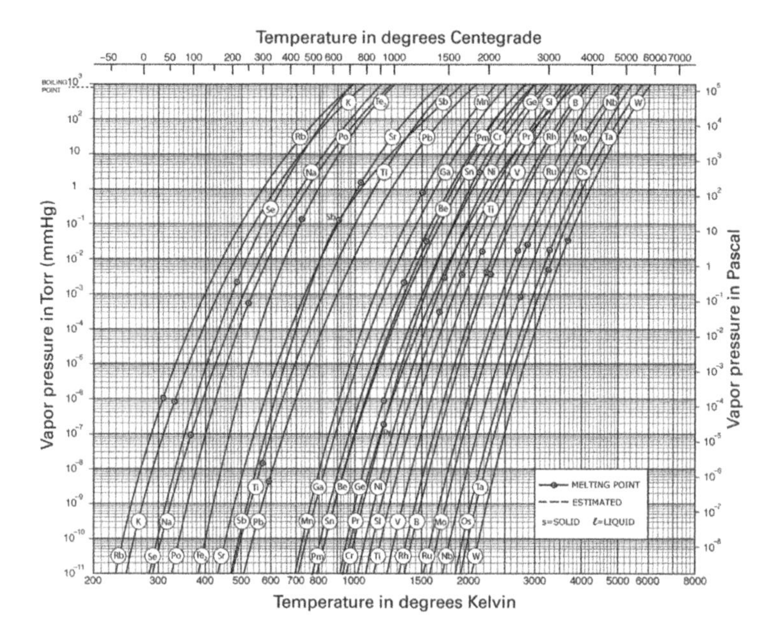

Vapor Pressure Curves of the Solid and Liquid Elements (Sheet C)

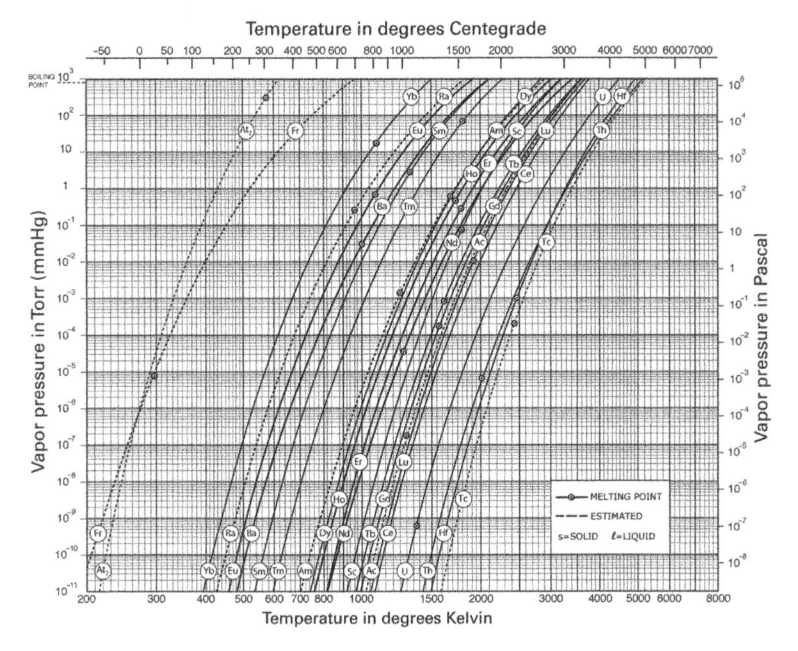

Source: Reproduced with permission from RCA Review, 30, p. 285, June 1969, Vapor Pressure Data for the Solid and Liquid Element, by R. E. Honig and D. A. Kramer. Copyright 1969, RCA Corp.

C.7 Outgassing Rates of Polymers

Material	q_1 (10^{-5} Pa-m/s)	α_1	q_{10} (10^{-5} Pa-m/s)	α_{10}
Araldite (molded)[a]	155.0	0.8	47.0	0.8
Araldite D[b]	253.0	0.3	167.0	0.5
Araldite F[b]	200.0	0.5	97.0	0.5
Kel-F[c]	5.0	0.57	2.3	0.53
Methyl Methacrylate[d]	560.0	0.9	187.0	0.57
Mylar (24-h at 95% RH)[e]	307.0	0.75	53.0	—
Nylon[f]	1600.0	0.5	800.0	0.5
Plexiglas[g]	96.1	0.44	36.0	0.44
Plexiglas[b]	413.0	0.4	240.0	0.4

(*Continued*)

(Continued)

Material	q_1 (10^{-5} Pa-m/s)	α_1	q_{10} (10^{-5} Pa-m/s)	α_{10}
Polyester–glass Laminate[c]	333.0	0.84	107.0	0.81
Polystyrene[c]	2667.0	1.6	267.0	1.6
PTFE[h]	40.0	0.45	26.0	0.56
PVC (24-h at 95% RH)[e]	113.0	1.0	2.7	—
Teflon[g]	8.7	0.5	3.3	0.2

For all q, $q_n = qt^{-\alpha_n}$, where n is in hours. See Fig. 4.6 for a detailed explanation.
[a] A. Schram, *Le Vide*, No. 103, 55 (1963).
[b] R. Geller, *Le Vide*, No.13, 71 (1958).
[c] B. B. Dayton, CVC Technical Report.
[d] J. Blears, E. J. Greer and J. Nightengale, *Adv. Vac. Sci. Technol.*, Vol. 2, E. Thomas, Ed., Pergamon Press, 1960, p. 473.
[e] D. J. Santeler, *Trans. 5th Symp. Vac. Tech. (1958)*, Pergamon Press, New York, 1959, p. 1;
[f] B. D. Power and D. J. Crawley, *Adv. Vac. Sci. Technol.*, Vol. 1, E. Thomas, Ed., Pergamon Press, New York, 1960, p. 207.
[g] G. Thieme, *Vacuum*, **13**, 137 (1963).
[h] B. B. Dayton, *Trans. 6th Natl. Vac. Symp. Vac. Technol. (1959)*, Pergamon Press, New York, 1960, p.101.
Source: Reproduced with permission from Vacuum, 25, p. 347, R. J. Elsey, Copyright 1975, Pergamon Press.

C.8 Austenitic Stainless Steels

	Performance—Recommended for								Tensile properties typical, at R. T. annealed		Impact strength at −196°C	
AISI Type	General forming	High temperature use	Cryogenic use	Resist carbide precipitation	High yield strength	Non-magnetic at cryogenic temperatures	Free machining	Welding ability	0.2% proof (yield) kgf mm^2	Ultimate TS kgf mm^2	Charpy-V J	Price index
302	×							E	22	60	70	100
303S, Se							×	P	22	60		105
304	×		×					E	20	50	70	100
304L	×		×	×				E	18	48	80	115
304N	×		×		×			E	0	55–75	min 63	110
304LN	×		×	×	×	×		E	28	55–75	min 55	125
310	×	>			×	×		G	23	65		210
316	×		×			~		E	22	50		130
316L	×		×	×		~		E	20	45	120	150
316N	×		×		×	×		E	30	60		140
316LN	×		×	×	×	×		E	30	60–80	min 55	160
317	×	>				~		E	24	60		200
321	×	>		×				E	21	50	65	130
347	×	>		×				G	22	50	65	150

Selection guide for vacuum and cryogenic equipment. × = recommended, P = poor,~ = Probable,–should be tested, E = excellent, G = good with precautions.
Source: Reproduced with permission from Vacuum, 26, p. 287, C. Geyari. Copyright 1976, Pergamon Press.

	Typical Composition (%)						
AISI Type	C max	Cr	Ni	Mo	N	Others	Description
302	0.12	17–19	8–10				General purpose. Good resistance to atm corrosion. Good mechanical properties
303 S, Se	0.15	17–19	8–10			S, Se ≥0.15	Free machining type. Good corrosion resistance
304	0.08	18–20	8–12				Low C variation of 302 Improved corrosion resistance after welding
304L	0.03	18–20	8–12				Extra low C prevents carbide precipitation
304N	0.06	18.5	9.5		0.25	Mn 2	Improved mechanical properties
304LN	0.03	18.5	9.5		0.18	Mn 2	Low C prevents carbide precipitation
310	0.025	24–26	19–22				High-scale resistance, superior corrosion resistance
316	0.08	16–18	10–14	2–3			Very good corrosion resistance in most media
316L	0.03	16–18	10–14	2–3			Extra low C variation of 316
316N	0.07	17–18	10–13	2.5–3	0.2	Mn 0.5–2	Improved mechanical properties
316LN	0.03	17.5	13	2	0.18	Mn 2	Improved mechanical properties. Extra low C prevents carbide precipitation
317	0.08	18–20	11–15	3–4			Higher alloy content improves creep and corrosion resistance of 316
321	0.08	17–19	9–12			Ti ≥ 5xC	Stabilized–Ti prevents carbide precipitation. Improved corrosion
347	0.08	17–19	9–13			Nb, Ta ≥10xC	Stabilized–Nb, Ta, prevents carbide precip. Improved corrosion resistance

Source: Reproduced with permission from Vacuum, 26, p. 287, C. Geyari. Copyright 1976, Pergamon Press.

Appendix D

Isotopes

D.1 Natural Abundances

Element	AMU	Relative Abundance	Element	AMU	Relative Abundance
H	1	99.985	Mg	24	78.60
	2	0.015		25	10.11
He	3	0.00013		26	11.29
	4	~100.0	Al	27	100.0
Li	6	7.42	Si	28	92.27
	7	92.58		29	4.68
Be	9	100.0		30	3.05
B	10	19.78	P	31	100.0
	11	80.22	S	32	95.06
C	12	98.892		33	0.74
	13	1.108		34	4.18
N	14	99.63		36	0.016
	15	0.37	Cl	35	75.4
O	16	99.759		37	24.6
	17	0.0374	Ar	36	0.337
	18	0.2039		38	0.063
F	19	100.0		40	99.600
Ne	20	90.92	K	39	93.08
	21	0.257		40	0.0119
	22	8.82		41	6.91
Na	23	100.0			

(*Continued*)

A Users Guide to Vacuum Technology, Fourth Edition. John F. O'Hanlon and Timothy A. Gessert.
© 2024 John Wiley & Sons, Inc. Published 2024 by John Wiley & Sons, Inc.

(Continued)

Element	AMU	Relative Abundance	Element	AMU	Relative Abundance
Ca	40	96.97	Zn	64	48.89
	42	0.64		66	27.82
	43	0.145		67	4.14
	44	2.06		68	18.54
	46	0.0033		70	0.617
	48	0.185	Ga	69	60.2
Sc	45	100.0		71	39.8
Ti	46	7.95	Ge	70	20.55
	47	7.75		72	27.37
	48	73.45		73	7.67
	49	5.51		74	36.74
	50	5.34		76	7.67
V	50	0.24	As	75	100.0
	51	99.76	Se	74	0.87
Cr	50	4.31		76	9.02
	52	83.76		77	7.58
	53	9.55		78	23.52
	54	2.38		80	49.82
Mn	55	100.0		82	9.19
Fe	54	5.82	Br	79	50.52
	56	91.66		81	49.48
	57	2.19	Kr	78	0.354
	58	0.33		80	2.27
Co	59	100.0		82	11.56
Ni	58	67.76		83	11.55
	60	26.16		84	56.90
	61	1.25	Rb	85	72.15
	62	3.66		87	27.85
	64	1.16	Sr	84	0.56
Cu	63	69.1		86	9.86
	65	30.9		87	7.02
				88	82.56

(Continued)

Element	AMU	Relative Abundance	Element	AMU	Relative Abundance
Y	98	100.0	Cd	106	1.22
Zr	90	51.46		108	0.87
	91	11.23		110	12.39
	92	17.11		111	12.75
	94	17.4		112	24.07
	96	2.8		113	12.26
Nb	93	100.0		114	28.86
Mo	92	15.86		116	7.85
	94	9.12	In	113	4.23
	95	15.70		115	95.77
	96	16.50	Sn	112	0.95
	97	9.45		114	0.65
	98	23.75		115	0.34
	100	9.62		116	14.24
Ru	96	5.47		117	7.57
	98	1.84		118	24.01
	99	12.77		119	8.58
	100	12.56		120	32.97
	101	17.10		122	4.71
	102	31.70		124	5.98
	104	18.56	Sb	121	57.25
Rh	103	100.0		123	42.75
Pd	102	0.96	Te	120	0.089
	104	10.97		122	2.46
	105	22.23		123	0.87
	106	27.33		124	4.61
	108	26.71		125	6.99
	110	11.81		126	18.71
Ag	107	51.82		128	31.79
	109	48.18		130	34.49

(Continued)

(Continued)

Element	AMU	Relative Abundance	Element	AMU	Relative Abundance
I	127	100.0	Nd	148	5.72
Xe	124	0.096		150	5.62
	126	0.090	Sm	144	3.16
	128	1.92		147	15.07
	129	26.44		148	11.27
	130	4.08		149	13.84
	131	21.18		150	7.47
	132	26.89		152	26.63
	134	10.44		154	22.53
	136	8.87	Eu	151	47.77
Cs	131	100.0		153	52.23
Ba	130	0.101	Gd	152	0.20
	132	0.097		154	2.15
	134	2.42		155	14.73
	135	6.59		156	20.47
	136	7.81		157	15.68
	137	11.32		158	24.87
	138	71.66		160	21.90
La	138	0.089	Tb	159	100.0
	139	99.911	Dy	156	0.052
Ce	136	0.193		158	0.090
	138	0.250		160	2.294
	140	88.48		161	18.88
	142	11.07		162	25.53
Pr	141	100.0		163	24.97
Nd	142	27.13		164	28.18
	143	12.20	Ho	165	100.0
	144	23.87	Er	162	0.136
	145	8.30		164	1.56
	146	17.18		166	33.41

(Continued)

Element	AMU	Relative Abundance	Element	AMU	Relative Abundance
Er	167	22.94	Os	188	13.3
	168	27.07		189	16.1
	170	14.88		190	26.4
Tm	169	100.0		192	41.0
Yb	168	0.140	Ir	191	37.3
	170	3.03		193	62.7
	171	14.31	Pt	190	0.012
	172	21.82		192	0.78
	173	16.13		194	32.8
	174	31.84		195	33.7
	176	12.73		196	25.4
Lu	175	97.40		198	7.21
	176	2.60	Au	197	100.0
Hf	174	0.18	Hg	196	0.15
	176	5.15		198	10.02
	177	18.39		199	16.84
	178	27.08		200	23.13
	179	13.78		201	13.22
	180	35.44		202	29.80
Ta	180	0.012		204	6.85
	181	99.988	Tl	203	29.50
W	180	0.135		205	70.50
	182	26.4	Pb	204	1.48
	183	14.4		206	23.6
	184	30.6		207	22.6
	186	28.4		208	52.3
Re	185	37.07	Bi	209	100.0
	187	62.9	Th	232	100.0
Os	184	0.018	U	234	0.0057
	186	1.59		235	0.72
	187	1.64		238	99.27

Source: Reprinted with permission from *Mass Spectroscopy for Science and Technology*, F. A. White, p. 339. Copyright 1968, John Wiley & Sons.

Appendix E

Cracking Patterns

E.1 Cracking Patterns of Pump Fluids

AMU	Welch 1407[a]	Fomblin Y-25[b]	DC-704[c]	DC-705[c]	Octoil-S[c]	Convalex-10[c,d]
18					1.86	
27					14.40	23.20
28			17.19	67.18	8.37	
29				2.56	28.53	23.20
30				3.52	1.10	20.20
31		31.49			1.27	6.00
32			2.15	10.08		
35						
36						6.50
37						
38					0.65	4.20
39				2.22	8.82	14.90
40				4.44	2.21	15.50
41	40			3.59	54.03	31.00
42					17.12	8.30
43	74		2.40	7.69	66.58	29.20
44			1.98	12.17	3.17	50.00
45					1.98	
47		20.67				

(*Continued*)

A Users Guide to Vacuum Technology, Fourth Edition. John F. O'Hanlon and Timothy A. Gessert.
© 2024 John Wiley & Sons, Inc. Published 2024 by John Wiley & Sons, Inc.

(Continued)

AMU	Welch 1407[a]	Fomblin Y-25[b]	DC-704[c]	DC-705[c]	Octoil-S[c]	Convalex-10[c,d]
50		15.30				5.40
51		27.66	2.61		0.84	11.30
52		2.92				
53	3				3.41	
54					4.34	
55	70		1.31	4.79	54.56	25.60
56	23				22.18	15.50
57	100		2.13		88.20	55.40
59					4.59	
60					2.0	
61						
62						
63						7.7
64						8.30
65						7.10
66		2.23				
67	40				4.52	3.60
68	14				4.23	3.00
69	91	100.00		3.59		11.30
70	31	3.39			56.83	18.50
71	83			2.90	46.92	9.50
72					3.14	
73			1.65		2.98	
76						3.40
77	2		2.20			25.60
78	0.7			4.62		4.80
81	22	3.10		1.88	3.80	3.60
82	10				3.91	2.40
83	30			2.90	19.78	11.30

(Continued)

AMU	Welch 1407[a]	Fomblin Y-25[b]	DC-704[c]	DC-705[c]	Octoil-S[c]	Convalex-10[c,d]
84	12				13.78	4.80
85		2.23		2.05	2.25	
87					1.82	
91		2.03	3.39	3.76		3.00
92						3.60
93	2				1.18	
94	7					3.60
95	3			2.05	2.87	2.40
96	10				2.11	
97		16.37		2.39	10.50	3.00
98					19.87	5.40
100		14.63				
101		14.63			2.00	
108						12.50
112					45.46	
113					25.01	61.30
119		13.88	3.87	3.24		
131		2.36				2.40
135		5.19	20.59	10.26		

Note: Only peaks up to 135 AMU are shown for the data taken from source *c*; the largest mass peak (100%) occurs at a higher mass number. The data are *not* renormalized for the range tabulated here.

[a] Data taken on UTI-100B quadrupole with $V_{EE} = -60$ V, only major peaks shown. Reproduced with permission from Uthe Technology Inc., 325 N. Mathilda Avenue, Sunnyvale, CA 94086.

[b] Sector data. Adapted with permission from *Vacuum*, **22**, p. 315, L. Holland, L. Laurenson, and P. N. Baker. Copyright 1972, Pergamon Press.

[c] Sector data. Reprinted with permission from *J. Vac. Sci. Technol.*, **6**, p. 871, G. M. Wood, Jr., and R. J. Roenig. Copyright 1969, The American Vacuum Society.

[d] Convalex-10, was discontinued by the distributor. It was identical to Santovac-5.

E.2 Cracking Patterns of Gases

AMU	Hydrogen[a] H_2	Helium[b] He	Neon[b] Ne	Carbon Monoxide[a] CO	Nitrogen[a] N_2	Oxygen[a] O_2	Argon[a] Ar	Carbon Dioxide[a] CO_2
1	2.7							
2	100	0.12						
3	0.31							
4		100						
6				0.0008				0.0005
7					0.0006			
8				0.0001		0.0013		0.0005
12				3.5				6.3
13								0.063
14				1.4	9			
15					0.026			
16				1.4		14		16
17						0.0052		
18						0.028	0.071	0.0088
19							0.016	
20			100				5.0	
21			0.33					
22			9.9					0.52
22.5								0.0047
23								0.0012
28				100	100			15
29				1.2	0.71			0.15
30				0.2	0.0014			0.029
32						100		
33						0.074		
34						0.38		
36						0.0023	0.36	
38							0.068	
40							100	
44								100
45								1.2

(Continued)

AMU	Hydrogen[a] H_2	Helium[b] He	Neon[b] Ne	Carbon Monoxide[a] CO	Nitrogen[a] N_2	Oxygen[a] O_2	Argon[a] Ar	Carbon Dioxide[a] CO_2
46								0.38
47								0.0034
48								0.0005

[a] Data taken on UTI-100C-02 quadrupole residual gas analyzer. Typical parameters, $V_{EE} = 70$ V, $V_{IE} = 15$ V, $V_{FO} = -20$ V, $I_E = 2.5$ mA, resolution potentiometer = 5.00. Reproduced with permission from Uthe Technology Inc., 325 N. Mathilda Avenue, Sunnyvale, CA 94086.
[b] Sector data. Reprinted with permission from E. I. du Pont de Nemours & Co., Wilmington, DE 19898.

E.3 Cracking Patterns of Common Vapors

AMU	Water Vapor[a] H_2O	Methane[b] CH_4	Acetylene[b] C_2H_2	Ethylene[b] C_2H_4	Ethane[b] C_2H_6	Cyclo-Propane[b] C_3H_6
1	0.1	3.8	3.8	6.4	3.2	1.4
2		0.64	1.2	1.1	0.93	32
3		0.009	0.002	0.022	0.15	0.10
6		0.0003	0.0006	0.0002		
7		0.0013		0.0018		
12		2.1	4.5	2.3	0.47	0.85
13		7.4	7.6	4.0	1.1	1.6
14		15	0.86	8.1	3.4	5.6
14.5					0.24	
15		83			5.7	8.1
16	3.07	100			0.53	2.0
17	27.01	1.3				0.07
18	100					
19	0.19					2.7
19.5						1.3
20						2.3
20.5						0.68
24			7.1	3.2	0.52	0.35

(Continued)

(Continued)

AMU	Water Vapor[a] H_2O	Methane[b] CH_4	Acetylene[b] C_2H_2	Ethylene[b] C_2H_4	Ethane[b] C_2H_6	Cyclo-Propane[b] C_3H_6
25			23.	12	3.5	2.1
26			100	61	24.	17
27			2.5	59	33	46
28				100	100	18
29				2.8	21	11
30					24	0.29
31					0.54	
36						1.4
37						11
38						15
39						69
40						30
41						100
42						90
43						18

[a] Sector data. Reproduced with permission from E. I. du Pont de Nemours & Co., Wilmington, DE 19898.
[b] Quadrupole data, same conditions as given in Appendix E1, footnote *a*. Reprinted with permission from Uthe Technology Inc., 325 N. Mathilda Avenue, Sunnyvale CA 94086.

E.4 Cracking Patterns of Common Solvents

AMU	Methyl Alcohol[a]	Ethyl Alcohol[a]	Acetone[a]	Isopropyl Alcohol[a]	Trichloro-Ethylene[a]	Gentron 142B[b]
2						3.4
12						2.9
13						3.2
14						9.2
15						16.0
18	1.9	5.5				
19		2.3		6.6		3.3

(Continued)

AMU	Methyl Alcohol[a]	Ethyl Alcohol[a]	Acetone[a]	Isopropyl Alcohol[a]	Trichloro-Ethylene[a]	Gentron 142B[b]
20						1.7
25						6.5
26		8.3	5.8			17.0
27		23.9	8.0	15.7		4.4
28	6.4	6.9				
29	67.4	23.4	4.3	10.1		
30	0.8	6.0				
31	100	100		5.6		17.0
32	66.7					1.0
35					39.9	5.2
36						1.8
37			2.1		12.8	1.6
38			2.3			0.9
39			3.8	5.7		
41			2.1	6.6		
42		2.9	7.0	4.0		
43		7.6	100	16.6		0.5
45		34.4		100		53.0
46		16.5				3.5
47					25.8	1.1
48						0.5
49						1.4
50						2.9
51						1.9
58			27.1			
59				3.4		
60					64.9	0.6
62					20.9	0.5
63						3.4
64						8.6
65						100.0
66						2.5

(Continued)

(Continued)

AMU	Methyl Alcohol[a]	Ethyl Alcohol[a]	Acetone[a]	Isopropyl Alcohol[a]	Trichloro-Ethylene[a]	Gentron 142B[b]
87						1.8
95					100.0	
97					63.9	
130					89.8	
132					84.8	
134					26.8	

[a] Sector data. Reprinted with permission from VG-Micromass Ltd., 3 Tudor Road, Altringham, Cheshire, TN34-1YQ, England.
[b] Quadrupole data, same conditions as given in Appendix E1, footnote *a*. Reproduced with permission from Uthe Technology Inc., 325 N. Mathilda Avenue, Sunnyvale, CA 94086.

E.5 Cracking Patterns of Semiconductor Dopants

AMU	Arsine AsH_3	Silane SiH_4	Phosphine PH_3	Disilane Si_2H_6	Diphospine P_2H_4	Diborane B_2H_6
1						21.1
2						134.7
3						0.35
10						9.72
11						39.4
12						26.4
13						34.9
14.		0.4				2.23
14.5		0.5				
15		0.4				
15.5		0.1	0.23			
16			0.62			
16.5			0.13			
17			0.48			
20						0.22
21						1.85
22						11.8

(Continued)

AMU	Arsine AsH_3	Silane SiH_4	Phosphine PH_3	Disilane Si_2H_6	Diphospine P_2H_4	Diborane B_2H_6
23						48.5
24						94.0
25						57.7
26						100.0
27						95.2
28		2				16.7
29		32				
30		100.0				
31		80.	26.7			
32		7.3	100.0			
33		1.5	25.4			
34		0.2	76.7			
56				33		
57				48		
58				82		
59				37		
60				100.0		
61				40		
62				42	100.0	
63				5.7	58.8	
64				4.0	70.6	
65					26.5	
66					1.5	
75	38.5					
76	100.0					
77	28.8					
78	92.3					

Quadrupole data, same conditions as given in Appendix E1.

[a] Reproduced with permission from Uthe Technology Inc., 325 N. Mathilda Avenue, Sunnyvale, CA 94086.

Appendix F

Pump Fluid Properties

F.1 Compatibility of Elastomers and Pump Fluids

Elastomer	Mineral Oil	Ester	Halo-carbon	Fluoro-carbon	Poly-siloxane
Elastomer	Oil	Ester	carbon	carbon	siloxane
Butyl	No	<100°C	No data	<90°C	No
Buna-N	<100°C	No	<100°C	<90°C	No
Buna-S	No	No	No	No	No
Neoprene	<120°C	No	<120°C	<90°C	No
EPR	Yes	<70°C[a]	Yes	<70°C[a]	No
Silicone	Yes	<175°C	No	<150°C	No
Viton	Yes	<145°C	Yes	<200°C	Yes
Teflon	Yes	<175°C	Yes	<200°C	Yes
Kalrez	Yes	<175°C	Yes	Yes	Yes

[a] No data available for $T > 70$°C.

A Users Guide to Vacuum Technology, Fourth Edition. John F. O'Hanlon and Timothy A. Gessert.
© 2024 John Wiley & Sons, Inc. Published 2024 by John Wiley & Sons, Inc.

F.2 Vapor Pressures of Mechanical Pump Fluids

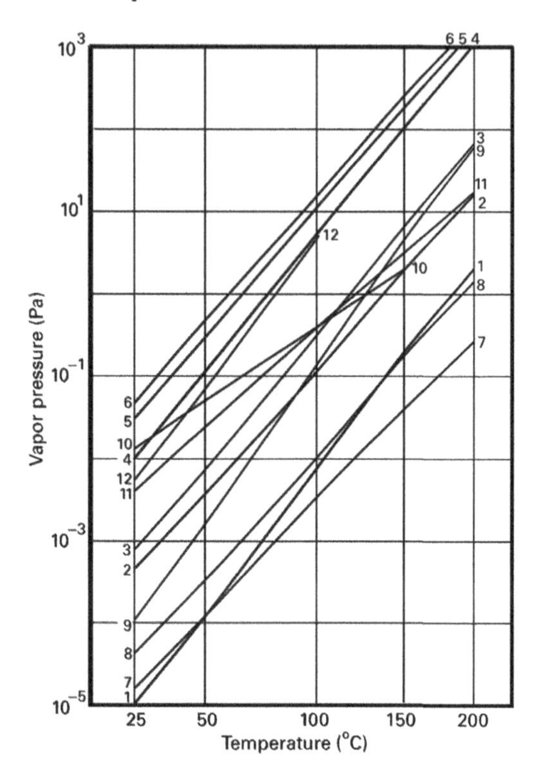

1) Fomblin Y-LVAC 25-6 [1]; (2) Fomblin Y-LVAC 14-6 [1]; (3) Fomblin Y-LVAC 06-6 [2]; (4) Halovac-190 [3]; (5) Halovac-125 [3]; (6) Halovac-100 [3]; (7) Krytox 1525 [4]; (8) Krytox 1514 [4]; (9) Krytox 1506 [4]; (10) Inland-77 [5]; (11) Inland-19 [5]; and (12) Balzers-Pfeiffer T-11 [6].

F.3 Vapor Pressures of Diffusion Pump Fluids

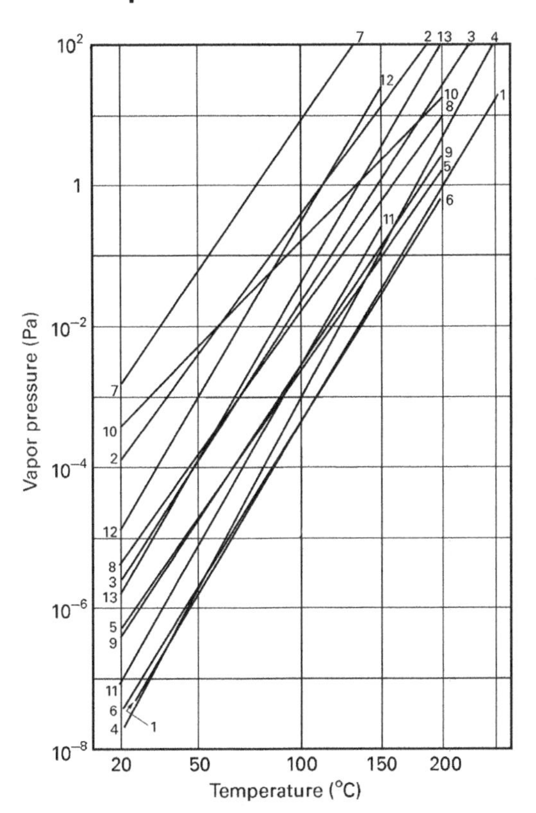

1) Santovac-5 [7]; (2) DC-702 [8]; (3) DC-704 [8]; (4) DC-705 [8]; (5) Krytox 1618 [4]; (6) Krytox 1625 [4]; (7) Butyl Phthalate; (8) Fomblin Y-HVAC 18/8 [2]; (9) Fomblin Y-HVAC 25/9 [2]; (10) Invoil-20 [5]; (11) Edwards L-9 [9]; (12) Octoil [10]; and (13) Octoil-S [10]

F.4 Kinematic Viscosities of Pump Fluids

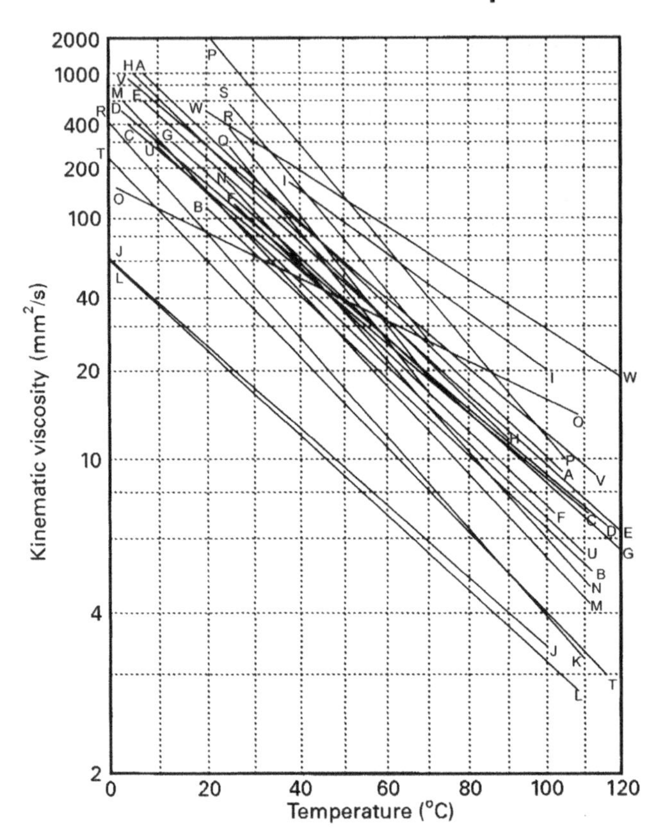

(A) Balzers P-3 [6], Kinney KV-100 [11]; (B) Convoil-20 [10]; (C) Inland-19 [5]; (D) Invoil-20 [5]; (E) Kinney OCR [11], Krytox 1525 and 1625 [4]; (F) Kinney Super X [11]; (G) Welch Duo-Seal 1407 [12], Krytox 1618 [3], Fomblin Y-HVAC 18/8 [2]; (H) Inland-77 [5]; (I) Synlube [13]; (J) Balzers-Pfeiffer T-11 [6]; (K) Octoil [10]; (L) Octoil-S [10]; (M) Fyrquel-220 [14]; (N) Kinlube 300 [11]; (O) Versilube F-50 [15]; (P) Santovac-5 [7]; (Q) Halovac 100 [3]; (R) Halovac 125 [3]; (S) Halovac 190 [3]; (T) Fomblin Y-LVAC 06/6 [2], Krytox 1506 [3]; (U) Fomblin Y-LVAC 14/6 [2], Krytox 1514 [4]; (V) Fomblin Y-LVAC 25/5, Y-HVAC 25/9 [2]; and (W) Dow Corning FS-1265 [8]. Fluids grouped under a common heading are not the same but have similar viscosities.

F.5 Viscosity Index, Viscosity and Temperature

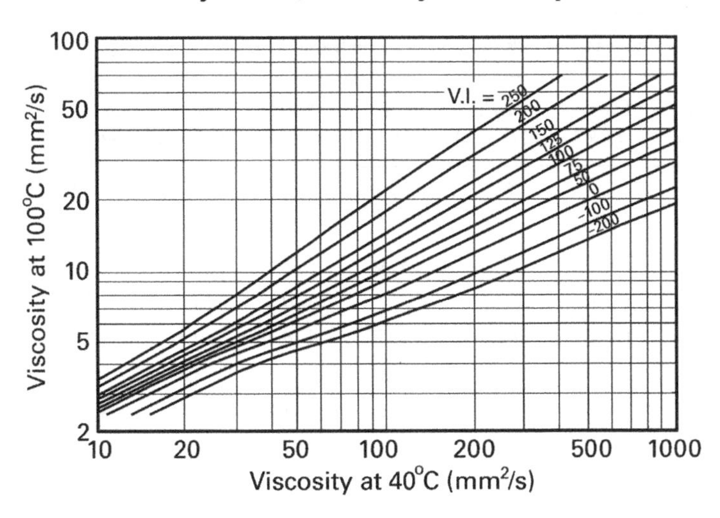

F.6 Kinematic Viscosity Conversion Factors

Kinematic Viscosity (mm^2/s)	Saybolt Universal Seconds (SUS) 100°F (37.8°C)	130°F (54.4°C)	210°F (98.8°C)	Redwood Seconds at 70°F (21.1°C)	140°F (60°C)	210°F (93.3°C)	Engler Degrees at all Temps
..5.0	42.3	42.4	42.6	37.9	38.5	38.9	1.40
..6.0	45.5	45.6	45.8	40.5	41.0	41.5	1.48
..7.0	48.7	48.8	49.0	43.2	43.7	44.2	1.56
..8.0	52.0	52.1	52.4	46.0	46.4	46.9	1.65
..9.0	55.4	55.5	55.8	48.9	49.1	49.7	1.75
10.0	58.8	58.9	59.2	51.7	52.0	52.6	1.84
12.0	65.9	66.0	66.4	57.9	58.1	58.8	2.02
14.0	73.4	73.5	73.9	64.4	64.6	65.3	2.22
16.0	81.1	81.3	81.7	71.0	71.4	72.2	2.43
18.0	89.2	89.4	89.8	77.9	78.5	79.4	2.64
20.0	97.5	97.7	98.2	85.0	85.8	86.9	2.87
22.0	106.0	106.2	106.7	92.4	93.3	94.5	3.10
24.0	114.6	114.8	115.4	99.9	100.9	102.2	3.34
26.0	123.3	123.5	124.2	107.5	108.6	110.0	3.58

(Continued)

(Continued)

Kinematic Viscosity (mm^2/s)	Saybolt Universal Seconds (SUS)			Redwood Seconds at			Engler Degrees at all Temps
	100°F (37.8°C)	130°F (54.4°C)	210°F (98.8°C)	70°F (21.1°C)	140°F (60°C)	210°F (93.3°C)	
28.0	132.1	132.4	133.0	115.3	116.5	118.0	3.82
30.0	140.9	141.2	141.9	123.1	124.4	126.0	4.07
32.0	149.7	150.0	150.8	131.0	132.3	134.1	4.32
34.0	158.7	159.0	159.8	138.9	140.2	142.2	4.57
36.0	167.7	168.0	168.9	146.9	148.2	150.3	4.83
38.0	176.7	177.0	177.9	155.0	156.2	158.3	5.08
40.0	185.7	186.0	187.0	163.0	164.3	166.7	5.34
42.0	194.7	195.1	196.1	171.0	172.3	175.0	5.59
44.0	203.8	204.2	205.2	179.1	180.4	183.3	5.85
46.0	213.0	213.4	214.5	187.1	188.5	191.7	6.11
48.0	222.2	222.6	223.8	195.2	196.6	200.0	6.37
50.0	231.4	231.8	233.0	203.3	204.7	208.3	6.63
60.0	277.4	277.9	279.3	243.5	245.3	250.0	7.90
70.0	323.4	324.0	325.7	283.9	286.0	291.7	9.21
80.0	369.6	370.3	372.2	323.9	326.6	333.4	10.53
90.0	415.8	416.6	418.7	364.4	367.4	375.0	11.84
100.0[a]	462.0	462.9	465.2	404.9	408.2	416.7	13.16

[a] At higher values use the same ratio as above for 100 mm^2/s.

References

1 Reproduced with permission from Kurt J. Lesker Company, 925-PA 51 Jefferson Hills, PA 15025.
2 Montedison, USA, Inc., 1114 Ave. of the Americas, New York, NY 10036.
3 Fluoro-Chem Corporation, 82 Burlews Court, Hackensack, NJ 07601.
4 Du Pont and Co., Chemicals and Pigments Department, Wilmington, DE 19898.
5 Inland Vacuum Industries, Inc., 35 Howard Ave., Churchville, NY 14428.
6 Balzers High Vacuum, Furstentum, Liechtenstein.
7 Monsanto Company, 800 N. Lindbergh Blvd. St. Louis, MO 63166.
8 Dow Corning Company, Inc., 2030 Dow Center, Midland, MI 48640.
9 Edwards High Vacuum, Ltd. Manor Royal, Crawley, West Sussex, England.

10 Veeco, Inc., 525 Lee Rd. Rochester, NY 14603.

11 Kinney Vacuum Co., 3529 Washington St., Boston, MA 02130.

12 Sargent-Welch Scientific Co., Vac. Prod. Div., 7300 N. Linder Ave. Skokie, IL 60077.

13 Synthatron Corp., 50 Intervale Rd., Parsippany, NJ 07054.

14 Stauffer Chemical Company, Specialty Chemical Division, Westport, CN 06880.

15 General Electric Company, Silicone Products Department, Waterford, NY 12188.

Index

A Users Guide to Vacuum Technology, Fourth Edition. John F. O'Hanlon and Timothy A. Gessert.
© 2024 John Wiley & Sons, Inc. Published 2024 by John Wiley & Sons, Inc.

m

n

o

p

q

r

Printed and bound by CPI Group (UK) Ltd, Croydon, CR0 4YY

16/04/2025

14658594-0002